S0-APC-858

# Methods in Enzymology

Volume 290

MOLECULAR CHAPERONES

# METHODS IN ENZYMOLOGY

EDITORS-IN-CHIEF

John N. Abelson Melvin I. Simon

DIVISION OF BIOLOGY
CALIFORNIA INSTITUTE OF TECHNOLOGY
PASADENA, CALIFORNIA

FOUNDING EDITORS

Sidney P. Colowick and Nathan O. Kaplan

*Methods in Enzymology*

*Volume 290*

# *Molecular Chaperones*

EDITED BY

*George H. Lorimer*

E. I. DUPONT DE NEMOURS AND COMPANY
WILMINGTON, DELAWARE

*Thomas O. Baldwin*

TEXAS A & M UNIVERSITY
COLLEGE STATION, TEXAS

ACADEMIC PRESS
San Diego London Boston New York Sydney Tokyo Toronto

This book is printed on acid-free paper. 

Academic Press
15 East 26th Street, 15th Floor, New York, New York 10010, USA
http://www.apnet.com

Academic Press Limited
24-28 Oval Road, London NW1 7DX, UK
http://www.hbuk.co.uk/ap/

International Standard Book Number: 0-12-182191-9

PRINTED IN THE UNITED STATES OF AMERICA
98 99 00 01 02 03 MM 9 8 7 6 5 4 3 2 1

# Table of Contents

# Contributors to Volume 290

Article numbers are in parentheses following the names of contributors.
Affiliations listed are current.

OSAMU ASAMI (4), *Toyota Central Research and Development Laboratories, Inc., Nagakute, Aichi 480-11, Japan*

ABDUSSALAM AZEM (22), *Department of Botany, Alexander Silbermann Institute of Life Sciences, Hebrew University of Jerusalem, 91904 Jerusalem, Israel*

KAREN BACOT (19), *Molecular Biology Division, Central Research and Development Department, E. I. duPont de Nemours and Company, Experimental Station, Wilmington, Delaware 19880-0402*

THOMAS O. BALDWIN (1), *Center for Macromolecular Design, Departments of Biochemistry and Biophysics, and Chemistry, Texas A&M University, College Station, Texas 77843-2128*

JAMES C. A. BARDWELL (5), *Department of Biology, University of Michigan, Ann Arbor, Michigan 48109*

JOACHIM BEHLKE (23), *Max Delbrück Center for Molecular Medicine, D-13122 Berlin, Germany*

WOLFGANG BERGMEIER (18), *Institüt für Biophysik and Physikalische Biochemie, Universität Regensburg, 93040 Regensburg, Germany*

SYLVIE BLOND (31), *Center for Pharmaceutical Biotechnology, University of Illinois at Chicago, Chicago, Illinois 60607-7173*

SUCHIRA BOSE (32, 33), *Department of Biochemistry, University of Bristol, School of Medical Sciences, Bristol BS8 1TD, Great Britain*

MICHAEL P. BOVA (30), *Jules Stein Eye Institute, University of California at Los Angeles School of Medicine, Los Angeles, California 90095-7008*

BILL T. BRAZIL (26), *Department of Biochemistry, University of Texas Health Science Center at San Antonio, San Antonio, Texas 78284-7760*

JOHANNES BUCHNER (27, 28, 32, 33), *Institüt für Biophysik und Physikalische Biochemie, Universität Regensburg, 93040 Regensburg, Germany*

HANS BÜGL (33), *Institüt für Biophysik und Physikalische Biochemie, Universität Regensburg, 93040 Regensburg, Germany*

STEVEN G. BURSTON (11), *Department of Genetics, Yale University School of Medicine, New Haven, Connecticut 06510*

S. CHEN (21), *Department of Crystallography, Birkbeck College, London WC1E 7HX, United Kingdom*

MATHIEU CHEVALIER (31), *Center for Pharmaceutical Biotechnology, University of Illinois at Chicago, Chicago, Illinois 60607-7173*

A. CLAY CLARK (8), *Department of Biochemistry and Molecular Biophysics, Washington University School of Medicine, St. Louis, Missouri 63110-3344*

NICHOLAS J. COWAN (20), *Department of Biochemistry, New York University School of Medicine, New York, New York 10016-6402*

DEBORAH L. DIAMOND (35), *Department of Biochemistry, Washington State University, Pullman, Washington 99164-4660*

RAMONA DICKSON (19), *Molecular Biology Division, Central Research and Development Department, E. I. duPont de Nemours and Company, Experimental Station, Wilmington, Delaware 19880-0402*

LINLIN DING (30), *Jules Stein Eye Institute, University of California at Los Angeles School of Medicine, Los Angeles, California 90095-7008*

YVES DUBAQUIÉ (17), *Basel Abteilung Biochemie, Biozentrum Universite Basel, Basel, Switzerland CH-4056*

MONIKA EHRNSPERGER (28), *Institüt für Biophysik und Physikalische Biochemie, Universität Regensburg, 93040 Regensburg, Germany*

EDWARD EISENSTEIN (9), *Center for Advanced Research in Biotechnology, University of Maryland Biotechnology Institute, Rockville, Maryland 20850*

ALEXEY N. FEDOROV (1), *Center for Macromolecular Design, Departments of Biochemistry and Biophysics, and Chemistry, Texas A&M University, College Station, Texas 77843-2128*

WAYNE A. FENTON (11), *Department of Genetics, Yale University School of Medicine, New Haven, Connecticut 06510*

MARK T. FISHER (9), *Department of Biochemistry and Molecular Biology, University of Kansas Medical Center, Kansas City, Kansas 66160-7421*

CARL FRIEDEN (8), *Department of Biochemistry and Molecular Biophysics, Washington University School of Medicine, St. Louis, Missouri 63110-3344*

MATTHIAS GAESTEL (28), *Max Delbrück Zentrum für Molekulare Medizin, 13122 Berlin, Germany*

H. F. GILBERT (3), *Verna and Marrs McLean Department of Biochemistry, Baylor College of Medicine, Houston, Texas 77030-3498*

PIERRE GOLOUBINOFF (18, 22), *Department of Botany, Alexander Silbermann Institute of Life Sciences, Hebrew University of Jerusalem, 91904 Jerusalem, Israel*

BORIS M. GOROVITS (25), *Department of Biochemistry, University of Texas Health Science Center at San Antonio, San Antonio, Texas 78284-7760*

HOLGER GRALLERT (27), *Institüt für Biophysik und Physikalische Biochemie, Universität Regensburg, 93040 Regensburg, Germany*

MICHAEL GROSS (24), *Oxford Centre for Molecular Sciences, New Chemistry Laboratory, Oxford OX1 3QT, United Kingdom*

BOYD HARDESTY (2), *Department of Chemistry and Biochemistry, University of Texas at Austin, Austin, Texas 78712*

SIMON J. S. HARDY (35), *Department of Biology, University of York, York YO1 5YW, England*

MASANA HIRAI (4), *Toyota Central Research and Development Laboratories, Inc., Nagakute, Aichi 480-11, Japan*

PAUL M. HOROWITZ (25, 26), *Department of Biochemistry, University of Texas Health Science Center at San Antonio, San Antonio, Texas 78284-7760*

ARTHUR L. HORWICH (11), *Howard Hughes Medical Institute, Yale University School of Medicine, New Haven, Connecticut 06510*

JOSEPH HORWITZ (30), *Jules Stein Eye Institute, University of California at Los Angeles School of Medicine, Los Angeles, California 90095-7008*

QING-LING HUANG (30), *Jules Stein Eye Institute, University of California at Los Angeles School of Medicine, Los Angeles, California 90095-7008*

HAJIME ISHIKAWA (16), *Department of Biological Sciences, Graduate School of Science, University of Tokyo, Tokyo 113, Japan*

URSULA JAKOB (27, 32), *Department of Biology, University of Michigan, Ann Arbor, Michigan 48109*

TSUTOMU KAJINO (4), *Toyota Central Research and Development Laboratories, Inc., Nagakute, Aichi 480-11, Japan*

MARTIN KESSEL (18), *Department of Membrane and Ultrastructural Research, Hebrew University-Hadassah Medical School, 91220 Jerusalem, Israel*

LASHAUNDA KING (31), *Center for Pharmaceutical Biotechnology, University of Illinois at Chicago, Chicago, Illinois 60607-7173*

GISELA KRAMER (2), *Department of Chemistry and Biochemistry, University of Texas at Austin, Austin, Texas 78712*

WIESLAW KUDLICKI (2), *Department of Chemistry and Biochemistry, University of Texas at Austin, Austin, Texas 78712*

GARRETT J. LEE (29), *Department of Biochemistry, University of Arizona, Tucson, Arizona 85721-0106*

W. THEODORE LEE (13), *Department of Microbiology, Ohio State University, Columbus, Ohio 43210*

SUSAN LINDQUIST (34), *Howard Hughes Medical Institute and Department of Molecular Genetics and Cell Biology, University of Chicago, Chicago, Illinois 60637-8049*

GEORGE H. LORIMER (10, 18), *Molecular Biology Division, Central Research and Development Department, E. I. duPont de Nemours and Company, Experimental Station, Wilmington, Delaware 19880-0402*

AUTAR K. MATTOO (7), *Vegetable Laboratory, USDA/ARS, Beltsville Agricultural Research Center (W), Beltsville, Maryland 20705-2350*

CHRISTIAN MAYR (32), *Institüt für Biophysik und Physikalische Biochemie, Universität Regensburg, 93040 Regensburg, Germany*

BRUCE A. MCFADDEN (12), *Department of Biochemistry and Biophysics, Washington State University, Pullman, Washington 99164-4660*

CHIE MIYAZAKI (4), *Toyota Central Research and Development Laboratories, Inc., Nagakute, Aichi 480-11, Japan*

MIZUE MORIOKA (16), *Department of Biological Sciences, Graduate School of Science, University of Tokyo, Tokyo 113, Japan*

NIKHIL D. PHADKE (5), *Department of Biology, University of Michigan, Ann Arbor, Michigan 48109*

FRANZISKA PIRKL (33), *Institüt für Biophysik und Physikalische Biochemie, Universität Regensburg, 93040 Regensburg, Germany*

SHEENA E. RADFORD (24), *Department of Biochemistry and Molecular Biology, University of Leeds, Leeds LS2 9JT, United Kingdom*

RAGULAN RAMANATHAN (8), *Center for Mass Spectrometry, Washington University School of Medicine, St. Louis, Missouri 63110-3344*

LINDA L. RANDALL (35), *Department of Biochemistry, Washington State University, Pullman, Washington 99164-4660*

PRASAD REDDY (9), *DNA Technologies Group, Biotechnology Division, National Institute of Standards and Technology, Gaithersburg, Maryland 20899*

CAROL V. ROBINSON (24), *Oxford Centre for Molecular Sciences, New Chemistry Laboratory, Oxford OX1 3QT, United Kingdom*

A. M. ROSEMAN (21), *Department of Crystallography, Birkbeck College, London WC1E 7HX, United Kingdom*

SABINE ROSPERT (17), *Basel Abteilung Biochemie, Biozentrum Universite Basel, Basel, Switzerland CH-4056*

HAYS S. RYE (11), *Howard Hughes Medical Institute, Yale University School of Medicine, New Haven, Connecticut 06510*

H. R. SAIBIL (21), *Department of Crystallography, Birkbeck College, London WC1E 7HX, United Kingdom*

GOTTFRIED SCHATZ (17), *Basel Abteilung Biochemie, Biozentrum Universite Basel, CH-4056 Basel, Switzerland*

ERIC C. SCHIRMER (34), *Howard Hughes Medical Institute and Department of Molecular Genetics and Cell Biology, University of Chicago, Chicago, Illinois 60637-8049*

HANS-JOACHIM SCHÖNFELD (23), *Pharmaceutical Research–Infectious Diseases, F. Hoffmann–La Roche Limited, CH-4070 Basel, Switzerland*

ROBERT K. SCOPES (14), *Center for Protein and Enzyme Technology, School of Biochemistry, La Trobe University, Bundoora, Victoria 3083, Australia*

JEFFREY W. SEALE (26), *Monsanto, St. Louis, Missouri 63189*

JOHN J. SIEKIERKA (6), *Discovery Research, The R. W. Johnson Pharmaceutical Research Institute, Raritan, New Jersey 08869-0602*

VIRGINIA F. SMITH (35), *Department of Biochemistry, Washington State University, Pullman, Washington 99164-4660*

F. ROBERT TABITA (13), *Department of Microbiology, Ohio State University, Columbus, Ohio 43210*

HIDEKI TAGUCHI (15), *Research Laboratory of Resources Utilization, R-1, Tokyo Institute of Technology, Yokohama 226, Japan*

MATTHEW J. TODD (10), *Molecular Biology Division, Central Research and Development Department, E. I. duPont de Nemours and Company, Experimental Station, Wilmington, Delaware 19880-0402*

TRACI B. TOPPING (35), *Department of Biochemistry, Washington State University, Pullman, Washington 99164-4660*

JOSE A. TORRES-RUIZ (12), *Department of Biochemistry, Ponce School of Medicine, Ponce, Puerto Rico 00732*

KAYE TRUSCOTT (14), *Center for Protein and Enzyme Technology, School of Biochemistry, La Trobe University, Bundoora, Victoria 3083, Australia*

SHIGEZU UDAKA (4), *Department of Brewing and Fermentation, Tokyo University of Agriculture, Setagaya-ku, Tokyo 156, Japan*

ELIZABETH VIERLING (29), *Department of Biochemistry, University of Arizona, Tucson, Arizona 85721-0106*

PAUL V. VIITANEN (18, 19), *Molecular Biology Division, Central Research and Development Department, E. I. duPont de Nemours and Company, Experimental Station, Wilmington, Delaware 19880-0402*

STEFAN WALKE (28), *Medical Research Council, Laboratory of Molecular Biology, Cambridge CB2 2QH, England*

Gregory M. F. WATSON (13), *Department of Microbiology, Ohio State University, Columbus, Ohio 43210*

TOM WEBB (19), *Molecular Biology Division, Central Research and Development Department, E. I. duPont de Nemours and Company, Experimental Station, Wilmington, Delaware 19880-0402*

TINA WEIKL (33), *Institüt für Biophysik und Physikalische Biochemie, Universität Regensburg, 93040 Regensburg, Germany*

CELESTE WEISS (18, 22), *Department of Botany, Alexander Silbermann Institute of Life Sciences, Hebrew University of Jerusalem, 91904 Jerusalem, Israel*

JONATHAN S. WEISSMAN (11), *Department of Pharmacology, University of California San Francisco School of Medicine, San Francisco, California 94143*

GREGORY WIEDERRECHT (6), *Immunology Research, Merck Research Laboratories, Rahway, New Jersey 07065*

YUKIO YAMADA (4), *Toyota Central Research and Development Laboratories, Inc., Nagakute, Aichi 480-11, Japan*

MASASUKE YOSHIDA (15), *Research Laboratory of Resources Utilization, R-1, Tokyo Institute of Technology, Yokohama 226, Japan*

THOMAS ZANDER (5), *Department of Biology, University of Michigan, Ann Arbor, Michigan 48109*

TONG ZHANG (2), *Department of Chemistry and Biochemistry, University of Texas at Austin, Austin, Texas 78712*

# Preface

The chapters contained in this volume represent a collection of methods currently used for the study of assisted folding and assembly of proteins. Much of the literature is published in short articles in journals that do not allow presentation of experimental details, and in this field, as in most, the devil is truly in the details.

Since the pioneering work of Anfinsen and his collaborators and contemporaries, the study of protein folding has been the province of biophysics and biochemistry. The discovery of molecular chaperones and the development of recombinant DNA methods have caused an enormous influx of cell biologists, genetists, and molecular biologists to the field, and, as a result, cultural and language differences often lead to misunderstandings of meaning and interpretation of experimental observations. The laws of thermodynamics pertain *in vivo* as well as *in vitro,* and while refolding of a protein following dilution of urea-containing buffers is hardly biological, the principles learned from such studies are applicable to the understanding of the biological processes as they occur within the living cell. For example, it is well known that to obtain high recovery of soluble protein on refolding from urea or guanidine, it is usually essential to dilute the protein solution into an aqueous buffer so that each molecule may refold effectively in isolation, thereby avoiding assembly of folding intermediates into large aggregates. Likewise, *in vivo,* overexpression of a protein from a plasmid or other vector often leads to aggregation and inclusion body formation as a consequence of the higher than normal steady-state concentration of folding intermediates of that protein.

As with all enzymes, a detailed biochemical description of the mechanism of action of chaperones and folding catalysts requires pure enzyme. Failure to remove bound, contaminating substrate polypeptides from chaperones can and does lead to errors in determination of stoichiometry and other enzymological parameters.

In assembling this volume, one of the major objectives was to solicit contributions from experts knowledgeable in the detailed "personalities" of the chaperones and related components of the protein metabolic machinery. The chemical transformations catalyzed by the chaperones will ultimately be understood in molecular detail. Until then, we hope and believe that the process of unraveling these details will be greatly facilitated through diligent application of the methods presented by the contributors to this volume.

We are deeply indebted to the contributors for their chapters and for their advice, which helped substantially in shaping the final volume. We also appreciate the efforts of the excellent staff of Academic Press, especially Ms. Shirley Light, for their assistance and gentle encouragement.

George H. Lorimer
Thomas O. Baldwin

# METHODS IN ENZYMOLOGY

VOLUME 91. Enzyme Structure (Part I)
*Edited by* C. H. W. HIRS AND SERGE N. TIMASHEFF

VOLUME 92. Immunochemical Techniques (Part E: Monoclonal Antibodies and General Immunoassay Methods)
*Edited by* JOHN J. LANGONE AND HELEN VAN VUNAKIS

VOLUME 93. Immunochemical Techniques (Part F: Conventional Antibodies, Fc Receptors, and Cytotoxicity)
*Edited by* JOHN J. LANGONE AND HELEN VAN VUNAKIS

VOLUME 94. Polyamines
*Edited by* HERBERT TABOR AND CELIA WHITE TABOR

VOLUME 95. Cumulative Subject Index Volumes 61–74, 76–80
*Edited by* EDWARD A. DENNIS AND MARTHA G. DENNIS

VOLUME 96. Biomembranes [Part J: Membrane Biogenesis: Assembly and Targeting (General Methods; Eukaryotes)]
*Edited by* SIDNEY FLEISCHER AND BECCA FLEISCHER

VOLUME 97. Biomembranes [Part K: Membrane Biogenesis: Assembly and Targeting (Prokaryotes, Mitochondria, and Chloroplasts)]
*Edited by* SIDNEY FLEISCHER AND BECCA FLEISCHER

VOLUME 98. Biomembranes (Part L: Membrane Biogenesis: Processing and Recycling)
*Edited by* SIDNEY FLEISCHER AND BECCA FLEISCHER

VOLUME 99. Hormone Action (Part F: Protein Kinases)
*Edited by* JACKIE D. CORBIN AND JOEL G. HARDMAN

VOLUME 100. Recombinant DNA (Part B)
*Edited by* RAY WU, LAWRENCE GROSSMAN, AND KIVIE MOLDAVE

VOLUME 101. Recombinant DNA (Part C)
*Edited by* RAY WU, LAWRENCE GROSSMAN, AND KIVIE MOLDAVE

VOLUME 102. Hormone Action (Part G: Calmodulin and Calcium-Binding Proteins)
*Edited by* ANTHONY R. MEANS AND BERT W. O'MALLEY

VOLUME 103. Hormone Action (Part H: Neuroendocrine Peptides)
*Edited by* P. MICHAEL CONN

VOLUME 104. Enzyme Purification and Related Techniques (Part C)
*Edited by* WILLIAM B. JAKOBY

VOLUME 105. Oxygen Radicals in Biological Systems
*Edited by* LESTER PACKER

VOLUME 106. Posttranslational Modifications (Part A)
*Edited by* FINN WOLD AND KIVIE MOLDAVE

VOLUME 107. Posttranslational Modifications (Part B)
*Edited by* FINN WOLD AND KIVIE MOLDAVE

Volume 266. Computer Methods for Macromolecular Sequence Analysis
*Edited by* Russell F. Doolittle

Volume 267. Combinatorial Chemistry
*Edited by* John N. Abelson

Volume 268. Nitric Oxide (Part A: Sources and Detection of NO; NO Synthase)
*Edited by* Lester Packer

Volume 269. Nitric Oxide (Part B: Physiological and Pathological Processes)
*Edited by* Lester Packer

Volume 270. High Resolution Separation and Analysis of Biological Macromolecules (Part A: Fundamentals)
*Edited by* Barry L. Karger and William S. Hancock

Volume 271. High Resolution Separation and Analysis of Biological Macromolecules (Part B: Applications)
*Edited by* Barry L. Karger and William S. Hancock

Volume 272. Cytochrome P450 (Part B)
*Edited by* Eric F. Johnson and Michael R. Waterman

Volume 273. RNA Polymerase and Associated Factors (Part A)
*Edited by* Sankar Adhya

Volume 274. RNA Polymerase and Associated Factors (Part B)
*Edited by* Sankar Adhya

Volume 275. Viral Polymerases and Related Proteins
*Edited by* Lawrence C. Kuo, David B. Olsen, and Steven S. Carroll

Volume 276. Macromolecular Crystallography (Part A)
*Edited by* Charles W. Carter, Jr., and Robert M. Sweet

Volume 277. Macromolecular Crystallography (Part B)
*Edited by* Charles W. Carter, Jr., and Robert M. Sweet

Volume 278. Fluorescence Spectroscopy
*Edited by* Ludwig Brand and Michael L. Johnson

Volume 279. Vitamins and Coenzymes, Part I
*Edited by* Donald B. McCormick, John W. Suttie, and Conrad Wagner

Volume 280. Vitamins and Coenzymes, Part J
*Edited by* Donald B. McCormick, John W. Suttie, and Conrad Wagner

Volume 281. Vitamins and Coenzymes, Part K
*Edited by* Donald B. McCormick, John W. Suttie, and Conrad Wagner

Volume 282. Vitamins and Coenzymes, Part L
*Edited by* Donald B. McCormick, John W. Suttie, and Conrad Wagner

Volume 283. Cell Cycle Control
*Edited by* William G. Dunphy

# [1] Protein Folding and Assembly in a Cell-Free Expression System

*By* ALEXEY N. FEDOROV and THOMAS O. BALDWIN

## Introduction

At the most fundamental level, the objective of all protein-folding studies is an understanding of how proteins fold and assemble into biologically active complexes within the living cell. The same physical and chemical principles that govern protein folding *in vitro* are, of course, operational *in vivo*. Using the knowledge gained from the studies of refolding of small model polypeptides, we can now probe the folding reactions *in vivo* to gain an understanding of these processes in the more complex environment. The involvement of ribosomes, chaperones, subcellular compartments, and other factors, as well as the added complexity of the vectorial component of folding imposed by the process of protein synthesis, make folding *in vivo* a more difficult problem to investigate. At present, one of the best available experimentally manipulatable strategies to probe the coupled processes of protein synthesis and folding is to use a cell-free expression system.

Cell-free expression systems are currently being used for a number of purposes. These purposes include the study of both co- and posttranslational events associated with biosynthetic protein folding,[1–7] including interactions with chaperones[2,4,7] and assembly of multicomponent complexes.[8] Other uses of cell-free expression systems include selection procedures with ribosome-bound polypeptides aimed at, for example, directed evolution.[9] Prob-

[1] A. N. Fedorov, B. Friguet, L. Djavadi-Ohaniance, Y. B. Alakhov, and M. E. Goldberg, *J. Mol. Biol.* **228,** 351 (1992).

[2] J. Frydman, E. Nimmesgern, K. Ohtsuka, and F. U. Hartl, *Nature* (*London*) **370,** 111 (1994).

[3] V. A. Kolb, E. V. Makeyev, and A. S. Spirin, *EMBO J.* **13,** 3631 (1994).

[4] W. Kudlicki, O. W. Odom, G. Kramer, and B. Hardesty, *J. Mol. Biol.* **244,** 319 (1994).

[5] A. N. Fedorov and T. O. Baldwin, *Proc. Natl. Acad. Sci. U.S.A.* **92,** 1227 (1995).

[6] W. Kudlicki, Y. Kitaoka, O. W. Odom, G. Kramer, and B. Hardesty, *J. Mol. Biol.* **252,** 203 (1995).

[7] K. Tokatlidis, B. Friguet, D. Deville-Bonne, F. Baleux, A. N. Fedorov, A. Navon, L. Djavadi-Ohaniance, and M. E. Goldberg, *Phil. Trans. R. Soc. Ser. B: Biol. Sci.* **348,** 89 (1995).

[8] L. A. Ryabova, D. Desplancq, A. S. Spirin, and A. Pluckthun, *Nature Biotechnol.* **15,** 79 (1997).

[9] L. C. Mattheakis, R. R. Bhatt, and W. J. Dower, *Proc. Natl. Acad. Sci. U.S.A.* **91,** 9022 (1994).



0076-6879/98 $25.00

ing of the structural properties[10] and epitope mapping[11] of nascent polypeptides has also been an exciting application of cell-free expression systems.

In this chapter, we describe several common procedures for use of cell-free expression systems, including strategies to obtain homogeneous populations of nascent polypeptides, separation of ribosomal complexes, induced release of nascent polypeptides from ribosomes, and immunoadsorption of synthesized polypeptides. A fundamental difficulty in the study of protein folding in a cell-free system pertains to the complexity of the reaction mixture and the fact that the polypeptide of interest comprises a small fraction of the total protein. As a result, the powerful biophysical methods, such as spectroscopy, that have yielded such detailed information about protein folding *in vitro* cannot be directly used in a cell-free expression system. Nonetheless, it is possible to investigate the basic structural properties of radiolabeled nascent and newly synthesized polypeptides without purification. Such factors as compactness, structural stability, cooperativity of folding, and some local structural properties of polypeptides can be studied by applying such techniques as size-exclusion chromatography under native and denaturing conditions, urea-gradient electrophoretic analysis, and limited proteolysis.[10] The latter method, which is easily applicable to study different target polypeptides in both free and ribosome-bound forms, is discussed. In the following section, an approach to study the kinetic mechanism of biosynthetic protein folding and assembly is presented. In the final section of this chapter, we discuss approaches to evaluate in a complex mixture the equilibrium binding affinity of specific proteins. This method, which is related to the enzyme-linked immunosorbent assay (ELISA) competitive binding technique,[12] allows analysis of protein–protein interactions and, more generally, protein–ligand interactions in complex mixtures even when the interacting components are present at low levels, as is the case in cell-free expression systems. In this chapter, we stress that the physical and chemical principles that allow detailed descriptions of molecular interactions in comparatively simple reactions also control reactions under the complex conditions of cell-free protein synthesis.

[10] A. N. Fedorov, D. A. Dolgikh, V. V. Chemeris, B. K. Chernov, A. V. Finkelstein, A. A. Schulga, Y. B. Alakhov, M. P. Kirpichnikov, and O. B. Ptitsyn, *J. Mol. Biol.* **225,** 927 (1992).

[11] B. Friguet, A. N. Fedorov, and L. Djavadi-Ohaniance, *J. Immunol. Methods* **158,** 243 (1993).

[12] B. Friguet, A. F. Chaffotte, L. Djavadi-Ohaniance, and M. E. Goldberg, *J. Immunol. Methods* **77,** 305 (1985).

## Basic Characteristics of Commercially Available Cell-Free Expression Systems

Cell-free expression systems from *Escherichia coli,* wheat germ, and rabbit reticulocyte are commercially available. These systems differ in the optimal temperature range for protein synthesis, the rate of translation, cell extract composition (including identity of chaperones), upper limit for the length of synthesized polypeptides, translation reinitiation patterns, and in other details. The *E. coli* system is active in the 25–37° temperature range with elongation rates at the optimal temperature (37°) of about 150–200 residues/min. Chaperones of *E. coli* are the best studied to date, consisting of the GroE family, DnaK family, SecB, trigger factor, and others.[13,14] The rabbit reticulocyte system is most active in the 30–37° temperature range with an elongation rate of around 200–300 residues/min at 37°. The chaperone families represented in rabbit reticulocyte lysate include heat shock protein 90 (HSP90), HSP70, and TRiC.[13,14] The wheat germ system is active from 22 to 27°, with an elongation rate of 50–70 residues/min over this range; synthesis declines sharply above this temperature range. The HSP90 and HSP70 families have been characterized from this source.[15,16] Expression systems can be used in two basic forms: either as translation systems supplemented with mRNA or as transcription–translation systems in which DNA is provided. The latter form is currently available not only for the *E. coli* system but also for others. The rabbit reticulocyte and wheat germ system are available in forms supplemented with phage RNA polymerase and other components for RNA transcription. The choice of a particular system should be based on careful consideration of its features and the specific details of the experiment. The wheat germ system is often the system of choice because it is reliable, as efficient as the other systems, and the least expensive. The overall efficiency of polypeptide synthesis in these expression systems employed in the conventional format is template dependent and usually in the range of 0.01–0.5 $\mu$g of product per 50 $\mu$l of the expression system, which corresponds to a 0.01–0.1 $\mu M$ concentration range of the synthesized polypeptide. It is essential to measure and consider the molar concentration of the product polypeptide if one hopes to understand the interaction between the product polypeptide and some other component of the reaction system.

[13] M. J. Gething and J. Sambrook, *Nature* (*London*) **355,** 33 (1992).

[14] F. U. Hartl, *Nature* (*London*) **381,** 571 (1996).

[15] M. V. Blagosklonny, J. Toretsky, S. Bohen, and L. Neckers, *Proc. Natl. Acad. Sci. U.S.A.* **93,** 8379 (1996).

[16] L. F. Stancato, K. A. Hutchison, P. Krishna, and W. B. Pratt, *Biochemistry* **35,** 554 (1996).

Numerous methods are available to measure polypeptide synthesis, but most are based on electrophoretic resolution of the products followed by quantitation of radioactivity in the separated components. Quantitation is usually accomplished by autoradiography or fluorography, or with radioactivity scanners that employ imaging technologies. The latter are available from Fuji (Tokyo, Japan), Molecular Dynamics (Sunnyvale, CA), Packard (Meriden, CT), and other companies. Total incorporation of radioactivity into product is usually determined using classic liquid scintillation counting of trichloroacetic acid (TCA)-precipitable material.[17,18]

## Buffers, Solutions, and Reagents Used in Cell-Free Protein Synthesis

All buffers and other solutions should be sterile and nuclease free. To this end, all buffers and solutions should be prepared with freshly deionized water. Buffer solutions may be stored at room temperature for periods of a few days. Longer storage should be at 4°.

- Buffer A: 10 m*M* Tris-HCl, 100 m*M* potassium-acetate, 10 m*M* magnesium-acetate (pH 7.6), 1 m*M* dithiothreitol (DTT; freshly prepared)
- Buffer A supplemented with 0.1% (v/v) Nonidet P-40
- Buffer A with 4 *M* urea: Prepare by mixing 1 vol of 2× buffer A with an equivalent volume of 8 *M* urea in water
- $MgCl_2$, 100 m*M*
- EGTA [ethylene glycol bis(β-aminoethyl ether)-*N*,*N*,*N*′,*N*′-tetraacetic acid; sodium salt] (pH 8.0), 0.2 *M*
- $CaCl_2$, 0.5 *M*
- Sucrose (10 and 29%, w/v) for isokinetic centrifugation: The sucrose solutions are prepared by mixing 1 vol of 2× buffer A with an equivalent volume of sucrose in water. (*Note:* Sucrose-containing solutions made with reagent-grade sucrose should be prepared fresh owing to the difficulty in maintaining sterility. Any microbial growth will lead to protease and RNase contamination)
- Glycerol (40%, v/v): Prepare with buffer A as described for sucrose solutions
- Micrococcal nuclease (nuclease S7; Boehringer GmbH, Mannheim, Germany): 15,000 units/ml
- Phenylmethylsulfonyl fluoride (PMSF, 40 m*M*): Prepare a fresh solution in 2-propanol
- Puromycin (50 m*M*; Sigma, St. Louis, MO)

[17] A. H. Erickson and G. Blobel, *Methods Enzymol.* **96,** 38 (1983).

[18] R. J. Jackson and T. Hunt, *Methods Enzymol.* **96,** 50 (1983).

Trypsin (0.5 mg/ml), treated to inhibit contaminant chymotryptic activity [trypsin-TPCK (tolylsulfonyl phenylalanyl chloromethyl ketone) treated; Worthington, Freehold, NJ]

Protein A– and protein G–Sepharose 4 Fast Flow (Pharmacia Biotech, Uppsala, Sweden)

## Synthesis of Nascent Polypeptides on Monoribosomes

It is often desirable to maintain conditions that favor monoribosomal complexes. More than one ribosome per mRNA molecule will necessarily lead to heterogeneity in nascent polypeptide length and confuse the analysis of rates of translation, and other details. A predominance of monoribosomal complexes is achieved by addition of excess mRNA relative to translationally active ribosomes in the expression system. In practice, this can be determined by titrating the expression system with increasing amounts of mRNA until a plateau in the rate of appearance of full-length product is achieved. Usually 3–10 $\mu$g of mRNA/50 $\mu$l of the translation system is enough for this purpose. The distribution of the nascent polypeptides between mono- and polyribosomes can be evaluated by centrifugation in a sucrose gradient; this method is described in the next section.

## Separation of Translation Mixture by Sedimentation Velocity Centrifugation in Sucrose Gradients

Depending on the specific experiment, it may be desirable to stop the translation reaction before resolution of the components. Also, the reaction mixture may be applied to the sucrose gradient without dilution. Other modifications of the following procedure may be incorporated without altering the overall approach. In general, the reaction mixture is diluted with buffer A to a total of 300 $\mu$l and applied gently to 12 ml of isokinetic sucrose gradient, 10 to 29% (w/v),[19] prepared with buffer A. Centrifugation is performed using a Beckman (Palo Alto, CA) SW 41-Ti swinging bucket rotor at 235,000 *g* for 2 hr at 4°. Fractions are collected and the absorbance at 260 nm measured to determine the positions of mono- and polyribosomes in the gradient. In the absence of a pump to drain the gradients, we have found that excellent results may be obtained simply by carefully withdrawing aliquots (100 $\mu$l to 1 ml, depending on the desired resolution) sequentially with a pipette from the top of the gradient. The distribution of nascent polypeptides in the gradient is evaluated by measuring incorporation of

[19] B. van der Zeijst and H. Bloemers, *in* "Handbook of Biochemistry and Molecular Biology: Physical and Chemical Data" (G. Fasman, ed.), p. 426. CRC Press, Cleveland, Ohio, 1976.

radioactivity. This is achieved either by measuring total trichloroacetic acid-precipitated radioactivity after heating to effect solubilization of nonpolypeptide radioactivity or by electrophoretic separation of components in the fractions followed by autoradiography.

### Conversion of Polyribosomes into Monoribosomes by Ribonuclease Treatment

For some experiments it is essential, following translation of the sample mRNA, to have a reaction mixture that is free of polyribosomes. For example, if one wants to analyze a heterogeneous mixture of nascent polypeptides with an antibody, it is critical that polypeptides be associated exclusively with monoribosomes. In addition to the use of excess mRNA to obtain a predominance of monoribosomes (discussed above), it is possible to treat the reaction mixture with nuclease to convert any residual polyribosomes to monoribosomes.

A sample of the reaction mixture is diluted by mixing 200 $\mu$l of the system with 300 $\mu$l of buffer A. To this mixture 5 $\mu$l of 0.5 *M* $CaCl_2$ and 100 units of calcium-dependent micrococcal nuclease (Boehringer GmbH) are added. The mixture is incubated for 20 min at 20°, after which 25 $\mu$l of 0.2 *M* EGTA (twofold molar excess relative to the $CaCl_2$) is added to block further nuclease activity. The distribution of the nascent polypeptides between mono- and polyribosomes should be evaluated. Also, the amount of nuclease and/or time of incubation can be adjusted, if necessary.

### Generation of Ribosome-Bound Polypeptides

When the mRNA template has an in-frame translation stop codon, predominantly free polypeptides are usually produced. However, if the mRNA is lacking an intact stop codon, then the ribosomes will stall at the 3′ end resulting in a ternary complex of mRNA, ribosome and nascent polypeptide. That is, mRNA ending with one or two in-frame nucleotides but without a stop codon will result in an excellent yield of homogeneous nascent polypeptides. The traditional way to prepare mRNA for this purpose is by transcription with phage RNA polymerase of a linearized plasmid that has been designed to include either one or two additional bases following the last sense codon. Expression of a template without a stop codon per se does not ensure exclusive presence of ribosome-bound polypeptides. It is necessary to separate the supernatant and ribosomal fractions, as from one-third to two-thirds of the total product polypeptide may dissociate on separation of the supernatant and ribosomal fractions. The purified complexes are stable for a period of hours.

Separation of ribosome-bound nascent polypeptides from free polypeptides and other soluble material components can be achieved either by centrifugation or size-exclusion chromatography. The advantages of size-exclusion chromatography are that the procedure is quick and does not require the use of a high-speed centrifuge. The most convenient approach is to use prepacked columns with the appropriate matrix. For example, Sephacryl S-300 chromatography columns (cDNA spun columns; Pharmacia Biotech) are equilibrated with buffer A supplemented with 0.1% (v/v) Nonidet P-40 [or 1% (v/v) Nonidet P-40, if nonspecific adsorption of the nascent polypeptides onto the beads is high]. A sample of the reaction mixture is adjusted to a final volume of 50–110 μl with buffer A, applied to the column, and centrifuged using an appropriate centrifuge with a swinging bucket rotor (e.g., Sorvall RC-5B centrifuge; Du Pont, Wilmington, DE). The ribosomal fraction is usually recovered without significant dilution.

Ribosomal pellets can be obtained using sedimentation of the ribosomal fraction by high-speed centrifugation. Sedimentation can be performed with or without a glycerol (or sucrose) cushion. Use of a cushion allows better separation of lighter nonribosomal complexes, including large chaperone particles (GroE and TriC), although it cannot eliminate them entirely. Samples are prepared by diluting 100–300 μl of expression system to 0.5 ml with buffer A, applied onto 0.5 ml of 40% (v/v) glycerol prepared with buffer A, and centrifuged using a Beckman Ti 50 fixed-angle rotor at 225,000 *g* for 2.5 hr at 4°. Alternatively, a 2 *M* sucrose cushion should be used if polypeptides in the reaction mixture are aggregation prone; at this concentration, sucrose will effect separation on the basis of particle densities. Resuspension of the glasslike ribosomal pellet in buffer A is a slow process that requires special attention. Using an automatic pipette of appropriate volume, the sample is carefully drawn into the pipette and expelled several times to assure complete dissolution and separation of physical aggregates. If the pellet is expected to be small, it is advisable to mark the expected position of the pellet (the side of the centrifuge tube to the outside of the rotor) before centrifugation. Smaller volumes of cell-free extracts, 10–50 μl, can be centrifuged using a TLA 100 fixed-angle rotor, TL-100 tabletop ultracentrifuge from Beckman. With this instrument, the extract is diluted to 100 μl with buffer A and applied onto 100 μl of 40% (v/v) glycerol made up with the same buffer. Centrifugation is carried out for 30 min at 150,000 *g* at 4°.

### Dissociation of Nascent Polypeptides from Ribosomes by Puromycin Treatment

Puromycin treatment is a standard approach to dissociate nascent polypeptides from ribosomes. To obtain complete dissociation of nascent polypeptides, 50 m*M* puromycin is diluted 1:50 into the expression system to yield a 1 m*M* final concentration and incubated at 37° for 20 min (or at 25–29° for 40 min). The reaction is pH dependent, being substantially slower at pH <7.0 than at higher values of pH.[20] Consequently, it is important that the pH of the mixture be kept above pH 7.0; the most common range is pH 7.4–7.6. The efficiency of the polypeptide dissociation is usually 75–90%. The final product is a polypeptide with puromycin covalently attached to the C terminus. It should be noted that puromycin itself is a large ($M_r$ 472) and hydrophobic molecule and potentially can affect properties of the polypeptide to which it is attached, such as folding, stability, and adsorption or interaction with chaperones. If puromycin is used in studies aimed at investigating protein folding, it is important to demonstrate that the compound has no effect on the property under investigation.

## Immunoadsorption of Synthesized Polypeptides

Immunoadsorption is a basic and widely used approach to analyze polypeptides generated by cell-free synthesis. Applications include epitope mapping,[11] determination of binding affinity, and evaluation of conformation of nascent or released polypeptides.[1,21] All of these approaches require access to the appropriate antibody reagents, usually monoclonal, that have been properly characterized. There are many articles available that describe these methods, which are not discussed here.

Antibodies can be conveniently used to separate the product polypeptides from other reaction components. This is most readily accomplished by attaching the antibody molecules to some chromatographic matrix. The attachment can be either direct, by physically attaching the antibody protein to the matrix, or indirect, by allowing the antibody to interact with column material to which protein A or protein G has been attached. For the former approach, immunoglobulin G (IgG) can be coupled directly to the activated matrix via primary amines or carbohydrate moiety. Attachment through the carbohydrate of the IgG yields a higher specific antigen binding capacity owing to the oriented coupling. For the latter method, protein A or protein G, each with different IgG-binding specificities and capacities dependent on the source of IgG, may be attached to the matrix. Methods to generate and use these materials are well described in the materials supplied by the manufacturers. Catalogs from Bio-Rad (Hercules, CA), Pharmacia, and other companies contain tables with the required information. The required

[20] S. Pestka, *Methods Enzymol.* **XXX,** 470 (1974).

[21] B. Friguet, A. N. Fedorov, A. Serganov, A. Navon, and M. E. Goldberg, *Anal. Biochem.* **210,** 344 (1993).

matrices can be obtained from a variety of suppliers and coupled with IgG according to instructions furnished by the manufacturer. After coupling of IgG to the beads, the suspension is washed three times with buffer A without DTT and stored at 4° as a 50% (v/v) slurry in buffer A.

*Immunoadsorption Procedures*

Immunoadsorption is normally accomplished simply by allowing the immobilized antibody to interact with the antigen, followed by separation of the resulting antibody–antigen complex from other components. Problems that often arise are the result of nonspecific binding, owing either to lack of specificity of the antibody or to nonspecific adsorption of the antigen to some surface associated with the immobilized antibody.[22] These problems can usually be alleviated by careful attention to experimental design and appropriate controls.

To achieve immunoadsorption of radiolabeled synthesized polypeptide, 2–10 $\mu$l of the reaction mixture is diluted with 150–300 $\mu$l of buffer A supplemented with 0.1% (v/v) Nonidet P-40 (NP-40). In this and following procedures buffer A is used without DTT. To the resulting solution, 40 $\mu$l of a 50% (v/v) slurry of IgG-coupled beads in buffer A is added and mixed for 10–30 min at 4°. Rotation or any other gentle mixing method should be used to avoid bubble formation. The mixture is diluted with 1 ml of the buffer A supplemented with 0.1% (v/v) NP-40 and centrifuged at maximum speed using a benchtop centrifuge for 15 sec to 1 min. The supernatant is carefully withdrawn by pipetting, and the pellet washed once with buffer A supplemented with 1% (v/v) NP-40 and then washed twice with buffer A containing 0.1% (v/v) NP-40. Nonspecific adsorption of the polypeptide onto the beads can be determined using experiments with unrelated antibodies bound to the matrix. If nonspecific adsorption is high, mixing of the reaction mixture with the beads can be achieved in buffer A with 1% (v/v) NP-40 supplemented also with a carrier protein (e.g., bovine serum albumin, hen egg albumin, or casein) at a 1-mg/ml final concentration to saturate the hydrophobic surfaces of the beads and thereby suppress nonspecific binding. The first wash of the pellet can be done in buffer A supplemented with 1% (v/v) NP-40 and 0.1% (w/v) sodium dodecyl sulfate (SDS). Subsequent washing procedures are then performed as described above. The bound polypeptides are released for electrophoretic analysis by addition of 20–50 $\mu$l of SDS protein electrophoresis sample buffer to the pellets, boiling for 2 min, and centrifugation.

[22] D. J. Anderson and G. Blobel, *Methods Enzymol.* **96,** 111 (1983).

### Proteolysis of Synthesized Polypeptides

Limited proteolysis is a widely used approach to probe the extent of structure formation in nascent and newly synthesized polypeptides. It is well known that the native structure of a protein is much less protease labile than is the same protein when fully unfolded. Limited proteolysis can be used to detect regions of local unfolding, and as a result, may be used to monitor the process of folding of polypeptides co- and posttranslationally.[2] Comparison of the kinetics of proteolysis of specific bonds within a newly synthesized polypeptide with the same region of the native and denatured protein can yield information about the relative conformational flexibility of the region in the polypeptide.[10] Many formats of these methods have been used. We describe here the basic approaches that have been most useful.

Stock trypsin solution in water at 0.5 mg/ml should be immediately distributed into aliquots and kept at −70°. Each aliquot is used only once and the remaining protease discarded. Immediately before use, an aliquot is thawed and kept on ice. Aliquots of the expression system or ribosomal fraction (50 $\mu$l) are mixed with equal volumes of buffer A or with buffer A containing 8 *M* urea. Trypsin is added at a final concentration of 10–15 $\mu$g/ml. If necessary, the concentration of trypsin can be adjusted in subsequent experiments. An effective method to stop the action of the protease is to withdraw 10- to 20-$\mu$l aliquots at various times, mix with an equal volume of hot 2× sodium dodecyl sulfate–polyacrylamide gel electrophoresis (SDS–PAGE) sample buffer, and immediately boil in water for 4 min in preparation for electrophoretic analysis. Care should be taken to use hot SDS buffer, because the protease may remain active briefly if the partially proteolyzed substrate unfolds rapidly and therefore is more protease labile. This is especially true for certain bacterial proteases such as subtilisin and the *Staphylococcus aureus* V8 protease. An alternative approach is to use a chemical inhibitor such as phenylmethylsulfonyl fluoride (PMSF). A 40 m*M* PMSF solution is added to 1–4 m*M* final concentration to block further trypsin activity. It should be noted that at 25° and pH 7.2, the second-order rate constant for reaction of PMSF with trypsin is 271 $M^{-1}$ $\text{min}^{-1}$,[23] so that at 1 m*M* PMSF, the half-time for inhibition of trypsin would be about 2 min. If the rate of proteolysis is slow relative to the rate of reaction of the inhibitor, this slow reaction will cause no difficulties, but if the rate of proteolysis is fast, an alternative method for stopping the protease must be used. For chymotrypsin under the same conditions, the second-order rate constant is 14,900 $M^{-1}$ $\text{min}^{-1}$, sufficiently fast to be effective under

[23] A. M. Gold, *Methods Enzymol.* **XI,** 706 (1967).

most conditions. Urea at a 4 *M* final concentration was selected to analyze proteolysis under "denaturing" conditions, as it is sufficient to destabilize the structure of most proteins and it is easy to compare the stability of a polypeptide under native conditions and in 4 *M* urea, as trypsin is exactly half as active at this urea concentration as in buffer.[24]

## Analysis of Kinetics of Protein Folding in Cell-Free Expression Systems

There is a growing number of proteins for which folding can be monitored during synthesis in a cell-free expression system. Folding events can be monitored by measuring enzymatic activity,[2,3,5] interaction with ligands,[25] conformation-dependent antibodies,[1] and specific spectroscopic probes incorporated cotranslationally into the nascent polypeptide chain.[4] The folding process of the polypeptide synthesized in a cell-free expression system can be broken down to the following events:

$$\text{Synthesis} \xrightarrow{k_1} \text{full-length intermediate} \xrightarrow{k_2} \text{monomer} \xrightarrow{k_3} \text{oligomer}$$

where full-length intermediate refers to the polypeptide on its release from the ribosome, monomer is the native form for monomeric polypeptides or association-competent form for the oligomeric proteins, and $k_1$, $k_2$, and $k_3$ are the rate constants of these processes. The C-terminal 20- to 30-amino acid residues of nascent polypeptides are sheltered by the ribosome during synthesis.[26,27] Consequently, folding can be completed only after the release of the entire polypeptide from the ribosome. Obviously, there will be a lag between the time of initiation of protein synthesis and the time of appearance of the first full-length polypeptide. This lag phase represents the time required for the ribosomes to complete the first cycle of translation and is determined by the rate constant of translation, $k_1$, and the length of the polypeptide. The kinetics of folding of the polypeptide and ultimately appearance of the native form will be a function of all of the rate constants of the processes outlined here. Further complicating matters, cotranslational folding processes and interactions with chaperones potentially can influence the order and rates of folding/association events. If the native form of the synthesized polypeptide is a monomer, its folding will be a first-order, or concentration-independent, reaction. If the synthesized polypeptide is

[24] J. I. Harris, *Nature* (*London*) **177,** 471 (1956).
[25] A. A. Komar, A. Kommer, I. A. Krasheninnikov, and A. S. Spirin, *FEBS Lett.* **326,** 261 (1993).
[26] L. I. Malkin and A. Rich, *J. Mol. Biol.* **26,** 329 (1967).
[27] L. A. Ryabova, O. M. Selivanova, V. I. Baranov, V. D. Vasiliev, and A. S. Spirin, *FEBS Lett.* **226,** 255 (1988).

involved as a component of an oligomeric structure, its association will be a high-order concentration-dependent reaction dependent on the number of subunits involved. Considering rates and efficiency of polypeptide synthesis in the conventional expression systems and the fact that association rate constants cannot exceed the diffusion limit of about $10^8$ $M^{-1}$ $sec^{-1}$, it is likely that at the low concentrations often achieved in cell-free expression systems, association will become the rate-limiting step for oligomeric protein formation.

There is no common protocol by which to study and analyze the kinetics of biosynthetic folding. To outline basic strategies for such studies, we describe an example. The biosynthetic folding of the $\beta$ subunit of bacterial luciferase, an $\alpha\beta$ heterodimer, has been studied in the *E. coli* translation system.[5] Of particular interest was a question concerning whether cotranslational folding of the $\beta$ subunit contributes to the rate of formation of the native heterodimer. The cell-free system was provided with prefolded $\alpha$ subunit at a concentration high enough to make folding of the $\beta$ subunit the rate-limiting step of the folding/association process. Without the high concentration of prefolded $\alpha$ subunit, the rate-limiting step in formation of native luciferase would be subunit association, as discussed above. Translation was performed at 29° because formation of the luciferase heterodimer declines with increasing temperature in this range. The kinetics of synthesis of full-length $\beta$ subunit were monitored by determining incorporation of [$^{35}$S]methionine into the polypeptide, using a $\beta$-particle radioactivity scanner. Formation of the enzymatically active heterodimer in aliquots of the translation mixture was followed by light emission, using the reduced flavin mononucleotide injection assay. We have established that it requires 12 min to complete synthesis of the polypeptide chain and that the lag between appearance of the entire $\beta$ subunit and formation of active heterodimer is about 3–4 min in the linear range of protein synthesis (Fig. 1A). In this experiment the lag reflects both folding/association events and time required for the polypeptide termination and release from the ribosome. To determine better the rate of folding and assembly of the $\beta$ subunit after release from the ribosome, translation in the linear range of protein synthesis was blocked by increasing the $Mg^{2+}$ concentration by 10 m*M*. It has been confirmed that this treatment completely blocked translation. Consequently, further increases in enzyme activity reflect folding/association events of the $\beta$ subunit released from the ribosome before the block was applied. As is seen in Fig. 1B, the luciferase activity reaches a plateau within 1 min after translation was blocked. At this stage, the conclusion would be simply that folding/association of the luciferase rapidly follows $\beta$ subunit synthesis. To be able to dissect some properties of biosynthetic folding of $\beta$ subunit, we have compared its kinetics with that of its refolding

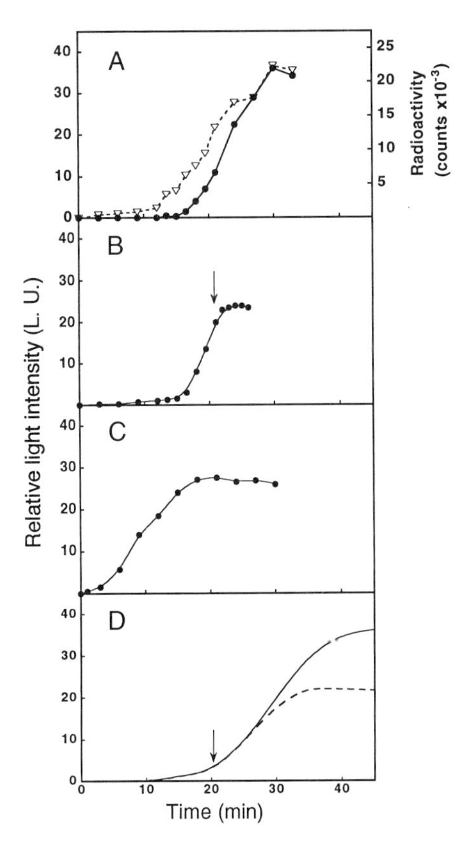


FIG. 1. (A) Kinetics of luciferase $\beta$ subunit synthesis and appearance of active luciferase heterodimer in an *E. coli* translation system. Translation was performed in a reaction mixture containing [$^{35}$S]methionine with mRNA encoding the luciferase $\beta$ subunit and supplemented with refolded luciferase $\alpha$ subunit at a final concentration of 10 $\mu$g/ml. At the indicated time points, two aliquots of the reaction mixture were withdrawn. One aliquot was used to measure bioluminescence activity (filled circles). Another aliquot was subjected to electrophoretic analysis, followed by determination of incorporation of radioactivity into full-length $\beta$ subunit (open triangles). (B) Kinetics of appearance of active luciferase in the translation mixture after blocking of synthesis. Cell-free synthesis was performed as described above. Translation was blocked 21 min after the initiation of the reaction (time indicated by arrow) by adding $MgCl_2$ at a final concentration of 10 m*M*. At the indicated time points, aliquots of the reaction mixture were withdrawn and bioluminescence activity was measured. (C) Renaturation of full-length $\beta$ subunit in the translation mixture. Urea-denatured $\beta$ subunit was added to the translation mixture prepared as described above, but without mRNA. Aliquots were removed and used to measure enzymatic activity. (D) Simulation of luciferase assembly reaction in the cell-free system. Kinetics of full-length $\beta$ subunit appearance are taken from (A). Simulated kinetics of active luciferase formation (solid line) were based on the assumption that $\beta$ subunit released from the ribosome (A, open symbols) folds at the same rate as urea-denatured polypeptide under the same conditions (C). Simulated kinetics of active luciferase formation following a block in synthesis at 21 min (time indicated by arrow) are shown by the dashed line. It is apparent that the simulations do not fit the data. Compare the solid line in (D) with the filled symbols in (A), and the dashed line in (D) with the data in (B).

from denaturant solution under the same conditions. To this end, urea-denatured $\beta$ subunit was renatured under conditions similar to those for biosynthetic folding. Urea-unfolded $\beta$ subunit was diluted at a concentration similar to its final concentration on synthesis into a translation mixture lacking mRNA but supplemented with $\alpha$ subunit. Refolding of the enzyme is a slow process, with luciferase regaining 50% of the final activity in about 9 min (Fig. 1C). Unlike renaturation experiments, in which all molecules start to refold immediately on dilution from denaturant into the native conditions, in the translation reaction there is a steady state appearance of the full-length intermediate. Consequently, a proper comparison of the kinetics of protein folding and assembly during renaturation and during biosynthesis cannot be done directly and requires simulation analysis using a modeling program such as Kinsim.[28] Modeling of the rate of $\beta$-subunit biosynthetic folding was performed as follows. It was assumed that $\beta$ subunit released from the ribosome folds at the same rate as the subunit diluted from denaturant into the cell-free system. In other words, it was assumed that whatever cotranslational folding events take place, they precede the rate-limiting step of the native heterodimer formation. The basic minimalist mechanism of luciferase subunit refolding and association is known. To simplify the analysis and fit the rate data, the renaturation kinetics presented in Fig. 1C were modeled by a polynomial fit. This approach avoids determination of the parameters of folding and association reactions under these particular conditions, including determination of the rate constants and consideration of potential involvement of chaperones. The rate of introduction of unfolded $\beta$ subunit was taken from the experimentally observed rate of polypeptide synthesis. Each increment of $\beta$ subunit was allowed to form active enzyme. The sum of these additions yields the predicted kinetics of luciferase formation, based on the initial assumption. As is seen in Fig. 1D, the lag between synthesis of $\beta$ subunit and appearance of active heterodimer in this simulation is around 10 min, much longer than the experimentally observed lag. Similar modeling of the experiment, in which translation was blocked during the accumulation phase, shows a large increase in luciferase activity after the block of $\beta$-subunit synthesis, contrary to the experimental observation. In this simulation, a plateau is reached after 17 min (Fig. 1D). The results of the simulations for both experiments are clearly contrary to the experimentally observed kinetics of luciferase biosynthetic folding. These observations demonstrate that the initial assumption used in the modeling, i.e., that the $\beta$ subunit released from the ribosome folds at the same rate as the $\beta$ subunit refolding from urea, is incorrect. In fact, $\beta$ subunit released from the ribosome interacts with $\alpha$

[28] B. A. Barshop, R. F. Wrenn, and C. Frieden, *Anal. Biochem.* **130,** 134 (1983).

subunit much faster than does the refolding subunit. Thus, for this protein, cotranslational folding does contribute significantly to the rate of formation of native structure. These results show that $\beta$ subunit is released from the ribosome in a form that is beyond the rate-limiting step encountered during refolding of the luciferase heterodimer.

## Determination of Affinity of Protein–Protein and Protein–Ligand Interactions in Cell-Free Systems

There are many approaches to the determination of binding affinity, but certain basic strategies may be discussed in detail. For any equilibrium binding studies, it is crucial to consider the temperature, the buffer composition and pH, and the concentrations of interacting species. Binding reactions may be accompanied by release of other molecules, such as salts. As a consequence, measured binding affinities are the composite of many solution parameters. This is especially true for measurement of binding affinities in a cell-free system, which contains a vast array of components that potentially could be involved in the interactions under investigation. The method that we describe below allows measurement of the apparent affinity of interacting species within complex mixtures. This approach is well suited to the study of interactions of newly synthesized polypeptides in cell-free expression systems.

The polypeptide of interest is incubated at a fixed concentration with different known concentrations of its partner, either protein or a ligand. After the reaction has reached equilibrium, the concentration of one of the polypeptide forms, either bound to its partner or free, should be determined by some available means. In most cases, antibodies against the polypeptide or its ligand coupled to an insoluble matrix can be used for this purpose. After a brief incubation of the matrix-coupled antibodies in the incubation mixture the matrix can be quickly separated by centrifugation and the amount of free or ligand-bound form of the polypeptide determined. Knowing the concentrations of the ligand added and the fractions of free or ligand-bound polypeptide, it is possible to determine the affinity (i.e., equilibrium association constant) of the protein–ligand interaction. A potential source of error in this method stems from the fact that the measurement can potentially perturb the equilibrium. That is, if the antibody binds only the free polypeptide and the dissociation rate constant is fast, then the equilibrium can reestablish itself and an erroneous measurement will result. This same difficulty arises from gel-shift and filter-binding assays that are commonly used to evalute protein–nucleic acid binding equilibria. If the complex under investigation is a high-affinity complex, then the dissociation rate constant will likely be slow, such that the equilibrium will not be perturbed by the measurement. Nonetheless, this is a potential source of error, and is discussed in more detail below.

### *Conditions for Determination of Affinity of Protein–Protein Interactions*

Conditions should be established such that only a small percentage of the polypeptide in the reaction mixture will be trapped by immunoadsorption. To this end, 2–10 μl of the translation mixture should be added to 150–300 μl of buffer A or any other buffer in which the equilibrium experiments will be performed. In this and following experiments buffer A should not contain DTT. To this mixture, 20–50 μl of IgG-coupled beads is added and mixed gently for 10–30 sec. The mixture should be immediately diluted with 1 ml of the buffer A and centrifuged at maximum speed in a benchtop centrifuge for 15 sec. The supernatant is carefully withdrawn by pipetting and the pellets washed twice with buffer A. The radioactivity incorporated into the polypeptide can be counted in the beads either directly, after elution of the polypeptide from the matrix, or following electrophoretic analysis. If nonspecific adsorption of radiolabeled polypeptide onto the beads is high, conditions of the incubation and washing procedures can be adjusted as discussed in the immunoadsorption procedure discussed above. If necessary, conditions under which an appropriate percentage of the total antigen is trapped should be adjusted.

After proper conditions have been established, a ligand is added to the translation mixture at various concentrations to measure their interaction affinity. After 10–60 min of incubation, immunobeads (Bio-Rad, Richmond, CA) should be added in the same manner as in the previous control experiment and the mixture should be treated exactly as in the control experiment. The amount of the polypeptide trapped at each ligand concentration should be determined.

### *Data Analysis*

The fraction of bound polypeptide at any given concentration of the ligand is equal to $Y = (R_0 - R_l)/(R_0 - R_c)$, where $R_0$ is the radioactivity recovered in the sample without ligand, $R_l$ is the radioactivity in the sample with the ligand, and $R_c$ is that in the control sample, i.e., where beads with any other unrelated antibody have been used. The reciprocal of the fraction of bound polypeptide, $1/Y$, should be plotted versus the reciprocal of the free ligand concentration, $1/L_f$. If the concentration of the ligand ($L_t$) is much higher than the total concentration of the polypeptide, then the free ligand concentration is essentially equal to its total concentration. The ligand concentration resulting in binding of half of the polypeptide, i.e., when $1/Y = 2$, provides a value for the equilibrium dissociation constant. The affinity corresponds to the reciprocal of this value.

Quantitation of the synthesized polypeptide in experiments of this type can be compromised by isotope dilution if there is appreciable free amino acid in the reaction mixture. Commercial expression systems are usually depleted but not entirely free of the amino acid to be used as a label. As a result, the actual specific radioactivity of the amino acid may be lower by a factor of 2 or more. The amount of the residual nonlabeled amino acid can be determined by quantitative amino acid analysis. Alternatively, mixtures of labeled amino acid can be used with 5- to 10-fold molar excess of nonlabeled amino acid, thus minimizing the effect of endogenous amino acid on the calculations.

A potential problem with this method, outlined above, is the essential condition to trap only a small percentage of the polypeptide on the matrix. Adsorption of too much of the polypeptide can lead to a shift of the equilibrium toward dissociation and, consequently, cause an error in the determination of the apparent affinity. In a system with a high dissociation rate constant, it is important that the mixing time of the reaction mixture with the immobilized antibody be as short as possible. If the dissociation rate constant is slow, however, as would be the case for a high-affinity complex, the equilibrium redistribution problem becomes insignificant.

This method has been applied to study the affinity of a monoclonal antibody toward synthesized polypeptide in a cell-free system.[1,21] A fixed amount of the reaction mixture was incubated with increasing amounts of free antibody. After the system had achieved equilibrium, the same antibody coupled to a resin was used to trap a small amount of the free polypeptide. The free polypeptide–free antibody interaction affinity was then determined, as well as for the nascent ribosome-bound polypeptide. The validity of this approach has been confirmed by comparison with the results obtained by ELISA techniques with purified polypeptides.

## Acknowledgments

We thank Dr. P. Swartz for critical reading of the manuscript. Supported by grants from the National Science Foundation (MCB-9513429) and the Office of Naval Research (N00014-96-1-87). Much of the work described here is based on previous work performed in collaboration with the laboratories of Drs. Yu. Alakhov, M. Goldberg, and O. Ptitsyn.

# [2] Preparation and Application of Chaperone-Deficient *Escherichia coli* Cell-Free Translation Systems

*By* Gisela Kramer, Tong Zhang, Wieslaw Kudlicki, and Boyd Hardesty

An important unanswered question for both protein folding and protein synthesis concerns how newly formed proteins acquire the conformation of their native state. Efficient refolding of many denatured proteins requires molecular chaperones. Do these chaperones play a role in folding of nascent proteins either as they are synthesized on ribosomes or immediately after they are terminated and released? We have shown that five bacterial chaperones (DnaJ, DnaK, GrpE, GroEL, and GroES) can promote folding and release of full-length rhodanese (E.C. 2.8.1.1) that was not properly folded and released from *Escherichia coli* ribosomes after its synthesis.[1] However, we question whether this is the major pathway by which nascent proteins are folded into their native conformation. Lorimer[2] estimated that GroEL could account for folding of no more than 5% of the proteins formed in rapidly growing *E. coli* cells. We have shown that *E. coli* salt-washed ribosomes, their 50S subunit, or even the 23S rRNA in its native state can efficiently refold denatured rhodanese independent of ATP.[3] Reports by Dasgupta and co-workers also indicate that prokaryotic and eukaryotic ribosomes have the ability to promote refolding of several other denatured enzymes.[4] Our results relate the ability of ribosomes to carry out refolding to the peptidyltransferase center and their ability to form peptide bonds.

To test the hypothesis that ribosomes themselves can promote folding of nascent proteins without the assistance of molecular chaperones, we have developed *in vitro* systems for protein synthesis that are deficient in molecular chaperones. Our aim is to develop a system that totally lacks these proteins.

The description of the experimental procedures addresses the following three aspects: (1) preparation of the cell-free systems, (2) methods to quantitate the chaperones that remain in these systems, and (3) *in vitro* synthesis of test proteins and their analysis, including their enzymatic activity.

[1] W. Kudlicki, O. W. Odom, G. Kramer, and B. Hardesty, *J. Biol. Chem.* **269,** 16549 (1994).
[2] G. H. Lorimer, *FASEB J.* **10,** 5 (1996).
[3] W. Kudlicki, A. Coffman, G. Kramer, and B. Hardesty, *Folding Design* **12,** 101 (1997).
[4] B. Dasgupta, A. Chattopadhyay, A. K. Bera, and C. Dasgupta, *Eur. J. Biochem.* **235,** 613 (1996).


METHODS IN ENZYMOLOGY, VOL. 290
0076-6879/98 $25.00

*Materials and Reagents*

All reagents, chemicals, and biochemicals are from Sigma (St. Louis, MO) unless specified otherwise.

*E. coli* strain A19 (originally obtained from H.-G. Wittmann and K. Nierhaus, Berlin) and strain PK101 (kindly provided by E. Craig, Madison, WI): Strain A19 (a K12 derivative) lacks one of the major ribonucleases; the genes for DnaK and DnaJ have been deleted from strain PK101[5]

Plasmids: pSP65 with the following coding sequences under the SP6 promoter: chloramphenicol acetyltransferase (CAT), dihydrofolate reductase (DHFR), rhodanese; pGEM with either CAT or rhodanese under the T7 promoter. In these constructs, the coding sequence for β-lactamase is on the opposite strand; thus only one protein is synthesized even though we do not linearize the plasmids

SP6 or T7 RNA polymerase (commercially available): We use our own preparations isolated by the procedure of Davanloo *et al.*[6] The enzyme protein is at 2–4 mg/ml in a solution with 50% (v/v) glycerol for storage at −20°

[$^{14}$C]Leucine from ICN (Costa Mesa, CA) or New England Nuclear (NEN)-Du Pont (Boston, MA): Dilute with unlabeled leucine to give 0.5 m*M* and a specific radioactivity of 40 Ci/mol

*E. coli* MRE 600 tRNA (Boehringer Mannheim, Indianapolis, IN): Resuspend in $H_2O$ at 10 mg/ml for use in coupled transcription/translation

For immunological analyses:

Molecular chaperones: Commercially available from Epicentre (Madison, WI)

Primary antibodies: Polyclonal rabbit anti-GroEL [kindly supplied by G. Merrill (Brooke Army Medical Center, Fort Sam Houston, San Antonio, TX)]; commercial mouse monoclonal anti-DnaK antibodies from Stressgen (Victoria, B.C., Canada)

Secondary antibodies to which horseradish peroxidase (HRP) is linked and 2,2′-azinobis(3-ethylbenzothiazoline)-6-sulfonic acid (ABTS), the substrate for the enzyme-linked immunosorbent assay (ELISA): Purchased from Zymed (South San Francisco, CA)

For Western blot: Polyvinylidene difluoride (PVDF) membrane is from Schleicher & Schuell (Keene, NH); the ECL (enhanced chemilumi-

[5] P. J. Kang and E. A. Craig, *J. Bacteriol.* **172,** 2055 (1990).

[6] P. Davanloo, A. Rosenburg, J. Dunn, and F. Studier, *Proc. Natl. Acad. Sci. U.S.A.* **81,** 2035 (1984).

nescence) detection system and Hyperfilm are from Amersham (Arlington Heights, IL)

## Procedures

### *Preparation of S30 Fraction*

The preparation of the *E. coli* S30 fraction is carried out essentially as described by Zubay.[7] *Escherichia coli* A19 cells are grown at 37°, PK101 cells at 30°, in Luria broth (LB) medium supplemented with 0.4% (w/v) glucose. The doubling time should be between 20 and 30 min. Cells are harvested by centrifugation at 5000 *g* for 15 min at 4° in midlog phase ($A_{600}$ 0.75), then washed by resuspension in a solution containing 10 m*M* Tris–acetate (pH 8.2), 14 m*M* magnesium acetate, 60 m*M* KCl, and 6 m*M* 2-mercaptoethanol (solution A). After centrifugation, the cells are spread in a thin layer on Saran Wrap, quickly frozen, and stored at −70°. About 5 g of cells is obtained per liter of culture.

To prepare the *E. coli* cell extract, the frozen cells are thawed, washed by resuspension in solution A, then centrifuged as before. The cell pellet is resuspended in solution A (about 1.2 ml/g cells). The cells are broken at room temperature in a French press (800 psi). Phenylmethylsulfonyl fluoride (PMSF; to 0.5 m*M*) and dithiothreitol (DTT; to 1 m*M*) are quickly added and the lysate is subjected to the French press for a second time under the same conditions, then centrifuged at 30,000 *g* for 30 min at 4°. The resulting supernatant fraction is removed, then incubated under protein-synthesizing conditions for 80 min at 37° as described by Zubay,[7] dialyzed overnight, quickly frozen in aliquots in liquid $N_2$, and stored at −70°.

### *Fractionation of S30 by Centrifugation*

About 3 ml of the S30 fraction is centrifuged at 100,000 *g* for 3 hr at 4°. Three distinct phases are discernible after this step: a clear postribosomal supernatant fraction that contains most of the soluble components, a gelatinous ribosomal pellet, and a relatively viscous, slightly amber-colored fluid between the ribosomes and the clear supernatant. We called this fraction the p-r (protein-rich) fraction because it contains about 60% of the total protein excluding the ribosomal proteins and about 65% of the EF-Tu (elongation factor Tu) that is present in the S30 fraction.[8] This fraction is carefully retrieved from above the ribosomal pellet after the supernatant

[7] G. Zubay, *Annu. Rev. Genet.* **7,** 267 (1973).

[8] W. Kudlicki, M. Mouat, J. P. Walterscheid, G. Kramer, and B. Hardesty, *Anal. Biochem.* **217,** 12 (1994).

fraction has been removed with a pipette. The p-r fraction comprises about 8–10% of the total volume of the S30 that is subjected to centrifugation. Finally, the ribosomes are resuspended in about half the original volume of an $NH_4Cl$ solution as described in the next section.

*Preparation of Salt-Washed Ribosomes*

The crude ribosomes from the previous steps are gently resuspended using a glass rod in a solution containing 20 m*M* Tris-HCl (pH 7.6), 10 m*M* magnesium acetate, 500 m*M* $NH_4Cl$, and 1 m*M* DTT and kept on ice for 30 min. The ribosomes are then sedimented by centrifugation at 100,000 *g* for 1.5 hr at 4°. The resulting supernatant (salt-wash fraction) is removed, then the ribosomal pellet is resuspended as described above but in a solution containing 30 m*M* rather than 500 m*M* $NH_4Cl$. The concentration of the ribosomes is adjusted to about 1500 $A_{260}$ units/ml. The ribosomes are stored in small aliquots at −70°.

*Chromatography of Protein-Rich Fraction*

Proteins of the p-r fraction are separated by size-exclusion chromatography on Sephadex G-150 exactly as described previously.[8] The matrix is loaded into a 1 × 73 cm column and equilibrated with a solution containing 20 m*M* Tris-HCl (pH 7.8), 10 m*M* magnesium acetate, 60 m*M* potassium acetate, and 1 m*M* DTT. About 130 $A_{280}$ units of the p-r fraction are chromatographed. The column is developed with the same solution at 7.5 ml/hr; 1-ml fractions are collected and assayed for their ability to support protein synthesis in a coupled transcription/translation system using salt-washed ribosomes. A plasmid of choice plus the respective RNA polymerase are used as detailed below. Fractions active for protein synthesis elute well behind the fractions that contain maximum concentrations of GroEL and DnaK (see Ref. 8). Fractions active for protein synthesis are pooled, concentrated by Centricon (Amicon, Danvers, MA) centrifugation if necessary, and stored at −70° in small aliquots.

*Immunological Analyses*

The ELISAs are carried out by the indirect method.[9] The wells are coated with the antigen for 45 min at room temperature, then for blocking treated with 30 μg of bovine serum albumin (BSA)/well for 30 min. The primary antibodies are added as antiserum in a 1:3000 to 1:5000 dilution. Incubation is for 90 min at room temperature. This is followed by a 60-min

[9] J. R. Crowther, "ELISA" (*Methods Mol. Biol.* **42**) (1995).

incubation with the HRP-linked secondary antibody. Color development is achieved by a 5-min incubation at room temperature with ABTS and $H_2O_2$ at pH 4.5. Color intensity is quantitated using an ELISA reader.

Western blotting is carried out by electrophoretic transfer of proteins from 10 to 12% sodium dodecyl sulfate–polyacrylamide gels (SDS–PAGE) (Laemmli[10]) onto a PVDF membrane using a semidry gel blotter. The membranes are probed with primary antibodies, then with secondary antibodies to which HRP is linked as described above. The position of HRP on the membrane is detected using the Amersham ECL kit, following manufacturer instructions.

As these are widely used methods, they are only outlined above; however, we will comment on a few points relevant to the topic discussed here. Because the titer of the antiserum used is quite high, GroEL can be quantitated by ELISA. A standard (pure GroEL as antigen) is included in all assays; linear response is obtained in the range of 5–30 ng. The DnaK antibodies used recognize only denatured proteins, thus they can be used only in Western blots. Here we use pure DnaK in the range of 20–100 ng as reference.

### *Cell-Free Coupled Transcription/Translation System*

*Basic Assay.* The standard assay contains the following components in a volume of 30 $\mu$l: 10 $\mu$l of LM-Mix (Table I), 2.5 $\mu$l of 120 m$M$ magnesium acetate, 5 $\mu$l of [$^{14}$C]leucine, 2 $\mu$l of tRNA, 0.5 $\mu$l of the appropriate RNA polymerase, 0.2 $\mu$l of pyruvate kinase at 1 mg/ml, and about 0.5 $\mu$g of plasmid. Rifampicin is added to give a 50 $\mu M$ final concentration. This compound blocks specifically the *E. coli* RNA polymerase. The S30 fraction (5 $\mu$l) is quickly thawed just before use and added last. Incubation is at 37° in a shaking water bath for 30 min. The reaction is stopped by adding about 100 $\mu$l of 1 $M$ NaOH. Incubation at 37° is continued for 5 min, then 2 ml of ice-cold 5% (w/v) trichloroacetic acid (TCA) is added. The precipitates are collected on glass fiber filters, then their radioactivity is determined by liquid scintillation counting. Background is determined with a sample from which the plasmid has been omitted.

*Chaperone-Deficient Assay.* In this assay system, the S30 fraction is replaced by salt-washed ribosomes (about 1.5 $A_{260}$ units) plus protein collected in fractions eluting after the main peak during the Sephadex G-150 chromatography of the p-r fraction. A typical elution profile from the chromatography of the p-r fraction from *E. coli* A19 is given in Ref. 8; protein synthesis was tested with aliquots of the fractions using a plasmid

[10] U. K. Laemmli, *Nature (London)* **227,** 680 (1970).

TABLE I
COMPOSITION OF LM-MIX[a]

| Component | Stock solution (*M*) | Concentration in LM-Mix (m*M*) |
|---|---|---|
| Tris–acetate | 2.0 | 165 |
| Potassium acetate | 2.0 | 216 |
| Ammonium acetate | 2.0 | 108 |
| Calcium acetate | 1.0 | 5 |
| PEG | (40%) | (6%) |
| DTT | 1.0 | 6 |
| ATP | 0.1 | 3.6 |
| GTP | 0.1 | 2.4 |
| CTP | 0.1 | 2.4 |
| UTP | 0.1 | 2.4 |
| Phosphoenolpyruvate | 0.5 | 90 |
| Glucose 6-phosphate | 0.1 | 1.0 |
| cAMP | 0.1 | 1.2 |
| EDTA | 0.5 | 1.5 |
| Folinic acid | (2.7 mg/ml) | (0.1 mg/ml) |
| Each of the 20 amino acids minus leucine | 0.1 | 1.5 |

[a] First, a solution containing a 5 m*M* concentration of each of the 19 amino acids is prepared from the 100 m*M* stock solutions. DTT is added to this mixture to 0.5 m*M* and the pH is adjusted to pH 7.0. All compounds listed above are then mixed together to give the concentrations indicated in the last column. Before the volume is finalized, the pH is checked and adjusted to pH 7.8 at room temperature. The LM-Mix is stored at $-70°$ in about 100-$\mu$l aliquots.

with the DHFR-coding sequence and salt-washed ribosomes from *E. coli* A19. The ribosomes by themselves do not incorporate leucine into polypeptides. Fractions from a corresponding chromatography on Sephadex G-150 of the p-r fraction isolated from *E. coli* PK101 were tested for synthesis and activity of CAT and for distribution of GroEL. The results are shown in Table II. Fractions from the main protein peak with high GroEL content are compared to those (fractions 30–36) with low GroEL concentrations. The specific enzymatic activity of CAT decreased only slightly when these chaperone-deficient fractions were used to supplement salt-washed ribosomes for synthesis of CAT.

*Note:* It is necessary to optimize the amounts of at least the following components that are used in the coupled transcription/translation assay: the S30 fraction or the salt-washed ribosomes, the plasmid, and the $Mg^{2+}$ concentration for each new preparation of S30 or salt-washed ribosomes.

TABLE II
SYNTHESIS OF ENZYMATICALLY ACTIVE CAT IN CHAPERONE-DEFICIENT SYSTEM[a]

| Fraction | $A_{280}$/ml | GroEL (ng/μl) | CAT synthesis (ng/30 μl) | CAT activity (U × $10^{-3}$/30 μl) | Specific activity (U/mg) |
|---|---|---|---|---|---|
| 20 | 2.05 | 376 | 94 | 11.46 | 121.9 |
| 22 | 2.13 | 377 | 123 | 14.77 | 120.1 |
| 30 | 1.62 | 36 | 155 | 18.81 | 121.4 |
| 34 | 1.34 | 25 | 156 | 16.47 | 105.6 |
| 36 | 1.18 | 22 | 124 | 13.87 | 111.9 |

[a] The protein-rich fraction isolated from PK101 cells was chromatographed on Sephadex G-150 as described.[8] Fractions of 0.5 ml were collected and their absorbance at 280 nm was measured. Distribution of GroEL was determined with a 0.01- to 0.1-μl aliquot from selected fractions by ELISA, using anti-GroEL polyclonal antibodies. The assay was carried out as described in text; the relative amount of GroEL was determined by comparison to the absorbance at 414 nm obtained with standard GroEL in the range of 3–20 ng. Synthesis of CAT was determined by coupled transcription/translation from a plasmid carrying the CAT-coding sequence using 1.5 μl of salt-washed ribosomes and 5 μl of each fraction plus all other components necessary for protein synthesis as described in text. Incubation was at 37° for 30 min. An aliquot was then taken from the reaction mixture to determine incorporation of [$^{14}$C]leucine into polypeptides. The results obtained were converted to picomoles of leucine incorporated per 30-μl reaction mixture; from these values, the amount of CAT synthesized (nanograms per 30-μl reaction mixture) was calculated; there are 13 leucine residues in the CAT sequence. Another aliquot was taken to determine CAT enzymatic activity, which was measured at 30° after a 10-min incubation. Specific enzymatic activity was calculated and is given as micromoles of product formed per minute per milligram of protein synthesized.

## *Analysis of Product Formed*

*Analysis of Released Polypeptides.* To analyze released polypeptides, the standard 30-μl reaction mixture is usually enlarged to 60 or 90 μl. After incubation, the reaction mixture is loaded over a 50-μl sucrose solution [0.5 *M* sucrose in 20 m*M* Tris–acetate (pH 7.8), 100 m*M* ammonium acetate, 10 m*M* magnesium acetate, 1 m*M* DTT; solution 1] and centrifuged in an airfuge (Beckman) at 140,000 *g* for 40 min at room temperature. After centrifugation, the supernatant (including the sucrose solution) is carefully removed, then the ribosomal pellet is resuspended in the original volume of solution 1. Aliquots (10–20 μl) of both the supernatant and the ribosomal fraction are taken to determine incorporation of leucine into polypeptides as described above.

Another aliquot (20 μl) is taken from each fraction to determine the amount and portion of full-length polypeptides. This is accomplished by SDS–PAGE electrophoresis carried out according to Laemmli,[10] followed by autoradiography. Usually the gel is exposed for about 48 hr to Hyperfilm

(Amersham) before the film is developed. To measure enzymatic activity of the test protein synthesized, an aliquot is removed from the supernatant fraction. No activity is observed with polypeptides (even full-length) remaining on the ribosomes except when at least a 23-amino acid extension is added to the C terminus of the enzyme.[11,12]

*Enzyme Assays.* The enzyme assays used are referenced and briefly outlined below. The incubation time and the size of the aliquot taken for the enzyme assay must be tested to ensure that initial rates are determined. The enzyme assays described here for rhodanese and DHFR require only about 10–20 ng of the synthesized polypeptides released into the supernatant. Even less is required to measure CAT activity. Background for the assay system is determined with a reaction mixture from which the plasmid was omitted during coupled transcription/translation. Virtually no background is detected when rhodanese or CAT is assayed. Significant endogenous activity for DHFR is present in the S30 fraction. It is barely detectable in the fractionated system.

Rhodanese activity is determined as described by Sörbo[13] with minor modifications (cf. Kudlicki *et al.*[1]). Rhodanese is a mitochondrial sulfur transferase that converts $CN^-$ to $SCN^-$ from $Na_2S_2O_3$. The amount of the product formed, thiocyanate, is determined by formation of a complex with ferric ions. Its absorbance is measured at 460 nm; $\varepsilon = 4200$. One unit of enzymatic activity is defined as 1 $\mu$mol of product formed per minute at 25°.

Dihydrofolate reductase activity is determined spectrophotometrically by the decrease in absorbance of reduced nicotinamide adenine dinucleotide phosphate (NADPH) at 340 nm, the substrate that reduces the cosubstrate, dihydrofolate (DHF). The enzymatic activity is measured in a solution containing 100 m$M$ imidazole chloride (pH 7.0), 10 m$M$ 2-mercaptoethanol, 75 $\mu M$ DHF, and 100 $\mu M$ NADPH (cf. Ma *et al.*[14]).[15] One unit of enzymatic activity is defined as the amount of enzyme required to reduce 1 $\mu$mol of DHF per minute at 30°, on the basis of a molar extinction coefficient of $6 \times 10^3$ at 340 nm for NADPH (note that 2 mol of NADPH is required to reduce 1 mol of DHF).

CAT activity is determined by transfer of the [$^{14}$C]acetyl group from acetyl-CoA to chloramphenicol.[16] The radioactive product formed is extracted from the reaction mixture with cold ethyl acetate and quantitated by liquid scintillation counting. One unit of enzyme activity is defined as

[11] W. Kudlicki, J. Chirgwin, G. Kramer, and B. Hardesty, *Biochemistry* **34,** 14284 (1995).

[12] E. Makeyev, V. Kolb, and A. Spirin, *FEBS Lett.* **378,** 166 (1996).

[13] B. H. Sörbo, *Acta. Chem. Scand.* **7,** 1129 (1953).

[14] C. Ma, W. Kudlicki, O. W. Odom, G. Kramer, and B. Hardesty, *Biochemistry* **32,** 7939 (1993).

[15] D. Baccanari, A. Phillips, S. Smith, D. Sinski, and J. Burchall, *Biochemistry* **14,** 5267 (1975).

[16] M. J. Sleigh, *Anal. Biochem.* **156,** 251 (1986).

the amount of enzyme that produces 1 μmol of acetylated chloramphenicol per minute at 30°.

## Comments

In the chaperone-deficient *E. coli* A19 system described here, using salt-washed ribosomes and the late-eluting fractions after Sephadex G-150 chromatography of the p-r fraction, the GroEL concentration is reduced to approximately 10% of the value determined for the coupled transcription/translation system with the S30 fraction. The corresponding chaperone-deficient system prepared from *E. coli* PK101 cells is devoid of detectable amounts of the molecular chaperones DnaJ and DnaK; however, the GroEL concentration is higher because the PK101 S30 fraction contains about three or four times more GroEL than the A19 S30 fraction.

Future experiments are aimed at eliminating GroEL completely from the protein-synthesizing system. These experiments will be successful only if GroEL is not tightly complexed with components necessary for the synthesis of proteins.

## Acknowledgments

The research presented in this manuscript was supported by a grant from the Foundation for Research. We thank Yolanda Easley for typing the manuscript.

# [3] Protein Disulfide Isomerase

*By* H. F. Gilbert

## Introduction

For proteins in which the native structure is stabilized by disulfide bonds, folding into the correct three-dimensional structure requires coupling tertiary structure formation to a chemical oxidation that cross-links specific cysteine residues (oxidative folding).[1,2] Spontaneous oxidative folding is often slow and may be complicated by competing aggregation. In the cell,

[1] T. E. Creighton, *Prog. Biophys. Mol. Biol.* **33,** 3231 (1978).

[2] H. F. Gilbert, *in* "Protein Folding" (R. Pain, ed.), pp. 104–136. Oxford IRL Press, Oxford, 1994.





0076-6879/98 $25.00

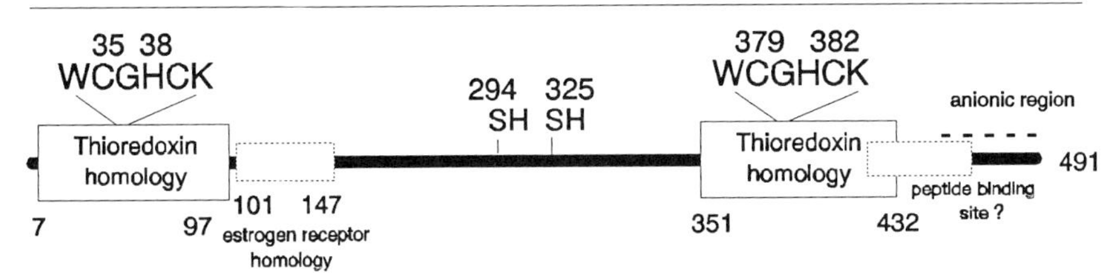


FIG. 1. Sequence organization of PDI. Sequence numbering is for the rat enzyme.

oxidative folding is assisted by enzymes that catalyze disulfide bond formation[3-6] and molecular chaperones that inhibit hydrophobic aggregation.[7]

Disulfide-containing proteins are mostly extracellular,[8] and in eukaryotes oxidative folding begins either during, or shortly after, translocation of the newly synthesized protein into the lumen of the endoplasmic reticulum (ER).[9,10] The ER maintains high concentrations of folding catalysts and molecular chaperones and provides a redox environment of glutathione (GSH) and its disulfide (GSSG) that is appropriate for disulfide formation.[11] Protein disulfide isomerase (PDI, EC5.3.4.1), a 55-kDa ER resident, is the most active and most abundant oxidative folding catalyst isolated to date.[3,4,12] PDI is a very abundant protein; its concentration in the ER has been estimated at near m*M*.[13]

Protein disulfide isomerase is a member of the thioredoxin superfamily, having two functional domains with significant sequence homology to the redox cofactor, thioredoxin.[14,15] These two domains, located near the N- and C-terminal regions of the protein (Fig. 1), are responsible for all of the disulfide isomerase activity.[16,17] PDI is a multifunctional protein with

[3] R. B. Freedman, *Cell* **57,** 1069 (1989).
[4] R. Noiva and W. J. Lennarz, *J. Biol. Chem.* **267,** 3553 (1992).
[5] D. A. Hillson, N. Lambert, and R. B. Freedman, *Methods Enzymol.* **107,** 281 (1984).
[6] R. Noiva, *Protein Expression Purif.* **5,** 1 (1994).
[7] M. J. Gething and J. Sambrook, *Nature (London)* **355,** 33 (1992).
[8] H. F. Gilbert, *Adv. Enzymol.* **63,** 69 (1990).
[9] L. W. Bergman and W. M. Kuehl, *J. Biol. Chem.* **254,** 8869 (1979).
[10] J. R. Huth, K. Mountjoy, F. Perini, E. Bedows, and R. W. Ruddon, *J. Biol. Chem.* **267,** 21396 (1992).
[11] C. Hwang, A. J. Sinskey, and H. F. Lodish, *Science* **257,** 1496 (1992).
[12] M. M. Lyles and H. F. Gilbert, *Biochemistry* **30,** 613 (1991).
[13] A. Zapun, T. E. Creighton, P. J. E. Rowling, and R. B. Freedman, *Proteins* **14,** 10 (1992).
[14] J. C. Edman, L. Ellis, R. W. Blacher, R. A. Roth, and W. J. Rutter, *Nature (London)* **317,** 267 (1985).
[15] R. B. Freedman, T. R. Hirst, and M. F. Tuite, *Trends Biol. Sci.* **19,** 331 (1994).
[16] K. Vuori, R. Myllyla, T. Pihlajaniemi, and R. I. Kivirikko, *J. Biol. Chem.* **267,** 7211 (1992).
[17] M. LaMantia and W. J. Lennarz, *Cell* **74,** 899 (1993).

a number of diverse functional roles in the cell. In addition to thiol–disulfide exchange, PDI catalyzes the GSH-dependent reduction of dehydroascorbate[18]; binds $Ca^{2+}$,[19,20] estrogen,[21] and thyroid hormones[22,23]; and serves as a subunit of prolyl hydroxylase[24] and an ER triglyceride transferase.[25] When present in stoichiometric excess, high concentrations of PDI exhibit chaperone activity and inhibit the aggregation of denatured proteins.[26,27] However, under conditions in which unfolded proteins are in stoichiometric excess, PDI can actually increase the formation of protein aggregates (antichaperone activity).[26,28]

This chapter provides a description of the purification of PDI from recombinant *Escherichia coli* and evaluates several assays for PDI activity. A summary of the properties of the purified enzyme is also presented, including those features of the protein that should be of interest in refolding strategies. It is relatively easy to isolate large quantities of PDI from natural sources (~100 mg from 500 g of bovine liver)[5] or by expression of mammalian PDI in *E. coli* (~10–20 mg/liter of culture).[29]

## Purification

Protein disulfide isomerase is a very abundant protein, comprising approximately 0.4% of total protein in liver.[5] From 500 g of bovine liver, 80–100 mg of PDI can be obtained in a relatively simple preparation that involves two ion-exchange columns. An excellent description of this purification appears in a previous volume of this series.[5]

[18] W. W. Wells, D. P. Xu, Y. F. Yang, and P. A. Rocque, *J. Biol. Chem.* **265,** 15361 (1990).
[19] P. N. Van Ngyen, K. Rupp, A. Lampen, and H. D. Soling, *Eur. J. Biochem.* **213,** 789 (1993).
[20] K. Rupp, U. Birnbach, J. Lundstrom, P. N. Van, and H. D. Soling, *J. Biol. Chem.* **269,** 2501 (1994).
[21] J. C. Tsibris, L. T. Hunt, G. Ballejo, W. C. Barker, L. J. Toney, and W. N. Spellacy, *J. Biol. Chem.* **264,** 13967 (1989).
[22] S.-Y. Cheng, Q.-H. Gong, C. Parkinson, E. A. Robinson, E. Apella, G. T. Merlino, and I. Pastan, *J. Biol. Chem.* **262,** 11221 (1987).
[23] K. Yamaughi, T. Yamamoto, H. Hayashi, S. Koya, H. Takikawa, K. Toyoshima, and R. Horiuchi, *Biochem. Biophys. Res. Commun.* **146,** 1485 (1987).
[24] J. Koivu, R. Myllyla, T. Helaakoski, T. Pihlajaniemi, K. Tasanen, and K. I. Kivirikko, *J. Biol. Chem.* **262,** 6447 (1987).
[25] J. R. Wetterau, K. A. Combs, S. N. Spinner, and B. J. Joiner, *J. Biol. Chem.* **265,** 9800 (1990).
[26] A. Puig and H. F. Gilbert, *J. Biol. Chem.* **269,** 7764 (1994).
[27] H. Cai, C.-C. Wang, and C.-L. Tsou, *J. Biol. Chem.* **269,** 24550 (1994).
[28] A. Puig, M. M. Lyles, R. Noiva, and H. F. Gilbert, *J. Biol. Chem.* **269,** 19128 (1994).
[29] H. F. Gilbert, M. L. Kruzel, M. M. Lyles, and J. W. Harper, *Protein Expression Purif.* **2,** 194 (1991).

*Recombinant Protein Disulfide Isomerase*

Recombinant PDI from several organisms has been expressed in *E. coli* both as a soluble cytosolic protein[17,29] or in the bacterial periplasm.[16] Functionally, the recombinant rat protein is identical to the protein isolated from bovine liver in all activity assays, including protein refolding,[29] insulin reduction,[29] and chaperone and antichaperone activities.[26] The following procedure for isolating recombinant rat PDI from *E. coli* has been used in our laboratory for several years to produce PDI that is >90% homogeneous by sodium dodecyl sulfate–polyacrylamide gel electrophoresis (SDS–PAGE). If desired, the protein can be purified to >99% homogeneity by high-performance liquid chromatography (HPLC) using gel filtration or DEAE ion-exchange columns.

*Bacterial Strains and Growth.* pETPDI.2 is a pET-8c-based plasmid (Novagen, Madison, WI) containing the coding sequence for rat PDI in an expression vector driven by a T7 RNA polymerase-dependent promoter and an ampicillin resistance marker. The ER signal sequence has been eliminated so that the PDI coding sequence begins with aspartate. The recombinant protein has Met-Asp at the N terminus; however, the *N*-formyl group is not present.[29] Expression is performed in *E. coli* strain BL21(DE3) (Novagen), which contains the T7 RNA polymerase gene under the control of an isopropylthiogalactoside (IPTG)-inducible *lacUV* promoter. Although it has not been studied systematically, PDI expression levels may decrease significantly on storage of cryopreserved *E. coli*; cells that are freshly transformed may yield higher levels of expression. Expression levels are generally high; however, an occasional clone (10–15%) may show low expression. It is advisable to test each of 6–10 individual colonies from the transformation for expression levels. After transformation with pETPDI.2, single colonies, selected on ampicillin plates, are introduced into 2 ml of LB medium containing ampicillin (50 μg/ml), grown to an absorbance at 600 nm of 0.4–0.6, induced with 0.4 m*M* IPTG, and grown until the absorbance at 600 nm is 1.0–1.4. Cells from 1 ml of culture are removed by centrifugation and lysed in 200 μl of 2× SDS–PAGE sample buffer containing 4% (w/v) SDS. After heating for 3–5 min at 100°, a 10- to 20-μl aliquot of each culture is subjected to SDS–PAGE. Expression levels are assessed by comparing the intensity of the PDI band (55 kDa) on a Coomassie-stained 10% (w/v) SDS polyacrylamide gel with that of an uninduced control. Clones showing high expression are used to inoculate two 1-liter cultures of LB medium containing ampicillin (50 μg/ml). Standard shaker cultures are grown at 37° to an $A_{600}$ of 0.4, induced with IPTG (0.4 m*M*), then grown to an $A_{600}$ of 1.2–1.4 (about 3–4 hr).

After centrifugation at 11,000 *g* for 15 min at 4° (in 500-ml bottles), packed cells from 500 ml of culture are resuspended in 35 ml of 20 m$M$ Tris-HCl (pH 7.6), 10% (w/v) sucrose, 0.2 $M$ NaCl, 2 m$M$ EDTA, 1 m$M$ phenylmethylsulfonyl fluoride (PMSF), and 25 $\mu M$ leupeptin. If desired, the preparation (or parallel preparations) may be frozen at $-80°$ at this point with no noticeable effect on subsequent steps. After incubation of the suspended cells with lysozyme (0.1 mg/ml) at 0° for 45 min, another aliquot of PMSF along with 0.1% (v/v) Triton X-100 (final concentration) is added, and each 35 ml is sonicated (five times, 30 sec each). After centrifugation (11,000 *g*, 30 min, 4°) to remove debris, the soluble extract is dialyzed against 25 m$M$ potassium phosphate buffer, pH 6.3 (twice, 4 liters each time).

*Chromatography on DEAE-Sephacel.* The dialyzed extract is applied to a 2.5 × 20 cm column of DEAE-Sephacel equilibrated with 0.025 $M$ sodium phosphate, pH 6.3. The column is washed until the $A_{280}$ is $<0.1$, then eluted with a linear gradient of 0–0.5 $M$ NaCl (1 liter of each). Elution is continued with 0.5 $M$ NaCl. This gradient is somewhat shallower than that originally reported and has been found to give better separation of PDI from contaminating proteins. PDI elutes as a broad peak near 0.3 $M$ NaCl (at a volume of approximately 1200–1350 ml after starting the gradient). The reason for the relatively broad peak is most likely due to aggregation heterogeneity in the PDI (see below). The DEAE column is the most critical step of fractionation. Because of the low specific activity of PDI, it is difficult to base purification on enzyme assays alone. Fractions are examined for PDI and impurities by SDS–PAGE, and fractions with the highest level of PDI and lowest levels of contaminating proteins are combined. A 22-kDa protein eluting at the trailing edge of the PDI peak is the most difficult impurity to remove, and fractions containing significant quantities of this protein should not be included. This protein is a C-terminal fragment of PDI resulting from initiation of translation at the second AUG in the sequence and is difficult to remove in subsequent steps.[30] PDI should be 75–80% homogeneous at this stage of the preparation.

*$Zn^{2+}$ Affinity Chromatography.* PDI binds $Zn^{2+}$ and other divalent metals, and the enzyme can be purified by chromatography on a $Zn^{2+}$ affinity column.[29] The chelate column (2.5 × 10 cm) is prepared by batch washing Pharmacia (Piscataway, NJ) chelating Sepharose Fast Flow (~50 ml) with 20 m$M$ sodium phosphate (pH 7.0) containing 1 $M$ NaCl, rinsing with water, washing with 400 ml of $ZnCl_2$ [5 mg/ml, adding a few drops of 1 $M$ HCl to avoid $Zn(OH)_2$ formation], and finally equilibrated with 20 m$M$ sodium phosphate (pH 7.0) containing 1 $M$ NaCl. The PDI-containing fractions from the DEAE column are combined, adjusted to pH 7.0, and additional

[30] H. F. Gilbert, unpublished data (1997).

salt is added (0.7 *M*) to bring the NaCl concentration to approximately 1 *M*. After applying PDI to the column and washing with 100–150 ml of 20 m*M* sodium phosphate with 1 *M* NaCl (pH 7.0), the enzyme is eluted in a sharp peak with 0.025 *M* sodium acetate, pH 4.5. $Zn^{2+}$ induces aggregation of PDI, particularly at low pH, so that the fractions from the $Zn^{2+}$ column (8 ml) are collected into tubes containing 0.16 ml of 1 *M* Tris base to raise the pH as the fractions are collected. The enzyme is generally >90% homogeneous by SDS–PAGE after this step. Fractions with an $A_{280} > 0.2$ are pooled and dialyzed against 0.1 *M* Tris-HCl, pH 8.0, containing 2 m*M* EDTA and 10 m*M* dithiothreitol. After further dialysis to remove the dithiothreitol, the solution is sterilized by filtering through a 0.2-$\mu$m (pore size) filter into sterile microcentrifuge tubes and stored at $-20°$ in appropriate-sized aliquots to avoid repeated freeze–thaw cycles. A preparation beginning with 2 liters of initial culture generally yields approximately 20–40 mg of purified PDI. At 280 nm, the absorbance of a 1.0-mg/ml solution of PDI is 1.0 as determined by quantitative amino acid composition analysis.

*High-Performance Liquid Chromatography.* If a higher state of purity (>90%) is required, PDI can be further purified by HPLC using a DEAE column (DEAE-5PW or comparable) at pH 6.3, 0.025 *M* sodium phosphate eluted with a 0.05–0.6 *M* NaCl gradient, or by HPLC gel filtration. At low pH (pH <7), PDI has a tendency to aggregate into a mixture of higher order oligomers. Dialysis and storage of the enzyme at pH 8.0 with EDTA decrease PDI aggregation. Preparations that have not been dialyzed at higher pH and reduced with dithiothreitol (DTT) may show multiple peaks of PDI on both DEAE and gel-filtration HPLC. Reduction of the protein with 10 m*M* DTT for 1 hr at pH 8.0, 2.0 m*M* EDTA will convert all aggregates to a single species.

*Protein Disulfide Isomerase Oligomerization.* PDI that is >90% homogeneous by SDS–PAGE (reducing or nonreducing) often gives rise to multiple peaks on HPLC ion-exchange or gel filtration. This behavior is not unique to recombinant PDI; PDI purified from bovine liver by the method of Hillson *et al.*[5] also exhibits such heterogeneity.[30,31] The mixture is dominated by two species that gel filter with apparent molecular weights of PDI dimers and tetramers; however, preliminary sedimentation equilibrium experiments suggest that at pH 7.0 (0.05 *M* HEPES), the lower molecular weight peak by gel filtration exists in concentration-dependent equilibrium between a 55-kDa monomer and a tetramer of 220 kDa.[30] Studies to define the relationships between the species observed by sedimentation equilibrium and the aggregates observed by gel filtration are in progress. Higher

[31] X. C. Yu, C. C. Wang, and C. L. Tsou, *Biochim. Biophys. Acta* **1207,** 109 (1994).

order aggregates (≥ octamer) may also be observed. PDI size heterogeneity has been attributed to C-terminal proteolysis[32]; however, the apparent molecular weight differences observed by gel-filtration HPLC are much too large not to be noticeable on SDS–PAGE, and mass spectrometry suggests that both species have the same (±20 amu) molecular weight.[30] PDI oligomers are not always cross-linked by intermolecular disulfides, but reduction with DTT (10 m$M$, pH 7.0) eliminates oligomers, converting them to PDI that appears dimeric by HPLC. Aggregation is enhanced by oxidation, lower pH, and 0.5–1 m$M$ concentrations of certain divalent metals ($Zn^{2+}$, $Cu^{2+}$, $Cd^{2+}$, and $Hg^{2+}$ but not $Ca^{2+}$, $Mg^{2+}$, $Co^{2+}$, or $Mn^{2+}$).[30] PDI that gel filters as a tetramer has approximately half the specific activity of PDI that gel filters as a dimer[30,31]; consequently, the accumulation of PDI oligomers reduces the specific activity, but by a relatively small amount (preparations that are 50% monomer will have 75% the specific activity of preparations containing only monomer). It is not clear how these oligomeric forms of PDI are related to the relatively small amount of large oligomers observed in PDI preparations by Pace and Dixon.[33] The implications of PDI aggregation on the *in vivo* behavior of the enzyme are not yet clear; however, the complex aggregation behavior can be minimized by dialysis of the enzyme at pH 8.0 in the presence of 2 m$M$ EDTA and 10 m$M$ dithiothreitol.

## Assay

Protein disulfide isomerase is not a particularly effective catalyst, and all assays for activity have a significant background reaction, making it difficult to asay the activity of the enzyme in crude homogenates. There are three assays of PDI enzyme activity commonly in use; none are entirely satisfactory. These include catalysis of the GSH-dependent reduction of insulin,[34,35] the thiol-dependent isomerization of scrambled RNase,[5,36] and the oxidative folding of reduced RNase in a glutathione redox buffer.[12] LaMantia and Lennarz[17] have reported an assay based on the increase in the activity of reduced, denatured bovine pancreatic trypsin inhibitor; however, no specific activity data were reported. For routine use in assaying enzyme preparations, the reduction of insulin is the easiest assay to perform, whereas the protein-refolding assays are more useful for mechanistic studies.

[32] C. H. Hu and C. L. Tsou, *Biochem. Biophys. Res. Commun.* **183,** 714 (1992).
[33] M. Pace and J. E. Dixon, *Int. J. Peptide Protein Res.* **14,** 409 (1979).
[34] N. Lambert and R. B. Freedman, *Biochem. J.* **213,** 235 (1983).
[35] N. A. Morjana and H. F. Gilbert, *Biochemistry* **30,** 4985 (1991).
[36] H. C. Hawkins, E. C. Blackburn, and R. B. Freedman, *Biochem. J.* **275,** 349 (1991).

*Insulin Reduction*

The PDI-catalyzed, GSH-dependent reduction of insulin can be observed continuously by coupling the formation of GSSG to NADPH oxidation through glutathione reductase.[34,35]

$$\text{Insulin-(SS)} + 2\text{GSH} \xrightarrow{\text{PDI}} \text{insulin-(SH)}_2 + \text{GSSG} \quad (1)$$

$$\text{GSSG} + \text{NADPH} + \text{H}^+ \xrightarrow[\text{reductase}]{\text{glutathione}} 2\text{GSH} + \text{NADP}^+ \quad (2)$$

Equilibrate 1 ml of 0.2 *M* phosphate, 5 m*M* EDTA, pH 7.5 at 25.0° in a spectrophotometer. The analytical wavelength is 340 nm. Add 3.7 m*M* GSH, 0.12 m*M* NADPH, and 16 units of glutathione reductase, and preincubate to reduce the contaminating GSSG in the GSH (15–30 sec). Because GSH will slowly oxidize to GSSG, GSH stock solutions are made fresh daily and stored on ice. Add 30 $\mu M$ bovine insulin, and record the rate of uncatalyzed insulin reduction for 2 min. Under the conditions specified, the background rate is <0.01 absorbance units (AU)/min. PDI is added to a final concentration of 10–50 $\mu$g/ml, and the rate of decrease of absorbance is measured for 1–2 min. After subtracting the background rate, the specific activity is calculated on the basis of a $\Delta\varepsilon_{340}$ of 6.23 m$M^{-1}$ cm$^{-1}$. The specific activity of homogeneous, monomeric PDI in this assay is 0.1 $\mu$mol of GSSG formed min$^{-1}$ (mg of protein)$^{-1}$. At a GSH concentration of 8 m*M*, the specific activity is higher, 0.2 $\mu$mol of GSSG formed min$^{-1}$ mg$^{-1}$, but the background reaction is also higher.[34] The apparent $K_m$ for insulin is 5 $\mu M$, whereas the apparent $K_m$ for GSH is 17 m*M*.[35]

This is by far the easiest assay for activity. The substrates (except for GSH and NADPH) can be prepared as stock solutions and frozen. The assay is quick and continuous. The major disadvantage is that the assay measures only one of the many activities of PDI, the ability to catalyze reduction of disulfide bonds. Other proteins such as thioredoxin are quite active in this assay.

*Continuous RNase Refolding*

In a glutathione redox buffer, PDI catalyzes the oxidative renaturation of reduced, denatured RNase. By carrying out the refolding of RNase in the presence of the RNase substrate, cCMP, the rate of cCMP hydrolysis at any time during the assay can be used to calculate the concentration of catalytically active RNase that is present.[12]

$$\text{RNase(SH)}_8 + 4\text{GSSG} \xrightarrow{\text{PDI}} \text{RNase(SS)}_4 + 8\text{GSH} \quad (3)$$

$$\text{cCMP} \xrightarrow{\text{RNase}} \text{CMP} \quad (4)$$

The hydrolysis of cCMP is followed by the change in absorbance at 296 nm, providing a relatively continuous assay of the active RNase concentration. The presence of cCMP and CMP has no effect on the rate of formation of the native, active enzyme.[37]

*Assay Procedure.* Tris–acetate buffer (0.2 $M$, pH 8.0), 2 m$M$ EDTA, a glutathione redox buffer (1 m$M$ GSH and 0.2 m$M$ GSSG), and 0.5–2 $\mu M$ PDI (0.03–0.12 mg/ml) are thermally equilibrated at 25° in a spectrophotometer along with a parallel background control that does not contain PDI. Using semimicrocuvettes, an assay volume of 0.4 ml can be used. The spectrophotometer is blanked on the redox buffer and PDI to subtract out any absorbance due to PDI. It is important to blank the absorbance on all components of the assay except cCMP because the initial absorbance after the addition of cCMP is used to determine the initial cCMP concentration. At PDI concentrations greater than 30 $\mu M$, the absorbance due to PDI at the analytical wavelength becomes too high to permit adequate subtraction of the background absorbance. cCMP (4.5 m$M$ final concentration) is added, followed by the addition of reduced, denatured RNase (8 $\mu M$ final concentration). The initial absorbance should be 0.85 $\pm$ 0.1. The absorbance is measured at 296 nm for 15–30 min at 0.5-min intervals. A 15- to 30-sec sampling is near optimal; a longer interval yields too few points for calculating the first derivative whereas a much shorter sampling time results in a small absorbance change and a noisy first derivative. A typical time course for the absorbance change during the assay in the presence and absence of PDI is shown in Fig. 2.

*Analysis of Data.* The concentration of refolded, catalytically active RNase at any time during the assay is proportional to the instantaneous velocity of cCMP hydrolysis. Because the cCMP concentration is near the $K_m$ and the product, CMP, is a competitive inhibitor of RNase, the observed velocity of cCMP hydrolysis depends not only on the concentration of active RNase, but also on the concentrations of the substrate (cCMP) and product (CMP). The depletion of cCMP and the inhibition by the product mean that the same molar amount of active RNase will generate a larger absorbance change during the initial part of the assay than in the final stages of the assay.

The procedure for determining the amount of active RNase at any time amounts to calculating the concentration of fully active RNase that would

[37] S. W. Schaffer, A. K. Ahmed, and D. B. Wetlaufer, *J. Biol. Chem.* **250,** 8483 (1975).

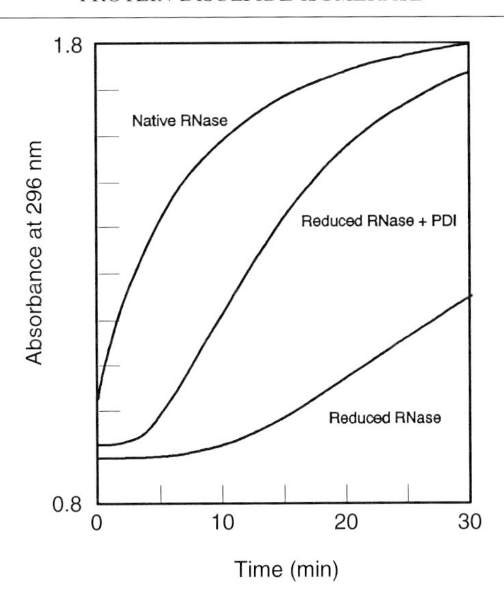


FIG. 2. Absorbance change during the PDI-catalyzed refolding of reduced RNase in the presence of cCMP. Time course for the absorbance change at 296 nm produced by RNase-catalyzed hydrolysis of cCMP at pH 8.0 (0.1 *M* Tris–acetate buffer), 25°. All samples contained a glutathione redox buffer (0.2 m*M* GSSG, 1.0 m*M* GSH). The concentration of RNase A (native or reduced, denatured) was 8.4 $\mu M$, and the concentration of PDI, when present, was 1.4 $\mu M$. There is no significant absorbance change in the absence of RNase A. (Adapted, with permission, from M. M. Lyles and H. F. Gilbert, *Biochemistry* **30,** 613 (1991). Copyright 1991 American Chemical Society.)

tions of cCMP and CMP that are present in the assay at that time.[12] At each time point, the velocity of cCMP hydrolysis ($v_t$) is evaluated numerically by determining the least-squares slope through the central data point and two points (taken at 15- to 30-sec intervals) on either side of the central point. The slope ($\Delta A$/min) is converted to m*M* cCMP hydrolyzed per minute by dividing by the $\Delta\varepsilon$ (0.19 m$M^{-1}$ cm$^{-1}$) for cCMP hydrolysis at pH 8.0.

At any time during the assay, the concentration of cCMP and CMP can be obtained from the observed absorbance ($A_t$), the extinction coefficients of cCMP ($\varepsilon_S$ = 0.19 m$M^{-1}$ cm$^{-1}$) and CMP ($\varepsilon_S$ = 0.38 m$M^{-1}$ cm$^{-1}$) at 296 nm, and the initial concentration of cCMP ([cCMP]$_0$).

$$[\text{cCMP}]_t = (\varepsilon_P[\text{cCMP}]_0 - A_t)/(\varepsilon_P - \varepsilon_S) \tag{5}$$

The concentration of the product, CMP, is given by

$$[\text{CMP}]_t = [\text{cCMP}]_0 - [\text{cCMP}]_t \tag{6}$$

The initial concentration of cCMP, $[\text{cCMP}]_0$, is calculated from the initial absorbance value and the extinction coefficient of cCMP. Because there is generally a significant lag time before RNase activity is observed, the first time point taken after addition of RNase provides a satisfactory measurement of the initial absorbance and the initial cCMP concentration if all other absorbance (including that of PDI or any additives to the assay) has been appropriately subtracted. At concentrations less than 60 $\mu M$, RNase has no significant absorbance at 296 nm, so that the absorbance due to the addition of RNase does not need to be subtracted to give an accurate initial cCMP concentration. At higher RNase concentrations (>30 $\mu M$), PDI catalyzes RNase aggregation, which results in an increase in absorbance due to light scattering.[38] For RNase concentrations less than 50 $\mu M$, it is possible to compensate for this effect by running a parallel control in which the cCMP is omitted and subtracting the absorbance at any time from that observed in the presence of cCMP. For routine use, however, such controls are not essential.

After calculating the instantaneous slope and the concentrations of cCMP and CMP at a given time, the concentration of fully active RNase ($E_t$, micromolar) that would yield the same slope at the same concentrations of cCMP and CMP is calculated from

$$E_t = v_t/\{V_{\max}[\text{cCMP}]_t/[[\text{cCMP}]_t + K_{\text{m}}(1 + [\text{CMP}]_t/K_i)]\} \qquad (7)$$

where $v_t$ is the velocity of cCMP hydrolysis observed at time $t$ [$v_t$(m$M$/min) = $\Delta A_{296}/(\varepsilon_{\text{P}} - \varepsilon_{\text{S}}) = \Delta A_{296}/0.19$], $V_{\max}$[39] is the activity of fully active RNase (0.20 ± 0.015 m$M$ cCMP/min/$\mu M$ RNase) at saturating cCMP, $K_{\text{m}}$ is the $K_{\text{m}}$ for cCMP (8.0 ± 0.5 m$M$) observed for native RNase under the conditions of the assay, and $K_{\text{i}}$ is the inhibition constant for CMP (2.1 ± 0.4 m$M$).

In our laboratory, the absorbance-vs-time data are collected directly into a personal computer, and all calculations are performed within a few seconds after the completion of data acquisition. The computer program (PC/DOS), designed for a direct interface to Beckman (Palo Alto, CA) DU7 or DU70 spectrophotometers, is available on request. We would also be willing to provide hardcopy source code for the subroutines needed for the specific calculations. The results of typical calculations of the active RNase concentration from the absorbance-vs-time data are shown in Fig. 3.

The regain of RNase activity is preceded by a significant lag due to the time required for the accumulation of inactive intermediates that precede

[38] M. M. Lyles and H. F. Gilbert, *J. Biol. Chem.* **269,** 30946 (1994).

[39] The value of $k_{\text{cat}}$ reported previously[12] was labeled with incorrect units. The specific activity of fully active RNase is 14.3 $\mu$mol of cCMP/min/mg RNase.

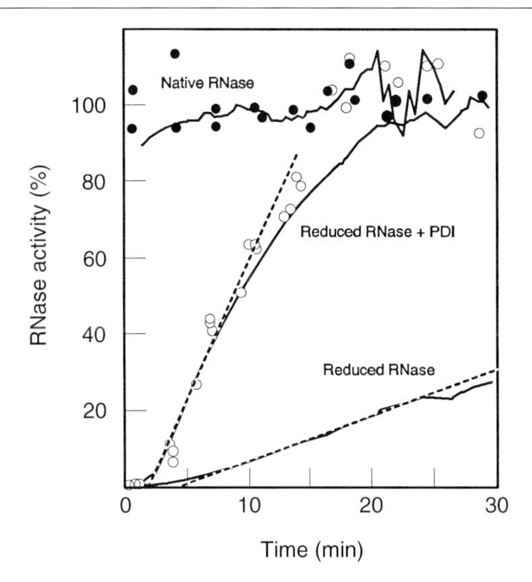


FIG. 3. Calculation of the concentration of active RNase from the change in absorbance at 296 nm due to cCMP hydrolysis. Comparison of the time course of the oxidative folding of reduced RNase determined by continuous assay and by a discontinuous assay. The concentration of active RNase as a function of time was determined from the change in absorbance at 296 nm due to cCMP hydrolysis as described in text. The total RNase concentration was 8.4 $\mu M$, and when present, the concentration of PDI was 1.4 $\mu M$. The curves are the results from the continuous assay and are not fits of the discontinuous assays of RNase activity shown by the symbols. The discontinuous measurement of the activity of RNase was determined in parallel experiments by withdrawing periodic aliquots from a preincubation and assaying for RNase A activity at pH 5.0. (Adapted, with permission, from M. M. Lyles and H. F. Gilbert, *Biochemistry* **30,** 613 (1991). Copyright 1991 American Chemical Society.)

the formation of active RNase. Such lags are observed in both continuous and discontinuous assays.[12] After the lag, RNase activity increases linearly for a time and then finally levels off as all of the RNase is converted to active enzyme. The slope of the initial linear increase in RNase activity following the lag is taken as the rate of the catalyzed reaction. The observed initial rate is then corrected for the background rate (in the absence of PDI) measured in a parallel assay. Under the conditions specified, the uncatalyzed reaction represents 15–20% of the catalyzed reaction. The $K_m$ for RNase is 7 ± 1 $\mu M$ and the turnover number is approximately 0.8 $\mu$mol of RNase refolded per minute per micromole of PDI.[38]

*Constraints on the Assay.* Although this may appear to be a laborious procedure, in practice it provides a much more convenient and complete assay for PDI activity than discontinuous methods, in which numerous aliquots must be removed and assayed individually for RNase [or bovine pancreatic trypsin inhibitor (BPTI)] activity during the time course of a single assay. The time course for the PDI-catalyzed regain of RNase activity is complex, so that a single point assay of the RNase activity regained after a fixed time of incubation is unsatisfactory because experimental conditions could affect the lag or rate. The assay is linear with PDI concentration over a rather limited range of concentrations (0.5–3 $\mu M$), and the rate of the reaction saturates with increasing PDI concentration as the PDI concentration exceeds that of the RNase substrate (Fig. 4).

The calculation of the concentration of active RNase at any time depends on knowing the $K_m$ for cCMP and the $K_i$ for CMP, values that will depend on pH, salt concentration, etc. Therefore, the assay can be used only at pH 8 under the conditions specified unless the $K_m$ and $K_i$ are determined independently. Because the initial cCMP concentration is used to calculate both the instantaneous cCMP and CMP concentrations, this value must be known with some certainty ($\pm 0.05$ AU). Rate measurements become much less precise as the substrate (cCMP) is depleted and the product inhibitor (CMP) increases. Generally, values observed when the absorbance is greater than 1.5 (with $cCMP_{total}$ at 4.5 m$M$) are too inaccurate to be useful. Decreasing the concentration of PDI or the concentration of RNase can be used to extend the time of the assay and to reduce the effects of substrate depletion. Enzyme activity is also a sensitive function of the concentration of GSH and GSSG[12]; consequently, GSH and solutions should be made fresh daily.

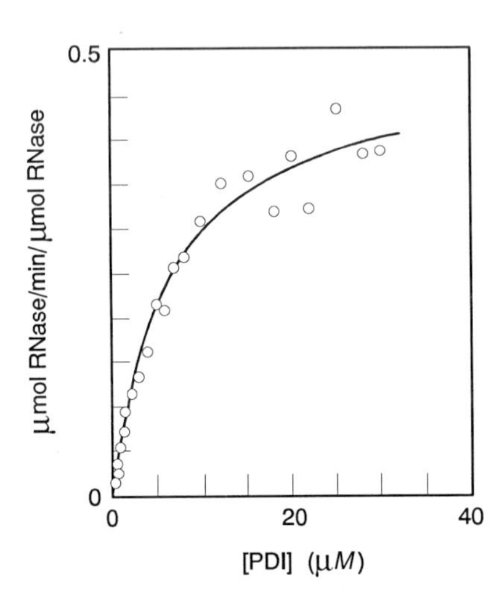


FIG. 4. Dependence of the velocity of RNase oxidative folding on the concentration of PDI. The experiments were performed at an RNase concentration of 8 $\mu M$ at pH 8.0 (0.1 $M$ Tris-HCl, 2 m$M$ EDTA), 25° using the continuous assay for RNase oxidative folding. (Data from M. M. Lyles and H. F. Gilbert, *J. Biol. Chem.* **269,** 30946 (1994), with permission.)

*Preparation of Substrate.* Reduced, denatured RNase is prepared by incubating 5 mg of native RNase overnight in 1 ml of 6 *M* guanidine hydrochloride (Gdn-HCl) in 0.1 *M* Tris–acetate (pH 8.0), 2 m*M* EDTA, and 0.14 *M* DTT.[40] The reduced, denatured enzyme is separated from excess DTT and Gdn-HCl by centrifugal gel filtration employing Bio-Gel P4 (Bio-Rad, Hercules, CA) equilibrated with 0.1% (v/v) acetic acid. The concentration of RNase is determined by absorbance at 277.5 nm, using an extinction coefficient of 9.3 $mM^{-1}$ $cm^{-1}$.[37] For each preparation of reduced RNase, the concentration of thiol groups per reduced RNase is determined using the method of Ellman.[41] Typically, 1 m*M* reduced RNase contains 7.8 ± 1.5 m*M* SH groups. The RNase (8 $\mu M$) introduces reducing equivalents (0.064 m*M*) into the assay. Compensating for this by increasing the GSSG concentration from 0.2 to 0.23 m*M* does not alter the rate significantly; however, when using more than 8 $\mu M$ RNase, additional GSSG should be included to compensate for that consumed by substrate oxidation.

*Scrambled RNase Refolding*

In the presence of low concentrations of thiols, PDI will catalyze the renaturation of RNase that has been reduced and then oxidized in the presence of denaturant (scrambled RNase). This assay and the preparation of the scrambled RNase substrate have been described in detail in an earlier volume of this series.[5] The advantage of using scrambled RNase rather than the fully reduced protein is that the scrambled RNase (sRNase) can be prepared in large quantities and stored. The disadvantage of this substrate is that it is a complex mixture of species including intermolecular disulfides, and activity varies with different sRNase preparations.[36] This is a discontinuous assay, and aliquots of the renaturation reaction are withdrawn and assayed individually for RNase activity at several different times during the time course of reactivation. The PDI activity is determined by replotting the RNase activity against time. The activity of PDI in this assay depends somewhat on the preparation of scrambled RNase. $K_m$ values vary between 2 and 5 $\mu M$ and $k_{cat}$ values vary between 0.2 and 0.5 $min^{-1}$.[42]

## Properties of Purified Enzyme

*Structure*

Each 55-kDa molecule (sequence molecular weight of rat PDI is 55,114) of PDI contains six cysteine residues. The molecule has two regions that

[40] T. E. Creighton, *J. Mol. Biol.* **113,** 329 (1977).
[41] G. L. Ellman, *Arch. Biochem. Biophys.* **82,** 70 (1959).
[42] H. C. Hawkins and R. B. Freedman, *Biochem. J.* **274,** 335 (1991).

are homologous to the redox active protein thioredoxin, one near the amino terminus and the other near the carboxyl terminus. Each thioredoxin domain contains two of the cysteines in the sequence WCGHCK[14,15] (Fig. 1). The functional homology to thioredoxin is illustrated by the observation that the two thioredoxin domains of PDI are substrates for thioredoxin reductase.[43] The cysteines in the two thioredoxin domains are responsible for the thiol–disulfide exchange activity.[16,17,38] The two central cysteines that are not in the thioredoxin domains do not contribute to the disulfide isomerase activity and are buried and inaccessible to chemical modification.[42] In addition to the thioredoxin homology domains at the N- and C-terminal active site regions, there are two more internally repeated homology domains of unknown function. A single region of homology to the estrogen receptor is located immediately after the N-terminal thioredoxin domain.[44] PDI also has a binding site for peptides and proteins.[35] Photoaffinity labeling of this site with a radiolabeled tripeptide photoaffinity reagent identified a single labeled peptide near the C terminus.[45] However, in the yeast enzyme, deleting this region had no effect on the activity of the enzyme.[17] The C terminus of PDI is very acidic; approximately half of the last 30 amino acids are aspartate or glutamate. The function of this anionic tail is unknown. The last four amino acids are KDEL, the consensus ER retention signal.[46]

As isolated, these cysteines of the thioredoxin sites are present as disulfides, and the oxidizing equivalents of these disulfides can be used stoichiometrically to oxidize substrate proteins.[47] The enzyme will also function catalytically in a glutathione redox buffer. Mutation of the cysteines at a given thioredoxin site is sufficient to inactivate the disulfide isomerase activity of that site[16,17,38]; PDI with all four thioredoxin domain cysteines mutated to serine has a low but detectable activity (0.5% of wild type).[38] The N- and C-terminal active sites do not contribute identically to PDI activity.[38] Inactivation of the N-terminal domain by mutagenesis results in a PDI with 33% of the $k_{cat}$ of the wild-type enzyme; however, inactivation of the C-terminal domain has no significant effect on $k_{cat}$. At saturating concentrations of RNase, the activity of wild-type PDI is somewhat less than that expected from the independent operation of the two individual active sites, suggesting that catalysis at one of the sites may interfere with catalysis at the other when substrate is saturating. However, at lower sub-

[43] J. Lundstrom and A. Holmgren, *J. Biol. Chem.* **265,** 9114 (1990).

[44] R. B. Freedman, H. C. Hawkins, S. J. Murant, and L. Reid, *Biochem. Soc. Trans.* **16,** 96 (1988).

[45] R. Noiva, R. B. Freedman, and W. J. Lennarz, *J. Biol. Chem.* **268,** 19210 (1993).

[46] H. R. B. Pelham, *Annu. Rev. Cell. Biol.* **5,** 1 (1989).

[47] M. M. Lyles and H. F. Gilbert, *Biochemistry* **30,** 619 (1991).

strate concentration ($k_{cat}/K_m$ conditions), the two active sites contribute equally to PDI catalysis.[16,38] Although the amino acid sequence in the immediate vicinity of the active site (WCGHCK) is absolutely conserved from yeast to human, mutations of residues other than cysteine have a relatively small overall effect on PDI activity.[48] Altering the active site of the N-terminal thioredoxin domain (in the presence of an inactive C-terminal domain) to that of thioredoxin (WCGPCK) does not have a significant effect on $k_{cat}$ or $K_m$ for the oxidative folding of RNase. However, altering the active site sequence of thioredoxin to that of PDI makes the thioredoxin a better oxidizing agent and increases the PDI activity of thioredoxin by 10-fold (at a constant substrate concentration).[49]

PDI as isolated from bovine liver or recombinant PDI from *E. coli* is not posttranslationally modified. Electrospray mass spectrometry of both proteins shows that the molecular weight of PDI is identical (±20 amu) to that expected from the sequence.[30] PDI has been reported to be phosphorylated by ATP[50]; however, the stoichiometry of ATP incorporation or the effect of phosphorylation on activity has not been reported. Whereas the bovine enzyme is not glycosylated, the enzyme from yeast is a glycoprotein.[51]

### *Function*

*Specificity and Catalysis.* Compared with other enzymes, PDI is not a particularly effective catalyst, with turnover numbers near 1 $min^{-1}$ and a $k_{cat}/K_m$ value (for RNase oxidative folding) of only $10^5$ $M^{-1}$ $min^{-1}$.[38] For RNase oxidative folding, the turnover number of 0.8 $min^{-1}$ is only 30- to 50-fold faster than the spontaneous renaturation of RNase in a glutathione redox buffer. However, PDI is a particularly effective catalyst of an intramolecular thiol–disulfide rearrangement that interconverts kinetically trapped, partially oxidized folding intermediates of bovine pancreatic trypsin inhibitor. The first-order rate constants at saturating levels of substrate are 4000 to 6000-fold faster than that of the uncatalyzed reaction. However, much of the rate acceleration is due to the very slow uncatalyzed reaction; the turnover number is still close to 1 $min^{-1}$.[52] The relatively low turnover number of PDI means that relatively high concentrations of PDI, approaching stoichiometric with substrate, are often required to observe significant catalysis.

[48] X. Lu, H. F. Gilbert, and J. W. Harper, *Biochemistry* **31,** 4205 (1992).
[49] J. Lundstrom, G. Krause, and A. Holmgren, *J. Biol. Chem.* **267,** 9047 (1992).
[50] E. Quemeneur and R. Guthapfel, *J. Biol. Chem.* **269,** 5485 (1994).
[51] T. Mizunaga, Y. Katakura, T. Miura, and Y. Maruyama, *J. Biochem.* (*Tokyo*) **108,** 846 (1990).
[52] J. S. Weissman and P. S. Kim, *Nature* (*London*) **365,** 185 (1993).

PDI also catalyzes nonspecific thiol–disulfide exchange reactions between a wide variety of substrates including organic thiols and disulfides, peptides, and proteins.[34] The lack of specificity of PDI may actually enhance its *in vivo* role as a catalyst of oxidative folding. Disulfide bonds in proteins occur in a wide variety of sequence motifs so that PDI must be able to react with cysteines and disulfides in a sequence-independent manner. This lack of specificity is also reflected in the PDI peptide/protein-binding site, which seems to have no discernible preference for a given sequence.[35,53]

*Chemical Reactivity and Redox Properties.* The active site disulfide of PDI is an extremely good oxidant. The equilibrium constant for oxidation of PDI by GSSG [Eq. (8)]

$$\mathrm{P}\begin{matrix}\diagup \mathrm{SH} \\ \diagdown \mathrm{SH}\end{matrix} + \mathrm{GSSG} \rightleftharpoons \mathrm{P}\begin{matrix}\diagup \mathrm{S} \\ \vert \\ \diagdown \mathrm{S}\end{matrix} + 2\mathrm{GSH} \tag{8}$$

has been estimated to be between 42 $\mu M$ (−0.108 mV)[12,42] and 3 m$M$ (−0.164 mV).[54] This is similar to that of DsbA (81 $\mu M$, −0.117 mV),[55,56] but considerably lower than that of *E. coli* thioredoxin, 14 $M$ (−0.269 mV).[57] The instability of the active site disulfide of PDI is also reflected in the kinetics of reduction.[58] The difficulty in forming intramolecular disulfide bonds at the thioredoxin sites suggests a low effective molarity for intramolecular reactions involving the second active cysteine. As a consequence, the PDI-catalyzed reduction[58] and oxidation[59] of model peptide substrates often occur through intermolecular reactions involving components of the redox buffer.

Disulfide formation during oxidative folding is not a straightforward and orderly process.[2] Stable intermediates with native-like structure may trap the protein in an incompletely oxidized state.[60,61] In addition, the early stages of oxidative folding of some proteins involve the formation of a

[53] R. Noiva, H. Kimura, J. Roos, and W. J. Lennarz, *J. Biol. Chem.* **266,** 19645 (1991).
[54] J. Lundstrom and A. Holmgren, *Biochemistry* **32,** 6649 (1993).
[55] A. Zapun, J. C. A. Bardwell, and T. E. Creighton, *Biochemistry* **32,** 5083 (1993).
[56] M. Wunderlich, R. Jaenicke, and R. Glockshuber, *J. Mol. Biol.* **233,** 559 (1993).
[57] A. Holmgren, *Methods Enzymol.* **107,** 295 (1984).
[58] H. F. Gilbert, *Biochemistry* **28,** 7298 (1989).
[59] N. J. Darby, R. B. Freedman, and T. E. Creighton, *Biochemistry* **33,** 7937 (1994).
[60] T. E. Creighton, *Science* **256,** 111 (1992).
[61] J. S. Weissman and P. S. Kim, *Science* **256,** 112 (1992).

large number of intermediates with nonnative disulfide pairings and with mixed disulfides between the protein and the glutathione redox buffer,[62–64] and the rearrangement of these nonnative disulfides often limits the rate of the overall process. Although protein disulfide formation is an oxidation, for many proteins, both catalyzed and uncatalyzed oxidative folding occurs fastest in a redox buffer that is slightly reducing.[65] Optimum refolding of RNase is observed with 0.2 m$M$ GSSG in the presence of 1 m$M$ GSH[12]; both higher and lower concentrations of GSH or GSSG yield lower folding rates. A somewhat reducing environment appears to be necessary to allow thiol-dependent reduction or rearrangement of disulfides that trap folding intermediates in incompletely oxidized states, but too high a GSH concentration diminishes the concentration of folding intermediates with disulfide bonds. GSSG is essential to provide oxidizing equivalents for disulfide formation, yet too high a concentration of GSSG may trap the protein in misoxidized forms including protein–glutathione mixed disulfides. The dependence of the rate of PDI-catalyzed oxidative folding of RNase on the composition of the redox buffer suggests that the reduced form of PDI is responsible for catalyzing rate-limiting disulfide rearrangements.[12] Kanaya *et al.*[66] have suggested that two cysteine residues of PDI, either at the same or different active sites, may cooperate in disulfide rearrangement of mutant lysozymes.

*Peptide–Protein Binding.* Protein disulfide isomerase has a peptide/protein-binding site that interacts relatively weakly with peptides and unfolded proteins. Peptides of a variety of sequences inhibit both the insulin reduction and RNase oxidative folding activities of PDI.[35] There appears to be little correlation of binding with charge or hydrophobicity; the only trend is that longer peptides are better inhibitors, consistent with the ability of longer peptides to bind in a larger number of ways. Cysteine-containing peptides bind approximately four- to eightfold tighter than peptides of the same length that do not contain cysteine. The $K_d$ values for peptides measured by inhibition are quite high, greater than 100 $\mu M$, suggesting that the peptide-binding site of PDI is relatively weak and nonspecific. A peptide photoaffinity label exhibits no peptide sequence specificity[53] and labels PDI at a site between residues 451 and 475[45]; however, deletion of this region in the yeast enzyme does not significantly affect activity.[17]

[62] D. M. Rothwarf and H. A. Scheraga, *Biochemistry* **32,** 2671 (1993).
[63] T. E. Creighton, *Proc. Natl. Acad. Sci. U.S.A.* **85,** 5082 (1988).
[64] B. Chatrenet and J.-Y. Chang, *J. Biol. Chem.* **268,** 20988 (1993).
[65] V. P. Saxena and D. B. Wetlaufer, *Biochemistry* **9,** 5015 (1970).
[66] E. Kanaya, H. Anaguchi, and M. Kikuchi, *J. Biol. Chem.* **269,** 4273 (1994).

*Other Functions.* Protein disulfide isomerase plays several different roles in the ER that do not involve its functions as a disulfide isomerase. PDI serves as a subunit of at least two enzymes, the $\beta$ subunit of the enzyme prolyl hydroxylase[24] and an ER triglyceride transferase.[25] For prolyl hydroxylase, the $\alpha$ subunit is not stable in the absence of PDI; however, mutational inactivation of the disulfide isomerase activity of PDI has no effect on the prolyl hydroxylase activity.[67] In the ER, PDI is present in large excess relative to the $\alpha$ subunit, suggesting that association with PDI may serve to stabilize the catalytically active $\alpha$ subunit and retain it in the ER.[68] PDI also binds thyroid hormones[22,23] and estrogen[21]; however, the functional significance of these activities is unknown. PDI catalyzes the glutathione-dependent reduction of dehydroascorbate to ascorbate.[18] Although PDI binds $Ca^{2+}$ with a high capacity and low affinity (approximately 19 binding sites per mole),[19] $Ca^{2+}$ binding does not appear to affect activity significantly.[20]

*Chaperone and Antichaperone Activities.* Protein disulfide isomerase is present in the ER at near millimolar concentrations, approaching or exceeding that of ER chaperones such as BiP, grp94, and calnexin.[13] When present in large stoichiometric excess relative to an unfolded protein substrate, PDI can exhibit chaperone activity, inhibiting aggregation and increasing the recovery of native protein.[26,27] The structural basis for the chaperone activity of PDI is not understood; however, PDI appears to have multiple sites for interaction with proteins/peptides and other hydrophobic molecules. In the oxidative folding of reduced lysozyme, the chaperone activity of PDI requires its disulfide isomerase activity and is observed only when PDI is present in solution before the denatured substrate is added (Fig. 5). Rapid PDI-catalyzed oxidation prevents lysozyme aggregation; however, PDI also displays chaperone activity for proteins that do not form disulfide bonds. When present at a concentration of 20–30 $\mu M$, PDI inhibits glyceraldehyde-3-phosphate dehydrogenase aggregation and protects the protein against thermal inactivation.[27] PDI chaperone activity does not require ATP.

Under certain conditions, PDI[26,28] and the ER chaperone, BiP,[69] display an unusual behavior—both proteins can specifically facilitate the formation and precipitation of large, insoluble protein aggregates, a behavior that has been termed "antichaperone" activity. Figure 5 shows the yield of native lysozyme as a function of the PDI concentration initially present when reduced lysozyme is diluted into a glutathione redox buffer to initiate

[67] K. Vuori, T. Pihlajaniemi, R. Myllyla, and K. I. Kivirikko, *EMBO J.* **11,** 4213 (1992).
[68] K. I. Kivirikko, R. Myllyla, and T. Pihlajaniemi, *FASEB J.* **3,** 1609 (1989).
[69] A. Puig and H. F. Gilbert, *J. Biol. Chem.* **269,** 25889 (1994).

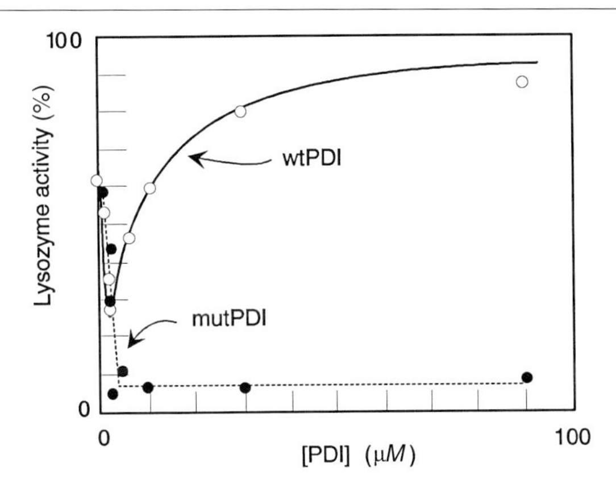


FIG. 5. The effect of PDI concentration on the yield of native lysozyme obtained during oxidative folding. Wild-type or mutant PDI was present in the refolding reaction (0.1 *M* HEPES, pH 7.0, 37°) along with a glutathione redox buffer (5 m*M* GSH, 0.5 m*M* GSSG) before lysozyme was added to a final concentration of 10 $\mu M$: (—) wild-type PDI; (---) mutant PDI with no disulfide isomerase activity, in which all four active site cysteines have been mutated to serine. (Adapted from A. Puig, M. M. Lyles, R. Noiva, and H. F. Gilbert, *J. Biol. Chem.* **269,** 19128 (1994), with permission.)

refolding. In the absence of PDI, reduced lysozyme at a final concentration of 10 $\mu M$ partitions almost equally between aggregation and refolding. Adding low concentrations of PDI (or BiP) actually decreases lysozyme refolding and promotes the formation of large, insoluble aggregates that contain the unfolded protein in association with the antichaperone. At a PDI-to-lysozyme ratio of approximately 5:1, almost all of the lysozyme and PDI is found in insoluble aggregates. Other folded proteins do not display this behavior. PDI-facilitated aggregation is not dependent on disulfide bond formation, disulfide isomerase activity, or the peptide-binding site.[28] Although the mechanism of this unusual activity is not known, the chaperone/antichaperone behavior of PDI may be compared with immunoprecipitation by a polyclonal antibody. Bivalent interaction between PDI and smaller protein aggregates may lead to an extensive network of cross-links at a stoichiometric ratio analogous to antigen:antibody equivalence. Lower or higher concentrations of PDI produce fewer aggregates, in analogy to antigen or antibody excess regions of immunoprecipitation curves.

The antichaperone activity of PDI and BiP may provide a mechanism for increasing the capacity of the ER for retaining unfolded proteins as aggregates when the concentration of unfolded proteins exceeds the monovalent binding capacity of ER chaperones.[70,71] PDI has been found to

[70] T. Marquardt and A. Helenius, *J. Cell Biol.* **117,** 505 (1992).

[71] I. Braakman, J. Helenius, and A. Helenius, *Nature (London)* **356,** 260 (1992).

associate with misfolded human lysozyme in the ER,[72] and pulse–chase experiments in HepG2 (human hepatocellular carcinoma) cells have suggested that the formation of large, disulfide-cross-linked aggregates is involved in the normal folding pathway for thyroglobulin.[73] Regardless of whether PDI chaperone and antichaperone activities play an active role *in vivo*, the behavior does have implications for using PDI as a "catalyst" for protein folding (see below).

*Stability.* Protein disulfide isomerase is a relatively stable protein. Guanidine and urea denaturation is fully reversible; however, denaturation behavior is complex, and Gdn-HCl specifically stabilizes an equilibrium folding intermediate.[74] In contrast to the observation that reduction of the disulfide stabilizes DsbA,[55,56] reduction of PDI does not have a large stabilizing or destabilizing effect. In addition, PDI is relatively stable to thermal denaturation; there is no significant change ($<5\%$) in secondary structure as determined by circular dichroism (CD) at temperatures of up to 55°, and the midpoint of the thermal denaturation occurs at approximately 67°.[75]

*Thioredoxin Family*

Protein disulfide isomerase is a member of the thioredoxin superfamily of oxidoreductase proteins. All members of this family have at least one domain with thioredoxin homology, all have a redox-active thiol/disulfide of the general sequence C*XX*C, and all catalyze various thiol–disulfide exchange reactions.[15] Thioredoxins, which are the best reducing agents of the family, are thiol–disulfide cofactors for a number of enzymes, including ribonucleotide reductase.[76] Glutaredoxins are cytoplasmic proteins that mediate the reduction of protein disulfides, including the catalytic disulfide of ribonucleotide reductase, by glutathione and the formation and reduction of protein glutathione mixed disulfides.[77] Glutaredoxins appear to be distinguished from thioredoxins by a specific binding site for glutathione.[78] DsbA, a soluble periplasmic protein from *E. coli* that catalyzes disulfide bond formation in secreted proteins,[79] is a member of a multiprotein system for transporting oxidizing equivalents to the bacterial periplasm and catalyzing

[72] M. Otsu, F. Omura, T. Yoshimori, and M. Kikuchi, *J. Biol. Chem.* **269,** 6874 (1994).
[73] P. S. Kim and P. Arvan, *J. Biol. Chem.* **268,** 4873 (1993).
[74] N. A. Morjana, B. J. McKeone, and H. F. Gilbert, *Proc. Natl. Acad. Sci. U.S.A.* **90,** 2107 (1993).
[75] N. A. Morjana and H. F. Gilbert, unpublished observations (1997).
[76] A. Holmgren, *J. Biol. Chem.* **264,** 13963 (1989).
[77] W. W. Wells, Y. Yang, T. L. Deits, and Z. R. Gan, *Adv. Enzymol.* **66,** 149 (1993).
[78] J. H. Bushweller, M. Billeter, A. Holmgren, and K. Wuthrich, *J. Mol. Biol.* **235,** 1585 (1994).
[79] J. C. A. Bardwell, J.-O. Lee, G. Jander, N. Martin, D. Belin, and J. Beckwith, *Proc. Natl. Acad. Sci. U.S.A.* **90,** 1038 (1991).

disulfide bond formation. Although DsbA is an excellent oxidant, it is a much less effective catalyst of the thiol–disulfide rearrangements than PDI.[80] An X-ray structure is available for DsbA that shows a thioredoxin-like fold and active site geometry.[81] A second, largely helical domain not present in thioredoxin is inserted into the sequence of the thioredoxin domain. Several hydrophobic patches and grooves are observed at interfaces between the two domains that may serve to mediate interactions between DsbA and its substrates. There are other proteins of the eukaryotic ER that are members of the thioredoxin family. Erp72 [also known as calcium-binding protein 2 (CaBP2)] has three thioredoxin domains.[19,82] A report has suggested that the molecule also had protease activity that degrades proteins of the ER, including PDI and calreticulin.[83] Erp61, which has two thioredoxin domains,[84] was originally reported to have phospholipase C activity, but subsequent reports suggest that this is not the case.[85] CaBP1, a hydroxyurea-induced protein, has two thioredoxin domains. Erp72 (CaBP2) and CaBP1 have been reported to have significant disulfide isomerase activity.[20]

### *Oxidative Folding in Vivo*

*Folding in Eukaryotes.* Disulfide bond formation is an early event in the synthesis of secreted proteins in eukaryotes. Disulfide bond formation can begin even before translation is complete,[9] but for some proteins disulfide formation is posttranslational.[10] In the one case where it has been studied, the order in which specific disulfide bonds are formed *in vivo* is the same as that observed in the PDI-catalyzed reaction *in vitro*.[86] The ER uses glutathione to maintain a redox environment ([GSH]/[GSSG] = 1–3)[11] that appears to be similar to the optimal redox requirements for PDI-catalyzed refolding of reduced RNase *in vitro* (1 m*M* GSH, 0.2 m*M* GSSG).[12]

There has been circumstantial evidence for some time that PDI was involved in disulfide bond formation in the ER,[34] but more direct evidence

[80] A. Zapun and T. E. Creighton, *Biochemistry* **33,** 5202 (1994).

[81] J. L. Martin, J. C. A. Bardwell, and J. Kuriyan, *Nature* (*London*) **365,** 464 (1994).

[82] R. A. Mazzarella, M. Srinivasan, S. M. Haugejorden, and M. Green, *J. Biol. Chem.* **265,** 1094 (1990).

[83] R. Urade, Y. Takenaka, and M. Kito, *J. Biol. Chem.* **268,** 22004 (1993).

[84] C. F. Bennet, J. M. Balcarek, A. Varrichio, and S. T. Crooke, *Nature* (*London*) **334,** 268 (1988).

[85] R. A. Mazzarella, N. Marcus, S. M. Haugejorden, J. M. Balcarek, J. J. Baldassare, B. Roy, L. J. Li, and M. Green, *Arch. Biochem. Biophys.* **308,** 454 (1994).

[86] J. R. Huth, R. Perini, O. Lockridge, E. Bedows, and R. W. Ruddon, *J. Biol. Chem.* **268,** 16472 (1993).

has been obtained. PDI is an essential gene product in yeast and a *pdi-1* null mutant is not viable.[17,87–89] A mutant PDI with no detectable disulfide isomerase activity could rescue the PDI null mutation, suggesting that some activity other than the disulfide isomerase activity is essential. However, the rescued strain was deficient in the ability to form disulfide bonds in carboxypeptidase Y, an ER-processed lysosomal protein. This suggests that PDI makes at least some contribution to disulfide bond formation in the ER.[17] A related molecule, Erp72, that has three thioredoxin domains but low disulfide isomerase activity also rescues a PDI null mutation in yeast.[90]

In prokaryotes such as *E. coli,* oxidative folding occurs in the periplasmic space, where it is catalyzed by DsbA,[79] a 21-kDa protein of the thioredoxin family. A similar molecule has been found in other bacteria.[91] Unlike the eukaryotic ER, the periplasm exchanges small molecules with the growth medium, and the redox environment cannot be maintained by small thiols and disulfides such as glutathione. In *E. coli,* oxidizing equivalents are provided to DsbA and possibly other periplasmic proteins by a transmembrane protein, DsbB.[92] A number of other components of the disulfide-forming process in *E. coli* and other prokaryotes are beginning to be elucidated.[93] The redox environment in the periplasm may place some constraints on the mechanism of disulfide bond formation in this compartment. For *E. coli* β-lactamase (one disulfide) and alkaline phosphatase (two disulfides), folding occurs fastest under very oxidizing conditions, suggesting that these periplasmic proteins and the catalysts of disulfide bond formation, DsbA, may have evolved to provide a simple oxidation mechanism that does not require the presence of thiols to catalyze disulfide rearrangements.[94]

### *Use of Protein Disulfide Isomerase in Protein Refolding*

Oxidative folding is a complex process that is frustrated by competing aggregation and kinetic traps that inhibit the regain of fully native structure.[95] A more complete understanding of the mechanisms by which folding catalysts and chaperones facilitate protein folding in the cell will undoubt-

[87] B. Scherens, E. Dubois, and F. Messenguy, *Yeast* **7,** 185 (1991).

[88] H. Tachikawa, T. Miura, Y. Katakura, and T. Mizunaga, *J. Biochem.* (*Tokyo*) **110,** 306 (1991).

[89] R. Farquhar, N. Honey, S. J. Murant, P. Bossier, L. Schultz, D. Montgomery, R. W. Ellis, R. B. Freedman, and M. F. Tuite, *Gene* **108,** 81 (1991).

[90] R. Gunther, M. Srinivasan, S. Haugejorden, M. Green, I. M. Ehbrecht, and H. Kuntzel, *J. Biol. Chem.* **268,** 7728 (1993).

[91] H. Yu, H. Webb, and T. R. Hirst, *Mol. Microbiol.* **6,** 1949 (1992).

[92] J. C. Bardwell and J. Beckwith, *Cell* **74,** 769 (1993).

[93] H. Loferer and H. Hennecke, *Trends. Biochem. Sci.* **19,** 169 (1994).

[94] K. A. Walker and H. F. Gilbert, *J. Biol. Chem.* **269,** 28487 (1994).

[95] R. Jaenicke, *Prog. Biophys. Mol. Biol.* **49,** 117 (1987).

edly assist in the practical problem of obtaining large quantities of correctly folded recombinant proteins; however, as yet, no universal refolding strategy, even using folding assistants, has emerged. Nevertheless, there are several considerations that may be generally useful, particularly when considering PDI-assisted refolding.

Using oxygen as the only oxidant can irreversibly trap the protein in a misfolded state if nonnative disulfides are formed. A thiol–disulfide redox buffer[65] provides a mechanism for reversible disulfide formation by thiol–disulfide exchange so that the stability of the native structure (including the native disulfides) can direct the folding process. For most eukaryotic proteins, the most suitable redox buffer contains a 2- to 10-fold excess of thiol relative to disulfide.[2]

Aggregation is the major competing side reaction encountered in protein refolding so that lowering the protein concentration and decreasing the temperature may increase refolding yields. For many proteins, aggregation occurs most readily from folding intermediates rather than from the unfolded protein,[96] and these intermediates may accumulate at intermediate concentrations of denaturant. For this type of protein, slowly reducing the denaturant concentration by dialysis virtually guarantees that the denaturant concentration will at some point reach just the right concentration to favor aggregation. Rapidly diluting the protein from denaturant into refolding conditions may provide a means to decrease the formation of these aggregation-prone intermediates.

For PDI-assisted folding, high concentrations of PDI (>10 $\mu M$), in stoichiometric excess relative to the unfolded protein may be required to increase refolding yields.[26] Aggregation and productive folding are competing fates; however, increasing the folding rate does not necessarily increase the yield. When unfolded lysozyme is diluted into folding conditions, some of the lysozyme rapidly (in seconds) partitions into intermediates that eventually fold into the native state (committed to fold) and into intermediates that eventually aggregate or misfold (committed to aggregate).[26,97] Catalytic concentrations of PDI do not affect this initial partitioning and only accelerate the renaturation of lysozyme that has already committed to fold. However, high concentrations of PDI, if present when the unfolded protein is diluted into folding conditions, can affect the initial partitioning, either by chaperone effects or by increasing the rate of disulfide formation sufficiently to compete with the commitment to aggregation.[26] Adding even a high concentration of PDI a few seconds after initiating folding may

[96] M. R. DeFilippis, L. A. Alter, A. H. Pekar, H. A. Havel, and D. N. Brems, *Biochemistry* **32,** 1555 (1993).

[97] M. E. Goldberg, R. Rudolph, and R. Jaenicke, *Biochemistry* **30,** 2790 (1992).

be ineffective.[98] In addition, catalytic concentrations of PDI may actually promote aggregation through the antichaperone activity of PDI.[26] Consequently, to observe a significant effect on protein refolding yields, PDI may have to be used at concentrations of 20–100 $\mu M$ (1–5 mg/ml).

It may be possible to limit aggregation by adding detergents or denaturing agents such as urea or guanidine. PDI maintains reasonable activity (about 50%) in 0.1% (w/v) Triton X-100,[34] and is stable and active in up to 3 *M* urea (pH 7.5) or 1.5 *M* Gdn-HCl.[74] PDI retains its catalytic activity when covalently coupled to Sepharose, and immobilized PDI can be successfully used in a column mode.[99] A problem with immobilized PDI involves removing the denaturant and keeping the unfolded protein soluble while maintaining a sufficiently high local PDI concentration to be effective. At least a partial solution to the problem is provided by running the column in 2 *M* urea. However, the effects of urea, detergents, or other additives must be explored on a trial-and-error basis.

## Acknowledgments

Work in the author's laboratory was supported by NIH Grants GM-40379 and HL-28521.

[98] H. Lilie, S. McLaughlin, R. Freedman, and J. Buchner, *J. Biol. Chem.* **269,** 14290 (1994).
[99] N. A. Morjana and H. F. Gilbert, *Protein Expression Purif.* **5,** 144 (1994).

# [4] Thermophilic Fungal Protein Disulfide Isomerase

*By* Tsutomu Kajino, Chie Miyazaki, Osamu Asami, Masana Hirai, Yukio Yamada, and Shigezo Udaka

## Introduction

The formation of disulfide bonds, which are important for the structure and function of proteins, comprises cotranslational or posttranslational modifications associated with the folding and assembly of secretory and cell surface proteins. The process is catalyzed within the lumen of the endoplasmic reticulum (ER) by protein disulfide isomerase (PDI, EC 5.3.4.1), which catalyzes thiol:protein disulfide interchange *in vitro*, with a broad protein substrate specificity. It is regarded as the *in vivo* catalyst for disulfide bond formation in the biosynthesis of secretory proteins.[1]

[1] D. A. Hillson, N. Lambert, and R. B. Freedman, *Methods Enzymol.* **107,** 281 (1984).



0076-6879/98 $25.00

Sequence analysis of PDIs from various eukaryotic sources revealed the presence of two consensus sequences (WCGHCK) that are closely related to the sequence WCGPCK in thioredoxin.[2] Thioredoxin has been detected in all classes of organisms[3] and is heat stable. It catalyzes the exchange of disulfide bonds in scrambled proteins,[4] but its activity is obviously less than that of PDI. A protein, DsbA, with disulfide formation activity has been isolated from *Escherichia coli.* It has a possible redox active site sequence, CPHC.[5] DsbA catalyzes the oxidation of reduced proteins but hardly catalyzes the exchange of preformed disulfide bonds.

PDIs from vertebrates and yeast are relatively heat labile. Even in the case of an algal enzyme,[6] which is most stable of the known PDIs, the stability against heat is not enough for industrial use of the enzyme. Hence, there is continuing interest in finding new, stable PDIs.

We have isolated and characterized a thermostable PDI from a thermophilic fungus, *Humicola insolens.*[7] The cDNA encoding the fungal PDI has been cloned and expressed in *Bacillus brevis.*[8]

## Assay

*Refolding of Scrambled Ribonuclease.* Refolding activity using scrambled ribonuclease (RNase) as a substrate is assayed by the method of Mizunaga *et al.*[9] with some modifications. One unit of PDI activity is defined as that catalyzing the reactivation of 1 unit of RNase per minute; 1 RNase unit is defined as the amount producing a change in $A_{260}$ of 1 absorbance unit (AU) per minute.[10]

*Refolding of Scrambled Lysozyme.* Various concentrations of PDI together with 5 $\mu M$ dithiothreitol (DTT) in 115 $\mu$l of 50 m$M$ sodium phosphate buffer (pH 7.5) are incubated at 30° for 5 min. After the addition of 5 $\mu$l of a disulfide-scrambled lysozyme solution (0.58 mg/ml), the mixture is incubated at 30° for 1 hr. Thirty microliters of 0.05% (w/v) glycol chitin in 0.1 $M$ acetate buffer (pH 5.5) is added and the mixture is then incubated

[2] T. Parkkonen, K. I. Kivirikko, and T. Pihlajaniemi, *Biochem. J.* **256,** 1005 (1988).

[3] A. Holmgren, *Annu. Rev. Biochem.* **54,** 237 (1985).

[4] V. P. Pigiet and B. J. Schuster, *Proc. Natl. Acad. Sci. U.S.A.* **83,** 7643 (1986).

[5] Y. Akiyama, S. Kimitani, N. Kusakawa, and K. Ito, *J. Biol. Chem.* **267,** 22440 (1992).

[6] D. D. Kaska, K. I. Kivirikko, and R. Myllyla, *Biochem. J.* **268,** 63 (1990).

[7] H. Sugiyama, C. Idekoba, T. Kajino, F. Hoshino, O. Asami, Y. Yamada, and S. Udaka, *Biosci. Biotech. Biochem.* **57,** 1704 (1993).

[8] T. Kajino, K. Sarai, T. Imaeda, C. Idekoba, O. Asami, Y. Yamada, M. Hirai, and S. Udaka, *Biosci. Biotech. Biochem.* **58,** 1424 (1994).

[9] T. Mizunaga, Y. Katakura, T. Miura, and Y. Maruyama, *J. Biochem.* **108,** 846 (1990).

[10] A. L. Ibbetson and R. B. Freedman, *Biochem. J.* **159,** 377 (1976).

at 40° for 30 min. Potassium ferricyanide (0.05%, w/v) in 0.5 *M* sodium carbonate is added, followed by incubation at 100° for 15 min in a test tube stoppered with aluminum foil. After cooling in ice, the optical density at 420 nm is measured.[11]

*Oxidation of Reduced Bovine Pancreatic Trypsin Inhibitor.* Various concentrations of PDI together with 5 $\mu M$ DTT in 99 $\mu$l of 50 m*M* sodium phosphate buffer (pH 7.5) are incubated at 30° for 5 min. After the addition of 1 $\mu$l of a disulfide-reduced bovine pancreatic trypsin inhibitor (BPTI) solution (1 mg/ml), the mixture is incubated at 30° for 1 hr. Then 2.5 $\mu$l of a trypsin solution (1 mg/ml) containing 1 m*M* HCl and 70 $\mu$l of 0.05 *M* Tris-HCl buffer (pH 8.2) containing 0.02 *M* $CaCl_2$ are added. After incubation of the mixture at 37° for 15 min and cooling to 25°, 30 $\mu$l of *N*-benzoyl-L-arginine *p*-nitroanilide (BAPA) (4.35 mg/ml) in 0.05 *M* Tris-HCl buffer (pH 8.2) containing 1% (v/v) dimethylsulfoxide (DMSO) is added. After incubation of the mixture at 25° for 10 min, 50 $\mu$l of 30% (v/v) acetic acid is added, and the absorbance at 410 nm is then measured.[12]

*Reduction of Insulin.* The reducing action of PDI is determined as insulin-glutathione transhydrogenase activity. A total of 950 $\mu$l of a mixture comprising 0.13 m*M* NADPH, 1 m*M* glutathione (reduced form), 0.13 m*M* insulin, 0.25 *M* sucrose, 20 m*M* KCl, 4 m*M* $MgCl_2$, 1 m*M* ethylenediaminetetraacetic acid (EDTA), 0.25 units of glutathione reductase, and 50 m*M* Tris (pH 7.5) is equilibrated in a quartz cuvette at 30° for 5 min. Various concentrations of PDI are added to the mixture and the decrease in $A_{340}$ is then measured for 10 min.[10]

## Preparation of Mycelial Extract

*Humicola insolens* isolated from geothermally heated soil is grown at 50° for 5 days in malt extract broth [4% (w/v) malt extract, 0.5% (w/v) Bacto-peptone (Difco, Detroit, MI), 0.1% (w/v) yeast extract; pH 5.5] on plastic plates (140 $cm^2$, Eiken Kizai, Tokyo, Japan). Mycelia are collected by filtration and stored in −80°.

The mycelium (100 g, wet weight) is disrupted with aluminum oxide and then extracted with 1 liter of extraction buffer A [20 m*M* sodium phosphate, 10 m*M* EDTA, 1 m*M* *N*-tosyl-L-lysylchloromethyl ketone (TLCK), 1 m*M* phenylmethylsulfonyl fluoride (PMSF), soybean trypsin inhibitor (SBTI, 50 $\mu$g/ml), aprotinin (50 $\mu$g/ml); pH 6.0]. All subsequent steps are performed at 4°. The mycelial debris is removed by centrifugation

[11] T. Imoto, *Agric. Biol. Chem.* **35,** 1154 (1971).

[12] T. E. Creighton, D. H. Hillson, and R. B. Freedman, *J. Mol. Biol.* **142,** 43 (1980).

at 12,000 *g* for 15 min and the supernatant is then adjusted to pH 7.5 with 10 *N* NaOH.

## Purification

*Anion-Exchange Chromatography.* The extract is put on a DEAE-Sephacel column (2.5 × 10 cm; Pharmacia Biotech, Uppsala, Sweden) equilibrated with buffer B (20 m*M* sodium phosphate, 10 m*M* EDTA; pH 7.5), and the column is washed with the same buffer. Proteins are eluted with a linear gradient of 0–0.5 *M* NaCl in buffer B at a flow rate of 100 ml/hr, and 2.7-ml fractions are collected.

*Lectin Affinity Chromatography.* The active fraction from the DEAE-Sephacel column is put on a concanavalin A (ConA)–Sepharose column (1 × 1 cm; Pharmacia Biotech) equilibrated with buffer C {20 m*M* Tris-HCl, 0.5 *M* NaCl, 0.1% (w/v) 3-[(3-cholamidopropyl)dimethylammonio]-1-propanesulfonate (CHAPS); pH 7.5}. After washing the column with buffer C, proteins are eluted with 0.5 *M* $\alpha$-methylmannoside in buffer C.

*High-Performance Liquid Chromatography.* The eluate from the ConA–Sepharose column is put on an Asahipak C4P-50 high-performance liquid chromatography (HPLC) column (5-$\mu$m particle size, 0.46 × 15 cm; Asahi Chemical Industry, Kanagawa, Japan) equilibrated with 10 m*M* ammonium acetate buffer (pH 7). Elution is performed with a 25–35% linear gradient of acetonitrile in the same buffer at a flow rate of 30 ml/hr, and 0.5-ml fractions are collected.

Table I shows the data for purification of PDI from the fungus, *Humicola insolens.*

TABLE I

PURIFICATION OF PROTEIN DISULFIDE ISOMERASE FROM FUNGUS *Humicola insolens*[a]

| Step | Protein (mg) | Enzyme activity (units) | Specific activity (units/mg) | Yield[b] (%) |
|---|---|---|---|---|
| Crude extract | 1600 | —[c] | — | — |
| DEAE-Sephacel | 32 | 200 | 6.3 | 100 |
| ConA–Sepharose | 0.18 | 110 | 610 | 55 |
| HPLC (C4P-50) | 0.041 | 34 | 829 | 17 |

[a] Protein was measured with a Bio-Rad (Hercules, CA) protein assay kit, using bovine serum albumin as a standard.

[b] The yield was calculated on the basis of a value of 100 for the DEAE-Sephacel step.

[c] The PDI activity of the crude extract could not be measured owing to the influence of impurities.

Peptide Sequence Analysis

The 12 peptide fragments of PDI digested with lysyl endopeptidase are separated by reversed-phase HPLC. The native PDI (N; Fig. 1) and the 12 peptide fragments (L1–L12; Fig. 1) are sequenced with an automated gas–liquid phase sequencer. Mapping positions of the six peptides L1–L6 are assigned on the basis of that of rat PDI.

```
-20                      M H K A Q K F A L G L L A A A A V A T A

  1  S D V V Q L K K D T F D D F I K T N D L V L A E F F A P W C
     (N)

 31  G H C K A L A P E Y E E A A T T L K E K N I K L A K V D C T
             (L1)                                        (L2)

 61  E E T D L C Q Q H G V E G Y P T L K V F R G L D N V S P Y K
                                         (L7)       ▲

 91  G Q R K A A A I T S Y M I K Q S L P A V S E V T K D N L E E
             (L8)

121  F K K A D K A V L V A Y V D A S D K A S S E V F T Q V A E K

151  L R D N Y P F G S S S D A A L A E A E G V K A P A I V L Y K
     (L3)

181  D F D E G K A V F S E K F E V E A I E K F A K T G A T P L I
                 (L9)

211  G E I G P E T Y S D Y M S A G I P L A Y I F A E T A E E R K

241  E L S D K L K P I A E A Q R G V I N F G T I D A K A F G A H

271  A G N L N L K T D K F P A F A I Q E V A K N Q K F P F D Q E

301  K E I T F E A I K A F V D D F V A G K I E P S I K S E P I P
                       (L10)

331  E K Q E G P V T V V V A K N Y N E I V L D D T K D V L I E F
         (L11)               (L4)

361  Y A P W C G H C K A L A P K Y E E L G A L Y A K S E F K D R
                                 (L12)

391  V V I A K V D A T A N D V P D E I Q G F P T I K L Y P A G A
                 (L5)

421  K G Q P V T Y S G S R T V E D L I K F I A E N G K Y K A A I
                                         (L6)

451  S E D A E E T S S A T E T T T E T A T K S E E A A K E T A T

481  E H D E L *
```

FIG. 1. Amino acid sequence of the PDI from *H. insolens* KASI. The amino acid sequence deduced from the nucleotide sequence is shown. Numbers indicate amino acid positions. The peptide sequences identified by amino acid sequence analysis are underlined. The consensus sequence of PDI is boxed. (▲) Potential glycosylation site; (*) a stop codon.

### cDNA Cloning

To amplify the DNA fragment around the N-terminal consensus region of fungal PDI, two oligonucleotides corresponding to the amino acid sequence are synthesized and used as primers for an RT-PCR (reverse transcriptase-mediated polymerase chain reaction). The DNA fragment amplified by RT-PCR is used as a unique probe for PDI to screen a cDNA library. We found a positive clone with an insert of about 1.8 kbp that hybridized with the probe. The insert contains a single open reading frame encoding a polypeptide of 505 amino acids. The deduced amino acid sequence contains all of the sequences found on amino acid sequence analysis and two WCGHCK sequences presumed to be the active site of PDI. From these properties, we conclude that the isolated cDNA encodes fungal PDI.

### Gene Expression in *Bacillus brevis*

The fungal PDI cDNA is expressed in a heterologous protein production system using *B. brevis* as a host.[13]

The fungal PDI expression plasmid, pNU211L4PDI (Fig. 2), contains the promoter region of the middle wall protein (MWP) of *B. brevis* 47, the L4 signal peptide derived from the MWP signal peptide, and a mature fungal PDI. In pNU211L4PDI, the complete L4 signal sequence is directly followed by the mature fungal PDI sequence. This vector is introduced into *B. brevis* 31-OK, a mutant that degrades secreted proteins less than the parent, by the Tris–polyethyleneglycol method.[14] The transformant is grown for 6 days at 30° in 3YC medium [3% (w/v) polypeptone P1 (Nihon Pharmaceuticals, Tokyo, Japan), 0.2% (w/v) yeast extract, 3% (w/v) glucose, 0.01% (w/v) $CaCl_2 \cdot 2H_2O$, 0.01% (w/v) $MgSO_4 \cdot 7H_2O$, $FeSO_4 \cdot 7H_2O$ (10 mg/liter), $MnSO_4 \cdot 4H_2O$ (10 mg/liter), $ZnSO_4 \cdot 7H_2O$ (1 mg/liter), erythromycin (10 mg/ml); pH 7.2]. The recombinant PDI secreted into the culture supernatant is purified to homogeneity by a two-step procedure, i.e., anion-exchange chromatography and hydrophobic interaction chromatography.

In this system, the maximum level of recombinant PDI in the culture supernatant reaches $1.36 \times 10^6$ units (2.1 g of bovine liver enzyme equivalent) per liter.

### Enzyme Properties

The properties of the native and recombinant fungal enzymes are compared with those of bovine liver PDI in Table II.

[13] S. Udaka and H. Yamagata, *Methods Enzymol.* **217,** 23 (1993).

[14] W. Takahashi, H. Yamagata, K. Yamaguchi, N. Tsukagoshi, and S. Udaka, *J. Bacteriol.* **156,** 1130 (1983).

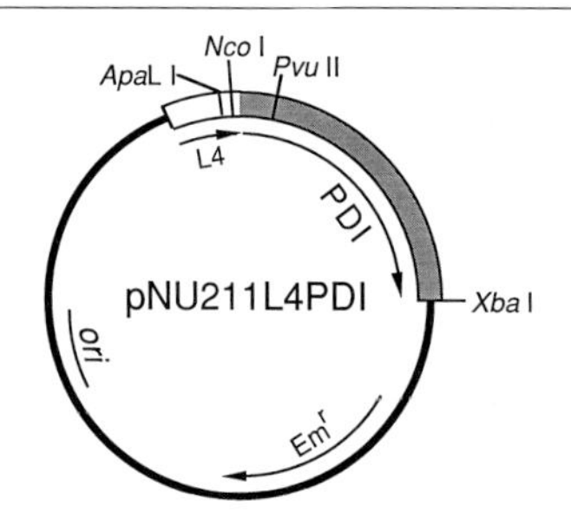


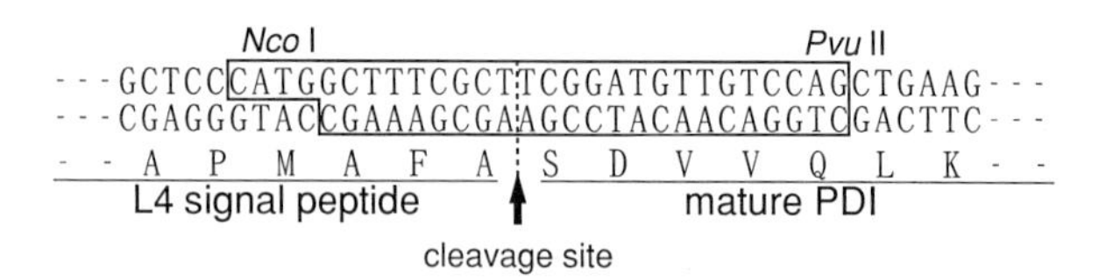


FIG. 2. Construction of an expression vector, pNU211L4PDI. The open bar denotes multiple promoters for the middle wall protein (MWP) gene of *B. brevis* 47 and a signal sequence, L4, which was constructed by insertion of three leucine residues into the hydrophobic region of the MWP signal peptide. The shaded bar represents the structural gene for mature PDI of *H. insolens* KASI. $Em^r$, Erythromycin resistance gene; *ori*, replication origin. Important restriction sites are indicated. *Bottom:* The nucleotide and amino acid sequences around the signal peptide cleavage site of the fused gene; the synthetic double-stranded DNA is boxed.

TABLE II

ENZYME PROPERTIES OF NATIVE AND RECOMBINANT FUNGAL AND BOVINE LIVER PROTEIN DISULFIDE ISOMERASES

| Property | Native PDI | Recombinant PDI | Bovine liver PDI |
|---|---|---|---|
| Molecular weight | 2 × (60 kDa) | 2 × (59 kDa) | 2 × (57 kDa) |
| p*I* | 3.5 | 3.5 | 4.2 |
| Glycosylation | Yes | No | No |
| Optimum temperature | 40° | 40° | 30° |
| Optimum pH | 8.0–9.0 | Not tested | 7.5–9.0 |
| $K_m$ ($\mu M$; scrambled RNase) | 3.3 | 4.2 | 1.1 |
| Stability | | | |
| Heat (>70°) | Stable | Stable | Less stable |
| $t_{1/2}$ (days, 25°) | 27 | Not tested | 9 |
| Acid (pH 5.0) | Stable | Stable | Unstable |
| Guanidine hydrochloride (0.2 *M*) | Stable | Stable | Unstable |
| Substrate specificity | Broad | Broad | Broad |

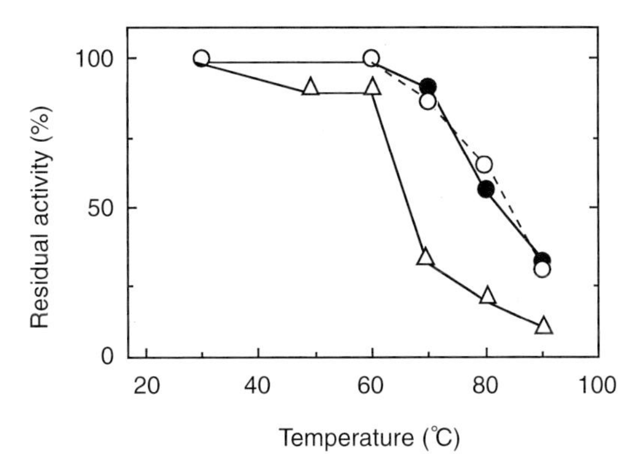


FIG. 3. Effect of temperature on PDI stability. The enzyme activity was assayed to determine the refolding of scrambled RNase. Samples were pretreated by heating at the specified temperatures for 30 min in 50 m$M$ sodium phosphate buffer (pH 7.5) before the enzyme assay. (△) PDI from bovine liver; (○) native PDI from the fungus *H. insolens;* (●) recombinant fungal PDI produced by *B. brevis.*

*Molecular properties and stability:* Molecular mass determined by sodium dodecyl sulfate–polyacrylamide gel electrophoresis (SDS–PAGE) under reducing and nonreducing conditions and by gel-permeation chromatography showed that the enzyme is a homodimer (120 kDa) of 60-kDa polypeptides, like the PDIs from other species. The recombinant PDI is also a homodimer, but it consists of 59-kDa polypeptides, which is 1 kDa smaller than in the case of the native PDI. The recombinant enzyme pro-

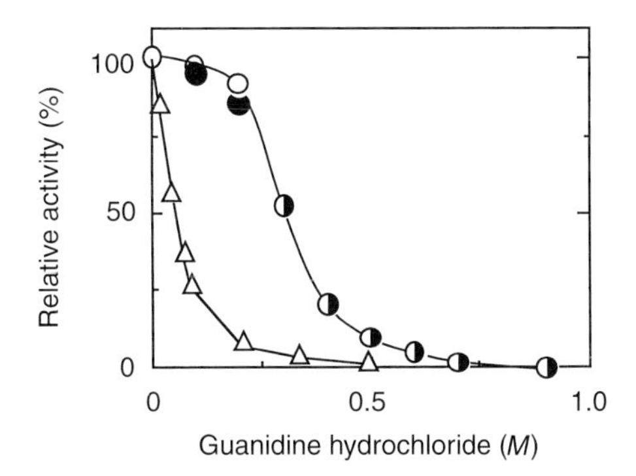


FIG. 4. Effect of guanidine hydrochloride on PDI activity. Various amounts of guanidine hydrochloride were added when PDI was incubated with scrambled RNase for the standard PDI assay procedure. Guanidine hydrochloride has no effect on the RNase assay. (△) PDI from bovine liver; (○) native PDI from the fungus *H. insolens;* (●) recombinant fungal PDI produced by *B. brevis.*

duced by the prokaryote, *B. brevis,* has no carbohydrate, and the difference in the molecular masses of the native and recombinant enzymes is consistent with the carbohydrate content of native PDI. This indicates that the recombinant PDI produced by *B. brevis* is a mature polypeptide.

Both the native and recombinant enzymes are acidic, with a p*I* value of 3.5, and their specific activities seem to be identical.

The native and recombinant fungal PDIs are considerably more heat stable than bovine liver PDI (Fig. 3). Yeast PDI is markedly inactivated by heat treatment at 54° for 15 min.[9] In the case of an algal PDI, heating to 62° reduces the activity to 50% of that of an unheated sample.[6] Of the known PDIs, fungal PDI is the most stable against heat treatment. These results also suggest that glycosylation of the enzyme is not essential for enzymatic activity and stability.

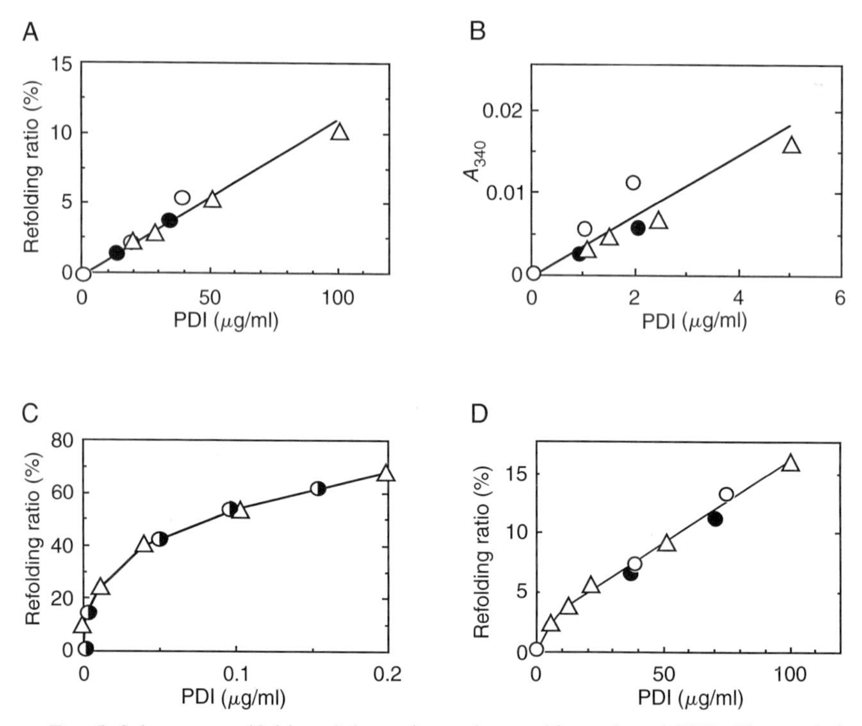


FIG. 5. Substrate specificities of the native and recombinant fungal PDIs. The catalytic properties of the native and recombinant fungal PDIs were compared with those of bovine PDI: reduced BPTI (A), insulin (B), scrambled RNase (C), and scrambled lysozyme (D). The enzyme activity was assayed as described in Assay. (△) PDI from bovine liver; (○) native PDI from the fungus *H. insolens;* (●) recombinant fungal PDI produced by *B. brevis.*

The fungal enzyme is stable in the presence of 0.2 *M* guanidine hydrochloride (Fig. 4), and has a longer half-life, 26 days, at room temperature and a wider stable pH range than bovine liver PDI.

*Specificity of the enzyme action:* Both the native and recombinant fungal PDIs are similar to bovine PDI in terms of the refolding of scrambled RNase and scrambled lysozyme, the oxidation of reduced BPTI, and the reduction of insulin (Fig. 5). The three enzymes showed no significant differences in substrate specificity. This suggests that the recombinant fungal PDI is also able to catalyze the reduction, oxidation, and isomerization of protein thiol or disulfide with a broad substrate specificity.

On the basis of these properties, the fungal PDI will be useful for the *in vitro* refolding and renaturing of scrambled proteins and for other industrial uses.

# [5] Disulfide Bond Catalysts in *Escherichia coli*

*By* THOMAS ZANDER, NIKHIL D. PHADKE, and JAMES C. A. BARDWELL

## Introduction

The proper oxidation status of cysteine residues is critical for protein stability and activity.[1–3] Although many proteins need to be reduced to be active, the presence of disulfide bonds is crucial for the folding and stability of other, mainly secreted, proteins. *Escherichia coli* mutants that are deficient in thioredoxin reductase and in the disulfide bond catalysts DsbA, B, C, and D show altered abilities in disulfide bond formation *in vivo,* demonstrating the important role of these folding catalysts in the cell.[1,4–8] One advantage of studying these catalysts is that disulfide exchange, both *in vitro* and in the cell, can be frozen at any instant in time using thiol-trapping agents and the reaction intermediates present within the substrate proteins and catalysts separated and quantified. This provides a unique

[1] J. C. A. Bardwell, *Mol. Microbiol.* **14,** 199 (1994).
[2] H. F. Gilbert, *Adv. Enzymol.* **63,** 69 (1990).
[3] N. Darby and T. E. Creighton, *Methods Mol. Biol.* **40,** 219 (1995).
[4] J. C. A. Bardwell, K. McGovern, and J. Beckwith, *Cell* **67,** 581 (1991).
[5] J. C. A. Bardwell, J.-O. Lee, G. Jander, N. Martin, D. Belin, and J. Beckwith, *Proc. Natl. Acad. Sci. U.S.A.* **90,** 1038 (1993).
[6] D. Missiakas, C. Georgopoulos, and S. Raina, *Proc. Natl. Acad. Sci. U.S.A.* **90,** 7084 (1993).
[7] A. I. Derman, W. A. Prinz, D. Belin, and J. Beckwith, *Science* **262,** 1744 (1993).
[8] D. Missiakas, F. Schwager, and S. Raina, *EMBO J.* **14,** 3415 (1995).



0076-6879/98 $25.00

opportunity to study the kinetics of folding catalysis *in vitro* and *in vivo*. Because proteins rely on disulfide bond formation for their stability the formation of disulfides is energetically coupled with protein folding and provides a powerful tool for studying the folding process itself.[3] However, accurately determining the rate of disulfide bond formation within cells presents an experimental challenge because the process normally occurs within seconds,[4,9] and artifactual oxidation or reduction of proteins can easily occur during sample processing.[10–12]

We describe general methods for rapidly blocking thiol–disulfide exchange and for analyzing cell extracts in ways allowing for the distinction between the oxidized and reduced forms of proteins. A sample labeling protocol specifically designed to detect defects in disulfide bond formation in *E. coli* is detailed as one specific example; however, modifications of this protocol should work with many species. In addition, the trapping and separation techniques described have many applications for the study of disulfide catalysis using purified proteins *in vitro*, and for the study of the folding process itself, in the absence of catalysts. These applications, additional methods, and the theory used to study disulfide exchange have been described in detail.[3] We also describe a generally applicable method to determine the equilibrium redox properties of disulfide catalysts.

## Trapping Reagents and Conditions

An ideal sulfhydryl trapping agent specifically modifies all sulfhydryls in a protein rapidly, in a stable way, and in a way that allows one to separate the modified forms of the protein from the disulfide-bonded form.[12] The following trapping conditions are rather general and can be used to block thiol–disulfide exchange within or between purified proteins,[3,11–15] in *in vitro* translation reactions,[16] or within cells that have been pulse labeled *in vivo*.[4,7]

### *Iodoacetate and Iodoacetamide*

Iodoacetate (IAA) reacts fairly specifically with exposed sulfhydryl groups, leaving behind the negatively charged carboxymethyl group

[9] T. Peters and L. K. Davidson, *J. Biol. Chem.* **257,** 8847 (1982).
[10] A. L. Derman and J. Beckwith, *J. Bacteriol.* **177,** 3764 (1995).
[11] J. S. Weissman and P. S. Kim, *Science* **253,** 1386 (1991).
[12] A. P. Pugsley, *Proc. Natl. Acad. Sci. U.S.A.* **89,** 12058 (1992).
[13] D. M. Rothwarf and H. A. Scheraga, *Biochemistry* **32,** 2671 (1993).
[14] T.-Y. Lin and P. S. Kim, *Biochemistry* **28,** 5282 (1989).
[15] J. W. Nelson and T. E. Creighton, *Biochemistry* **33,** 5974 (1994).
[16] A. Zapun, J. C. A. Bardwell, and T. E. Creighton, *Biochemistry* **32,** 5083 (1993).

($CH_2COO^-$) irreversibly linked to the sulfur in cysteine. This alteration of the net charge of a protein assists in the separation of proteins on urea polyacrylamide gels or native gels, based on the number of free cysteine residues present in the protein at the time of trapping. Because IAA reacts with the thiolate ion, trapping is pH dependent. At a concentration of 0.1 *M* at pH 8.7 the half-time of reaction between IAA and protein thiols should be about 1–3 sec,[3] generally fast enough for quantification of the relative amount of oxidized versus reduced protein.[14,18–20] At pH 8.7 IAA also reacts slowly with histidines, methionines, and amino groups. To avoid this, quench reactions should be kept short, on the order of 2 min. A quench carried out at pH 7.5 or with lower concentrations of IAA will be somewhat slower, but allows the protein to remain in the presence of the trapping agent for longer periods without significant side reactions. The two major disadvantages of IAA is that buried thiols will not react to an appreciable extent[3] and that IAA, even at pH 8.7, may not be fast enough to stop all intramolecular disulfide exchange reactions.[11,14,18–19,21]

Trapping can be accomplished by addition of 0.2 vol of 1 *M* IAA stock solution directly to growing cells or to thiol disulfide exchange reactions. Following a 2-min incubation at 25° for solutions at pH 8.7 or >15 min for solutions at pH 7.3, the excess IAA can be removed by desalting on a disposable 10DG column (Bio-Rad, Hercules, CA), and should be, if the samples are to be stored before analysis. If the trapping is carried out in whole cells or within cell extracts the trapping temperature should be lowered to 4° to help slow proteolysis. A closely related blocking agent, iodoacetamide (IAM), can be used interchangeably with IAA, except that it leaves behind an uncharged $CH_2CONH_2$ group. The lack of charge on this trapping agent increases its ability to pass through cellular membranes.

*Stock Solutions*

Prepare stock solutions from white crystals just before use.

Iodoacetic acid (1.0 *M*): Dissolve 184 mg of iodoacetic acid in 0.5 ml of 1 *M* KOH, then add 0.5 ml of 1 *M* Tris-HCl, pH 7.3 or 8.7

Iodoacetamide (0.5 *M*): Dissolve 92 mg of iodoacetamide in 1 ml of 0.5 *M* Tris-HCl, pH 7.3 or 8.7

*Double Trapping with Both Iodoacetate and Iodoacetamide.* Using both the acidic iodoacetate, and the neutral iodoacetamide added sequentially,

[17] T. E. Creighton, C. J. Bagley, L. Cooper, N. J. Darby, R. B. Freedman, J. Kemmink, and A. Sheikh, *J. Mol. Biol.* **232,** 1176 (1993).

[18] T. E. Creighton, *J. Mol. Biol.* **87,** 579 (1974).

[19] T. E. Creighton, *J. Mol. Biol.* **87,** 603 (1974).

[20] T. E. Creighton, *J. Mol. Biol.* **96,** 767 (1975).

[21] G. H. Snyder, *Biochemistry* **26,** 688 (1987).

it is possible to count the integral number of cysteine residues and disulfide bonds present in protein and classify their accessibility.[3,22,23] One protocol involves first adding a negative charge to free sulfhydryl groups with iodoacetic acid in the presence or absence of urea; the second step consists of reducing all disulfides with dithiothreitol (DTT) and then blocking all remaining sulfhydryl groups using the uncharged iodoacetamide. Free sulfhydryls are first trapped by incubation with 30 m*M* IAA in 50 m*M* Tris (pH 8.2), 1 m*M* EDTA in either the presence or absence of 8 *M* urea for 15 min at 37°. The reaction is stopped by the addition of 0.01 vol of a carrier protein (alkylated lysozyme, 5 mg/ml) followed by precipitation with 10 vol of a cold acetone–1 *N* HCl mix (98:2, v/v). The precipitate is recovered by centrifugation (3000 *g*, 5 min, 4°). The precipitates are washed three times by resuspension in a cold acetone–1 *N* HCl–$H_2O$ mix (98:2:10, v/v/v) and recentrifugation. The next step consists of dissolving the pellet in 8 *M* urea, 3.5 m*M* DTT, 50 m*M* Tris (pH 8.2), 1 m*M* EDTA and incubating for 30 min at 37° to reduce the remaining disulfides. The newly generated free sulfhydryls are then alkylated by addition of 10 m*M* IAM and incubated for 10 min at 37°.

When IAA reacts with a cysteine residue it leaves behind the charged carboxymethyl group; IAM leaves behind an uncharged group. This procedure allows the classification of all cysteines into three categories: reactive sulfhydryls, nonreactive sulfhydryls, and disulfide bonded. The total number of IAA molecules incorporated can be determined because migration in urea gels is dependent on charge. This method has the advantage that the effect of protein conformation is minimized. To calibrate the relationship between the mobility and the number of IAA molecules introduced a marker lane in which the protein of interest is modified with various mixtures of IAM and IAA is used.[22–25] The number of distinct bands present in this marker lane should correspond to one more than the number of half-cystines present in the molecule. This process allows the quantitative analysis of the disulfide status of proteins that contain up to 18 intramolecular disulfides within the complex mixture of proteins present in *in vivo* labeling reactions.[22]

### *N*-Ethylmaleimide

*N*-Ethylmaleimide (NEM) is a highly specific sulfhydryl reagent: blocking of sulfhydryls occurs 1000 times more rapidly than the side reaction

[22] N. Takahashi and M. Hirose, *Anal. Biochem.* **188,** 359 (1990).
[23] R. A. Reisfield, U. J. Lewis, and D. E. Williams, *Nature* (*London*) **195,** 281 (1962).
[24] A. Zapun, D. Missiakas, S. Raina, and T. E. Creighton, *Biochemistry* **34,** 5075 (1995).
[25] T. E. Creighton, *Nature* (*London*) **284,** 487 (1980).

with amino groups at pH 7.[26] *N*-Ethylmaleimide reacts more rapidly than iodoacetate with exposed sulfhydryl groups; however, it suffers from the disadvantage that it reacts even more slowly than iodoacetate with sulfhydryls that are buried or partially exposed, which leads to reactions that are not completed for 100 min or more.[27] The pH of the reaction with *N*-ethylmaleimide is critical. The rate of reaction drops off rapidly below pH 6. Above neutrality the rate of undesirable side reactions is increased and the reagent itself breaks down.[27] Adducts of *N*-ethylmaleimide and thiol are stable at pH values below 9.5, but unstable above pH 10.[28] The adduct it leaves behind is a rather bulky but uncharged $C_6NO_2H_8$ group. Stock solutions are 1 *M* in ethanol and can be stored at $-20°$. Crystals that form on storage should be dissolved before use. Blocking reactions are for 20 min using 20 m*M* *N*-ethylmaleimide at pH 7.

### *2-Aminoethyl Methanethiosulfonate*

The reaction rate of the blocking reagent 2-aminoethyl methanethiosulfonate[13] (AEMTS) is about seven orders of magnitude greater than that of IAA. It is sufficiently rapid to gain access to essentially buried protein thiols owing to local fluctuations or "breathing" of the protein structure during the blocking period. The use of AEMTS should be contemplated if rapid intramolecular disulfide exchange reactions are anticipated. This blocking reagent leaves behind the positively charged cysteamine moiety, which facilitates the separation, in a predictable manner, of proteins that contain different numbers of blocked thiols. Unfortunately, AEMTS is not commercially available; however, it can be synthesized.[29] Before use it should be freshly dissolved in 100 m*M* Tris, 2 m*M* EDTA, pH 8.0. AEMTS appears to be specific; no side reaction modifications are observed using up to 100 m*M* AEMTS at pH 5 and above.[13] At a minimum, it should be used in at least fivefold molar excess relative to free thiols. Following a 2-min incubation at 25° (or 5 min at 4°) at pH 8 this blocking agent can, if desired, be removed by desalting on a disposable 10DG column (Bio-Rad).

### *Acid*

Acid quenching is fast,[14] occurring with a rate constant greater than $10^9$ $sec^{-1}$ $M^{-1}$, and is effective even when the thiol groups are relatively

[26] G. E. Means and R. E. Feeney, "Chemical Modification of Proteins." Holdan-Day, San Francisco, 1971.
[27] J. F. Riordan and B. L. Vallee, *Methods Enzymol.* **XI,** 541 (1967).
[28] J. Nishiyama and T. Kuninori, *Anal. Biochem.* **200,** 230 (1992).
[29] G. L. Kenyon and T. W. Bruice, *Methods Enzymol.* **47,** 407 (1977).

inaccessible. However, pH quenching does not stop thiol–disulfide exchange; it just slows it down by reducing the concentration of the reactive thiolate ion, necessitating the timely processing of samples.[14,18] This reversibility is also an advantage, because it allows one to isolate individual folding intermediates and reinitiate thiol–disulfide exchange by raising the pH.[11] Acid quenching can be accomplished by addition of various acids, i.e., a 1/4 vol of 1 *M* HCl, or an equal volume of 10% (w/v) trichloroacetic acid.

*AMS (4-acetamido-4'-maleimidylstilbene-2,2'-disulfonic acid)*

AMS (Molecular Probes, OR) reacts specifically and irreversibly with sulfhydryl groups conjugating a large 490 Dalton, negatively charged moiety to the free cysteines. This major change in molecular weight can be used to clearly separate proteins on SDS polyacrylamide gels based on the number of AMS moieties added onto the protein at the time of trapping. In addition to the large increase in molecular weight, AMS also has the added advantage of being more cysteine specific and reacting faster with cysteines as compared with the haloacetates.

*Combination Quench Techniques Using Acid and AMS*

Internal disulfide rearrangement reactions and disulfide catalysts such as DsbA are capable of rapid disulfide exchange reactions, which are difficult to quench using IAA. Acid quenching is sufficiently fast but necessitates acidic pH conditions for sample processing. This is possible if high-performance liquid chromatography (HPLC) separation techniques are used but not if any of the electrophoresis techniques discussed below are used. A combination of the acid and chemical quench techniques[30,31] can make use of the best properties of both techniques. It involves first mixing the sample to be quenched with an equal volume of ice-cold 20% (w/v) trichloroacetic acid. This both rapidly quenches disulfide exchange and also denatures proteins present, including any disulfide catalysts, inhibiting their function. The precipitates of denatured proteins are pelleted by centrifugation (17,000 *g* for 20 min at 4°). The pellet is washed twice with 100% ethanol, dried in a dessicator, and then redissolved in freshly prepared 10 m*M* AMS in a 50-m*M* Tris-HCl (pH 8.1), 0.1% SDS, 1 m*M* EDTA solution. After 30 min of reaction, these samples can be resolved in nonreducing SDS gels. 90% acetone has been used in place of ethanol in the washing step, however, this can lead to poor recovery. Carrier proteins can be used to increase the size of the pellet, and a cysteine-free carrier protein has been successfully

[30] J. C. Joly and J. R. Swartz, *Biochemistry* **36**(33), 10067 (1997).
[31] T. Kobayashi, *et al., PNAS,* in press.

used in this procedure. Because AMS is a light-sensitive reagent, all steps involving AMS are carried out in the dark.

*Safety*

All of the thiol-trapping agents described above are rather toxic and should be handled accordingly.

*In Vivo* Labeling and Trapping

See Bardwell *et al.*[4] for details.

1. Dilute 1 : 100 an overnight culture of *E. coli* cells that had been grown in M63 minimal medium[32] supplemented with glucose (0.4%, w/v) and amino acids (20 $\mu$g/ml each except for methionine and cysteine), into the same medium.
2. Let the culture grow until the cells are in the logarithmic phase ($OD_{600}$ 0.3).
3. Label the cells for 40 sec by addition of [$^{35}$S]methionine (50 $\mu$Ci/ml; 29 TBq/mmol).
4. Chase with unlabeled methionine should be initiated by addition of cold methionine to a final concentration of 20 m*M*.
5. Immediately after addition of the cold methionine, transfer 1 ml of cells into a prechilled Eppendorf tube that contains 200 $\mu$l of ice-cold 100 m*M* IAM, vortex, and transfer to an ice–water bath.
6. After 1, 5, and 15 min, transfer additional 1-ml aliquots of the pulse–chased culture into Eppendorf tubes containing the trapping reagent.
7. After the samples have been incubated for 15 min on ice, recover the cells by centrifugation for 5 min at 10,000 rpm in a cooled Eppendorf centrifuge. Remove the supernatant.
8. The samples can now either be further processed by cellular fractionation, immunoprecipitation, and gel electrophoresis or frozen at −70° in a dry ice–ethanol bath before further processing.

*Comments*

This simple protocol is adequate for trapping disulfide bond formation in many periplasmic proteins; if more rapid trapping is desired use one of the alternate trapping agents and conditions previously described.

Immunoprecipitation

Specific visualization of the substrates for disulfide catalysts requires separation of the substrates from the total pattern of proteins synthesized.

[32] T. J. Silhavy, M. J. Berman, and L. W. Enquist, "Experiments with Gene Fusions." Cold Spring Harbor Press, Cold Spring Harbor, New York, 1984.

This can be easily accomplished by immunoprecipitation. Perhaps the most suitable protein for measuring the rates of disulfide bond formation in *E. coli* is the outer membrane protein OmpA.[4] OmpA has the fortunate quality that it is stable *in vivo* in the absence of its disulfide bond. The oxidized and reduced forms of OmpA are easily distinguished because the reduced form can be cleaved by trypsin to a discrete fragment.

1. Resuspend the frozen pellets from the pulse chase (sample protocol above) in 0.5 ml of ice-cold spheroplast buffer (100 m*M* Tris, pH 8.0; 100 m*M* KCl; 0.5 m*M* EDTA; 0.5 *M* sucrose).

2. Add EDTA and lysozyme to final concentrations of 7 m*M* and 60 μg/ml, respectively, and incubate for 15 min on ice.

3. Divide each sample into two equal aliquots, and lyse the cells by the addition of Triton X-100 to 1% (v/v). To one aliquot, add trypsin to a final concentration of 10 μg/ml. Proteolysis is carried out on ice for 20 min and stopped by addition of phenylmethylsulfonyl fluoride (1 m*M* final concentration) and trypsin inhibitor (2.5 mg/ml, final concentration).

4. Add an equal volume of KISDS buffer [100 m*M* Tris-HCl (pH 8.0), 300 m*M* NaCl, 4% (v/v) Triton X-100, 2 m*M* EDTA, 0.06% (w/v) SDS] and freeze–thaw three times to disrupt the spheroplasts completely. Pellet the debris for 5 min at 10,000 rpm in an Eppendorf centrifuge and pipette the soluble cell extract into a fresh tube.

5. OmpA is immunoprecipitated by incubation with rabbit polyclonal anti-OmpA antibody (obtained from C. Kumamoto, Tufts Medical School, Boston, MA) overnight at 4°.

6. Fifty microliters of immunoadsorbent *Staphylococcus aureus* (10%, w/v; Sigma, St. Louis, MO) is added, incubated for 20 min on ice with occasional inversion.

7. Pellet *Stapylococcus aureus* by centrifuging at 8000 *g* for 30 sec at 4°, resuspend in 750 μl of HS buffer [1% (v/v) Triton X-100; 1 *M* NaCl; 50 m*M* Tris-HCl, pH 8.0; 1 m*M* EDTA], by vortexing for 1 min.

8. Repeat the preceding wash step. Complete resuspension is important; do not spin longer or faster than required to pellet the *Staphylococcus aureus* suspension.

9. Remove the salt by washing with 750 μl of 50 m*M* Tris-HCl, pH 7.5.

10. Resuspend the pellet in 50 μl of SDS–PAGE sample buffer [2.5 m*M* DTT, 2% (w/v) SDS, 12.5 m*M* Tris-HCl (pH 8.0), 15% (v/v) glycerol]. Boil for 5 min to release the antigen from antibody and *Staphylococcus aureus,* spin *Staphylococcus aureus* down, and transfer the supernatant into a new Eppendorf tube. Store the supernatant at −20° before electrophoresis.

11. Of the various electrophoresis conditions described below for separating the oxidized and reduced forms of OmpA, a nonreducing 9% (w/v)

acrylamide–Tricine–SDS gel[33] has been found to be the one that gives the highest separation between oxidized and reduced forms. Split samples into two aliquots; to one aliquot add 2 m*M* DTT to illustrate the migration position of reduced OmpA, boil the samples for 5 min, and apply reduced sample to one-half of the gel and nonreduced sample to the other half. The reduced nonproteolyzed form of OmpA will run slightly slower than the oxidized form, and the trypsin fragment of the reduced form will run much faster than the oxidized form.[4]

## Gel Separation of Oxidized and Reduced Protein

Methods used for the separation of trapped species with purified proteins, such as reversed-phase HPLC and ion-exchange chromatography,[3] are not generally suitable for studies of disulfide bond formation *in vivo,* in which the limited amount of the proteins of interest is contaminated with much higher amounts of nonspecific cellular components.

Gel electrophoresis is one of the most useful techniques for analysis of disulfide bond formation *in vivo* because it readily allows the rapid, simultaneous separation of multiple samples and the detection of the small amounts of proteins produced by *in vivo* labeling reactions. Visualization can be achieved by immunoprecipitation before gel electrophoresis or by Western blotting.[5] Native gels, urea gels, and nonreducing SDS acrylamide gels have all been used to resolve the oxidized and reduced forms of proteins. The best choice for any particular protein should be determined experimentally.

### *Nonreducing Sodium Dodecyl Sulfate Gel Electrophoresis*

Nonreducing SDS gel electrophoresis is a simple modification of the conventional Laemmli gel system,[34] in which samples on adjacent gel lanes alternatively contain or do not contain a reducing agent. This gel system separates reduced from oxidized proteins primarily on the basis of the conformational state of the SDS-solubilized proteins. The oxidized forms of proteins are often slightly more compact and thus migrate slightly faster than the reduced form.[4] Small proteins with multiple disulfide bonds that create large loop sizes will generally show large differences in mobility between oxidized and reduced forms. Subtle mobility differences found with larger proteins with small numbers of disulfides can be emphasized by running very long protein gels in apparatus normally used to electrophorese

[33] H. Schagger and G. von Jagow, *Anal. Biochem.* **166,** 368 (1987).
[34] D. M. Bollab and S. J. Edelstein, "Protein Methods." Wiley-Liss, New York, 1991.

DNA sequence samples, modified only by the substitution of thicker (0.8 mm) spacers. Large sheets of Mylar obtained relatively inexpensively from local plastic suppliers can be cut cleanly with a sharp paper cutter to construct inexpensive spacers of any length (1/32-inch-thick stock is equivalent to 0.8 mm). Small differences in mobility between adjacent lanes can be better visualized by minimizing the width of the teeth on the comb used to cast the stacking gel.

Reducing agents tend to diffuse into adjacent lanes during electrophoresis and cause reduction of proteins in these lanes. Diffusion of 2-mercaptoethanol is relatively rapid and can pass through several lanes during prolonged electrophoresis. This necessitates the grouping of samples that have been treated with 2-mercaptoethanol on one portion of the gel. Substitution of 2 m*M* DTT, which diffuses slower than the 700 m*M* 2-mercaptoethanol normally present in SDS sample buffer, will greatly reduce this problem. The transition between oxidized and reduced protein will show up as a band containing a step or spur connecting the oxidized and reduced versions of the same protein. This feature can be helpful in spotting small differences in mobility because it usually occurs within a band instead of between lanes. This phenomenon is also useful in identifying the oxidized and reduced forms of a protein that exhibits a major shift in mobility, such as a protein linked by intermolecular disulfides.[35] SDS–PAGE gels run in the presence of Tricine instead of glycine[33] run more slowly but may give better resolution of oxidized and reduced forms.

### *Native Gel Systems*

One simple native gel system is identical to conventional SDS–PAGE except that SDS is eliminated from all solutions.[34] Performing the electrophoresis at a slightly elevated pH (pH 9.5) has the advantage that more thiol groups will be ionized, and thus cause a mobility shift. Here separation is proportional to net charge and hydrodynamic volume.[3] Differences in charge between oxidized and reduced proteins can be introduced by the use of trapping agents such as IAA and AEMTS that leave a charged group covalently linked to the free sulfhydryl. Native gels are primarily useful for analyzing small purified proteins or extracts that consist entirely of soluble proteins, such as periplasmic extracts. The presence of membranes and other cellular components in an extract tends to make the bands smear.

### *Urea Polyacrylamide Gels*

All of the preceding separation techniques are easiest to interpret if the protein of interest has only one possible disulfide resulting in only two

[35] R. J. Allore and B. H. Barber, *Anal. Biochem.* **137,** 523 (1984).

bands, oxidized and reduced. Proteins that have more than one disulfide will have multiple intermediates in the protein-folding pathway and may migrate as multiple, poorly resolved bands on native and nonreducing gels. These different forms can be separated by urea PAGE. Oxidized and reduced forms of the same protein are separated on urea gels almost entirely on the basis of charge introduced by the trapping agent.

*Urea Polyacrylamide Gel Protocol.* In the urea polyacrylamide gel protocol,[22] proteins should be electrophoresed on a discontinuous acrylamide slab gel. Both stacking and resolving gels contain 8 *M* urea. The stacking gel contains 0.12 *M* Tris-HCl buffer, pH 6.8 and the resolving gel contains 0.037 *M* Tris-HCl buffer, pH 8.8. The reservoir contains 0.025 *M* Tris, 0.192 *M* glycine (pH 8.3). The concentration of acrylamide in the separating gel should be 9% (w/v); the stacking gel uses 2.5% (w/v). Electrophoresis should be carried out at 4°. For additional gel protocols see Ref. 3.

## Enzymatic Assays to Detect Defects in Disulfide Bond Formation in *Escherichia coli*

Protein folding in the cell is a competition between two processes: correct folding leads to active protein and incorrect folding leads, at a minimum, to inactive protein but more often to degraded or aggregated protein *in vivo*. As a result, one can get an idea of the effectiveness of protein-folding catalysts simply by comparing the enzymatic activity of their substrate proteins between wild-type strains and strains that are deficient in these folding catalysts.

Of the large number of substrates whose activity is affected by mutations in the disulfide catalysts[1] only two are described here in detail: alkaline phosphatase and the MalF–$\beta$-galactosidase #102 fusion protein. Other genetic tests that can be used to distinguish between wild-type strains and those that form disulfides slowly are acid phosphatase levels,[37] sensitivity to 7 m*M* DTT,[6] sensitivity to benzylpenicillin (15 $\mu$g/ml),[6] and motility in LB plates containing 0.3% (w/v) agar.[38]

### *Alkaline Phosphatase*

Disulfide bond formation in both the periplasm and cytoplasm can be assayed by measuring specific alkaline phosphatase activity. Two intrachain disulfide bonds are necessary for the activity of alkaline phosphatase as phosphomonoesterase. Mutations that decrease the rate of disulfide bond

[36] B. J. Davis, *Ann. N.Y. Acad. Sci.* **121,** 404 (1964).
[37] P. Belin, E. Quéméneur, and P. L. Boquet, *Mol. Gen. Genet.* **242,** 23 (1994).
[38] F. E. Dailey and H. C. Berg, *Proc. Natl. Acad. Sci. U.S.A.* **90,** 1043 (1993).

formation in the periplasm will result in a loss of activity of this periplasmic enzyme.[4,6,37] Mutations that increase the rate of disulfide bond formation in the cytoplasm will result in activation of artificially cytoplasmically located alkaline phosphatase.[7]

*Alkaline Phosphatase Activity Assay.* Grow cultures overnight in NZ medium. Dilute cultures 1:100 (v/v) into M63 salts[32] supplemented with 0.4% (w/v) glucose and a 50-$\mu$g/ml concentration of all amino acids except cysteine and methionine. When the $OD_{600}$ is between 0.3 to 0.6, wash the cells twice by centrifuging at 10,000 *g* for 5 min at 25° and resuspending in an equal volume of TS buffer [10 m*M* Tris (pH 8), 150 m*M* NaCl] at room temperature. Do not leave the cells on ice at any time, as this can result in artificially high levels of alkaline phosphatase, although this problem can be prevented by addition of 10 m*M* IAM to the TS buffer.[10] Use 800 $\mu$l to determine the $OD_{600}$ of the cell suspension following the wash. To determine alkaline phosphatase activity, mix 100 $\mu$l of cells with 900 $\mu$l of 1 *M* Tris-HCl, pH 8.0, containing 0.1 m*M* $ZnCl_2$. Add 25 $\mu$l of 0.1% (w/v) SDS and 25 $\mu$l of chloroform; vortex. Incubate for 5 min at 28° and add 100 $\mu$l of 0.4% (w/v) *p*-nitrophenyl phosphate (PNPP in 1 *M* Tris-HCl, pH 8.0). Record the starting time. Incubate at 28° until a light yellow color develops. Stop the reaction by adding 120 $\mu$l of 2.5 *M* $K_2HPO_4$; record the time. Spin for 5 min at 10,000 *g* in an Eppendorf microcentrifuge to remove cells. Read the $OD_{420}$ and calculate the alkaline phosphatase (AP) activity by the following equation:

$$\text{AP activity units} = \frac{OD_{420} \times 1000}{\text{Time} \times OD_{600} \times \text{ml of cell culture used}}$$

The steady level of active alkaline phosphatase found in *dsb*$^-$ strains is a complex function of growth conditions, and especially of availability of oxygen and cystine. To obtain reproducible values it is important to grow the cultures in exactly the same way each time, in the absence of added cystine and with the same rate of aeration. The accuracy of this procedure is improved by using *phoR*$^-$ strains, in which the synthesis rate of alkaline phosphatase is high and constitutive.[4]

Disulfides are not normally formed in cytoplasmic proteins; as a result, if alkaline phosphatase is forced to stay in the cytoplasm by elimination of its secretion signal, it fails to form its disulfides and is inactive.[7,30] However, a mutation in the gene for thioredoxin reductase leads to an increased rate of disulfide bond formation in the cytoplasm and up to 25% of the total alkaline phosphatase becoming active.[7] The formation of disulfides in the cytoplasm can therefore be monitored and quantified either by measuring the specific activity of the alkaline phosphatase or by immunoprecipitation and separation of reduced and oxidized forms of alkaline phosphatase by nonreducing SDS–PAGE. Antibody to alkaline phosphatase can be obtained from 5 Prime → 3 Prime (Boulder, CO).

*MalF–β-Galactosidase #102 Fusion Protein*

MalF–β-galactosidase #102 fusion protein, an artificial substrate for detection of disulfide bond formation *in vivo,* is particularly useful because it shows a large difference in activity between strains with defects in disulfide bond formation and wild-type strains; a big enough difference to allow the selection of mutants that show decreased levels of disulfide bond formation.[4,5] In this disulfide indicator protein, β-galactosidase, a normally cytoplasmic protein, is fused to the large periplasmic domain of the *E. coli* integral membrane protein MalF. In strains that are wild type for the *dsb* genes the fusion has virtually no β-galactosidase activity. The Dsb proteins are apparently able to oxidize and thus inactivate the portion of this fusion protein that is exposed to the periplasm. However, this fusion protein has substantial β-galactosidase activity if the strain in which this fusion protein is expressed contains a null mutation in the genes for the disulfide catalyst DsbA or DsbB. This increase in β-galactosidase activity can be selected for using lactose minimal plates[32] or can be screened for using LB plates that incorporate X-Gal (5-bromo-4-chloro-3-indolyl-β-D-galactopyranoside; 60 μg/ml). The MalF–β-galactosidase #102 fusion protein is expressed from λ integrated by homology at the maltose locus. Because the fusion is under the maltose promoter, one needs to induce the maltose promoter to see substantial levels of β-galactosidase. This can be done either by introducing the *malT*$^c$ mutation, which constitutively expresses the maltose genes, or by growing cells in the presence of 0.4% (w/v) maltose. It is also helpful to stabilize the λ MalF–β-galactosidase #102 fusion lysogen in the chromosome by introduction of the *recA* mutation into the strain. A semiquantitative test for Dsb function *in vivo* has been developed by exploiting the observation that partially active *dsb* mutants isolated in a strain that contains the MalF–β-galactosidase #102 fusion protein vary in their ability to overcome millimolar concentrations of DTT added to the growth medium.[39]

## Determination of Redox Equilibrium with Glutathione

The relative oxidizing power of disulfide catalysts or indeed of any disulfide bond can, in principle, be measured against glutathione (GSH) as a standard.[2,40–42] This redox scale compares the ability of oxidized gluta-

[39] U. Grauschopf, J. Winther, P. Korber, T. Zander, P. Dallinger, and J. C. A. Bardwell, *Cell* **83,** 947 (1995).

[40] M. Wunderlich and R. Glockshuber, *Protein Sci.* **2,** 717 (1993).

thione (GSSG) to oxidize protein thiols as measured by the equilibrium constant $K_{ox}$ for the following reaction:

$$\text{Protein}_{red} + \text{GSSG} \rightleftharpoons \text{Protein}_{ox} + 2\text{GSH} \tag{1}$$

$$K_{ox} = \frac{[\text{Protein}_{ox}][\text{GSH}]^2}{[\text{Protein}_{red}][\text{GSSG}]} \tag{2}$$

$K_{ox}$ can be measured by any technique that detects changes in the concentration of the reduced protein. One convenient method exploits the quenching effect of disulfides on the fluorescence of adjacent tryptophan residues.[40,41] A number of disulfide exchange enzymes (thioredoxin, DsbA, calf liver protein disulfide isomerase, and the periplasmic domain of TlpA) exhibit a much higher fluorescent yield of reduced as compared with oxidized proteins.[40,41] The fluorescence emission properties of oxidized proteins, and proteins reduced by the presence of 10 m*M* DTT, can be measured by exciting a 1 $\mu M$ solution of the purified protein at 295 nm in a scanning fluorometer and recording emission spectra between 300 and 400 nm. A significant difference in fluorescent yield between reduced and oxidized protein allows their relative quantification. The wavelength that shows the biggest difference in fluorescence between fully oxidized and fully reduced proteins (wavelength $Q$) should be used to monitor the oxidation status of the protein. This quantification is greatly simplified if the protein of interest contains only one disulfide bond. As the number of sulfhydryl groups that participate in thiol–disulfide exchange increases, analysis of the multiple redox states present at equilibrium becomes increasingly difficult.[2,41] Our analysis applies only to the simplest case, in which the protein contains one disulfide bond. It is also important to show that the protein does not show a significantly altered secondary structure on reduction as judged by near- and far-UV circular dichroism measurements,[40,41] because unfolding transition equilibria will complicate the measurement of redox equilibria.

The specific fluorescence of a protein is used to measure the equilibrium concentrations of the oxidized and the reduced form of the protein in the presence of different ratios of oxidized and reduced glutathione. The protein (1 $\mu M$) is incubated at 30° under a nitrogen atmosphere in the presence of degassed glutathione redox buffers that contain constant concentrations of 0.01 m*M* GSSG, 100 m*M* sodium phosphate (pH 7.0) and 1 m*M* EDTA and various concentrations of GSH ranging from 0.005 to 125 m*M* until equilibrium is reached. $R$, the relative amount of reduced protein present at equilibrium, can be calculated from the measured fluorescence intensity using the following formula:

[41] H. Loferer, M. Wunderlich, H. Hennecke, and R. Glockshuber, *J. Biol. Chem.* **270,** 26178 (1995).

[42] R. P. Szajewski and G. M. Whitesides, *J. Am. Chem. Soc.* **102,** 2011 (1980).

$$R = (F - F_{ox})/(F_{red} - F_{ox})$$

where $F$ is the measured fluorescence intensity at the wavelength $Q$, and $F_{red}$ and $F_{ox}$ are the fluorescence intensities of completely reduced or oxidized protein, respectively. The reactions are judged to be in equilibrium when further incubation does not result in a change in intensity and when the intensity is the same independent of whether one approaches equilibrium starting with reduced or oxidized protein.

### *Determination of GSH and GSSG Ratios*

Degassing buffers, flushing with nitrogen, and then incubating under nitrogen atmosphere will reduce air oxidation of GSH. However, to account for air oxidation and correct for the small amount of GSSG contaminating commercial GSH, GSH and GSSG concentrations actually present in the various glutathione mixtures at equilibrium should be measured in conjunction with fluorescence measurements.[41] The equilibrium concentrations of GSH can be determined by the method of Ellman[43]; an $\varepsilon_{412}$ of 14,140 $M^{-1}$ $cm^{-1}$ can be used for the thionitrobenzoate anion. The equilibrium concentrations of GSSG can be quantified enzymatically using yeast glutathione reductase (EC 1.6.4.2; Boehringer GmbH, Mannheim, Germany). NADPH is added to a final concentration of 200 $\mu M$ to a degassed reaction at 25° that contains the GSSG to be assayed, 100 m$M$ sodium phosphate (pH 7.0), and 1 m$M$ EDTA. Reactions are started by addition of 1 U glutathione reductase. After a 10-min incubation the rate of the decrease in absorption at 340 nm is recorded ($\varepsilon_{NADPH}^{340\ nm}$ 6220 $M^{-1}$ $cm^{-1}$). A linear calibration curve should be obtained with GSSG concentrations ranging from 1 to 200 $\mu M$.

### *Calculation of $K_{ox}$*

Once the proportion of the protein present in the reduced form at equilibrium ($R$) and the GSH and GSSG concentrations actually present in the various redox buffers have been determined the redox equilibrium constant, $K_{ox}$, of the protein can be calculated. This is done by plotting $R$ against the $[GSH]^2/[GSSG]$ ratio and performing a nonlinear regression fit using the following formula:

$$R = \frac{[GSH]^2/[GSSG]}{K_{ox} + [GSH]^2/[GSSG]}$$

[43] P. W. Riddles, R. L. Blakeley, and B. Zerner, *Methods Enzymol.* **91,** 49 (1983).

This analysis contains the simplifying assumption that the mixed disulfide between glutathione and the protein is present at equilibrium in negligible quantities. This has been shown to be true for DsbA.[16] To check if this is true for the protein to be tested it may be necessary to quench the protein–glutathione equilibrium reactions by addition of HCl (to about pH 2). Separation, identification, and quantification of all the thiol and disulfide species to determine their concentration at equilibrium may be attempted by reversed-phase HPLC analysis.[3,15] This type of analysis provides an alternative method for determining $K_{ox}$ values. If the $K_{ox}$ value of the tested protein is high, equilibrium constants may be measured by using DTT redox buffer instead of glutathione redox buffers.[2,40] A number of other techniques used in the investigation of disulfide catalysts are not discussed here but are explained in detail in Refs. 15–17, 24, and 42–48.

## Acknowledgments

To assemble the methods described here, we were helped by the knowledge and experience of J. Beckwith, A. Derman, R. Glockshuber, and T. Creighton, all of whom we gratefully acknowledge. We also thank R. Jaenicke for his continuous interest in this project. This work was supported by grants from the Deutsche Forschungsgemeinshaft (DFG) and the Bundesministerium für Forschung und Technologie (BMFT) to J.C.A.B. J.C.A.B. was an Alexander von Humboldt fellow.

[44] M. Wunderlich and R. Glockshuber, *J. Biol. Chem.* **268,** 24547 (1993).
[45] M. Wunderlich, A. Otto, R. Seckler, and R. Glockshuber, *Biochemistry* **32,** 12251 (1993).
[46] M. Wunderlich, R. Jaenicke, and R. Glockshuber, *J. Mol. Biol.* **233,** 559 (1993).
[47] M. Wunderlich, A. Otto, K. Maskos, M. Mücke, R. Seckler, and R. Glockshuber, *J. Mol. Biol.* **247,** 28 (1995).
[48] A. Zapun and T. E. Creighton, *Biochemistry* **33,** 5202 (1994).

# [6] Yeast Immunophilins: Purification and Assay of Yeast FKBP12

*By* GREGORY WIEDERRECHT and JOHN J. SIEKIERKA

## Introduction

The immunophilins constitute a class of ubiquitously expressed proteins named for their ability to bind the immunosuppressive drugs cyclosporin A (CsA), FK506 (Prograf), and rapamycin (RAPA).[1,2] The genes for five

[1] S. L. Schreiber and G. R. Crabtree, *Immunol. Today* **13,** 136 (1992).
[2] J. J. Siekierka, *Immunol. Res.* **13,** 110 (1994).



0076-6879/98 $25.00

CsA-binding proteins and three FK506/RAPA-binding proteins have so far been characterized in *Saccharomyces cerevisiae*.[3–10] The protein products of these genes possess properties essentially identical to those of the mammalian equivalents and bind either CsA or FK506/RAPA with high affinity and possess peptidylprolyl *cis–trans*-isomerase (PPIase, EC 5.2.1.8) activity.

Yeast FKB1 (Fpr1), FKB2 (Fpr2), and Fpr3 gene products (yFKBP12, yFKBP13, and yFKBP47, respectively) exhibit PPIase activity and are localized to different cellular compartments. FKB1 (Fpr1) is largely cytoplasmic, FKB2 is microsomal, and Fpr3 is nucleolar.[3–5] Although it is likely that yFKBPs play a role in catalyzing protein conformational changes *in vivo*, little is known regarding the precise biochemical events regulated by these proteins. Surprisingly, individual or triple disruption of these genes does not compromise viability or lead to any overt phenotypic defects. Presumably this is due to redundant or overlapping functions with the CsA-binding class of immunophilins. There is experimental evidence indicating that the CsA-binding class of immunophilins may play a chaperonin-like function during heat shock stress responses.[11] A fourth yFKBP, having a molecular mass of 62 kDa, has been biochemically identified and appears to associate with both heat shock protein (Hsp) 70 and Hsp 82, suggestive of a role in the heat shock response.[12]

In this chapter we describe the purification and functional assay of yFKBP12 from *S. cerevisiae* and the expression and purification of recombinant yFKBP12. FKBP12 can be assayed in three ways: by peptidylprolyl isomerase (PPIase) activity, by direct binding of the immunosuppressive ligand FK506, and by inhibition of calcineurin phosphatase activity by FK506 · yFKBP12 complexes. The assay methodologies presented here are

[3] G. Wiederrecht, L. Brizuela, K. Elliston, N. H. Sigal, and J. J. Siekierka, *Proc. Natl. Acad. Sci. U.S.A.* **88,** 1029 (1991).

[4] J. B. Nielsen, F. Foor, J. J. Siekierka, M. J. Hsu, N. Ramadan, N. Morin, A. Shafiee, A. M. Dahl, L. Brizuela, G. Chrebet, K. Bostian, and S. Parent, *Proc. Natl. Acad. Sci. U.S.A.* **89,** 7471 (1992).

[5] B. Benton, J.-H. Zhang, and J. Thorner, *J. Cell. Biol.* **127,** 623 (1994).

[6] B. Haendler, R. Keller, P. C. Hiestand, H. P. Kocher, G. Wegmann, and N. R. Movva, *Gene* **83,** 39 (1989).

[7] P. L. Koser, D. Sylvester, G. P. Livi, and D. J. Bergsma, *Nucleic Acids Res.* **18,** 1643 (1990).

[8] M. M. McLaughlin, M. J. Bossard, P. L. Koser, R. Cafferkey, R. A. Morris, L. M. Miles, J. Strickler, D. J. Bergsma, M. A. Levy, and G. P. Livi, *Gene* **111,** 85 (1992).

[9] G. Frigerio and H. R. B. Pelham, *J. Mol. Biol.* **233,** 183 (1993).

[10] L. Franco, A. Jimenez, J. Demolder, F. Molemans, W. Fiers, and R. Contreras, *Yeast* **7,** 971 (1991).

[11] K. Sykes, M.-J. Gething, and J. Sambrook, *Proc. Natl. Acad. Sci. U.S.A.* **90,** 5853 (1993).

[12] P.-K. Tai, H. Chang, M. W. Albers, S. L. Schreiber, D. O. Toft, and L. E. Faber, *Biochemistry* **32,** 8842 (1993).

generally applicable to the characterization of other members of this class of proteins, as has been previously demonstrated for the immunophilin yFKBP13.[4]

## Purification of FKBP12 from *Saccharomyces cerevisiae*

FKBP12 from *S. cerevisiae* (yFKBP12) is purified essentially as described.[13] The protease-deficient *S. cerevisiae* strain EJ926 is grown at 30° with shaking (250 rpm) in YEPD (yeast extract–peptone–dextrose) medium to a density of about 4.5 $A_{600}$ units. The cells are pelleted by centrifugation at 5000 rpm for 30 min at 25° in a JA-10 (Beckman, Palo Alto, CA) rotor. The yeast paste is resuspended 1:2 (w/v) in buffer containing 50 m*M* Tris-HCl (pH 7.8), 10% (w/v) glycerol, 10 m*M* $MgCl_2$, 1 m*M* phenylmethylsulfonyl fluoride (PMSF), 400 $\mu M$ diisopropyl fluorophosphate, 70 $\mu$g of leupeptin per gram of yeast, and 50 $\mu$g of pepstatin A per gram of yeast. The yeast suspension is frozen by dripping into liquid nitrogen. The resulting "popcorn-like" pellets are collected and ground under liquid nitrogen with a pestle in a large mortar. The mortar is precooled to −80° and cooled further by the addition of liquid nitrogen to the cavity. The yeast suspension is pulverized to the consistency of talcum powder or processed flour and is kept frozen by the continuous addition of liquid nitrogen to the mortar. The finely ground material is thawed and insoluble debris is removed by centrifugation at 17,000 rpm for 30 min at 4° in a JA-20 (Beckman) rotor.

The purification of yFKBP12 shares some steps with the procedure used to purify human FKBP12 from Jurkat cells.[14] The extract is subjected to a 56° heat treatment for 20 min. Denatured proteins are removed by centrifugation at 17,000 rpm for 30 min at 4° in a JA-20 rotor. The resulting heat-treated extract is dialyzed overnight at 4° against buffer containing 10 m*M* potassium phosphate (pH 7.2), 1 m*M* EDTA, 5 m*M* 2-mercaptoethanol, and 0.5 m*M* PMSF. The dialyzed protein is applied to a column (1.5 × 40 cm) of Affi-Gel Blue and eluted with a linear gradient (180 ml total) of 10–200 m*M* potassium phosphate, pH 7.2, containing 5 m*M* 2-mercaptoethanol, 1 m*M* EDTA, and 0.5 m*M* PMSF. Fractions (2.5 ml) are assayed for binding of [$^3$H]dihydro-FK506 using an LH-20 assay.[15] Unbound drug remains bound to the LH-20 resin while the yFKBP12 · [$^3$H]dihydro-FK506

[13] J. J. Siekierka, G. Wiederrecht, H. Greulich, D. Boulton, S. H. Y. Hung, J. Cryan, P. J. Hodges, and N. H. Sigal, *J. Biol. Chem.* **265,** 21011 (1990).

[14] J. J. Siekierka, S. H. Y. Hung, M. Poe, C. S. Lin, and N. H. Sigal, *Nature* (*London*) **341,** 755 (1989).

[15] R. Handschumacher, M. Harding, J. Rice, R. Druggs, and D. Speicher, *Science* **226,** 544 (1984).

complex flows through the column. The active fractions are dialyzed against buffer containing 20 m*M* sodium phosphate (pH 6.8), 50 m*M* $Na_2SO_4$, 5 m*M* 2-mercaptoethanol, and 1 m*M* EDTA (TSK buffer). The dialyzed protein is applied to a TSK-125 column (21.5 mm × 60 cm; Bio-Rad, Hercules, CA) at a flow rate of 4 ml/min. Fractions (3 ml) are collected and active fractions, centered around fraction 55, are combined, concentrated to about 5 ml, and applied to an FK506 affinity resin prepared as described.[13] Nonspecifically bound proteins are eluted with buffer containing 10 m*M* potassium phosphate (pH 7.2), 250 m*M* NaCl, and 0.02% (v/v) Tween 20. yFKBP12 is eluted from the column with 4 *M* guanidine hydrochloride. The protein is renatured by dialysis against buffer containing 10 m*M* potassium phosphate (pH 7.2), and 1 m*M* EDTA. Approximately 1.4 mg of yFKBP12 can be obtained from 6.9 g of protein in the crude extract.

## Expression and Purification of Recombinant Yeast FKBP12

The open reading frame (ORF) encoding *S. cerevisiae* FKBP12 can be cloned as an *Nco*I–*Bam*HI fragment into the bacterial expression vector pET3d (Novagen, Madison, WI) and transformed into BL21(DE3) bacteria.[3] A 10-ml overnight culture of the yFKBP12-expressing bacteria is grown at 37° in LB medium containing ampicillin (100 μg/ml). The 10-ml overnight culture is inoculated into a 2-liter Fernbach flask containing 1 liter of M9 medium and ampicillin (100 μg/ml). The flask is incubated at 37° with shaking at 250 rpm. When the optical density is between 0.6 and 0.8 $A_{600}$ units, isopropyl-β-D-thiogalactopyranoside (IPTG) is added to a final concentration of 0.5 m*M* and induction is allowed to proceed overnight.

Yeast FKBP12 is exported to the periplasmic space, greatly simplifying purification.[3] The cells are pelleted by centrifugation (6000 rpm, 15 min, 4°) in a JA-10 (Beckman) rotor and resuspended in 20 ml of 20 m*M* Tris (pH 7.4). The cell suspension, in a 50-ml polypropylene tube, is flash-frozen by immersion of the polypropylene tube in liquid nitrogen. Freezing in liquid nitrogen breaks open the periplasm, releasing its contents. After thawing, all subsequent steps are performed at 4°. Debris is removed by centrifugation at 17,000 rpm in a JA-20 (Beckman) rotor. The supernatant, containing the periplasmic contents, is concentrated to 10 mg/ml and yFKBP12 is purified by size-exclusion chromatography. Using a 5-ml injection loop, up to 50 mg of the extract can be applied per injection to a TosoHaas TSK-Gel G2000SW HPLC column (21.5 mm × 60 cm; Analytical Sales and Services, Manwati, NJ) at a flow rate of 4 ml/min. The column is protected by a TSK SW guard column (21.5 mm × 7.5 cm). The running buffer is the TSK buffer described above and 0.75-min (3-ml) fractions are collected. Protein is followed by absorbance at 280 nm. Because most of the protein in the periplasm is yFKBP12, the major protein (UV-absorbing) peak centered at fraction 57 is yFKBP12. Recombinant yFKBP12 is homogeneous as determined by Coomassie blue staining of a 12% (w/v) denaturing gel (Novex) and is stable indefinitely when stored at −80° or on ice in the presence of 0.02% (w/v) sodium azide. The FK506-binding activity of the recombinant protein is measured using the modified LH-20 binding assay.[14]

## Peptidylprolyl Isomerase Activity of Yeast FKBP12 and FKBP13

The peptidylprolyl isomerase (PPIase) assay is performed essentially as described.[16] The peptide substrate is *N*-succinyl-Ala-Leu-Pro-Phe-*p*-nitroanilide (Bachem, Torrance, CA). The 3.2-mg/ml peptide stock, in dimethyl sulfoxide (DMSO), is stored shielded from light. Yeast FKBP12 is diluted from the concentrated stock to 0.15 mg/ml in 100 m*M* Tris, pH 7.5. A fresh 5-mg/ml stock of bovine pancreatic chymotrypsin (Sigma, St. Louis, MO) is prepared in water and can be maintained at room temperature during the day that the assay is performed. Do not place the chymotrypsin stock on ice. The release of *p*-nitroanilide from the *trans* form of the peptide by chymotrypsin is quantitated by following absorbance at 405 nm at 3-sec intervals for 1.5 min. Reactions (total volume, 1.5 ml) are performed at room temperature in a 3-ml quartz cuvette containing a stir flea (1.5 × 8 mm) set to mix such that no bubbles are created. Conversion to the *trans* form of the peptide is accelerated by yFKBP12. A "blank" reaction is performed to measure the uncatalyzed isomerization reaction. The blank contains 1432.5 μl of 100 m*M* Tris (pH 7.5), 6 μ*M* chymotrypsin (45 μl of the 5-mg/ml stock), and the reaction is initiated by the addition of 22.5 μl of the peptide solution (72 μ*M* final concentration) to the mixing solution in the cuvette. The contents of the enzyme-catalyzed reaction are 1412.5 μl of 100 m*M* Tris (pH 7.5), 6 μ*M* chymotrypsin, and 1-μg/ml yFKBP12 (10 μl of the 0.15-mg/ml yFKBP12 stock). As with the uncatalyzed reaction, the catalyzed reaction is initiated by the addition of peptide substrate. Most of the peptide preexists in the *trans* conformation and so the initial rapid increase in absorbance represents the hydrolysis of the *trans* peptide by chymotrypsin. The slow increase in absorbance represents the conversion of the *cis* peptide to the *trans* peptide. The data are fit to a simple first-order rate law and the first-order rate constant, $k$ ($sec^{-1}$), calculated after subtracting out the blank.

[16] T. Sewell, E. Lam, M. Martin, J. Leszyk, J. Weidner, J. Calaycay, P. Griffin, H. Williams, S. Hung, J. Cryan, N. Sigal, and G. Wiederrecht, *J. Biol. Chem.* **269,** 21094 (1994).

### Calcineurin Phosphatase Inhibition by yFKBP12 · Drug Complex

Calcineurin phosphatase activity is measured by the release of $^{33}P$ from the serine residue of the $^{33}P$-labeled RII peptide (DLDVPIPGRFDRRVS VAAE) substrate, which is the phosphorylation site of the RII subunit of cAMP-dependent protein kinase. FK506, and related analogs, are not inhibitors of calcineurin (CaN) alone. Rather, it is the complex of FK506 with its abundant intracellular receptor, FKBP12, forming a unique entity, that is the actual inhibitor of CaN. In cells, the ability of FK506 to inhibit a CaN-dependent function is a reflection of the ability of the drug to get into the cell, the abundance of the various FKBPs, the concentration of CaN, and the efficacies with which the various FKBP · FK506 complexes can inhibit CaN. The ability of the yFKBP · FK506 complex to inhibit CaN is the result of a complex equilibrium, which can be represented as:

$$\text{yFKBP12} + \text{FK506} + \text{CaN} \rightleftharpoons \text{yFKBP12} \cdot \text{FK506} + \text{CaN} \rightleftharpoons \text{yFKBP12} \cdot \text{FK506} \cdot \text{CaN}$$

To measure the ability of the yFKBP12 · FK506 complex to inhibit CaN, it is important to minimize the equilibrium effect made on that measurement by the dissociation of the yFKBP12 · FK506 complex into its drug and protein components. Therefore, if drug is being titrated in an experiment designed to measure its ability to inhibit CaN, it is important to ensure that all of the added drug is bound by yFKBP12. Thus, drug titrations are performed in the presence of an excess of yFKBP12 (50 $\mu M$) to force equilibrium to the right and toward the formation of the yFKBP12 · FK506 complex. When the assays are set up in this manner, the amount of drug added is an approximation of the amount of yFKBP12 · drug complex formed. Likewise, if yFKBP12 is being titrated, precautions must be taken to ensure that all of the added protein binds to FK506. Protein titrations are performed in the presence of near-saturating drug concentrations (30 $\mu M$), which will also force equilibrium to the right and toward the formation of the yFKBP12 · drug complex. Therefore, the amount of yFKBP12 added will approximate the amount of yFKBP12 · drug complex formed. The crystal structure of the bovine FKBP12 · FK506 · CaN complex indicates that the FKBP12 · FK506 complex blocks the approach of the phosphopeptide substrate to the CaN active site.[17] This and other evidence (S. Salowe, unpublished results, 1997) suggest that the FKBP12 · FK506 complex is a competitive inhibitor of CaN activity. Thus, when measuring the ability of

[17] J. P. Griffith, J. L. Kim, E. E. Kim, M. D. Sintchak, J. A. Thomson, M. J. Fitzgibbon, M. A. Fleming, P. R. Caron, K. Hsiao, and M. A. Navia, *Cell* **82,** 507 (1995).

FKBP · drug complexes to inhibit CaN, more accurate $K_i$ measurements are obtained when the substrate concentration is maintained at a value less than the $K_m$, which for bovine CaN is 40 $\mu M$.[18]

The assay presented here is a modified version of previous CaN assays[18–20] resulting in greatly increased speed and throughput. The major changes to the assay include the substitution of a [$^{33}$P]phospho-RII peptide for a [$^{32}$P]phospho-RII peptide and the conversion to a 96-well microtiter plate format to aid in drug screening. In both assays, CaN activity is measured by the release of free, radiolabeled phosphate from the RII phosphopeptide. Previously, free phosphate was separated from the phosphopeptide using Dowex AG 50W-X8 minicolumns. In the assay presented here, a 96-well membrane-filtration microplate format is used to separate the released radiolabeled phosphate from the substrate.

*Preparation of Phospho-RII Peptide*

The $^{33}$P-labeled RII peptide substrate is prepared as follows. To 1.5 mg of synthetic RII peptide (DLDVPIPGRFDRRVSVAAE; Peptides International) is added 40 $\mu$l of 500 m$M$ morpholinepropanesulfonic acid (MOPS, pH 7.0), 4 $\mu$l of 500 m$M$ magnesium acetate, 10 $\mu$l of 100 m$M$ ATP, pH 7.5 (Pharmacia, Piscataway, NJ), and approximately 17 $\mu$l of [$\gamma$-$^{33}$P]ATP (Amersham, Arlington Heights, IL). The volume of [$\gamma$-$^{33}$P]ATP is adjusted such that the final specific activity of the ATP in the reaction mixture is about 600 cpm/pmol. $H_2O$ is added to bring the volume to 950 $\mu$l. The lyophilized bovine protein kinase catalytic subunit (Sigma) is activated by the addition of 200 $\mu$l of dithiothreitol (DTT, 6 mg/ml) to 1000 units of the lyophilized protein. Activation proceeds for 10 min at room temperature. Fifty microliters of the activated protein kinase is then added to the peptide solution. Phosphorylation of the RII peptide is performed at 30° for 50 min. The phosphorylated peptide is purified by applying the reaction mixture to a 1.0-ml Dowex AG1-X8 (Bio-Rad) column previously equilibrated in 30% (v/v) acetic acid. The flowthrough is collected and the column washed twice with 500 $\mu$l of 30% (v/v) acetic acid. The flowthrough and washes (total volume, 2 ml) are evaporated to dryness overnight in a SpeedVac (Savant, Hicksville, NY) without the application of heat. The dry [$^{33}$P]phosphopeptide may appear yellow to brown in color. The [$^{33}$P]phosphopeptide is resuspended in 1 ml of 60 m$M$ Tris that has not

[18] M. J. Hubbard and C. B. Klee, *in* "Molecular Neurobiology: A Practical Approach" (J. Chad and H. Wheal, eds.), pp. 135–157. IRL Press, Oxford, 1991.

[19] A. Manalan and C. Klee, *Proc. Natl. Acad. Sci. U.S.A.* **80,** 4291 (1983).

[20] J. Liu, J. D. Farmer, W. S. Lane, J. Friedman, I. Weissman, and S. L. Schreiber, *Cell* **66,** 807 (1991).

been adjusted for pH. On resuspension of the [$^{33}$P]phosphopeptide, the resulting pH should be about pH 7. The final concentration of the labeled phosphopeptide will be between 300 and 400 pmol of [$^{33}$P]peptide/$\mu$l. The $^{33}$P-phosphorylated peptide is stable at $-20°$.

*Calcineurin Phosphatase Reaction*

When measuring the ability of an FK506-like compound to inhibit CaN, the CaN phosphatase reaction contains the following components: 40 m*M* Tris-HCl (pH 8.0), 100 m*M* NaCl, 6 m*M* magnesium acetate, 0.1 m*M* $CaCl_2$, bovine serum albumin (BSA, 0.1 mg/ml; Sigma) 0.5 m*M* DTT, 20 $\mu M$ [$^{33}$P]phospho-RII peptide, 190 n*M* calmodulin (CaM; Sigma) 3 n*M* CaN (Sigma), 50 $\mu M$ recombinant yFKBP12, and drug. It is important that the highest purity bovine serum albumin be used because of the presence of trace amounts of FK506 binding activity in less pure preparations that will affect measurements in the presence of drug. In setting up the reaction, a 2× phosphatase buffer stock is prepared consisting of 80 m*M* Tris (pH 8.0), 200 m*M* NaCl, 12 m*M* magnesium acetate, 0.2 m*M* $CaCl_2$, BSA (0.2 mg/ml), and 1 m*M* DTT. A 24× calmodulin stock is prepared in 20 m*M* Tris (pH 7.5) and 30% (v/v) glycerin. A 600× CaN stock is also prepared in 20 m*M* Tris (pH 7.5) and 30% (v/v) glycerin. A 2× phosphatase cocktail (total volume, 3.6 ml) sufficient for performing a reaction in each well of a 96-well microtiter plate is prepared by mixing 100 $\mu M$ yFKBP12 (from a 15-mg/ml solution), 300 $\mu$l of the 24× calmodulin stock, and 2× phosphatase buffer, prepared as described above, making up the remaining volume.

The total volume of the CaN phosphatase reaction is 60 $\mu$l. The reaction mixture is prepared by adding 30 $\mu$l of the 2× phosphatase cocktail to each well of a 96-well polypropylene plate (Nunc, Roskilde, Denmark). This is followed by the addition of 10 $\mu$l of FK506, or related analog, to each of the 96 wells. The drug, diluted from a 10-mg/ml stock in DMSO, is in 20 m*M* Tris (pH 8.0) and 0.1% (w/v) BSA before its addition to the well. Siliconized pipette tips are used at all times when handling FK506 or related compounds. The final drug concentration during titrations is normally between 0.3 n*M* and 50 $\mu M$. Next, 10 $\mu$l of the 600× CaN stock is diluted 100-fold into 990 $\mu$l of 20 m*M* Tris (pH 8.0)–0.1% (w/v) BSA and 10 $\mu$l of this 100-fold diluted CaN is added to each of the 96 wells.

The final 10 $\mu$l of the reaction mixture to be added consists of the phosphopeptide substrate and $H_2O$. Calculate the volume of peptide to be added such that its final concentration will be 20 $\mu M$, one-half of the $K_m$ value. However, before adding the phosphopeptide, add the appropriate volume of water to each well of the polypropylene plate and vortex briefly on a microtiter plate vortexer. Then, incubate the plate at 30° for 30 min.

This incubation allows time for the yFKBP12 · drug complexes to form the ternary complex with CaN and CaM. The dephosphorylation reaction is initiated by the addition of the phosphopeptide and the reaction is allowed to proceed at 30° for 10 min. Terminate the reaction by adding 100 $\mu$l of ice-cold trichloroacetic acid (TCA) containing 0.1 *M* potassium phosphate, pH 7.0. The total volume in each well of the microtiter plate is now 160 $\mu$l.

*Assay of Phosphate Released*

Prepare an Immobilon-P plate (MultiScreen-IP, Immobilon-P 96-well filtration plate; Millipore, Bedford, MA) according to manufacturer instructions. Apply one-fourth (40 $\mu$l) of the reaction mixture to the Immobilon-P filtration plate that has been placed on top of a MultiScreen vacuum manifold (Millipore MultiScreen manifold). Apply medium vacuum and wash each well of the Immobilon-P plate twice with 40 $\mu$l of distilled $H_2O$. Allow 1 to 2 min for all of the liquid to flow through. Collect the flowthrough, containing the released [$^{33}$P]phosphate, and washes in a Microlite-2 96-well plate (Dynatech, McLean, VA). The phosphorylated peptide adheres to the Immobilon-P plate, which is discarded. MicroScint 40 (150 $\mu$l; Packard, Downers Grove, IL) scintillation cocktail is added to each well of the Microlite-2 plate, which is then sealed with a plastic sheet (Packard Topseal-S microplate heat sealing film) using a Micromate 496 sealer (Packard). Each well of the Microlite-2 plate is counted on a TopCount 96-well microplate scintillation counter (Packard). The counts are corrected for decay and the picomoles of phosphate released calculated from the known specific radioactivity of the substrate. Results are calculated as a percentage of the uninhibited (no-drug-added) control.

Figure 1 shows the results obtained measuring the ability of yFKBP12 to mediate the inhibitory effects of FK506 and a related analog, L-685,818 ('818), toward CaN. For comparative purposes, the abilities of the same two compounds, when complexed to human FKBP12, to inhibit CaN is also shown. In Fig. 1A, the two compounds are titrated in the presence of either 50 $\mu M$ yFKBP12 or human FKBP12 (hFKBP12). In Fig. 1B, yeast and human FKBP12 are titrated in the presence of a constant high (30 $\mu M$) concentration of drug. Both methods give similar results. In the presence of FK506, yFKBP12 and hFKBP12 are equipotent inhibitors of bovine CaN. The $IC_{50}$ values for CaN inhibition by the yFKBP12 · FK506 and hFKBP12 · FK506 complexes are both about 10 n*M*. In T cells, '818 is not immunosuppressive and antagonizes the immunosuppression caused by FK506.[21] The antagonism mediated by '818 results from its ability to displace

[21] F. Dumont, M. Staruch, S. Koprak, J. Siekierka, C. Lin, R. Harrison, T. Sewell, V. Kindt, T. Beattie, M. Wyvratt, and N. Sigal, *J. Exp. Med.* **176,** 751 (1992).

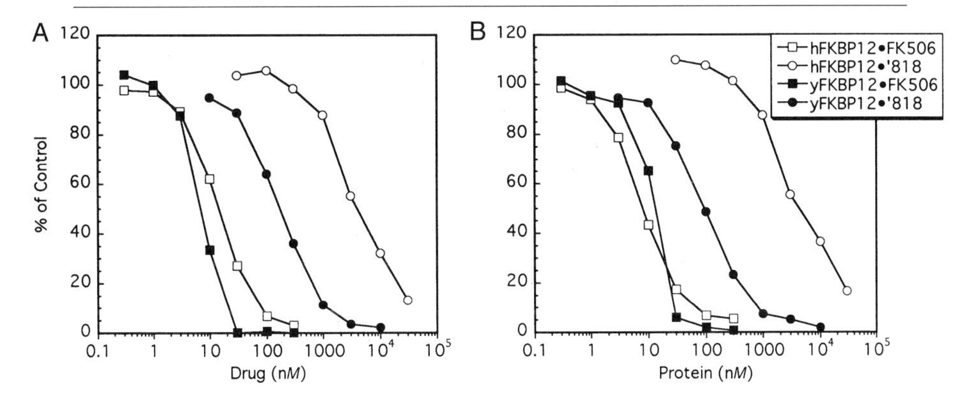


FIG. 1. Inhibition of bovine CaN by the FK506 and L-685,818 ('818) complexes with yFKBP12 and hFKBP12. (A) Either FK506 or L-685,818 is titrated into the assay in the presence of a high (50 $\mu M$) concentration of yFKBP12 or hFKBP12. The $IC_{50}$ values (in parentheses) of the indicated FKBP12 · drug complexes are as follows: hFKBP12 · FK506 (14.3 n*M*); hFKBP12 · '818 (2.9 $\mu M$); yFKBP12 · FK506 (7.4 n*M*); and yFKBP12 · '818 (94 n*M*). (B) Either yFKBP12 or hFKBP12 is titrated into the CaN assay in the presence of a high concentration of FK506 or L-685,818. The $IC_{50}$ values (in parentheses) of the indicated FKBP · drug complexes are as follows: hFKBP12 · FK506 (7.6 n*M*); hFKBP12 · '818 (3.4 $\mu M$); yFKBP12 · FK506 (10.4 n*M*); and yFKBP12 · '818 (176 n*M*).

FK506 from FKBP12. However, the hFKBP12 · '818 complex is a poor inhibitor of CaN (Fig. 1A and B). Both compounds bind well to FKBP12 because they are identical in that portion of the molecule required for binding. However, in that face of the molecule that binds CaN, the presence of a hydroxyl group on '818 at the C-18 position of the macrolactam ring inhibits binding to CaN, either through steric interference or charge effects, when complexed with hFKBP12. The concentration of hFKBP12 · '818 complex required to inhibit CaN cannot be attained in T cells as it can *in vitro*. As shown in Fig. 1, the yFKBP12 · '818 complex is much more potent than the hFKBP12 · '818 at inhibiting CaN. The three-dimensional structures of the '818 complex with yeast and human FKBP12 are almost identical.[22] However, among the 26 amino acids on that portion of FKBP12 likely to be in close proximity to CaN in the presence of drug, there are 10 residue differences between yeast and human FKBP12. Thus, among these 10 residues there are amino acids that are compensating for the inhibitory effects of the C-18 hydroxyl group.

Care must be exercised in the interpretation of assay results when the $IC_{50}$ approaches that of the concentration of CaN, which in this assay is 3 n*M*. Thus, development of a more potent FK506 analog that, when com-

[22] J. Rotonda, J. Burbaum, H. Chan, A. Marcy, and J. Becker, *J. Biol. Chem.* **268,** 7607 (1993).

plexed with FKBP12, inhibits CaN in this assay with an $IC_{50}$ of 3 n$M$, suggests that CaN is merely being titrated by the FKBP12 · drug complex. Thus, for such analogs to be accurately assayed, it becomes important to lower the CaN concentration further so that it is significantly below the $IC_{50}$ of the FKBP12 · drug complex.

# [7] Peptidylprolyl *cis–trans*-Isomerases from Plant Organelles

*By* AUTAR K. MATTOO

Correct protein folding is central in determining the secondary and tertiary structure, substrate specificity, and functional activity of a protein, thereby regulating cellular metabolism and organismic physiology.[1] It is apparent that living cells utilize diverse and intricate processes to assemble nascent chains of polypeptides into functional proteins. Protein folding involves conformational changes that give each protein its structural and functional signature. Proper configuration of disulfide bonds catalyzed by protein disulfide isomerases,[2] as well as *cis–trans* isomerization of the peptide bonds adjacent to proline residues catalyzed by peptidylprolyl isomerases,[3] are two reactions known to regulate protein folding *in vitro*. Both these enzymes (EC 5.1.) are ubiquitous in nature and are distributed in the cytosol and most subcellular organelles. This chapter particularly deals with plant peptidylprolyl *cis–trans*-isomerases (PPIases). Two groups of PPIases are distinguishable on the basis of their binding affinity to immunosuppressants, either cyclosporin A (CsA), or FK506 and rapamycin.[4] Cyclophilins (immunophilins) specifically bind CsA whereas FK506-binding proteins (FKBPs) bind FK506 and rapamycin. The binding of the immunosuppressants to PPIases results in inhibition of the enzymatic activity. PPIases catalyze slow steps in the folding/assembly of proteins, the exact mechanism of the reactions around specific peptide bonds being largely unknown. The demonstration of specific PPIase activities in plant organelles[5] has now been confirmed via cloning of several distinct cDNAs for cytosolic and

[1] S. L. Rutherford and C. S. Zuker, *Cell* **79,** 1129 (1994).
[2] R. Noiva and W. J. Lennarz, *J. Biol. Chem.* **267,** 3553 (1992).
[3] K. Lang, F. X. Schmid, and G. Fischer, *Nature* (*London*) **329,** 268 (1987).
[4] S. L. Schreiber, *Science* **251,** 283 (1991).
[5] A. Breiman, T. W. Fawcett, M. L. Ghirardi, and A. K. Mattoo, *J. Biol. Chem.* **267,** 21293 (1992).


METHODS IN ENZYMOLOGY, VOL. 290
0076-6879/98 $25.00

TABLE I
COMPOSITION OF BUFFERS EMPLOYED FOR CHLOROPLAST ISOLATION AND FRACTIONATION

| Component | Extraction buffer | Wash buffer | Buffer A[a] (*M*) | Resuspension buffer (*M*) |
|---|---|---|---|---|
| Sorbitol | 0.4 *M* | — | 0.1 | 0.4 |
| $MgCl_2$ | 0.01 *M* | 0.01 *M* | 0.01 | 0.01 |
| NaCl | 0.01 *M* | 0.01 *M* | 0.01 | 0.01 |
| Tricine-OH, pH 7.8 | 0.1 *M* | 0.1 *M* | 0.01 | 0.1 |
| PVP-40[b] | 0.2% (w/v) | 0.2% (w/v) | — | — |
| Sodium ascorbate[b] | 0.2% (w/v) | 0.2% (w/v) | — | — |

[a] For example, to make buffer A add (final volume per liter) sorbitol (18.22 g), $MgCl_2$ (2.03 g), NaCl (0.584 g), and Tricine (1.792 g).
[b] Add immediately before use. PVP-40, polyvinylpyrrolidine-40.

plastidic PPIases. We compile here methods to isolate, assay, and characterize PPIases from plant organelles.

## Isolation and Subfractionation of Chloroplasts and Mitochondria

### *Plant Materials*

For chloroplast preparations, leaves are harvested from plants grown for 10–15 days in either growth chambers or greenhouses at 25° under 50–100 $\mu$mol $m^{-2}$ $sec^{-1}$ PAR (photosynthetically active radiation) or natural light, respectively. For preparation of mitochondria, we prefer to use etiolated seedlings grown for 5–8 days in growth chambers at 25° in the dark.

### *Isolation of Chloroplast Stroma and Thylakoids*

Intact chloroplasts are isolated from fully developed leaves by the method of Bartlett *et al.*[6] This method, which uses sorbitol as the osmoticum in the extraction medium and Percoll gradients to purify crude chloroplast fraction, works well with pea, spinach, barley, and corn leaves.

1. Leaves are first gently washed sequentially with distilled water, 70% (v/v) ethanol, and distilled water, then cut into small pieces with a razor blade or scissors directly into an ice-cold homogenizer container.
2. Leaf pieces are homogenized [twice, 15 sec each time; Polytron (Brinkmann, Westbury, NY) setting 3] in ice-cold extraction buffer (Table I) using 4 ml of buffer/g (fresh weight) of tissue.

[6] S. G. Bartlett, A. R. Grossman, and N.-H. Chua, *in* "Methods in Chloroplast Molecular Biology" (M. Edelman, R. B. Hallick, and N.-H. Chua, eds.), p. 985. Elsevier Science, Amsterdam, 1982.

3. The homogenate is filtered through 2 layers of Miracloth (Calbiochem, La Jolla, CA) into centrifuge tubes or bottles that are then centrifuged (5000 *g* for 15 sec, 4°). The supernatant is discarded.

4. The green pellet is gently resuspended in the extraction medium (1/10 the original volume) and purified on 10–80% Percoll gradients [centrifuge at 14,600 *g* in a swing-out (such as HB-4) rotor for 10 min at 4°; see Bartlett *et al.*[6] for additional details]. The lower green band containing intact chloroplasts is collected. *Note:* This purification step is intended to collect chloroplast stroma; however, if only the chloroplast membranes are to be analyzed, this step can be omitted.

5. Lysis of chloroplasts is achieved by resuspension in a lysis buffer [50 m*M* Tris-HCl (pH 7.4), 50 m*M* NaCl, 20 m*M* $MgCl_2$, 0.1 m*M* EDTA, 10% (v/v) glycerol, 2 m*M* phenylmethylsulfonyl fluoride (PMSF), leupeptin (2 $\mu$g/ml), and 5 m*M* 2-mercaptoethanol], followed by either vortexing or use of a Teflon–glass hand homogenizer. *Note:* Addition of EDTA, phenylmethylsulfonyl fluoride, leupeptin, and 2-mercaptoethanol is highly recommended because these help in minimizing proteolytic degradation during chloroplast lysis, sample handling, and storage.

6. Membranes are separated from the stroma by centrifugation of the lysed mixture at 7000 *g* for 10 min at 4°. The supernatant is saved as soluble (stroma) fraction.

7. The membrane pellets are washed twice in lysis buffer supplemented with 300 m*M* NaCl and once with 10 m*M* Tris-HCl, pH 8.0. The washed chloroplast membranes (thylakoids) are immediately processed for protein solubilization and fractionation.

### *Fractionation of Thylakoids into Grana and Stroma Membranes*

The following fractionation of chloroplast membranes into stromal and granal subfractions is based on previously described methods.[7,8] The one routinely used in the author's laboratory is described here.

1. The chloroplast membrane pellet from step 6 in the previous section is resuspended in buffer A (Table I) and the chlorophyll content of a small aliquot is determined by the method of Arnon.[9] The suspension is appropriately diluted with buffer A to obtain a 0.8-mg/ml chlorophyll concentration. *Note:* A uniform suspension of thylakoids can be obtained manually by using a Teflon–glass homogenizer.

[7] K. J. Leto, E. Bell, and L. McIntosh, *EMBO J.* **4,** 1645 (1985).
[8] A. K. Mattoo and M. Edelman, *Proc. Natl. Acad. Sci. U.S.A.* **84,** 1497 (1987).
[9] D. I. Arnon, *Plant Physiol.* **24,** 1 (1949).

2. Add an equal volume of 1% (w/v) digitonin (Sigma, St. Louis, MO) to the suspension such that final concentrations are 0.4-mg/ml chlorophyll and 0.5% (w/v) digitonin. Incubate for 30 min in the dark on ice with gentle stirring. Centrifuge at 1000 *g* for 3 min at 4° [2690 rpm, JA-17 Beckman, Fullerton, CA) rotor; or 2900 rpm, SS-34 Sorvall–du Pont (Norwalk, CT) rotor]. Discard the pellet. *Note:* A stock solution of digitonin is made by adding an appropriate volume of water to the powder in a reagent bottle, which is then autoclaved to dissolve the detergent.

3. Centrifuge the supernatant at 40,000 *g* for 30 min at 4° (17,000 rpm, JA-17 rotor; or 18,300 rpm, SS-34 rotor). Save both the supernatant (for stromal membrane isolation) and the pellet (for grana isolation).

4. The supernatant is recentrifuged at 144,000 *g* for 90 min at 4° (34,600 rpm in a Beckman SW50.1 rotor or 41,000 rpm in a TLS55 rotor). Discard the supernatant. The pellet is gently suspended in resuspension buffer (Table I), 10 $\mu$l/g (fresh weight) of leaves. The chlorophyll content is then determined as before, and the suspension stored at −70°. This fraction contains stromal membranes.

5. For grana preparation, the pellet from step 3 is resuspended in buffer A, 20 $\mu$l/g (fresh weight) of the original tissue, and the chlorophyll concentration is determined.

6. Dilute the suspension (from step 5) with buffer A to 1 mg of chlorophyll per milliliter. Add 75 $\mu$l of 20% (w/v) Triton X-100/ml extract to obtain a Triton-to-chlorophyll ratio of 15. Incubate in the dark for 1 min while stirring on ice.

7. Centrifuge at 1000 *g* for 3 min at 4° (2690 rpm in the JA-17 rotor or 2900 rpm in the SS-34 rotor) and discard the pellet. Recentrifuge the supernatant at 40,000 *g* for 30 min at 4° (17,000 rpm in the JA-17 rotor or 18,300 rpm in the SS-34 rotor) and discard the supernatant. The pellet (containing grana membranes) is gently suspended in resuspension buffer (Table I), 40 $\mu$l/g (fresh weight) of leaves. The chlorophyll content is then determined and the suspension is stored at −70°.

*Notes.* The purity and quality of grana and stroma membranes is determined by electron microscopy as well as sodium dodecyl sulfate-polyacrylamide gel electrophoresis (SDS–PAGE).[10] The fractionation method employed provides submembrane fractions that have less than 2% contamination of one type with the other, based on visual examination in the electron microscope. The grana membranes appear as double-membrane

[10] F. E. Callahan, W. P. Wergin, N. Nelson, M. Edelman, and A. K. Mattoo, *Plant Physiol.* **91,** 629 (1989).

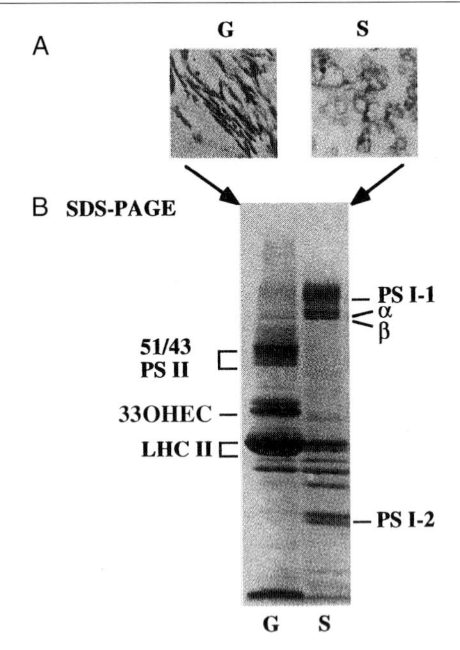


FIG. 1. (A) Electron micrographs illustrating ultrastructural characteristics of the isolated granal (G) and stromal (S) lamellae. (B) SDS–PAGE protein profiles of granal (G) versus stromal (S) lamellae. The positions of subunits 1 and 2 of photosystem I (PS I-1; PS I-2), $\alpha$ and $\beta$ subunits of proton ATPase, 51/43 chlorophyll-binding proteins of photosystem II (PS II), 33-kDa polypeptide of the oxygen-evolving complex (33OHEC), and light-harvesting chlorophyll *a/b* apoprotein (LHC II) are indicated.

sheets (Fig. 1A) and are enriched in photosystem II proteins (Fig. 1B, lane G), whereas the stroma membranes are seen as single-membrane vesicles (Fig. 1A) and are enriched in photosystem I proteins and the proton ATPase complex (Fig. 1B, lane S).

### *Isolation of Mitochondria*

Etiolated plant tissue is preferred[11] because it decreases interference from chloroplasts; however, green leaf tissue may be used following the modifications introduced by Galun *et al.*[12] The method described here is

[11] D. R. Pring and C. S. Levings III, *Genetics* **89,** 121 (1978).

[12] E. Galun, P. Arzee-Gonen, R. Fluhr, M. Edelman, and D. Aviv, *Mol. Gen. Genet.* **186,** 50 (1982).

TABLE II
COMPOSITION OF BUFFERS EMPLOYED FOR ISOLATION AND SUBFRACTIONATION OF MITOCHONDRIA

| Component[a] | Buffer B | Buffer C | Buffer D | SP buffer | Solubilization buffer |
|---|---|---|---|---|---|
| Sucrose | 0.5 *M* | 0.4 *M* | 0.3 *M* | 0.3 *M* | — |
| Tris-HCl, pH 7.5 | 0.05 *M* | 0.05 *M* | 0.05 *M* | — | 0.01 *M* |
| Sodium phosphate | — | — | — | 0.02 *M* | — |
| EDTA (disodium) | 0.005 *M* | 0.005 *M* | — | 0.0025 *M* | — |
| BSA | 0.1% | 0.1% | — | — | — |
| EGTA | — | — | — | 0.0025 *M* | — |
| PMSF | — | — | — | — | 0.001 *M* |
| NaCl | — | — | — | — | 0.15 *M* |
| *N*-Octylglucoside | — | — | — | — | 1.8% |

[a] BSA, Bovine serum albumin; EGTA, ethylene glycol-bis($\beta$-aminoethyl ether)-*N,N,N',N'*-tetraacetic acid; PMSF, phenylmethylsulfonyl fluoride.

based on Pring and Levings,[11] Galun *et al.,*[12] and Breiman.[13] All procedures are carried out at 4°.

1. Etiolated tissue is weighed and homogenized with ice-cold buffer B (4 ml/g tissue) (Table II) in either a Polytron (setting 3, 15 sec) or a blender (Waring; three-fourths maximum speed, 10 sec). Green leafy tissue can be homogenized in the same manner except that buffer B is supplemented with 1 m*M* 2-mercaptoethanol.

2. The homogenate is filtered through two layers of cheesecloth and two layers of Miracloth. *Note:* Cheesecloth and Miracloth are previously wetted with cold distilled water and briefly squeezed before use for filtration. The volume of the filtrate is measured before centrifugation (Beckman JA-17 rotor, 3500 rpm, 10 min or Sorvall-Du Pont GSA rotor, 3100 rpm, 10 min, 4°). The pellet is discarded.

3. The supernatant is centrifuged [GSA rotor, 10,000 rpm (16,300 *g*), 10 min] to pellet mitochondria. The pellet is then gently suspended in buffer C to one-fifth the original volume of the filtered homogenate and centrifuged [GSA rotor, 3500 rpm (2000 *g*), 10 min]. The pellet is discarded.

4. The supernatant is recentrifuged [GSA rotor, 8500 rpm (11,800 *g*), 15 min] and the pellets collected. These are washed once and then resuspended in buffer D (using 1/10 the original weight of the plant material), 10 ml/100 g tissue (Table II), and centrifuged at 12,000 *g* for 15 min (GSA

[13] A. Breiman, *Theor. Appl. Genet.* **73,** 563 (1987).

rotor, 8700 rpm, or the SS-34 rotor, 10,000 rpm; 4°) to obtain the crude mitochondrial fraction. This pellet is suspended in a small volume of SP buffer (Table II), and then more SP buffer is added to obtain a mitochondrial suspension containing total protein equivalent to 40 mg/ml. Protein concentration is measured by Rose Bengal[14] or Bradford[15] methods.

### *Fractionation of Mitochondria into Matrix and Membranes*

Mitochondria obtained as in step 4 of the previous section are further fractionated to obtain the soluble (matrix) and membrane fractions using the methods described by Hack *et al.*[16] and Breiman *et al.*[5]

1. Mitochondrial suspension is sonicated (we use the microprobe in a Virsonic 300-VirTis (VirTis, Gardiner, NY) at 60% of full power) six times for 5 sec at 25-sec intervals. The tubes are kept in a mixture of methanol–ice.
2. The sonicated mixture is centrifuged (15,600 *g*, 10 min, 4°; 11,400 rpm in the SS-34 rotor) to pellet unbroken mitochondria and any aggregated material. The supernatant is centrifuged (230,000 *g*, 70 min, 4°) to obtain mitochondrial matrix. The pellets are saved.
3. The 230K pellet is suspended in SP buffer, divided into two portions, and each centrifuged at 230,000 *g* for 70 min at 4°. The supernatants are discarded and both pellets saved.
4. One of the washed pellets is resuspended in 0.1 *M* sodium bicarbonate, pH 11, to solubilize peripheral proteins (Fujika *et al.*[17]). After 40 min on ice, the pellets are collected by centrifugation (230,000 *g*, 70 min, 4°).
5. The other pellet is resuspended in the solubilization buffer (Table II) and incubated on ice for 60 min. The solubilized mitochondrial membrane fraction (supernatant) is collected by centrifugation at 15,600 *g* for 5 min at 4° (11,400 rpm in the SS-34 rotor).

## Cyclosporin A Binding to Peptidylprolyl Isomerases

Both chloroplasts and mitochondria contain PPIases that bind the drug CsA and lose a major proportion of PPIase activity in the process (Breiman *et al.*[5]). Therefore, CsA-binding ability is a good measure of the presence of PPIases in these tissue extracts. The method of choice is that of Handschumacher *et al.*[18]

[14] J. I. Elliott and J. M. Brewer, *Arch. Biochem. Biophys.* **190,** 351 (1978).
[15] M. M. Bradford, *Anal. Biochem.* **72,** 248 (1976).
[16] E. Hack, C. Lin, H. Yang, and H. T. Horner, *Plant Physiol.* **95,** 861 (1991).
[17] Y. Fujika, S. Fowler, H. Shio, A. L. Hubbard, and P. B. Lazarow, *J. Cell Biol.* **93,** 103 (1982).
[18] R. E. Handschumacher, M. W. Harding, J. Rice, R. J. Drugge, and D. W. Speicher, *Science* **226,** 544 (1984).

*Preparation of Chloroplast Membrane Samples*

1. Chloroplast membranes (1 mg of chlorophyll per milliliter) are solubilized in 2% (w/v) Triton X-100 for 30 min on ice. Insoluble material is pelleted by centrifugation at 2000 *g* for 10 min at 4° and discarded. The supernatant is saved.

2. A 2-cm column of DEAE-Toyopearl 650S (Supelco, Bellefonte, PA) is prepared and equilibrated with 50 m*M* Tris-HCl, pH 7.2, and 0.2% (w/v) Triton X-100.

3. The supernatant from step 1, containing 2 mg equivalent of chlorophyll, is applied to the DEAE-Toyopearl column. The column is washed with 10 ml of the equilibration buffer as in step 2.

4. PPIases bind to the column and are eluted with 20 ml of 0.5 *M* NaCl. Fractions of 0.5 ml are collected and analyzed for chlorophyll content and CsA binding. Maximum CsA binding occurs in fractions 4 through 6.

*Note.* The abundance of chlorophylls associated with isolated thylakoids interferes with the PPIase assay. Therefore, to circumvent this problem, it is necessary to use detergent solubilization and DEAE-Toyopearl fractionation, as described above, before assaying for PPIase activity associated with chloroplast membranes.

*Mitochondrial Membrane Samples*

The mitochondrial membranes, isolated as described above, are solubilized in 1.8% (w/v) *N*-octylglucoside for 60 min on ice. The insoluble material is pelleted by centrifugation at 15,600 *g* for 10 min at 4°. The mitochondrial membrane-associated PPIase as well as CsA-binding activity are removable by treatment of membranes with 0.1 *M* sodium carbonate, pH 11, which solubilizes the peripheral proteins.

*Radioactive Cyclosporin A Tagging Assay*

One method to identify PPIases is to determine the ability of subcellular fractions to bind radioactive CsA.[18] To determine specific binding, different concentrations of nonradioactive CsA are added to the reaction mixture to dilute the specific radioactivity of [$^3$H]CsA. The amount of specific binding is calculated as the difference in the binding at 500 n*M* CsA versus that at 5 n*M*.

1. Minicolumns (1.7-ml) of Sephadex LH-20 resin (Pharmacia, Piscataway, NJ) are equilibrated with 20 ml of a solution containing 20 m*M* Tris-HCl (pH 7.2), 5 m*M* 2-mercaptoethanol, and 0.02% (w/v) sodium azide, and kept ready in the cold room. The void volume of the column is determined using 0.2% (w/v) Blue Dextran dissolved in the equilibration buffer.

2. Several tubes are prepared to contain the following reaction mixture (100-μl final volume): 10 m*M* *N*-2-hydroxyethylpiperazine-*N'*-2-ethanesulfonic acid (HEPES, pH 8), 100 m*M* NaCl, 0.015% (w/v) Triton X-100, 5 m*M* 2-mercaptoethanol, 0.5 μ*M* CsA, and [$^3$H]CsA ($10^5$ cpm; specific activity, 11 Ci/mmol; Amersham, Arlington Heights, IL). Different aliquots of the plant extract to be tested are then added.

3. The reaction mixtures are incubated at 25° for 60 min and then applied to the LH-20 columns.

4. Each column is washed with 400 μl of 10 m*M* HEPES (pH 8.0), 100 m*M* NaCl, 0.015% (w/v) Triton X-100 and 5 m*M* 2-mercaptoethanol.

5. Elution of the bound material is achieved by applying 500 μl of the buffer to the column as in the previous step.

Relative binding of CsA by plant organellar fractions is summarized in Table III. Maximum binding is seen with DEAE-fractionated thylakoids, grana membranes binding twice as much as either the stroma-exposed membranes or stroma (soluble) fraction. The CsA binding by total mitochondrial extract equals that bound solely by either chloroplast stroma or stroma-exposed membranes. Within mitochondria, membranes bind less than twice the amount of CsA bound by the matrix (Table III). A major difference in CsA binding between mitochondrial membranes and stroma-exposed chloroplast membranes is that CsA is peripherally bound to mitochondrial membranes (the binding can be released by treatment with 0.1 *M* sodium carbonate, pH 11) whereas binding associated with chloroplast membranes is not removable by washing at alkaline pH.

### *Peptidylprolyl Isomerase Assay*

The assay is based on measuring the interconversion of *cis* to *trans* isomerization of the Ala-Pro peptide bond in the substrate peptide *cis*-*N*-succinyl-Ala-Ala-Pro-Phe-*p*-nitroanilidine (Sigma) (Fischer *et al.*[19]). In the presence of PPIase, the Ala-Pro bond takes the *trans* configuration, which then allows chymotrypsin (Boehringer Mannheim, Indianapolis, IN) to cleave the test peptide. The resultant 4-nitroanilide is quantified spectrophotometrically at 390 nm.

1. The test peptide (20–60 μ*M* final concentration) is added to a solution containing 40 m*M* HEPES (pH 8.0) and 0.015% (w/v) Triton X-100 and then mixed with appropriate volumes of the plant extract or subcellular fractions (50–200 μg of protein). Cyclosporin A (10–20 μ*M*) and rapamycin

[19] G. Fischer, B. Wittmann-Liebold, K. Lang, T. Kiefhaber, and F. X. Schmid, *Nature* (*London*) **337,** 476 (1989).

TABLE III
IMMUNOPHILIN-BINDING PROPERTIES OF PLANT PPIases[a]

| Characteristic | Mitochondria | | | | Chloroplasts | | | |
|---|---|---|---|---|---|---|---|---|
| | Matrix | Membranes | Washed membranes[b] | Stroma | DEAE-fractionated thylakoids | Grana | Stroma lamellae | Washed stroma lamellae[c] |
| CsA binding[d] (cpm/mg protein) | 78,800 | 31,900 | 2,100 | 85,917 | 533,140 | 184,000 | 81,952 | 91,667 |
| CsA inhibition of PPIase activity[e] (%) | 86 | 100 | | 100 | 66 | ND | ND | ND |
| Rapamycin inhibition of PPIase activity[f] (%) | 0 | 39 | | | | | | |
| Cross-reactivity to anti-RBP[g] | + | + | – | – | – | – | – | – |

[a] Adapted from A. Breiman, T. W. Fawcett, M. L. Ghirardi, and A. K. Mattoo, *J. Biol. Chem.* **267,** 21293 (1992) with permission.
[b] Washed with 0.1 *M* $Na_2CO_3$, pH 11.
[c] Washed at pH 9.
[d] Given as the difference in binding at 500 and 5 n*M*.
[e] CsA was tested at 14 $\mu M$.
[f] Rapamycin was tested at 30 $\mu$M.
[g] Measured on immunoblots; ND, not detectable.

(20–40 $\mu M$), inhibitors of PPIase, are added in two other tubes containing this mixture, to provide an index of the specificity of the enzymatic reaction. The final volume of the mixture is brought to 1.5 ml.

2. Each reaction mixture is held at 10° until transferred to a spectrophotometer cuvette. Chymotrypsin (final concentration, 20 $\mu M$) is added to the cuvette to initiate the enzymatic reaction and the change in absorbance at 390 nm is monitored until the curve plateaus.

3. First-order rate constants are calculated from semilog plots derived for each curve. The difference between the catalyzed and uncatalyzed first-order rate constants, derived from the kinetics of the $A_{390}$ change, is multiplied by the amount of substrate in each reaction. One unit of PPIase activity is expressed as nanomoles (substrate hydrolyzed) per second per milligram of protein.

*Notes*

1. Under equilibrium conditions, 88% of the test peptide is present in the *trans* form and the remaining 12% in the *cis* form. It is this lower percent form that, when acted on by PPIase, is converted to the *trans* form, which is then cleaved by chymotrypsin.

2. When *X*aa in the test substrate succinyl-Ala-*X*aa-Pro-Phe-*p*-nitroanilide was replaced with different amino acids, the specificity found followed the order: Ala > Val > Leu > Ile > Glu > Phe (Luan *et al.*[20]).

3. PPIase activity, in the presence and absence of CsA, as a function of protein concentration, is shown for solubilized pea mitochondria (Fig. 2) and partially purified solubilized thylakoids (Fig. 3).

## Plant Mitochondria But Not Chloroplasts, Containing FK506-Binding Proteins

In addition to the family of CsA-binding cyclophilins present in plant organelles, another distinct class of PPIases, the FKBPs, show characteristic binding to the other immunosuppressants, FK506 and rapamycin.[4] In plants, the rapamycin-binding proteins seem selectively present in the mitochondria but not in the chloroplasts (Table III; Fig. 4). Two methods are used to identify the FKBPs. One method involves tagging the protein with radioactive FK506[21,22] in a manner similar to that described above for the CsA-

[20] S. Luan, W. S. Lane, and S. L. Schreiber, *Plant Cell* **6,** 885 (1994).

[21] J. J. Siekierka, S. H. Y. Hung, M. Poe, C. S. Lin, and N. H. Sigal, *Nature (London)* **341,** 755 (1989).

[22] M. W. Harding, A. Galat, D. E. Uehling, and S. L. Schreiber, *Nature (London)* **341,** 758 (1989).

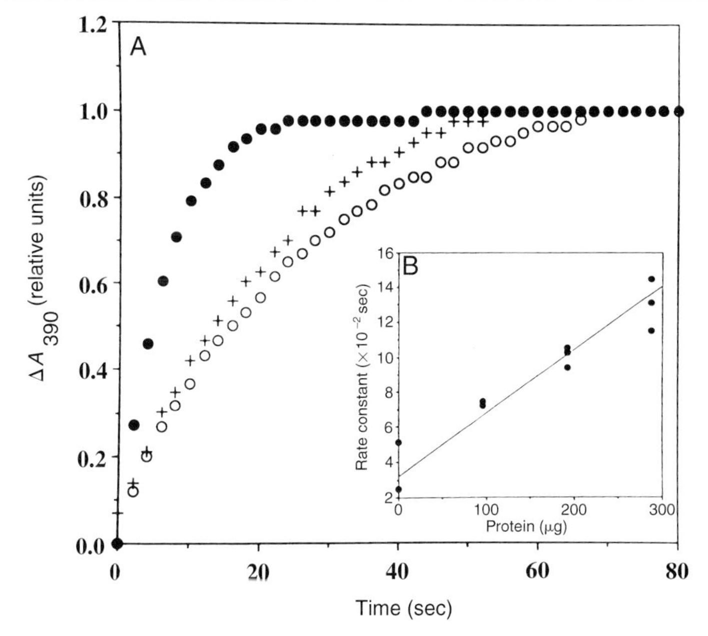


FIG. 2. (A) Kinetics of solubilized pea mitochondrial PPIase activity with *N*-succinyl-Ala-Ala-Pro-Phe-*p*-nitroanilidine as substrate (filled circles). Uncatalyzed, nonenzymatic reaction (unfilled circles) and the effect of 14 $\mu M$ CsA (marked with plus sign) are also shown. (B) Linear dependence of the enzyme activity as a function of protein concentration. (From A. Breiman, T. W. Fawcett, M. L. Ghirardi, and A. K. Mattoo, *J. Biol. Chem.* **267,** 21293 (1992) with permission.)

binding assay. Alternatively, FKBPs or rapamycin-binding proteins (RBPs) can be identified on immunoblots using antibodies to homologs of FKBP or RBP as described in Breiman *et al.*[5]

*Radioactive Assay*

Microtiter plates (96-well) previously coated with Sigmacote (Sigma) are used because of the ease of handling multiple samples. Aliquots of the test sample are mixed with 40–60 n$M$ [$^3$H]dihydro-FK506[21] (specific activity, ~50 Ci/mmol) or 32-[1-$^{14}$C]FK506[22] (specific activity, >2.5 × $10^6$ cpm $\mu g^{-1}$) in a solution containing 50 m$M$ HEPES (or 20 m$M$ sodium phosphate), pH 7.5, and 0.5% (w/v) bovine serum albumin in the presence and absence of a severalfold excess of cold FK506. The mixtures are incubated for 15–30 min at 25° and then assayed by the LH-20 resin assay as described above.

*Notes*

1. The appropriate concentrations of the buffer and protein equivalent of the plant extract to be used are determined in preliminary experiments.

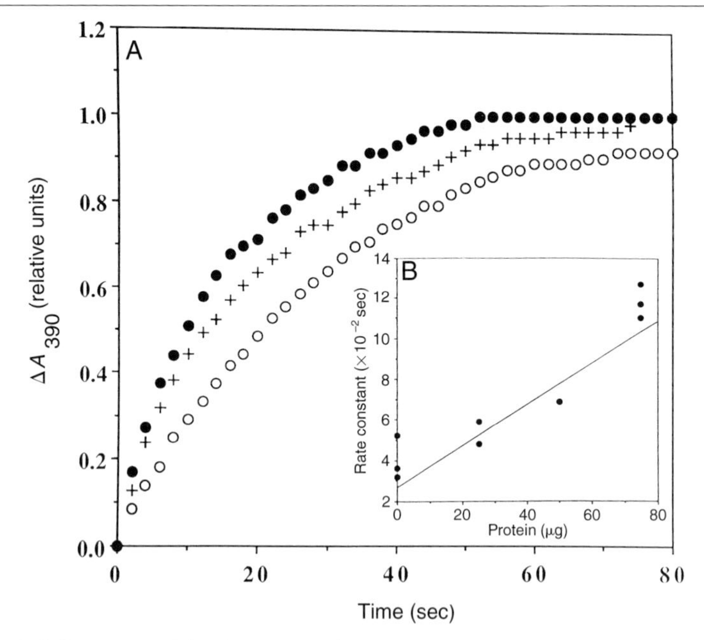


FIG. 3. (A) PPIase activity (filled circles) of chloroplast membranes following detergent solubilization and fractionation on a DEAE-Toyopearl 650S column. Nonenzymatic activity (unfilled circles) and the effect of 14 $\mu M$ CsA (marked with plus signs) are shown as well. (B) Linear dependence of the rate constant as a function of protein concentration. (From A. Breiman, T. W. Fawcett, M. L. Ghirardi, and A. K. Mattoo, *J. Biol. Chem.* **267,** 21293 (1992) with permission.)

2. Specific binding is dependent on the displacement assay using excess unlabeled FK506. The titration with cold ligand at a single concentration of the radiolabeled ligand is determined[21,22] to optimize the success of identifying the respective proteins.

*Immunological Identification*

Antibodies raised against FKBPs or RBPs (even from other systems) are useful to test for cross-reactive homologs from plant extracts on immunoblots.[23]

1. Soluble proteins are prepared for gel electrophoresis by mixing samples with sample application buffer[24] (2:1, v/v), then vortexed and heated at 90° for 5 min. For membrane samples, we use a 1:1 (v/v) ratio of sample

[23] H. Towbin, T. Staehelin, and J. Gordon, *Proc. Natl. Acad. Sci. U.S.A.* **76,** 4350 (1979).

[24] A. K. Mattoo, U. Pick, H. Hoffman-Falk, and M. Edelman, *Proc. Natl. Acad. Sci. U.S.A.* **78,** 1572 (1981).

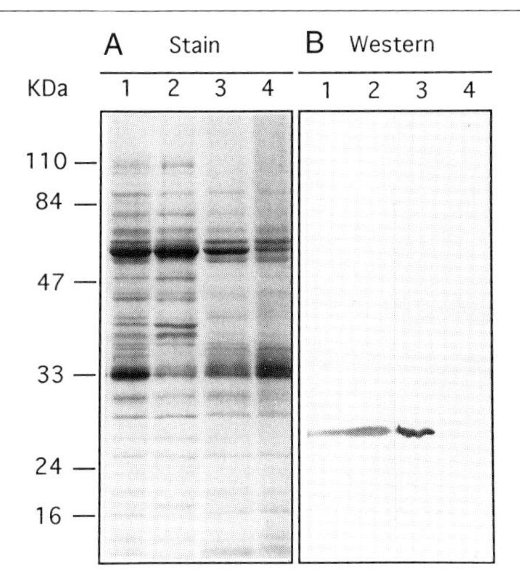


FIG. 4. Plant mitochondria contain a rapamycin-binding protein. Samples (20 $\mu$g of protein) of solubilized whole mitochondria (lanes 1), mitochondrial matrix (lanes 2), mitochondrial membranes (lanes 3), and mitochondrial membranes washed with 0.1 *M* $Na_2CO_3$, pH 11 (lanes 4) were fractionated by SDS–PAGE, then either stained (A) or immunoblotted and reacted with an antibody against yeast RBP [(B) Western blot.] (From A. Breiman, T. W. Fawcett, M. L. Ghirardi, and A. K. Mattoo, *J. Biol. Chem.* **267,** 21293 (1992) with permission.)

to the buffer and incubate the mixture at room temperature for 1 hr with intermittent vortexing.[10]

2. Gradient (10 to 20%, w/v) sodium dodecyl sulfate-polyacrylamide gels are prepared as described.[25] Samples (20 $\mu$g protein equivalent) are loaded and gels are run at a constant voltage of 150 V (or 30 mA) until the 6-kDa prestained marker (Life Technologies, Rockville, MD) is electrophoresed out of the gel.

3. The proteins on the gels are electrophoretically transferred to nitrocellulose paper (0.1-$\mu$m pore size; Schleicher & Schuell, Keene, NH) using a standard apparatus (Hoefer, San Francisco, CA).

4. Transfer buffer consists of 25 m*M* Tris, 0.192 *M* glycine, 0.02% (w/v) SDS, and 20% (v/v) methanol.

5. The transfer is carried out at a constant voltage of 50 V at 20° for 12–16 hr.

6. The nitrocellulose paper (blot) is briefly rinsed with phosphate-buffered saline (PBS), air dried, and then incubated for 2 hr with PBS containing 1% (w/v) milk powder (Carnation) and 0.5% bovine serum albumin (BSA).

[25] J. B. Marder, A. K. Mattoo, and M. Edelman, *Methods Enzymol.* **118,** 384 (1986).

7. The blots are washed twice (5 min each time) in PBS containing 0.05% (w/v) Tween 20 (PBST) and incubated for 1–2 hr with the primary antibody appropriately diluted in 1% (w/v) BSA in PBST.

8. The blots are washed three times (5 min each) in PBST and then incubated for 1 hr with a 1:1000 or 1:2000 dilution of goat anti-rabbit IgG conjugated to horseradish peroxidase (Bio-Rad, Hercules, CA) or phosphatase (Pierce, Rockford, IL).

9. The blots are washed as in step 8 in PBS lacking Tween 20 before detection of cross-reactive bands with enzyme-linked assays using peroxidase (4-chloro-1-naphthol as the chromogenic substrate) or phosphatase [with 5-bromo-4-chloro-3-indolylphosphate toluidinium (BCIP) and nitroblue tetrazolium (NBT) concentrate; Kirkegaard & Perry Laboratories, Gaithersburg, MD].

Pea mitochondrial matrix PPIases are not inhibited by rapamycin to the same degree as by CsA. In fact, a significant loss in PPIase activity in the presence of rapamycin was seen only with mitochondrial membranes (Table III). Nonetheless, both matrix and membranes of pea mitochondria show the presence of a 25-kDa RBP using antibodies against the conserved region (12–26 amino acids) of the yeast RBP[26] (Fig. 4). Interestingly, mitochondrial membranes washed free of peripheral proteins do not retain either PPIase activity, CsA binding or the 25-kDa RBP (Table III and Fig. 4).

*Notes*

1. Samples prepared for electrophoresis should be at room temperature and centrifuged for 30 sec (12,000 *g*, Eppendorf microcentrifuge) before loading on the gel.

2. During electrotransfer of proteins, the temperature can be maintained by using a water reflux system attached to a cold water tap.

3. All manipulations of blots are carried out at room temperature with mild shaking. The containers should have a sufficient volume of the buffers/reagents so that the blots are free floating.

4. Using Tween 20 in PBS for washes, and in diluting the primary antibody, helps reduce the background on immunoblots.

## Other Potential Characteristics of Peptidylprolyl Isomerases

Plants have, in addition to soluble PPIases, characteristically distinct classes of PPIases in mitochondria and chloroplasts. To enable their proper

[26] Y. Koltin, L. Faucette, D. J. Bergsma, M. A. Levy, R. Cafferkey, P. L. Koser, R. K. Johnson, and G. P. Livi, *Mol. Cell. Biol.* **11,** 1718 (1991).

TABLE IV
CHARACTERISTICS OF PLANT PEPTIDYLPROLYL ISOMERASES[a]

| Source | Form[b] | Molecular weight | Amino acid residues | p*I* | Net positive charge | Ref. |
|---|---|---|---|---|---|---|
| Rice | | | | | | |
| 1 | S | 18,361 | 172 | 8.34 | 3 | 29 |
| 2 | S | 18,331 | 172 | 8.34 | 3 | 29 |
| 3 | S | 19,200 | 179 | 7.76 | 1 | 29 |
| Maize | S | 18,349 | 172 | 8.87 | 4 | 30 |
| Tomato | S | 17,910 | 171 | 8.62 | 4 | 30 |
| *Brassica* | S | 18,172 | 168 | 8.34 | 3 | 30 |
| *Arabidopsis* | S | 18,373 | 172 | 7.83 | 1 | 31 |
| | — | 18,378 | 172 | 8.87 | 4 | 32 |
| | S | 18,161 | 169 | 8.38 | 2 | 33 |
| | C | 28,208 | 260 | 8.70 | 5 | 31 |
| Fava bean | C | 26,547 | 248 | 8.51 | 3 | 20 |
| Pea | M | 25,000 | — | — | — | 5 |

[a] Based on deduced amino acid sequences from the indicated references. Sequences were analyzed using Protein Analysis program (version 8) of Wisconsin Sequence Analysis Package, Genetics Computer Group, Madison, Wisconsin, 1994.
[b] S, Soluble (cytosolic) form; C, chloroplastic form; M, mitochondrial.

transport and sorting for final destination to specific organelles, these organellar proteins, encoded by nuclear genes, are made as precursor proteins whose leader sequences would be processed after they are specifically targeted to either the mitochondria or the chloroplasts.[27,28] The presence or absence of a leader sequence in a protein is taken as an indication of its possible destination within the cell. On the basis of this assumption, several PPIases have been cloned from a number of plants (Table IV[5,20,29–33]). Some of these, from *Arabidopsis* and fava beans, are made as precursors (have leader sequences), localized to the chloroplasts, and have significantly larger molecular weights as compared to the cytosol-localized PPIases. None of the mitochondrial PPIases have as yet been cloned. Overall, the

[27] K. Keegstra and L. J. Olsen, *Annu. Rev. Plant Physiol. Plant Mol. Biol.* **40,** 471 (1989).
[28] B. S. Glick, E. M. Beasley, and G. Schatz, *Trends Biochem. Sci.* **17,** 453 (1992).
[29] W. G. Buchholz, L. Harris-Haller, R. T. DeRose, and T. C. Hall, *Plant Mol. Biol.* **25,** 837 (1994).
[30] C. S. Gasser, D. A. Gunning, K. A. Budelier, and S. M. Brown, *Proc. Natl. Acad. Sci. U.S.A.* **87,** 9519 (1990).
[31] V. Lippuner, I. T. Chou, S. V. Scott, W. F. Ettinger, S. M. Theg, and C. S. Gasser, *J. Biol. Chem.* **269,** 7863 (1994).
[32] G. Hayman and J. A. Miernyk, Accession No. U07276 (unpublished) (1994).
[33] D. Bartling, A. Heese, and E. W. Weiler, *Plant Mol. Biol.* **19,** 529 (1992).

fact that multiple genes encode PPIases is consistent with the biochemical evidence for the presence of distinct PPIases in plant cells. Table IV summarizes the predicted physical characteristics of different PPIases whose amino acid sequences were deduced from their nucleotide sequences.

## Acknowledgments

I thank Drs. Adina Breiman, Timothy Fawcett, Maria-Lousia Ghirardi, Franklin Callahan, and Marvin Edelman for collaboration and productive association. I thank Dr. Dingbo Zhou for help in compiling Fig. 1 and Table IV. I thank Drs. Marvin Edelman, Roshni Mehta, and Mark Swegle for constructive comments and for going through the manuscript. Mention of a trade mark or proprietary product does not constitute a guarantee or warranty of the products by the USDA and does not imply its approval to the exclusion of other products that may also be suitable.

# [8] Purification of GroEL with Low Fluorescence Background

*By* A. Clay Clark, Ragulan Ramanathan, and Carl Frieden

## Introduction

The molecular chaperonin GroEL from *Escherichia coli* is a member of the heat shock protein 60 (Hsp60) class of chaperones. Such chaperonins assist in protein folding reactions both *in vivo*[1] and *in vitro*[2] by binding unfolded proteins, presumably within the central cavity of the tetradecameric double toroid structure.[3] This process decreases the concentration of aggregation-prone species, thus decreasing the rate of the off-pathway reaction, and favors partitioning to the native conformation. Proteins also may be released from the apical domain-binding region of GroEL into the central cavity on binding the cochaperonin GroES to the complex.[4] This sequestration into the central cavity allows the bound protein to fold in dilute protein concentrations, again preventing aggregation. On hydrolysis of ATP, the complex dissociates, and the protein may partition between folding to the native conformation, rebinding to GroEL, or forming aggregates.

[1] P. V. Viitanen, A. A. Gatenby, and G. H. Lorimer, *Protein Sci.* **1,** 363 (1992).
[2] J. P. Hendrick and F. U. Hartl, *FASEB J.* **9,** 1559 (1995).
[3] W. A. Fenton, Y. Kashi, K. Furtak, and A. L. Horwich, *Nature* (*London*) **371,** 614 (1994).
[4] R. J. Ellis, *Folding Design* **1,** R9 (1996).



0076-6879/98 $25.00

The mechanism of GroEL-mediated protein folding has been examined using a variety of proteins as model substrates (for reviews see Refs. 2 and 5–7). However, most studies have employed enzymatic activity measurements of the folding protein to monitor the interactions with GroEL. Because the return of activity corresponds to late processes in folding, interactions between GroEL and refolding proteins that occur early in the refolding reaction may not be observed, thus making it difficult to determine well-defined kinetic mechanisms.

A suitable probe for monitoring protein folding in the absence of GroEL should report on the formation of intermediate conformations that occur during refolding as well as on the formation of the native conformation. Ideally, the same probe also should function as a reporter for the interactions of the refolding protein in the presence of GroEL. Fluorescence emission of tryptophan residues is frequently used in refolding studies because it reports on intermediate structures that occur transiently during the refolding process. Because GroEL contains no tryptophan in its primary sequence,[8] it should be possible, in theory, to follow the changes in the fluorescence emission of the refolding protein in the presence of GroEL. For this assay to be useful, however, the background fluorescence due to GroEL should be low. Until recently, GroEL purified by standard protocols contained fluorescent material, and the background fluorescence emission was too large to correct easily. We describe here a protocol for the purification of GroEL that consists of simple chromatographic procedures that completely remove species responsible for the contaminating fluorescence. We assume that the contaminating fluorescence is due to tryptophan-containing peptides because the samples used in fluorescence measurements are excited at 295 nm; however, the chromatographic procedures also remove many species with $M_r < 350$. We show that GroEL purified by this method has a low level of fluorescence emission and contains a maximum tryptophan level of 0.04 mol/mol of GroEL protomer.

*Overexpression of GroEL*

There are several plasmid expression systems that harbor the genes encoding both GroEL and GroES.[9–17] We use a system in which the genes

[5] J. Buchner, *FASEB J.* **10,** 10 (1996).

[6] A. R. Clarke, *Curr. Opin. Struct. Biol.* **6,** 43 (1996).

[7] G. H. Lorimer, *FASEB J.* **10,** 5 (1996).

[8] S. M. Hemmingsen, C. Woolford, S. M. van der Vies, K. Tilly, D. T. Dennis, C. P. Georgopoulos, R. W. Hendrix, and R. J. Ellis, *Nature (London)* **333,** 330 (1988).

[9] G. N. Chandrasekhar, K. Tilly, C. Woolford, R. Hendrix, and C. Georgopoulos, *J. Biol. Chem.* **261,** 12414 (1986).

[10] A. J. Jenkins, J. B. March, I. R. Oliver, and M. A. Masters, *Mol. Gen. Genet.* **202,** 446 (1986).

are controlled by the RecA promoter, which is induced by the addition of nalidixic acid to the medium.[16] This section includes a brief description on growth and induction of *E. coli* cells harboring the inducible plasmid pHOG1.[16] However, the protocols for the purification of GroEL described in the subsequent sections of this chapter do not depend on the system one chooses to express the protein.

Stock cultures of *E. coli* BL21(DE3) harboring the plasmid pHOG1[16] are streaked for single colonies onto plates that contain Luria broth[18] (LB) and carbenicillin (50 $\mu$g/ml). The plates are incubated overnight at 37°. A single colony is then used to inoculate a liquid culture containing 5 ml of LB and ampicillin (50 $\mu$g/ml). The liquid culture is incubated overnight with shaking at 37°.

The cultures are grown in Fernbach flasks containing Terrific broth[18] (TB; see Table I) using the following protocol. First, 500 $\mu$l of the 5-ml culture (grown overnight) is added to a 250-ml flask that contains 50 ml of TB and ampicillin (50 $\mu$g/ml; 50 $\mu$l of a 50-mg/ml stock ampicillin solution, see Table I). The flask is shaken at 37° for about 2 hr, and the absorbance at 600 nm ($A_{600}$) is measured. This culture is used to inoculate each Fernbach flask.

Fernbach flasks containing TB should be made and sterilized in advance as described in Table I. To each Fernbach flask containing sterile TB is added 60 ml of a 10× salt solution (Table I), 600 $\mu$l of stock ampicillin (Table I), and 60 $\mu$l of a 30% (v/v) solution of antifoam C (Table I). The Fernbach flasks should then be inoculated from the 50-ml culture flask so that the initial $A_{600}$ equals 0.02. The volume of the inoculum will depend on the $A_{600}$ measured for the 50-ml culture. The Fernbach flasks are then incubated at 37° while being shaken at 300 rpm. Care should be taken to ensure that the liquid is rotating in the flask rather than splashing, so that the cultures are properly aerated. The absorbance of the culture should be monitored over time until an $A_{600}$ of approximately 4 is reached (about 4 hr). At this point the production of GroESL is induced by the addition to each Fernbach flask of 3 ml of stock nalidixic acid (Table I). The cultures

[11] O. Fayet, T. Ziegelhoffer, and C. Georgopoulos, *J. Bacteriol.* **171,** 1379 (1989).
[12] P. Goloubinoff, A. A. Gatenby, and G. H. Lorimer, *Nature* (*London*) **337,** 44 (1989).
[13] Y. Makino, H. Taguchi, and M. Yoshida, *FEBS Lett.* **336,** 363 (1993).
[14] A. Azem, S. Diamant, and P. Goloubinoff, *Biochemistry* **33,** 6671 (1994).
[15] C. Weiss and P. Goloubinoff, *J. Biol. Chem.* **270,** 13956 (1995).
[16] A. C. Clark, E. Hugo, and C. Frieden, *Biochemistry* **35,** 5893 (1996).
[17] N. Murai, Y. Makino, and M. Yoshida, *J. Biol. Chem.* **271,** 28229 (1996).
[18] J. Sambrook, E. F. Fritsch, and T. Maniatis, "Molecular Cloning: A Laboratory Manual." Cold Spring Harbor Laboratory Press, Cold Spring Harbor, New York, 1989.

TABLE I
STOCK SOLUTIONS FOR GROWTH OF *Escherichia coli* BL21 (DE3)/pHOG1[a]

1. Terrific broth[b] (for 10 Fernbach flasks and four 250-ml flasks)
   a. Combine:

      Bacto-tryptone, 80 g
      Yeast extract, 160 g
      Glycerol, 26.7 ml
      $ddH_2O$[c] to 6 liters

   b. Aliquot 550 ml to each Fernbach flask, 50 ml to each 250-ml flask, and autoclave to sterilize
2. 10× salts[b]
   a. Combine:

      $KH_2PO_4$, 23.1 g
      $K_2HPO_4$, 125.4 g
      $ddH_2O$ to 1 liter

   b. Filter sterilize (0.2-$\mu$m pore size filter)
3. Ampicillin (50 mg/ml)
   a. Combine:

      Ampicillin, 500 mg
      $ddH_2O$, 10 ml

   b. Filter sterilize (0.2-$\mu$m pore size filter)
4. Antifoam C [30% (v/v) solution]
   a. Combine:

      Antifoam C (Sigma, St. Louis, MO), 30 ml
      $ddH_2O$, 70 ml

   b. Autoclave to sterilize
5. Nalidixic acid (10 mg/ml)
   a. Combine:

      Nalidixic acid, 350 mg
      $ddH_2O$, 35 ml
      NaOH (50%, w/w), 270 $\mu$l

   b. Filter sterilize (0.2-$\mu$m pore size filter)

[a] All solutions except stock nalidixic acid can be made in advance. The ampicillin solution is stored at −20°.

[b] J. Sambrook, E. F. Fritsch, and T. Maniatis, "Molecular Cloning: A Laboratory Manual." Cold Spring Harbor Press, Cold Spring Harbor, New York, 1989.

[c] $ddH_2O$, Doubly distilled water.

are grown for an additional 3 hr following induction (37°, 300 rpm), then the cells are harvested by centrifugation for 20 min at 4°. After removing the supernatant, the cell pellet can be used immediately to purify GroEL (see below) or stored as a cell paste at −80°.

*Lysis of Cells and Ammonium Sulfate Fractionations*

The *E. coli* cell pellet is resuspended in ice-cold lysis buffer (Table II, buffer 1; approximately 2 ml/g of cells) and lysed in a French pressure cell (16,000–18,000 psi). The lysate is centrifuged to remove cell debris (14,000 rpm for 30 min at 4°), and the supernatant is transferred to a new container. The cell pellet is then washed with 50 ml of lysis buffer, centrifuged (14,000 rpm for 30 min at 4°), and the supernatant combined with the first. Ammonium sulfate is then added to 30% (w/v) saturation (16.4 g/100 ml of solution), the solution stirred for 30 min at 4°, and then centrifuged to remove the precipitate (14,000 rpm for 25 min at 4°). Ammonium sulfate should be added in about four aliquots over several minutes. The supernatant is transferred to a new container and ammonium sulfate added (in several aliquots) to 65% (w/v) saturation (21.4 g/100 ml of solution). After stirring for 30 min at 4°, the solution is centrifuged for 30 min (14,000 rpm at 4°) to collect the precipitate, and the precipitate is dissolved in DEAE–low salt buffer (Table II, buffer 2A). The solution is transferred to dialysis membrane [50,000 molecular weight cutoff (MWCO)]. The total volume for dialysis should be approximately 80 ml. The protein is then dialyzed against DEAE–low salt buffer at 4° (two changes of 2 liters each, minimum of 5 hr per dialysis).

## Column Chromatography

*General Features*

The GroEL is purified by elution from three columns. The use of each column is described in detail below; however, an overview of the column chromatography procedures is given in Table III. The first column is an ion-exchange column (DEAE–Sepharose Fast-Flow), and the protein is eluted by a linear gradient of KCl from 100 m*M* to 1.2 *M*. This column also separates GroEL from GroES. The second column (Bio-Gel A-5m; Bio-Rad, Hercules, CA) is a size-exclusion column. Following the first two column treatments, the GroEL is >90% pure as judged by sodium dodecyl sulfate-polyacrylamide gel electrophoresis (SDS–PAGE) (Fig. 1) but has a large background fluorescence emission (see below). The third column is a dye column (Reactive Red 120–agarose) that removes the contaminating

TABLE II
BUFFERS FOR GroEL PURIFICATION[a]

1. Lysis buffer [50 m*M* potassium phosphate (pH 7.2[b]), 1 m*M* EDTA,[c] 1 m*M* PMSF, 1 m*M* DTT]

   $KH_2PO_4$, 955 mg
   $K_2HPO_4$, 3.125 g
   DTT,[d] 77 mg
   $Na_2EDTA$ (0.5 *M*, pH 8.0), 1 ml
   PMSF[e] (200 m*M*), 2.5 ml
   dd$H_2O$ to 500 ml

2. DEAE–Sepharose Fast-Flow column buffers
   A. DEAE–low salt buffer [20 m*M* Bis–Tris (pH 6.5), 100 m*M* KCl, 1 m*M* EDTA, 1 m*M* DTT]
      i. Combine:

         Bis–Tris, 4.2 g
         KCl, 7.4 g
         $Na_2EDTA$ (0.5 *M*, pH 8.0), 2 ml
         DTT, 154 mg
         dd$H_2O$ to 1 liter

      ii. Adjust pH to 6.5 with 12 *N* HCl
   B. DEAE–high salt buffer [20 m*M* Bis–Tris (pH 6.5), 1.2 *M* KCl, 1 m*M* EDTA, 1 m*M* DTT]
      i. Combine:

         Bis–Tris, 4.2 g
         KCl, 89.4 g
         $Na_2EDTA$ (0.5 *M*, pH 8.0), 2 ml
         DTT, 154 mg
         dd$H_2O$ to 1 liter

      ii. Adjust pH to 6.5 with 12 *N* HCl
   C. DEAE wash buffer 1

      NaOH (0.5 *M*), 150 ml

   D. DEAE wash buffer 2 [50 m*M* potassium phosphate (pH 7.2), 2 *M* KCl, 5 *M* urea]

      $KH_2PO_4$, 955 mg
      $K_2HPO_4$, 3.125 g
      KCl, 74.5 g
      Ultrapure urea, 150 g
      dd$H_2O$ to 500 ml

   E. Storage buffer [20 m*M* Bis–Tris (pH 6.5), 100 m*M* KCl, 0.02% (w/v) $NaN_3$]
      i. Combine:

         Bis–Tris, 4.2 g
         KCl, 7.4 g
         $NaN_3$, 200 mg
         dd$H_2O$ to 1 liter

      ii. Adjust to pH 6.5 with 12 *N* HCl

(*continues*)

TABLE II (*continued*)

3. Bio-Gel A-5m column buffers
   A. Elution buffer [50 m*M* Tris-HCl (pH 7.6), 100 m*M* KCl, 2 m*M* $MgCl_2$, 1 m*M* DTT, 0.5 m*M* $Na_2ATP$]
      i. Combine:

         Tris base, 6.1 g
         KCl, 7.4 g
         $MgCl_2 \cdot 6H_2O$, 0.41 g
         DTT, 154 mg
         $Na_2ATP$, 276 mg
         $ddH_2O$ to 1 liter

      ii. Adjust to pH 7.6 with 12 *N* HCl (about 3.2 ml)
   B. Storage buffer [50 m*M* Tris-HCl (pH 7.6), 5 m*M* $NaN_3$]
      i. Combine:

         Tris base, 6.1 g
         $NaN_3$, 330 mg
         $ddH_2O$ to 1 liter

      ii. Adjust to pH 7.6 with 12 *N* HCl (about 3.2 ml)
4. Reactive Red–agarose column buffers
   A. Elution buffer [20 m*M* Tris-HCl (pH 7.5), 5 m*M* $MgCl_2$]
      i. Combine:

         Tris base, 2.42 g
         $MgCl_2 \cdot 6H_2O$, 1.02 g
         $ddH_2O$ to 1 liter

      ii. Adjust to pH 7.5 with 12 *N* HCl (about 1.5 ml) (*Note:* At room temperature, pH should be about pH 7.1)
   B. Wash buffer 1 [20 m*M* Tris-HCl (pH 7.5), 5 *M* urea]
      i. Combine:

         Tris base, 2.42 g
         Urea, 300 g
         $ddH_2O$ to 1 liter

      ii. Adjust to pH 7.5 with 12 *N* HCl (*Note:* At room temperature, pH should be about pH 7.1)
   C. Wash buffer 2/storage buffer [20 m*M* Tris-HCl (pH 7.5), 1.5 *M* NaCl, 0.02% (w/v) $NaN_3$]
      i. Combine:

         Tris base, 2.42 g
         NaCl, 87.8 g
         $NaN_3$, 200 mg
         $ddH_2O$ to 1 liter

      ii. Adjust to pH 7.5 with 12 *N* HCl (about 1.5 ml) (*Note:* At room temperature, pH should be about pH 7.1)

TABLE II (*continued*)

5. GroEL storage buffer [50 m*M* Tris-HCl (pH 7.6), 1 m*M* DTT]
   a. Combine:

      Tris base, 6.1 g
      DTT, 154 mg
      $ddH_2O$ to 1 liter

   b. Adjust to pH 7.6 with 12 *N* HCl (about 3.3 ml) (*Note:* At room temperature, pH should be about pH 7.3)

[a] All buffers should be filtered with a 0.2-μm pore size filter. Owing to the volumes involved in each case, we prefer a 0.2-μm pore size bottletop vacuum filter unit (VacuCap 90; Gelman Sciences).
[b] The pH of the phosphate buffers may need to be adjusted slightly.
[c] Bis–Tris, Bis(2-hydroxyethyl)iminotris(hydroxymethyl)methane; Tris, tris(hydroxymethyl)aminomethane; DTT, dithiothreitol; PMSF, phenylmethylsulfonyl fluoride; EDTA, ethylenediaminetetraacetic acid; DEAE–Sepharose, diethylaminoethyl–Sepharose; $ddH_2O$, doubly distilled water.
[d] Buffers can be made in advance; however, DTT, PMSF, and $Na_2ATP$ should be added to the respective buffers just before use.
[e] PMSF (200 m*M*): 131.3 mg of PMSF plus 3.5 ml of 95% (v/v) ethanol.

peptides responsible for the background fluorescence. The flow rates of the first two columns may be controlled using a pump.

### *DEAE–Sepharose Fast-Flow Ion-Exchange Column*

The column (33.5 × 2.6 cm) is preequilibrated with DEAE–low salt buffer (Table II, buffer 2A; 4 liters at a flow rate of 5 ml/min) and used at room temperature (see Table III). The dialyzed protein (see above) is loaded at a flow rate of 7 ml/min, and the column is washed with DEAE–low salt buffer (400 ml) at a flow rate of 5 ml/min. The protein is eluted by a linear salt gradient from 100 m*M* KCl to 1.2 *M* KCl (Table II, buffers 2A and 2B). The gradient (1 liter of each buffer) is run at a flow rate of 5 ml/min, and 20-ml fractions are collected.

The column fractions can be monitored by SDS–PAGE.[19] We find it convenient to monitor every other fraction using Tris–glycine SDS–polyacrylamide gels that have a 4–20% (w/v) acrylamide gradient (Bio-Rad Ready gels). This allows one to observe both GroEL and GroES on the same gel (see also Fig. 1). GroEL elutes from this column at KCl concentrations between 330 and 430 m*M*. The fractions containing GroEL, as determined from the SDS–PAGE analysis, are pooled (approximately 200-ml total volume) and the protein precipitated by the addition of ammo-

[19] U. K. Laemmli, *Nature* (*London*) **227,** 680 (1970).

TABLE III
COLUMN CHROMATOGRAPHY FOR PURIFICATION OF GroEL

1. DEAE–Sepharose Fast-Flow (Sigma, St. Louis, MO; 33.5 × 2.6 cm; run at room temperature)
   a. Column is preequilibrated with 4 liters of DEAE–low salt buffer (Table II)
   b. Protein is loaded at a flow rate of 7 ml/min
   c. Column is washed with 400 ml of DEAE–low salt buffer (flow rate of 5 ml/min)
   d. Protein is eluted by a linear KCl gradient [1 liter each of DEAE–low salt and DEAE–high salt buffers (see Table II) at a flow rate of 5 ml/min)]; 20-ml fractions are collected
   e. Every second fraction is analyzed by SDS–PAGE [4–20% (w/v) acrylamide gradient], and those fractions that contain GroEL are pooled (GroEL elutes between 330 and 430 m*M* KCl)
   f. Ammonium sulfate is added to 65% (w/v) saturation (39.8 g/100 ml of solution), and the precipitate is collected by centrifugation (14,000 rpm for 30 min at 4°)
   g. Pellets are resuspended in Bio-Gel A-5m elution buffer (Table II) and dialyzed (50,000-MWCO membrane) against the same buffer (2 × 2 liters of buffer) at 4°
   h. Column is washed with 150 ml of wash buffer 1, 500 ml of wash buffer 2, and reequilibrated with 4 liters of storage buffer (Table II)
2. Bio-Gel A-5m column (Bio-Rad; 96 × 6 cm; run at room temperature)
   a. Column is preequilibrated with 12 liters of elution buffer (Table II)
   b. Protein (about 80 ml from dialysis) is loaded at a flow rate of 5 ml/min
   c. Protein is eluted at a flow rate of 5 ml/min, and 20-ml fractions are collected
   d. Each fraction is analyzed by "micro"-Bradford analysis for presence of protein
   e. Every third fraction with protein is analyzed by SDS–PAGE [4–20% (w/v) acrylamide gradient], and those fractions that contain GroEL are pooled (GroEL elutes between 1100 and 1700 ml)
   f. Ammonium sulfate is added to 65% (w/v) saturation (39.8 g/100 ml of solution), and the precipitate is collected by centrifugation (14,000 rpm for 30 min at 4°)
   g. Pellets are resuspended in DEAE–low salt buffer (Table II) and dialyzed (50,000-MWCO membrane) against the same buffer (2 × 2 liters of buffer)
   h. Protein is dialyzed (2 × 2 liters of buffer) against Reactive Red elution buffer (Table II)
   i. Column is washed with 12 liters of $ddH_2O$ and 12 liters of A-5m storage buffer (Table II)
3. Reactive Red 120–agarose (type 3000-CL) column (Sigma; 33.5 × 5 cm; run at 4°)
   a. Column is preequilibrated with 8 liters of elution buffer (Table II)
   b. About 400 mg of protein (total) is loaded onto the column (concentration of 2–10 mg/ml)
   c. Column is equilibrated for approximately 15 min
   d. Protein is eluted at a flow rate of 4 ml/min and 20-ml fractions are collected
   e. Elution profile is generated by monitoring $A_{280}$ of each fraction
   f. Tryptophan fluorescence emission of fractions containing protein is measured to determine fractions to pool (excitation, 295 nm; emission, from 310 to 400 nm)
   g. Fractions with fluorescence emission profile as shown in Fig. 3 are pooled and concentrated (Amicon concentrator, YM30 membrane)
   h. Protein is dialyzed against GroEL storage buffer, glycerol is added to 10% (v/v), and aliquots of the protein are stored at −80°

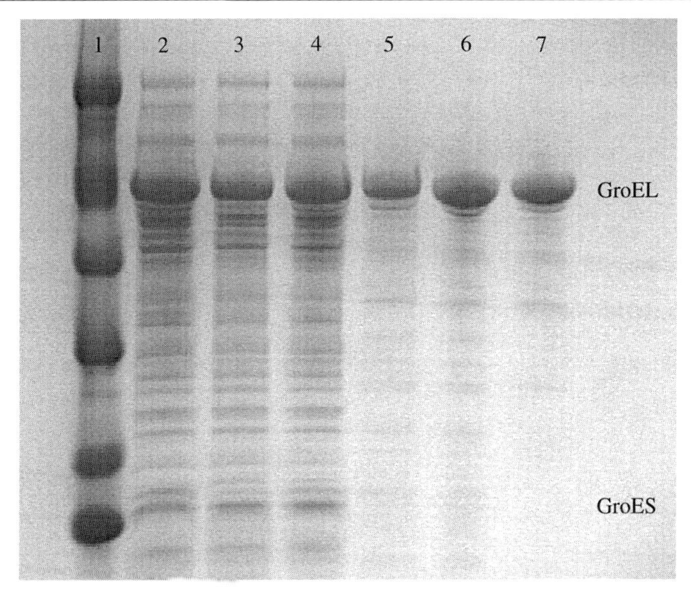


FIG. 1. SDS–PAGE of GroEL fractions. Lane 1, molecular weight markers (top to bottom; 97K, 66K, 46K, 30K, 21.5K, 14.3K); lane 2, crude lysate; lane 3, supernatant from 30% (w/v) ammonium sulfate treatment; lane 4, sample loaded onto the DEAE–Sepharose column; lane 5, sample loaded onto the Bio-Gel A-5m column; lane 6, sample loaded onto the Reactive Red 120–agarose column; lane 7, approximately 3.5 $\mu$g of purified GroEL (eluate from the Reactive Red 120–agarose column). The gel was a 4–20% (w/v) Tris–glycine acrylamide gel (Bio-Rad Ready gels) and was run at 100 V for approximately 1.5 hr. The gel was stained with Coomassie Brilliant Blue R-250 (Sigma), and the molecular weight markers (Rainbow RPN 756) were from Amersham (Arlington Heights, IL).

nium sulfate to 65% (w/v) saturation (39.8 g of ammonium sulfate per 100 ml of solution). After stirring for 30 min at 4°, the precipitate is collected by centrifugation (14,000 rpm for 30 min at 4°), and the ammonium sulfate pellet is resuspended in Bio-Gel A-5m elution buffer (Table II, buffer 3A) so that the total volume is about 60 ml. This solution is then dialyzed (50,000-MWCO membrane) at 4° against Bio-Gel A-5m elution buffer (two changes of 2 liters each with a minimum of 5 hr for each dialysis).

It should be noted that GroES elutes from the DEAE–Sepharose Fast-Flow columns at KCl concentrations between 170 and 210 m*M*. Those fractions also can be pooled, the GroES can be precipitated by the addition of ammonium sulfate to 65% (w/v) saturation, and the pellet can be stored at −80° pending further purification.

Following use, the DEAE–Sepharose Fast-Flow column should be washed with 150 ml of DEAE–Sepharose wash buffer 1 (Table II, buffer

2C) and then with 500 ml of DEAE–Sepharose wash buffer 2 (Table II, buffer 2D). The column should be reequilibrated with 4 liters of DEAE–Sepharose storage buffer (Table II, buffer 2E). Because the protein solution loaded onto this column is relatively crude, one should wash the column soon (within a few days) after use.

*Bio-Gel A-5m Size-Exclusion Column*

The Bio-Gel A-5m sizing column (96 × 6 cm) is preequilibrated with Bio-Gel A-5m elution buffer (Table II, buffer 3A; 12 liters at a flow rate of 5 ml/min). This column is also run at room temperature (see Table III). The dialyzed protein solution (about 80 ml) is loaded onto the column at a flow rate of 5 ml/min, the protein is eluted at the same flow rate, and 20-ml fractions are collected. The fractions are analyzed initially by a "micro"-Bradford analysis. In this assay, 75 $\mu$l of a 1 : 4 dilution of Bradford reagent[20] is added to each well of a 96-well microtiter plate; 25 $\mu$l from each column fraction is then added. The solutions in the microtiter plate that correspond to fractions that contain protein will turn blue, and those fractions should be further analyzed by SDS–PAGE (we find it convenient to analyze every third fraction that contains protein). GroEL elutes from this column between about 1100 and 1700 ml of elution buffer and, on the basis of SDS–PAGE analysis, those fractions that contain GroEL are pooled. The protein is then precipitated by the addition of ammonium sulfate to 65% (w/v) saturation (39.8 g per 100 ml of solution). After stirring for 30 min at 4°, the precipitate is collected by centrifugation (14,000 rpm for 30 min at 4°) and the ammonium sulfate pellet resuspended in DEAE–Sepharose low salt buffer (Table II, buffer 2A). This solution is first dialyzed (50,000-MWCO membrane) at 4° against the same buffer (two changes of 2 liters each with a minimum of 5 hr for each dialysis), then against Reactive Red 120–agarose elution buffer (Table II, buffer 4A; two changes of 2 liters each).

The Bio-Gel A-5m column should be washed with 12 liters of doubly distilled $H_2O$, then with 12 liters of Bio-Gel A-5m storage buffer (Table II, buffer 3B). Because the elution buffer for this column contains ATP, one should wash the column immediately after use to prevent bacterial growth.

*Reactive Red 120-Agarose (Type 3000-CL) Dye Column*

Following elution from the first two columns, the GroEL is >90% pure as judged by SDS–PAGE analysis (Fig. 1). However, tryptophan fluorescence emission spectra of the GroEL pooled from the Bio-Gel A-5m

[20] M. M. Bradford, *Anal. Biochem.* **72,** 248 (1976).

column fractions demonstrate a significant contamination by tryptophan-containing peptides (see below). The purpose of the third column is to remove the contamination to obtain GroEL with low background fluorescence.

The column (33.5 × 5 cm) is preequilibrated with 8 liters of Reactive Red 120–agarose elution buffer (Table II, buffer 4A) and run at 4° (Table III). GroEL is loaded onto the column at a concentration of 2–10 mg/ml (maximum total protein of about 400 mg), and the column is equilibrated for about 15 min. The protein is eluted at a flow rate of 4 ml/min, and 20-ml fractions are collected. It should be noted that the total quantity of GroEL loaded onto the column is about half (in our case) of the total amount of GroEL obtained from the *E. coli* cell paste. Thus, depending on the expression system used to obtain the starting material, one may not use all of the GroEL at this point. The excess protein can be stored at −80° (see below) and loaded onto the column when needed.

For this column, an elution profile can be generated by determining the absorbance at 280 nm ($A_{280}$) for each fraction. As shown in Fig. 2, a single peak eluted from the column, and SDS–PAGE analysis of the column fractions (not shown) demonstrated that the single peak corresponded to

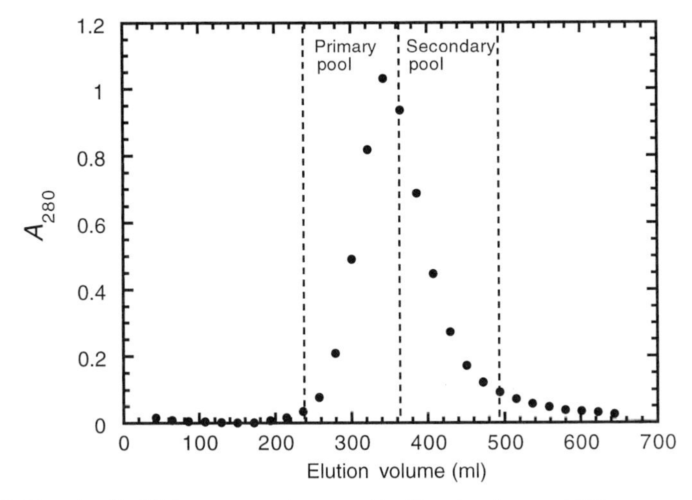


FIG. 2. Reactive Red 120–agarose column profile. The absorbance at 280 nm was measured for each fraction, and the fluorescence emission was measured for those fractions that contained GroEL, as described in text. The fractions within the elution volumes labeled "Primary Pool" (240 to 360 ml) had little tryptophan fluorescence emission and were pooled. The fractions within the elution volumes labeled "Secondary Pool" contained an unacceptable amount of tryptophan fluorescence between 310 and 350 nm, as described in text. These fractions were pooled and stored at −20° so that the GroEL could be repurified.

the elution of GroEL. One should not assume, however, that the GroEL contained in these fractions is free of contaminating peptides. The fluorescence emission spectra of the column fractions are monitored in order to determine the fractions to pool. Typically, a sample from each fraction that contains GroEL is excited at 295 nm, and the fluorescence emission is measured over the range of 310 to 400 nm. This procedure is critical for obtaining GroEL with low background fluorescence. The fluorescence emission spectra of the GroEL fractions that are free of tryptophan-containing contaminating peptides should resemble that shown in Fig. 3 for the final product (Fig. 3 inset, spectrum B). That is, there should be no peak in fluorescence emission from 310 to 350 nm (see Fig. 3, inset). For the $A_{280}$ elution profile shown in Fig. 2, this corresponds to the fractions designated as the primary pool. The fractions within the elution volumes of 240 and 360 ml were considered free of tryptophan contamination, on the basis of the fluorescence emission spectra, and were pooled for further use. The fractions in the secondary pool (Fig. 2) also were pooled because they contained a significant amount of GroEL; however, these fractions contained an unacceptable amount of fluorescence, on the basis of their fluo-

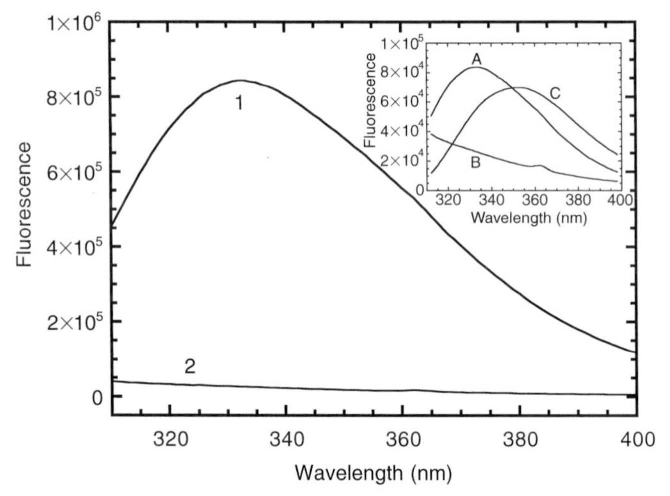


FIG. 3. Fluorescence emission spectra of 1 $\mu M$ GroEL before (1) and after (2) elution from the Reactive Red 120–agarose column. The excitation wavelength was 295 nm, and the fluorescence emission was monitored from 310 to 400 nm. *Inset:* Fluorescence emission of (A) 0.1 $\mu M$ GroEL before elution from the Reactive Red 120–agarose column, (B) 1.0 $\mu M$ GroEL after elution from the Reactive Red 120–agarose column, and (C) 4.0 $\mu M$ L-tryptophan. The samples were in a buffer of 50 m$M$ potassium phosphate (pH 7.2), 100 m$M$ KCl, 1 m$M$ DTT, 22°, and the spectra were recorded on a PTI Alpha scan spectrofluorometer (Photon Technologies, Inc., South Brunswick, NJ).

rescence emission spectra, which demonstrated an emission peak at approximately 332 nm (similar to that shown in Fig. 3, inset, spectrum A). These fractions can be pooled, stored at $-20°$, and later repurified with this column.

The GroEL from the primary pool can be concentrated [Amicon (Danvers, MA) concentrator with a YM30 membrane] and dialyzed (50,000-MWCO membrane) against GroEL storage buffer (Table II, buffer 5). Glycerol is then added to the concentrated, dialyzed sample to a final concentration of 10% (v/v), and aliquots are stored at $-80°$. Alternatively, the GroEL can be stored at $-80°$ as a 65% (w/v, saturation) ammonium sulfate pellet. The purified GroEL should have an absorbance maximum at approximately 276 nm, and the concentration can be determined using $\varepsilon_{280}$ $1.22 \times 10^4$ $M^{-1}\text{cm}^{-1}$.[21]

The Reactive Red 120–agarose column should be washed with 2 liters of Reactive Red wash buffer 1 (Table II, buffer 4B) then with 8 liters of Reactive Red wash buffer 2 (Table II, buffer 4C). The column is also stored in this buffer. As with the first two columns, the wash procedures should be performed soon after use.

It should be noted that the GroEL used in experiments with substrate proteins can be "recycled" simply by dialyzing (50,000-MWCO membrane) the mixture against Reactive Red 120–agarose elution buffer and repeating the protocols described here for the Reactive Red 120–agarose column. This procedure works well for mixtures of GroEL and dihydrofolate reductase (DHFR) because the DHFR diffuses through the dialysis membrane. In theory, this "recycling" protocol should work for larger proteins as well. One may need to adjust the pore size of the dialysis membrane, but the contaminating protein should be removed by the column nonetheless.

## Characterization of Purified GroEL

We have used four methods to assess the quality of the GroEL purified by the protocol described here: (1) SDS–PAGE (Fig. 1), (2) fluorescence emission (Fig. 3), (3) electrospray ionization (ESI) mass spectrometry (Fig. 4), and (4) ATPase activity. The results from these experiments are described in the following sections. On a routine basis, however, only the SDS–PAGE and fluorescence emission analyses are used.

### *SDS–PAGE*

The data shown in Fig. 1 demonstrate that GroEL is the predominant protein in the crude lysate of the induced cultures (Fig. 1, lane 2) and that

[21] M. T. Fisher, *Biochemistry* **31,** 3955 (1992).

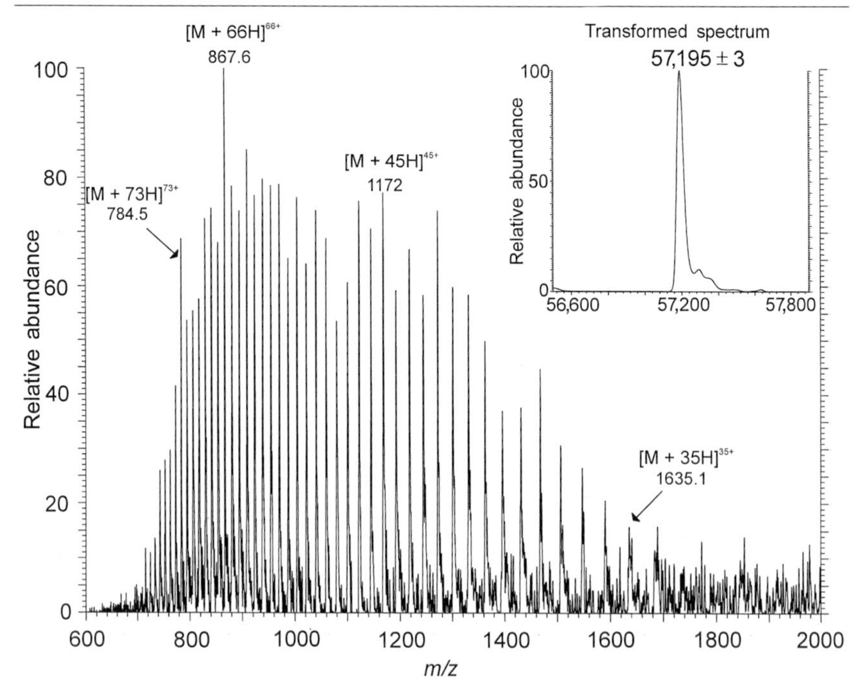


FIG. 4. Electrospray ionization mass spectrum of GroEL eluted from Reactive Red 120–agarose column. The GroEL was concentrated and the buffer was exchanged by several cycles of dilution and concentration with a solution of water–formic acid, pH 5, by ultrafiltration on a Centricon-30 concentrator. Electrospray ionization (ESI) mass spectra were obtained by using a VG ZAB-T four-sector tandem mass spectrometer (BEBE configuration) [M. L. Gross, *Methods Enzymol.* **193,** 131 (1990)] equipped with a VG electrospray source (Micromass, Manchester, UK). A Harvard model 22 syringe pump (Harvard Apparatus, South Natick, MA) was used to infuse the spray solution (water–acetonitrile–formic acid, 50:49:1, v/v/v) to the spray needle at a rate of 10 $\mu$l/min. A 20-$\mu$l aliquot of GroEL was mixed with 180 $\mu$l of the spray solution, and a 20-$\mu$l aliquot of the sample solution was introduced into the flow of the spray solution via a Rheodyne 7125 loop injector. The concentration of GroEL (tetradecamer) in each injection was approximately 4 $\mu M$. The spray needle was maintained at 8000 V, and the counterelectrode (pepper pot) potential was 5000 V. Sampling cone, skimmer lens, skimmer, hexapole, and ring electrode were typically 4177, 4125, 4119, 4117, and 4116 V, respectively. Nitrogen was used as both bath gas and nebulizer gas with a flow rates of approximately 300 and 12 liters/hr, respectively. The drying gas temperature was maintained at 80°. The mass spectrometer was calibrated from 100 to 2600 Da by using a solution of CsI. All experiments were done with a mass resolving power of 1300 and a scan speed of 15 sec/decade. Ten scans were signal averaged and processed by using a VG opus operating system and DEC-alpha work station. The raw ESI mass spectra were transformed by using the maximum entropy algorithm [A. G. Ferrige, M. J. Seddon, and S. Jarvis, *Rapid Commun. Mass Spectrom.* **5,** 374 (1991)].

separation from other proteins occurs primarily during elution from the DEAE–Sepharose column (Fig. 1, lane 5). One may also observe the separation of GroES from GroEL at this stage by a comparison of the data shown in Fig. 1 (lane 4 compared with lane 5). A comparison of the GroEL eluted from the Bio-Gel A-5m column (Fig. 1, lane 6) with that eluted from the Reactive Red 120–agarose column (Fig. 1, lane 7) suggested that few additional proteins were removed by the latter column, but that the Reactive Red 120–agarose column is critical for obtaining pure protein. Fluorescence emission spectra (see below) demonstrated that the protein eluted from the Bio-Gel A-5m column (Fig. 1, lane 6) contained a significant amount of tryptophan-containing contaminating peptides whereas the pure GroEL (Fig. 1, lane 7) did not.

*Fluorescence Emission*

In addition to other methods, the amount of fluorescence emission is another good measure of the GroEL purity. Fluorescence emission spectra of GroEL before and after elution from the Reactive Red 120–agarose column are shown in Fig. 3. The fluorescence emission spectrum of the sample eluted from the Bio-Gel A-5m column but not the Reactive Red 120–agarose column (see Fig. 1, lane 6) demonstrates a peak at approximately 332 nm (Fig. 3, spectrum 1). The fluorescence emission spectrum of the GroEL sample eluted from the Reactive Red 120–agarose column (see Fig. 1, lane 7) demonstrates a significantly lower fluorescence emission and no fluorescence emission peak over the range of 310 to 350 nm (Fig. 3, spectrum 2). The GroEL concentrations for both samples were 1 $\mu M$. The inset to Fig. 3 shows a comparison of the fluorescence emission spectra for 1 $\mu M$ GroEL eluted from the Reactive Red 120–agarose column (Fig. 3, inset, spectrum B) and 0.1 $\mu M$ GroEL eluted from the Bio-Gel A-5m column (Fig. 3, inset, spectrum A). The spectra also are compared to a standard of L-tryptophan (4 $\mu M$; Fig. 3, inset, spectrum C), which has a fluorescence emission maximum at approximately 350 nm. The GroEL eluted from the Bio-Gel A-5m column had a fluorescence emission maximum at approximately 332 nm, demonstrating that the environment of the bound contaminating peptides is more hydrophobic than that of tryptophan in bulk solution (Fig. 3, inset, spectrum C). The GroEL eluted from the Reactive Red 120–agarose column had no peak in fluorescence emission between 310 and 350 nm (Fig. 3, inset, spectrum B). These data suggested that the sample from the Reactive Red 120–agarose column contained approximately 30-fold less fluorescence emission (at 332 nm) than the sample from the Bio-Gel A-5m column.

If one assumes that the fluorescence observed in the GroEL is due to tryptophan contamination, then one may quantitate the amount of tryptophan in the samples before and after elution from the Reactive Red 120–agarose column. This may be done by incubating the samples in buffer containing 6 *M* guanidine hydrochloride (GdnCl) and comparing the fluorescence emission of the denatured protein with that of several standards of L-tryptophan (0–3 $\mu M$) in the same buffer. The samples are excited at 295 nm, and the fluorescence emission is measured at 350 nm. The choice of buffer is not critical; however, for this experiment we routinely use solution conditions of 50 m*M* potassium phosphate (pH 7.2), 6 *M* GdnCl, 100 m*M* KCl, 1 m*M* dithiothreitol (DTT), 22°, and a concentration of GroEL (tetradecamer) of 0.3 $\mu M$. If the standard curve of L-tryptophan is linear, one can calculate the amount of tryptophan in GroEL. For the samples described here, the data demonstrate that before elution from the Reactive Red 120–agarose column, GroEL contained 0.98 ± 0.3 mol of tryptophan per mole of GroEL protomer. Following elution from the Reactive Red 120–agarose column, GroEL contained approximately 0.04 mol of tryptophan per mole of GroEL protomer. Therefore, although the sample from the Bio-Gel A-5m column appeared >90% pure by SDS–PAGE analysis (Fig. 1, lane 6), the GroEL had bound to the tetradecamer approximately 14 mol of tryptophan. It is not yet clear what effect, if any, the tightly bound contaminants may have on the measurements of equilibrium binding constants.

### *Electrospray Ionization Mass Spectrometry*

The molecular mass of GroEL was determined by electrospray ionization (ESI) mass spectrometry for samples eluted from the Bio-Gel A-5m column and from the Reactive Red 120–agarose column. Under the ESI conditions used (see caption to Fig. 4) monomer species with charge states ranging from +32 to +81 were observed. Under these denaturing conditions, we saw no evidence for multimers of GroEL. The relative molecular weight of the GroEL protomer, determined from the transformed spectra (Fig. 4, inset) for both GroEL samples (pre- and post-Reactive Red 120–agarose), was 57,195.2 ± 3, in excellent agreement with that determined previously by ESI mass spectrometry.[22] These data show that the covalent structure of GroEL is not altered by elution from the Reactive Red 120–agarose column. An examination of the mass spectra for these samples indicated that the primary sources of contamination for the GroEL eluted from the Bio-Gel A-5m column were species with relative molecular weights less than 350. Several of these species were completely removed by elution

[22] C. V. Robinson, M. Gross, S. J. Eyles, J. J. Ewbank, M. Mayhew, F. U. Hartl, C. M. Dobson, and S. E. Radford, *Nature* (*London*) **372,** 646 (1994).

from the Reactive Red 120–agarose column, suggesting that the sources of tryptophan contamination in the GroEL may be limited to peptides less than three residues.

### *ATPase Activity*

The ATPase activity of the purified GroEL was determined using a coupled ATPase assay system.[23] In this assay, the ATPase activity of GroEL is coupled to the activities of pyruvate kinase (PK) and lactate dehydrogenase (LDH). The ADP generated during the GroEL-mediated hydrolysis of ATP is converted back to ATP by PK, and the pyruvate produced by PK (with the concomitant hydrolysis of phosphoenolpyruvate) is used by LDH to generate lactate. In the LDH reaction, NADH is oxidized to $NAD^+$; therefore, a decrease in absorbance is monitored at 340 nm. One mole of NADH is oxidized via the coupled enzyme assay for each mole of ATP hydrolyzed by GroEL, and the $A_{340}$ decreases linearly until all of the NADH is oxidized. This allows for a rapid determination of the GroEL ATPase activity. Using this assay, we determined the $k_{cat}$ of the purified GroEL (that is, GroEL eluted from the Reactive Red 120–agarose column) to be 19.5 ± 2.9 $min^{-1}$ at 22° (50 m*M* potassium phosphate, 100 m*M* KCl, 1 m*M* DTT, 5 m*M* $MgCl_2$, 0.5–4 m*M* MgATP). This is in good agreement with previous results that were determined by measuring the release of inorganic phosphate.[24] In that study, the $k_{cat}$ was determined to be approximately 15 $min^{-1}$ at ATP concentrations between 0.3 and 0.8 m*M* and at a temperature of 25°.

## Conclusion

We have described a protocol for the purification of GroEL that consists of three simple chromatographic procedures that completely remove tryptophan-containing contaminating peptides and several other peptides of $M_r <$ 350. The GroEL purified by this protocol contains a low level of background fluorescence emission, but the ATPase activity of the purified GroEL is similar to that determined for GroEL purified from other protocols. We also have shown by ESI mass spectrometry that before elution from the Reactive Red 120–agarose column, the fluorescence contamination appeared to be due to single amino acid residues or small peptides (di- or tripeptides) that were tightly bound to GroEL. This is consistent with our quantitation of approximately 14 mol of bound tryptophan per mole of

[23] D. A. Harris, *in* "Spectrophotometry and Spectrofluorimetry: A Practical Approach" (C. L. Bashford and D. A. Harris, eds.), p. 86. IRL Press, Washington, DC, 1987.

[24] O. Yifrach and A. Horovitz, *Biochemistry* **34,** 5303 (1995).

GroEL tetradecamer for samples not eluted from the Reactive Red dye column. The purified GroEL contained only 0.04 mol of tryptophan per mole of GroEL protomer.

It should be noted that several other protocols for the purification of GroEL also have been described.[21,25–29] These protocols (along with the one described here) rely to a greater or lesser extent on that published originally by Hendrix[30] and later refined by Chandrasekhar *et al.*[9] In general, most purification protocols for GroEL use an ion-exchange column and a sizing column as the first two chromatography procedures. The GroEL may then be eluted from an Affi-Gel Blue column,[21] a hydroxyapatite column,[25] or a gel-filtration column, either in the presence[28] or absence[27] of 20% (v/v) methanol. In one procedure,[26] the contaminating proteins are removed by incubating GroEL with GroES and ATP; this is then followed with separation by gel filtration. The GroEL purified by that protocol was reported to contain 0.26 mol of tryptophan per mole of GroEL protomer.[26] Similarly, GroEL purified from an Affi-Gel Blue column[21] was reported to contain 0.2–0.3 mol of tryptophan per mole of GroEL protomer. In another procedure,[29] the GroEL oligomer is disassembled by elution from a QAE–Sepharose column, in the presence of ATP, and reassembled by the addition of ammonium sulfate at 0° to a final concentration of 1 *M*. In that procedure, the fluorescence emission of the reassembled GroEL oligomer was reported to contain tryptophan contamination at levels comparable to that described here for GroEL eluted from the Reactive Red 120–agarose column (0.04 mol of tryptophan per mole of GroEL protomer). The amount of tryptophan contamination in GroEL purified by the other protocols has not been reported.

## Acknowledgments

The authors thank Dr. Michael Gross and the Washington University Resource for Biomedical and Bio-organic Mass Spectrometry Facility for the use of the ESI mass spectrometer. R. R. gratefully acknowledges support from NIH Grant 1PO1CA49210 and Grant 2P41RR00954 to Dr. M. L. Gross. This work was supported by NIH Grant DK13332 (to C. F.).

[25] M. J. Todd, P. V. Viitanen, and G. H. Lorimer, *Biochemistry* **32,** 8560 (1993).
[26] T. Mizobata and Y. Kawata, *Biochem. Biophys. Acta* **1209,** 83 (1994).
[27] S. G. Burston, N. A. Ranson, and A. R. Clarke, *J. Mol. Biol.* **249,** 138 (1995).
[28] J. S. Weissman, C. M. Hohl, O. Kovalenko, Y. Kashi, S. X. Chen, K. Braig, H. R. Saibil, W. A. Fenton, and A. L. Horwich, *Cell* **83,** 577 (1995).
[29] J. Ybarra and P. M. Horowitz, *J. Biol. Chem.* **270,** 22962 (1995).
[30] R. W. Hendrix, *J. Mol. Biol.* **129,** 375 (1979).

# [9] Overexpression, Purification, and Properties of GroES from *Escherichia coli*

*By* EDWARD EISENSTEIN, PRASAD REDDY, and MARK T. FISHER

## Introduction

The GroES and GroEL chaperonins were first discovered as host factors in *Escherichia coli* that were necessary for the assembly of the dodecameric lambda (λ) head coat protein E.[1] These factors are located on the *groE* operon, which contains two genes, encoding small (*groES*) and large (*groEL*) polypeptides.[2] The GroES and GroEL chaperonins are heat shock proteins (Hsps),[3] which are essential for cell viability.[4] Highly homologous counterparts of the bacterial chaperonins are found in both the mitochondria and chloroplasts of eukaryotic cells, suggesting an important and ubiquitous role in highly conserved physiological processes. However, the functions of the GroES and GroEL heat shock proteins are not limited to processes that alleviate the biochemical "shock" a cell experiences when exposed to an inhospitable environment; they also play key roles in normal operations in cells.[5] One of the major roles of heat shock proteins is to cooperate as molecular chaperones to modulate the interactions of nonnative intermediates and facilitate intracellular protein folding.[6]

Although both GroES and GroEL are required for cell viability, and several targets have been implicated from inventive genetic and biochemical studies,[7,8] the identity of authentic, *in vivo* substrates for the chaperonins remains elusive. Nonetheless, early work on the ability of the chaperonins to increase the yield of ribulose-bisphosphate carboxylase (Rubisco) refolding has spawned numerous attempts to improve the yields of other proteins

[1] C. Georgopoulos, R. W. Hendrix, S. R. Casjens, and A. D. Kaiser, *J. Mol. Biol.* **76,** 45 (1973).

[2] S. M. Hemmingsen, C. Woolford, S. M. van der Vies, K. Tilly, D. T. Dennis, C. P. Georgopoulos, R. W. Hendrix, and R. J. Ellis, *Nature (London)* **333,** 330 (1988).

[3] F. D. Niehardt, T. A. Phillips, R. A. VanBogelen, M. W. Smith, Y. Georgalis, and A. P. Subramanian, *J. Bacteriol.* **145,** 513 (1981).

[4] O. Fayet, T. Ziegelhoffer, and C. P. Georgopoulos, *J. Bacteriol.* **171,** 1379 (1989).

[5] R. I. Morimoto, A. Tissieres, and C. Georgopoulos, "The Biology of Heat Shock Proteins and Molecular Chaperones." Cold Spring Harbor Laboratory Press, Cold Spring Harbor, New York, 1994.

[6] E. A. Craig, B. D. Gambill, and R. J. Nelson, *Microbiol. Rev.* **57,** 402 (1993).

[7] A. Gragerov, E. Nudler, N. Komissarova, G. A. Gaitanaris, M. E. Gottesman, and V. Nikiforov, *Proc. Natl. Acad. Sci. U.S.A.* **89,** 10341 (1992).

[8] A. L. Horwich, K. Brooks Low, W. A. Fenton, I. N. Hirshfield, and K. Furtak, *Cell* **74,** 909 (1993).



0076-6879/98 $25.00

during *in vitro* folding.[9,10] The promiscuous nature of the interaction of chaperonins with a wide spectrum of polypeptides,[11,12] and their key role in facilitating intracellular protein folding, has been illustrated vividly by attempts to highly express recombinant proteins in *E. coli.* In many situations in which the overexpression of recombinant proteins leads to insoluble aggregates (inclusion bodies), the coexpression of both GroES and GroEL results in the recovery of significant fractions of the recombinant protein is soluble cell extracts.[13–17] Investigations into the role of chaperonins in protein folding, in addition to studies on their biochemical and biophysical attributes, have been stimulated greatly by early functional and structural observations. Biochemical studies have indicated that nonnative proteins bind to GroEL in or near the edge of the central cavity,[18–20] and that their affinity is modulated by ATP binding and hydrolysis.[10,21–23] In addition, GroES binding to a GroEL–nucleotide complex further decreases the affinity of GroEL for nonnative chains.[24–26] GroES binding inhibits the ATPase activity of GroEL,[27,28] but increases the sigmoidality of ATP hydro-

[9] P. Goloubinoff, A. A. Gatenby, and G. H. Lorimer, *Nature (London)* **337,** 44 (1989).

[10] P. Goloubinoff, J. T. Christeller, A. A. Gatenby, and G. H. Lorimer, *Nature (London)* **342,** 884 (1989).

[11] T. K. Van Dyk, A. A. Gatenby, and R. A. LaRossa, *Nature (London)* **342,** 451 (1989).

[12] P. V. Viitanen, A. A. Gatenby, and G. H. Lorimer, *Protein Sci.* **1,** 363 (1991).

[13] S. C. Lee and P. Olins, *J. Biol. Chem.* **267,** 2849 (1992).

[14] R. M. Wynn, J. R. Davie, R. P. Cox, and D. T. Chuang, *J. Biol. Chem.* **267,** 12400 (1992).

[15] N. Carrillo, E. A. Ceccarelli, A. R. Krapp, S. Boggio, R. G. Ferreyra, and A. M. Viale, *J. Biol. Chem.* **267,** 15537 (1992).

[16] A. Esher and A. A. Szalay, *Mol. Gen. Genet.* **238,** 65 (1993).

[17] J. Tremblay, G. A. Helmkamp, and L. R. Yarbrough, *J. Biol. Chem.* **271,** 21075 (1996).

[18] K. Braig, M. Simon, F. Furaya, J. F. Hainfeld, and A. Horwich, *Proc. Natl. Acad. Sci. U.S.A.* **90,** 3978 (1993).

[19] N. Ishii, H. Taguchi, H. Sasabe, and M. Yoshida, *J. Mol. Biol.* **236,** 691 (1994).

[20] S. Chen, A. M. Roseman, A. S. Hunter, S. P. Wood, S. G. Burston, N. A. Ranson, A. R. Clarke, and H. Saibil, *Nature (London)* **371,** 261 (1994).

[21] J. Martin, T. Langer, R. Boteva, A. Schramel, A. L. Horwich, and F. U. Hartl, *Nature (London)* **352,** 36 (1991).

[22] J. Buchner, M. Schmidt, M. Fuchs, R. Jaenicke, R. Rudolph, F. X. Schmid, and T. Kiefhaber, *Biochemistry* **30,** 1586 (1991).

[23] M. T. Fisher, *Biochemistry* **31,** 3955 (1992).

[24] R. A. Staniforth, S. G. Burston, T. Atkinson, and A. R. Clarke, *Biochem. J.* **300,** 651 (1994).

[25] M. T. Fisher, *J. Biol. Chem.* **269,** 13629 (1994).

[26] Z. Lin and E. Eisenstein, *Proc. Natl. Acad. Sci. U.S.A.* **93,** 1977 (1996).

[27] G. N. Chandrasekhar, K. Tilly, C. Woolford, R. Hendrix, and C. Georgopoulos, *J. Biol. Chem.* **261,** 12414 (1986).

[28] P. V. Vittanen, T. H. Lubben, J. Reed, P. Goloubinoff, D. P. O'Keefe, and G. H. Lorimer, *Biochemistry* **29,** 5665 (1990).

lysis by the chaperonins.[29,30] Finally, GroES is thought to stabilize large conformational changes in GroEL promoted by nucleotide binding, which may regulate polypeptide substrate binding and dissociation.[31,32]

Architecturally, the GroE proteins possess unusual sevenfold symmetrical quaternary structures. GroEL is composed of 14 chains of 57.3 kDa, assembling as 2 stacked, 7-membered rings with a conspicuous central pore or cavity.[33] However, GroES consists of seven polypeptides of 10.3 kDa.[27] The X-ray crystal structures of each chaperonin protein have been determined.[34,35] The structural properties, conformational changes, and function of GroEL are discussed elsewhere in this volume. Crystal structures have been solved for two highly homologous chaperonin 10 (GroES) heptamers from *Mycobacterium leprae* and *E. coli*.[35,36] The crystal structures reveal that the GroES heptamer is a dome-shaped structure with a centrally located hole. The oligomer is 70 Å in width and 30 Å in height. The basic structural motif of the individual groES subunit is a two-layer antiparallel $\beta$ barrel composed of nine $\beta$ strands arranged in an orthogonal manner. In addition, two $\beta$ turns extend from the main structure. One turn, composed of residues 41–52, forms the roof of the structure. The other loop (residues 17–35) is highly flexible and is thought to be of critical importance for the interaction with the GroEL apical domains.[37] Mutations in the structural gene encoding this loop region block $\lambda$ phage growth.[1,38] Owing to their high flexibility, a majority of these loops are not resolved in either the *M. leprae* or *E. coli* structures. However, one loop in the *E. coli* structure is resolved owing to crystal contacts with an adjacent GroES molecule.

Because the chaperonins are being used increasingly for studies aimed at the *in vitro* and *in vivo* refolding of recombinant polypeptides, reliable, efficient, convenient, and verifiable methods for the preparation of bio-

[29] T. E. Gray and A. R. Fersht, *FEBS Lett.* **292,** 254 (1991).

[30] O. Kovalenko, O. Yifrach, and A. Horovitz, *Biochemistry* **33,** 14974 (1994).

[31] S. Chen, A. M. Roseman, A. S. Hunter, S. P. Wood, S. G. Burston, N. A. Ranson, A. R. Clarke, and H. Saibil, *Nature* (*London*) **371,** 261 (1994).

[32] A. M. Roseman, S. Chen, H. White, K. Braig, and H. Saibil, *Cell* **87,** 241 (1996).

[33] R. W. Hendrix, *J. Mol. Biol.* **129,** 375 (1979).

[34] K. Braig, Z. Otwinowski, R. Hedge, D. C. Boisvert, A. Joachimiak, A. L. Horwich, and P. B. Sigler, *Nature* (*London*) **371,** 578 (1994).

[35] J. F. Hunt, A. J. Weaver, S. J. Landry, L. Gierasch, and J. Deisenhofer, *Nature* (*London*) **379,** 37 (1996).

[36] S. C. Mande, V. Mehra, B. R. Bloom, and W. G. J. Hol, *Science* **271,** 203 (1996).

[37] S. J. Landry, J. Zeilstra-Ryalls, O. Fayet, C. P. Georgopoulos, and L. M. Geirasch, *Nature* (*London*) **364,** 455 (1992).

[38] J. Zeilstra-Ryalls, O. Fayet, L. Barid, and C. P. Georgopoulos, *J. Bacteriol.* **175,** 1134 (1993).

chemical quantities of the chaperonins are essential. The focus of this chapter is on methods for the expression and purification of GroES, and various properties of the co-chaperonin that can be used to assess its integrity and activity in chaperonin-facilitated protein folding.

## Overexpression

GroES has been expressed from its endogenous promoter on multicopy plasmids[27] as well as from heterologous promoters on high copy number plasmids.[9,24,39–41] These recombinant constructions have enabled the expression of GroES protein from between a few percent to almost 40% of the soluble protein. Interestingly, the level of expression plays an important role in choosing a procedure to purify the co-chaperonin to homogeneity. The high levels of expression combined with the relatively cumbersome methods for assaying GroES activity have resulted in protocols that depend primarily on visualization of a highly expressed polypeptide band in the 10- to 15-kDa range in polyacrylamide gels in the presence of sodium dodecyl sulfate (SDS). By monitoring GroES in this way during the purification protocols, the purity of the preparation can be assessed so that the requirement for additional steps can be evaluated.

### *Vectors and Strains for GroES Expression*

The source of the *E. coli groES* gene described in these methods was pOF39[4] or pGroESL,[9] a derivative of pOF39. The 291-bp *groES* gene can be manipulated easily through PCR (polymerase chain reaction) techniques to amplify the coding sequence while at the same time creating convenient restriction enzyme sites to facilitate cloning in a variety of expression vectors. The four vectors described here exhibit a high level of expression that facilitates the purification of milligram quantities of GroES for biochemical and biophysical properties, the production of mutational variants of GroES without contamination of wild-type chains, and the rapid production of co-chaperonin for quick evaluation of the impact of chaperonins on the refolding of a particular protein of interest. A plasmid containing GroES (along with GroEL) under the control of the T7 promoter has been constructed by cloning an *Eco*RI–*Hin*dIII fragment from pOF39 into pSPT18, yielding pT7GroE (Pharmacia, Piscataway, NJ).[39] This construction enables the accumulation of GroES to 4–5% of the total cellular protein.

[39] C. E. Kalbach and A. A. Gatenby, *Enzyme Microbiol. Tech.* **15,** 730 (1993).

[40] J. Zondlo, K. E. Fisher, Z. Lin, K. R. Ducote, and E. Eisenstein, *Biochemistry* **34,** 10334 (1995).

[41] M. Kamireddi, E. Eisenstein, and P. Reddy, *Protein Expression Purif.* **11,** 47 (1997).

A vector for the expression of GroES from the *trc* promoter in pEGS1 has been constructed by cloning the *groES* gene into pKK233-2 (Pharmacia) for induction by the addition of isopropylthiogalactoside (IPTG), allowing accumulation of GroES to almost 20% of the soluble protein. Although there have not been any reports of strain lethality on high level expression of GroES, this vector has been propagated in a *lacIQ* strain.[40]

The expression of a histidine-tagged variant of GroES from pEGSHis1 can facilitate the purification of mutant chains from the low levels of contaminating, constitutively expressed wild-type polypeptide. This plasmid vector combines the *groES* gene, a small fragment containing the $(His)_6$-encoding sequence of pET-15b (Novagen, Madison, WI) and the *trc* promoter of pKK233-2. This construct results in the expression of GroES to about 10% of the soluble protein, but with the 20-residue segment M-G-S-S-$(H)_6$-S-S-G-L-V-P-R-G-S-H covalently linked to the usual N-terminal methionine.

Finally, a vector has been constructed for the expression of GroES from the $\lambda P_L$ promoter in pRE1[41] by temperature shift of *E. coli* strain MZ1 from 32 to 42°.[41] This system results in the expression of GroES to almost 40% of the soluble protein, which enables the most rapid of the purification procedures described.

*Cell Growth and Protein Expression*

Expression of wild-type GroES from the T7 promoter has been achieved in *E. coli* strain BL21 (DE3), which possesses a stable chromosomal copy of the T7 RNA polymerase gene under control of the *lacUV5* promoter.[42] Cultures can be grown in either LB or minimal M9 medium, supplemented with ampicillin (100 $\mu$g/ml). Addition of 0.5 m*M* IPTG is made when the cells reach an $A_{600}$ of about 1.0, resulting in the expression of T7 RNA polymerase that transcribes the *groEL* and *groES* genes from pT7GroE. Cells can be grown another 4–8 hr for accumulation of the chaperonins. Expression of wild-type GroES from *E. coli* strain JV30 containing pEGS1 can be achieved by growing cells in Super LB medium at 37° in the presence of ampicillin (100 $\mu$g/ml) to midlog phase, followed by the addition of 1 m*M* IPTG. This is usually reached in 2–3 hr when using a fresh, stationary-phase inoculum diluted 1:100 in culture medium. Cell growth and GroES induction can be continued for another 4–16 hr, as accumulation of GroES has no obvious deleterious effect on cell growth.

The expression of histidine-tagged GroES in *E. coli* strain JV30 containing pEGSHis1 can be achieved by growing cells at 37° in Super LB medium in the presence of ampicillin (100 $\mu$g/ml) to an $A_{600}$ of ~0.6–0.8,

[42] F. W. Studier and B. A. Moffatt, *J. Mol. Biol.* **189,** 113 (1986).

at which time IPTG is added to 1 m$M$ to induce the synthesis of histidine-tagged GroES from the *trc* promoter. For this vector, however, growth should be continued for only about two additional hours to minimize the occurrence of histidine-tagged GroES production in inclusion bodies. Under the recommended conditions, histidine-tagged GroES constitutes about 10% of the soluble protein in native cell lysates, with about half of the total GroES synthesized in an insoluble form.

The expression of GroES from *E. coli* strain MZ1 containing the pRE–GroES recombinant plasmid can be achieved by growing cells in LB medium containing ampicillin (50 $\mu$g/ml) at 32° to an $A_{650}$ of 0.40. The culture is then diluted with an equal volume of fresh LB medium equilibrated at 60° and immediately placed in a 42° water bath shaker for 5 hr for optimal yield of soluble expressed protein.

Cells prepared from any of these systems can be stored at −70 to −80° for several weeks with no apparent effect on the purity or yield of GroES prepared in the manner described in the following sections.

## Purification

A number of procedures have been developed for the purification of the GroES cochaperonin, some in combination with fractionation of GroEL. Many are modifications of one of the earliest reports from Chandrasekhar and co-workers,[27] which involved six purification steps including ammonium sulfate fractionation, several column chromatography steps, and a final glycerol gradient sedimentation step. Although these modifications result in the production of a highly purified fraction of GroES, they have met with only limited use for *in vitro* refolding or biophysical experiments owing to the low yield of purified material from small volumes of starting material. Depending on the level of expression of soluble GroES polypeptide, different protocols have achieved varying success in the preparation of homogeneous material. The purification of native, wild-type GroES from either the T7 or the *trc* promoter systems can be achieved by a combination of classic purification steps and by taking advantage of the high level of solubility of GroES at low pH values. Alternatively, nickel-chelate chromatography can be used for histidine-tagged variants of GroES yielding milligram quantities of mutational variants of GroES. When GroES is expressed at levels of up to 40% of the soluble protein, a simple heat treatment is generally sufficient to recover milligram quantities of the cochaperonin for use in refolding assays. The procedures outlined here, which take advantage of the unusual stability of GroES at extremes of pH and temperature, have repeatedly yielded large quantities of highly purified material that has been useful in determining some of the molecular characteristics and properties of the co-chaperonin. The relative ease of purifying GroES by these methods should facilitate its use in experimental studies of *in vitro* protein refolding.

*Conventional Purification of Native GroES from T7 and trc Vectors*

Cells grown as described above are resuspended in 3 vol (ml buffer/g cell) of lysis buffer [0.1 *M* Tris-HCl (pH 8.0), 10 m*M* ethylenediaminetetraacetic acid (EDTA), 1 m*M* dithiothreitol (DTT)] by stirring at 4° about 30 min, or until most of the clumps of cells have been broken. Immediately before lysis, the suspension is made 0.2 m*M* in phenylmethylsulfonyl fluoride (PMSF). Lysates are generally prepared by two passages through a French press at 8,000–10,000 psi. After disruption, cell debris can be removed by centrifugation. To recover more (~20%) GroES, the cell pellet can then be redissolved in 2 vol of lysis buffer supplemented with 1 *M* NaCl and stirred gently at 4° for 30 min. The debris is again removed by centrifugation and the supernatants containing GroES can be combined for 35–55% ammonium sulfate fractionation.

A saturated solution of ammonium sulfate is prepared by the addition of 710 g of solid ammonium sulfate to 1 liter of 0.1 *M* Tris base, without pH adjustment. This solution is then used to make the extract 35% (v/v) saturated in ammonium sulfate by adding 559 ml of saturated ammonium sulfate per liter of extract. The saturated solution is added dropwise and the mixture is stirred for an additional 20 min at 4° before removing the precipitated material by centrifugation.

The 35% (v/v) solution is then made 55% (v/v in saturation with ammonium sulfate by the addition of 895 ml of saturated ammonium sulfate solution per liter of 35% (v/v) saturated extract. The mixture is stirred for 20 min at 4°, and the pellet obtained by centrifugation, which contains the GroES polypeptide, is resuspended in 50 m*M* Tris-HCl, pH 8.5, containing 1 m*M* DTT, 1 m*M* EDTA, and 0.2 m*M* PMSF, and dialyzed against the same buffer. The extract is then applied to a DEAE–Sepharose Fast Flow (Pharmacia) column, equilibrated with dialysis buffer, with the matrix amounting to about 1 ml of resin to 1 g of cells that were originally lysed. The column is washed until the absorbance of the flowthrough material reaches zero, and then eluted with a linear gradient of 10 column volumes from 0 to 0.6 *M* KCl gradient in equilibration buffer. The fractions containing GroES are assessed by sodium dodecyl sulfate-polyacrylamide gel electrophoresis (SDS–PAGE), pooled, and dialyzed against 50 m*M* Tris-HCl, pH 8.0, containing 0.1 m*M* EDTA, 0.1 m*M* DTT, and 0.2 m*M* PMSF.

A Q-Sepharose Fast Flow (Pharmacia) column (about 80% the volume of the DEAE column) is equilibrated with dialysis buffer, the GroES fractions are applied, and the column is washed with equilibration buffer until the absorbance reaches zero. The column is then eluted with a linear gradient of 10 column volumes from 0 to 0.4 *M* potassium phosphate in equilibration buffer. The fractions containing GroES are first identified by SDS–PAGE. The fractions are then evaluated spectrophotometrically for contamination by determining the tyrosine-to-tryptophan ratio by second derivative spectroscopy.[43,44] Because GroES contains only one tyrosine and is devoid of tryptophan, fractions exhibiting a value greater than 1.1 for the ratio of the peak-to-trough difference in the second derivative spectrum for 287–283 nm compared with 294–291 nm are pooled. The pooled fractions are concentrated by ultrafiltration to 20 mg/ml, and then dialyzed against 50 m*M* potassium acetate, pH 5.0, containing 0.1 m*M* EDTA, 0.1 m*M* DTT, and 0.2 m*M* PMSF for 16–20 hr.

This dialysis step results in extensive precipitation of contaminating material, but little GroES. In fact, at the lower levels of protein expression obtained with the pGroESL vector, only one contaminating protein species is observed by SDS–PAGE. The next step utilizes cation-exchange chromatography. A column of S-Sepharose Fast Flow (Pharmacia), about 60% the volume of the original DEAE column, is equilibrated with dialysis buffer, the GroES fractions are applied, and the column is eluted with a linear gradient of 10 column volumes from 0 to 0.2 *M* KCl in equilibration buffer. The fractions containing GroES are assessed by SDS–PAGE, pooled, and dialyzed against 50 m*M* Tris-HCl, pH 7.5, containing 0.1 m*M* DTT and 0.1 m*M* EDTA, and concentrated by ultrafiltration. When preparing GroES using the pGroESL expression vector, the chaperonin elutes from the S-Sepharose column as a single peak yielding a highly purified product.

If homogeneous GroES is not obtained in the cation-exchange step, a gel-filtration step can be added to resolve GroES from minor contaminants. A useful high-performance liquid chromatography (HPLC) size-exclusion column is a Bio-Sil SEC-250 column, which provides baseline separation of the GroES pool. The pure fractions are pooled and dialyzed against 50 m*M* Tris-HCl, pH 7.6, containing 0.1 m*M* DTT and 0.1 m*M* EDTA and stored at $-80°$ in the presence of 20% (v/v) glycerol. Various extinction coefficients for GroES have been reported. The value of 0.24 $cm^2\ mg^{-1}$ heptamer reported here is based on comparisons with theoretical methods.[45] This estimate for the extinction coefficient yields 5–10 mg of GroES per gram of cells, with some variability due to the level of expression and the recovery of GroES at the individual steps in the purification.

[43] A. P. Demchenko, *in* "Ultraviolet Spectroscopy," p. 123. Springer-Verlag, Berlin, 1986.

[44] R. Ragone, G. Colonna, C. Balestrieri, L. Servillo, and G. Irace, *Biochemistry* **23,** 1871 (1984).

[45] S. C. Gill and P. H. von Hippel, *Anal. Biochem.* **182,** 319 (1989).

### *Affinity Purification of Histidine Tagged GroES and Cleavage of Polyhistidine-Containing Peptide*

Purification of GroES by capitalizing on the expression of a histidine-tagged fusion protein has been achieved using nickel-conjugated His-bind resin (Novagen) essentially as suggested by the manufacturer with the following modification. Cells were lysed with a French press, and before application of the crude lysate to the His-bind column, the extract was made 1 *M* in NaSCN to facilitate the separation of wild-type chains that had assembled into heptamers with histidine-tagged chains by dissociation of oligomers to folded monomers.[40] The binding step and subsequent wash step were performed with buffers containing 1 *M* NaSCN and, on elution, the fractions containing histidine-tagged GroES were immediately made 100 m*M* with EDTA, which was found to minimize aggregation.

Cleavage of histidine-tagged GroES yielded histidine-cleaved GroES (HC-GroES) with 17 of the 20 N-terminal amino acids removed. Three residues (Gly-Ser-His) remain because they lie within the thrombin cleavage site of the histidine-tagged fusion sequence. Digestion is achieved by first dialyzing the protein against 20 m*M* Tris-HCl (pH 8.4), 150 m*M* NaCl, 50 m*M* EDTA, followed by a change into the same buffer with a reduced (1 m*M*) level of EDTA. Cleavage of the histidine-containing peptide is accomplished using 10 units (U) or thrombin per milligram of histidine-tagged GroES (Haematologic Technologies, Essex Junction, VT) for 6 hr at room temperature. The resulting mixture was dialyzed against wash buffer (Novagen) and reapplied to the His-bind column to bind and remove uncleaved chains and the histidine peptide. The column wash, containing unbound HC-GroES, was dialyzed against 50 m*M* Tris-HCl, pH 8.0, containing 0.1 m*M* EDTA and DTT for a final chromatography step employing Poros HQ50 with elution by a linear 0.5 *M* NaCl gradient. This method enables the production of about 10–15 mg of HC-GroES per liter of culture, or about 1–3 mg of HC-GroES per gram of cells.

### *Purification of Native GroES from Hyperexpressed Cultures by Heat Treatment*

Each gram of cells expressing GroES from the $\lambda P_L$ promoter should be resuspended in 8 ml of lysis buffer consisting of 50 m*M* Tris-HCl, pH 8.0, containing 1 m*M* EDTA, 1 m*M* DTT, and 0.1 m*M* PMSF. Cell lysates are prepared by passage of the suspension through a French press at 8000–10,000 psi. After removal of the debris by centrifugation, the extract is heated in a water bath to 80 ± 2°, and maintained at this temperature with gentle stirring for 20 min. After clarification of the solution by centrifugation, it is applied to a DE-52 column containing about 20 vol of resin per gram of cells initially lysed. GroES is then eluted from the column with a linear gradient of 10 column volumes from 0 to 0.6 *M* NaCl in lysis buffer. The fractions containing homogeneous GroES are assessed by SDS–PAGE, pooled, concentrated, and stored at −80°. The yield of homogeneous GroES purified by this procedure is about 15 mg/g cells.

## Properties

Application of the methods and the materials described in the preceding sections results in the recovery of a highly pure preparation of GroES in good yield, and generally in sufficient quantities for studies on the *in vitro* reconstitution of enzyme activity or for biochemical studies of its role in chaperone-mediated protein folding. However, because GroES possesses no enzymological activity per se, and a precise and quantitative role for the co-chaperonin in facilitated protein refolding has been elusive, it is difficult to evaluate unambiguously the fraction of "functional" or "active" GroES obtained in a preparation. Alternatively, the functional attributes and properties of GroES have been defined only qualitatively. GroES function is assessed largely by its effect on GroEL-mediated protein refolding, by its effect on some of the properties of GroEL, and by the unusually dynamic quaternary structure of the co-chaperonin. The following four assays can be employed to obtain a general picture of the structural and functional integrity and competence of purified GroES, and could form the basis for an evaluation of the effects of mutations on the properties of the co-chaperonin.

### *Formation of Stable GroEL–GroES Complexes in the Presence of Adenosine Nucleotides*

Although there have been many studies on the formation of chaperonin complexes between GroES and GroEL using biochemical and spectroscopic approaches, a rapid and convenient gel-filtration assay to corroborate the strong (nanomolar) binding of GroES with GroEL in the presence of magnesium nucleotides has been reported.[46] The principle of the assay rests on the vast difference in molecular weights for GroEL (805,000) relative to that for GroES (73,000), which results in their different elution times from a gel-filtration column. In addition, the fact that the two chaperonins form a stable complex in the presence of nucleotides facilitates the coelution of GroES with GroEL in the void volume of a size-exclusion column, and

[46] P. V. Viitanen, T. H. Lubben, J. Reed, P. Goloubinoff, D. P. O'Keefe, and G. P. Lorimer, *Biochemistry* **29,** 5665 (1990).

the corresponding disappearance of GroES from the included volume as it elutes later from the column. Use of an HPLC greatly increases the time resolution of the experiment, so that it can be performed in a matter of an hour or so.

An analytical HPLC size-exclusion column can be equilibrated in 50 m*M* Tris-HCl, pH 7.6, containing 10 m*M* $MgCl_2$, 10 m*M* KCl, and 0.25 m*M* ATP. Roughly equimolar mixtures of GroEL and GroES are incubated in 50 m*M* Tris-HCl, pH 7.6, containing 10 m*M* $MgCl_2$, 10 m*M* KCl, and 0.5 m*M* ATP. After a 10-min incubation at room temperature, an aliquot of the mixture is applied to the size-exclusion column and the elution profile is followed by protein absorbance at 220 nm. The presence of any GroES bound to GroEL eluting in the void volume can be estimated by comparing this fraction with those from control experiments of uncomplexed GroEL and GroES using SDS–PAGE. The presence of a polypeptide eluting in the void volume that comigrates with authentic GroES can be taken as evidence for complex formation with the GroEL chaperonin. Specificity can be verified by control experiments performed in the absence of nucleotide. Depending on the initial concentration of the chaperonin proteins and the size of the size-exclusion column, it may be necessary to concentrate the void volume fractions to visualize stained bands for the chaperonin components. This procedure has been used to verify that several mutants of GroES form strong complexes with GroEL, and that even a variant of GroES with an N-terminal, 20-amino acid peptide containing a polyhistidine sequence can bind strongly to the chaperonin.[40]

### *Effect of GroES on ATPase Activity of GroEL*

The functional integrity of GroES can also be observed by its effect on the GroEL $K^+$-dependent ATP hydrolysis reaction. The GroEL-dependent ATP hydrolysis rate is inhibited by GroES.[28,30,46] The observed ATP hydrolysis rate for the GroEL–GroES complex decreases to about 50% of the value observed with GroEL alone.[47] This reduction in rate has been attributed to the direct consequence of the asymmetric binding interaction between GroEL and GroES. Normally, the ATP hydrolysis rate of GroEL, monitored by the hydrolysis of [$\gamma$-$P^{32}$]ATP, is halted by adding EDTA, excess ADP, or cold ATP as a chase. Curiously, in the presence of GroES, ATP hydrolysis is not immediately quenched but continues until an equivalent of six or seven ATPs per tetradecamer is hydrolyzed.[48] The delay in quenching has been proposed to be due to the quantized ATP hydrolysis of the GroES-bound GroEL heptamer. In the latest interpretation of the

[47] M. J. Todd, P. V. Viitanen, and G. H. Lorimer, *Biochemistry* **23,** 8560 (1993).
[48] M. J. Todd, P. V. Viitanen, and G. H. Lorimer, *Science* **265,** 659 (1994).

chaperonin cycle, GroES facilitates hydrolysis of ATP in one GroEL toroid. It is proposed that ATP will bind to the non-GroES-bound toroid and, following hydrolysis, ADP and GroES from the opposite toroid dissociate.[49] The question of whether ATP hydrolysis on the opposite ring also requires GroES binding (formation of a transient symmetric complex) is under investigation.

With increasing ATP concentrations, the initial ATP hydrolysis rate of GroEL shows a distinct sigmoidal response with Hill coefficients of about 2. When GroES is present, the sigmoidality becomes more pronounced and the Hill coefficient increases to 4.[30] In addition, with increasing ATP concentrations, adding GroES to GroEL increases the sigmoidal character of the substrate dissociation rates.[26] GroES also increases the polypeptide release rates from ADP or AMP–PNP GroEL complexes but at lower overall rates than are observed with ATP–GroEL complexes. The increase in the sigmoidal character of the ATP hydrolysis or polypeptide release rates almost certainly reflects a concerted conformational change in the chaperonin complex. At a molecular level, the changes are largely undefined although Lys-34 on GroES has been implicated as playing a critical role in modulating the allosteric transition between GroEL and GroES. Specific mutations in this residue (change to alanine or glutamate) enhance the sigmoidal transition above that observed with the wild-type GroES.[30]

*Refolding Activities*

An intriguing feature that has emerged from numerous studies of chaperonin-assisted protein folding is the differing requirements for GroES and nucleotides that support polypeptide refolding.[50] These results have led to the notion that the detailed mechanism for protein folding by chaperonins may vary with respect to the target protein under study. In spite of this apparent complexity, however, there are two main, general regimens of protein folding in which chaperonin components exert experimentally measurable effects.[51] The first general scenario involves solution conditions under which the protein of interest is capable of spontaneously refolding, although perhaps in low yield. The chaperonin requirement for reconstitution of activity under these conditions is highly variable. In some cases, only GroEL and ADP are required, whereas in others GroEL and ATP are required. Even though refolding is observed with GroEL and ATP

[49] S. G. Burston, N. A. Ranson, and A. R. Clarke, *J. Mol. Biol.* **249,** 138 (1995).
[50] R. Jaenicke, *Curr. Topics Cell. Regul.* **34,** 209 (1993).
[51] M. Schmidt, J. Buchner, M. J. Todd, G. H. Lorimer, and P. Viitanen, *J. Biol. Chem.* **269,** 10304 (1994).

alone, adding GroES to this mixture usually results in (1) an increase the refolding rate, (2) an increase in yield of recoverable activity, or (3) an increase in both rate and yield. The second regimen in which chaperonins play a key role in refolding is under conditions in which the polypeptides are unable to refold spontaneously, and are prone to aggregation. Under these experimental conditions, high-level refolding is virtually always dependent on GroEL, GroES, and ATP. However, results suggest that careful elucidation of optimal refolding conditions will affect the requirement for chaperonins in reconstitution assays. Nevertheless, the functional behavior of GroES purified using the protocols described here can be readily assessed under both limiting regimes of chaperonin-mediated protein folding. For this review, the functional role of GroES during chaperonin-assisted folding was assessed by focusing on the refolding of dimeric mitochondrial malate dehydrogenase (mMDH) and dodecameric *E. coli* glutamine synthetase (GS, glutamate–ammonia ligase).[23,25,52]

Using multisubunit proteins allows one to evaluate the complete function of GroES within the chaperonin cycle. For example, it is now established that a number of the more commonly used monomeric substrates can, under certain conditions, attain an active fold inside the large cavity created by the GroEL–GroES complex.[53–55] Here, acquisition of native monomeric structure would not depend on a complete cycle of GroES binding and dissociation from GroEL. Because enzyme activity assays are commonly used to assess successful chaperonin-assisted folding, one is unable to distinguish between activities contributed from free enzyme species and those from chaperonin-trapped, yet active, species. To complicate matters further, some monomeric substrates have a tendency to rebind to the GroE chaperonin complex even after acquiring a native fold.[56–58] With oligomeric substrates, all aspects of GroES interactions with GroEL are tested. GroES accelerates the ATP-dependent dissociation and subsequent folding of subunits from GroEL–substrate complexes. Because assembly

[52] R. A. Staniforth, A. Cortes, S. G. Burston, T. Atkinson, J. J. Holbrook, and A. R. Clarke, *FEBS Lett.* **344,** 129 (1994).

[53] M. Mayhew, A. C. R. da Silva, J. Martin, H. Erdjument-Bromage, P. Tempst, and F. U. Hartl, *Nature (London)* **379,** 420 (1996).

[54] J. S. Wiessman, H. S. Rye, W. A. Fenton, J. M. Beecham, and A. L. Horwich, *Cell* **84,** 481 (1996).

[55] M. K. Hayer-Hartl, F. Weber, and F. U. Hartl, *EMBO J.* **15,** 6111 (1996).

[56] P. V. Viitanen, G. K. Donaldson, G. H. Lorimer, T. H. Lubben, and A. A. Gatenby, *Biochemistry* **30,** 9716 (1991).

[57] K. E. Smith and M. T. Fisher, *J. Biol. Chem.* **270,** 21517 (1995).

[58] A. C. Clark, E. Hugo, and C. Frieden, *Biochemistry* **35,** 5893 (1996).

reactions are independent of the chaperonin complex,[25,59,60,61] both the associative and dissociative properties of GroES play key roles in the formation of assembly-competent subunits before oligomer assembly. Interestingly, a majority of the *E. coli* substrates whose folding may require the aid of chaperonins *in vivo* are multisubunit proteins.[8,62]

GroES plays a critical role during the chaperonin-assisted folding and assembly of mMDH and GS. GS is representative of a class of proteins that, under specific solution conditions, can undergo some spontaneous folding (30% of the total yield) without the aid of chaperonins. Under these solution conditions, GroEL and ATP alone can enhance folding yields. When functional GroES is included in the reaction, GS monomers are released more efficiently and a dramatic increase in the GS assembly rate to its dodecameric form results.[23,25,61] The substrate mMDH represents the other class of substrates whose folding and assembly are poor without the chaperonins ($<$10% of the original activity). mMDH absolutely requires GroEL, GroES, and ATP to attain its active state. Curiously, mMDH folding rates appear to be enhanced in the presence of chaperonins.[63] Evidence suggests that GroEL may provide the sufficient binding strain to favor the unfolding of misfolded states.[64–66] Because of this interaction, the concentrations of on-pathway intermediates increase[63] and hence the folding and assembly of native proteins increase. Here again, GroES rebinding and release from GroEL allow for multiple cycling of mMDH subunits onto and off GroEL.[60,63,67] The above-mentioned test proteins are readily obtained from commercial sources (Sigma, St. Louis, MO).

The substrates are typically unfolded between concentrations of 50–100 $\mu M$ with 8 *M* urea, 10 m*M* DTT, 50 m*M* Tris-HCl for a minimum of 2–4 hr at 0°. Refolding conditions are initiated by rapidly diluting away the denaturant 100-fold into buffers containing a 2-fold molar excess of GroEL relative to the refolding protein.[23,62] Because transient folding intermediates that bind to chaperonins have short lifetimes, it is best to dilute the refolding protein rapidly into a buffer containing enough GroEL to capture these

[59] X. Zheng, L. E. Rosenberg, K. Kalousek, and W. A. Fenton, *J. Biol. Chem.* **268,** 7489 (1993).
[60] N. A. Ranson, S. G. Burston, and A. R. Clarke, *J. Mol. Biol.* **266,** 656 (1997).
[61] M. T. Fisher and X. Yuan, *J. Biol. Chem.* **269,** 29598 (1994).
[62] C. Georgopoulos, D. Ang, K. Liberek, and M. Zyliczin, "Stress Proteins in Biology and Medicine" (R. I. Morimoto, A. Tissieres, and C. Georgopoulos, eds.), p. 202. Cold Spring Harbor Laboratory Press, Cold Spring Harbor, New York, 1990.
[63] N. A. Ranson, N. J. Dunster, S. G. Burston, and A. R. Clarke, *J. Mol. Biol.* **250,** 581 (1995).
[64] R. Zahn, S. Perret, G. Stenberg, and A. R. Fersht, *Science* **271,** 642 (1996).
[65] R. Zahn, S. Perret, and A. R. Fersht, *J. Mol. Biol.* **261,** 43 (1996).
[66] F. J. Corrales and A. R. Fersht, *Proc. Natl. Acad. Sci. U.S.A.* **93,** 4509 (1996).
[67] A. D. Miller, K. Maghlaoui, G. Albanese, D. A. Kleinjan, and C. Smith, *Biochem. J.* **291,** 139 (1993).

intermediates efficiently. Rapid dilution also prevents nonproductive aggregation and decreases denaturant concentrations to favor proper folding. Best results are obtained if a small, concentrated volume of denatured protein is placed in the bottom of a 15-ml disposable centrifuge tube, followed by rapid dilution of this volume 100-fold with refolding buffer containing the excess GroE proteins.[10,21–24] Rapidly introducing the refolding buffer with simultaneous vortexing yields the most consistent results for replicate runs as assessed by scatter about the mean in the final yield. To enhance further optimal folding yields, the refolding protein concentration should be varied over a wide range while maintaining at least a twofold molar excess concentration of GroEL.[10,22,23,57,68,69] After forming the GroEL–protein complex, a twofold molar excess of GroES relative to GroEL is added along with sufficient amounts of ATP (2–5 m*M*) to initiate release and refolding. The best protein-folding yields are usually obtained when the GroEL–substrate complex is first formed without nucleotide and with GroES present. The nucleotide-free form of GroEL has the highest binding affinity for partially folded or unfolded intermediates[24] and is most efficient in capturing most or all of the transient folding intermediates. Folding is then initiated by adding GroES and ATP to the GroEL–protein substrate complex. GroES concentrations should be maintained at or above 2 $\mu M$ oligomer to ensure that the predominant steady-state population of GroES is a heptamer[40] (see the next section). This specific sequential order of addition procedure avoids the depletion of potential polypeptide binding sites. Because GroES and unfolded polypeptide compete for the same site,[70] forming the ATP–GroEL–GroES complex before forming a GroEL–protein complex reduces the concentration of the potential polypeptide-binding sites on GroEL.

*Characterization of the Dynamic Quaternary Structure of GroES*

A surprising feature of GroES is the relative ease it displays in dissociating from its heptameric quaternary structure to monomers at submicromolar concentrations.[40] In addition, this attribute of the co-chaperonin was found not only for wild-type GroES, but also for several functional mutant variants, as well as for a construct that contained up to 20 amino acids at the N terminus of the protein.[40] This is consistent with the nature of the subunit contacts revealed by the X-ray crystal structure of GroES[35] and by biochemical studies on a proteolytic cleavage product of GroES that is missing seven C-terminal amino acids and fails to assemble into a hepta-

[68] W. Zhi, S. J. Landry, L. M. Gierasch, and P. A. Srere, *Protein Sci.* **1,** 522 (1992).
[69] M. T. Fisher, *J. Biol. Chem.* **268,** 13777 (1993).
[70] W. A. Fenton, Y. Kashi, K. Furtak, and A. L. Horwich, *Nature* (*London*) **371,** 614 (1994).

meric quaternary structure.[71] Although the functional implication of the monomer–heptamer equilibrium for GroES in chaperonin function is unclear at this time, its prevalence does provide a means to evaluate the integrity of GroES preparations. Several convenient methods can be used to corroborate the dynamic quaternary structure of the co-chaperonin.

Two simple spectroscopic measurements to follow the dissociation of GroES heptamers have capitalized on the changes in fluorescence seen in tryptophan-containing mutants of GroES, or on changes in light scattering at 550 nm on dilution into neutral buffer of a relatively concentrated stock of GroES (150–200 *M* heptamer) to submicromolar concentration.[40] Changes in the fluorescence and light scattering occur exponentially as a function of time, yielding half-times between 2 and 4 min for dissociation for wild-type GroES.

A simple, qualitative measure of the dynamic nature of the oligomeric structure of GroES can be made by measuring the formation of hybrid heptamers between wild-type GroES and a variant containing a 20-amino acid, N-terminal peptide that contains a $(His)_6$ sequence for Ni-chelate affinity chromatography. Simply by mixing equivalent concentrations of histidine-tagged GroES with wild-type protein for several minutes, eight discrete hybrids can be resolved on native polyacrylamide gels, demonstrating the rapid equilibrium between monomer and heptamers.

The most rigorous method by which to characterize the monomer–heptamer equilibrium of GroES is sedimentation equilibrium in an analytical ultracentrifuge. Analysis of the concentration distribution of GroES at sedimentation equilibrium for a model describing a monomer–heptamer equilibrium yields dissociation constants in the range of $10^{-38}$–$10^{-32}$ $M^{-6}$, which corresponds to concentrations of 50% monomer and heptamer between 0.7 and 8.0 $\mu M$ GroES monomer.[40]

## Conclusions

The methods described in this chapter have been used in a number of laboratories for the high-level expression, efficient purification, and biochemical characterization of the co-chaperonin GroES. The protocols described are the most readily adaptable, and the assays presented are the most versatile for implementation using a variety of equipment. Finally, the methodology proposed for the verification of the integrity and molecular properties of wild-type GroES should find wide use in the characterization of a number of mutational and proteolytically cleaved variants of the co-

[71] J. W. Seale and P. M. Horowitz, *J. Biol. Chem.* **270,** 30268 (1995).

chaperonin to elucidate a quantitative role for GroES in chaperone-facilitated protein folding.

Acknowledgments

This research was supported by NIH Grants GM49316 and RR08937 to E. E., GM 49309 to M. T. F., and by funds from the NIST to P. R.

# [10] Criteria for Assessing the Purity and Quality of GroEL

*By* MATTHEW J. TODD and GEORGE H. LORIMER

It is not difficult to purify GroEL to a level of purity that would be regarded as acceptable by many investigators. Typically some combination of anion-exchange and gel-filtration chromatography will yield a preparation that, when a few micrograms are electrophoresed on a sodium dodecyl sulfate (SDS)–polyacrylamide gel, will give a single Coomassie-stained band at about 60 kDa. Although such preparations are quite satisfactory for many purposes, some caution must be exercised in interpreting experimental results obtained with such preparations, most especially as they relate to the fundamental mechanism of the chaperonins. For example, experiments have been reported that claim to show that the addition of unfolded protein specifically induces the dissociation of GroES from the stable, asymmetric $GroEL_{14}$–$ADP_7$–$GroES_7$ complex.[1] However, the authors admit that the preparations of GroEL used in these experiments were severely contaminated by an ensemble of unfolded proteins.[2] One must therefore wonder how, in the first place, it was possible to create such a stable complex in the presence of reagents (unfolded proteins) that supposedly induce its dissociation. In a similar vein, conclusions concerning the stoichiometry of protein binding to GroEL, the influence of unfolded proteins on the basal ATPase activity of GroEL, and other similar parameters are likely to be compromised when the experiments are performed with preparations of GroEL already containing significant quantities of contaminating protein.

[1] J. Martin, M. Mayhew, T. Langer, and F. U. Hartl, *Nature* (*London*) **366,** 228 (1993).
[2] M. K. Hayer-Hartl and F. U. Hartl, *FEBS Lett.* **320,** 83 (1993).



0076-6879/98 $25.00

GroEL binds to a wide variety of unfolded proteins with dissociation constants varying from micromolar to picomolar.[3] For proteins with molecular masses >25,000 Da, the stoichiometry of interaction is generally one unfolded protein per heptameric ring. A standard preparation of GroEL, contaminated by 1 mol of a single, 40-kDa protein, contains 4.8% by weight of contaminant. When the impurity is a single species, it can easily be detected at this level by Coomassie-stained SDS–polyacrylamide gel electrophoresis (PAGE). However, when the contaminating protein is not a single species, but a whole ensemble of unfolded protein, ranging in molecular mass from 10,000 to 70,000 Da, detection of the impurities by standard Coomassie-stained SDS–PAGE may not work, leaving the investigator with the belief that the preparation is "pure." The following considerations reveal that this is rarely the case.

GroEL contains no tryptophan residues.[4] Thus, the presence of tryptophan in a preparation of GroEL can be taken as an indicator of contamination. Tryptophan is one of the rarer of the natural amino acids. On average, proteins contain 1.1% tryptophan.[5] Thus, an average protein with a molecular mass of 40,000 Da (or 350–400 amino acid residues), is expected to contain about 4 tryptophan residues. Consider now the ensemble of contaminating proteins in the standard preparation of GroEL of the sort yielding a single band on Coomassie-stained SDS–PAGE. Such preparations typically contain 4.2–5.6 mol of tryptophan per mole of $GroEL_{14}$.[2] This translates collectively into 1.0–1.4 mol of contaminating protein per mole of $GroEL_{14}$. Having survived passage through several chromatography columns, this contaminating material is evidently quite tightly bound. The danger that such contaminating protein could interfere with the binding of a specific substrate protein to GroEL should therefore be kept in mind.

Thus, standard preparations of GroEL are generally contaminated by an ensemble of unfolded protein. The contaminating protein may have been associated with GroEL *in vivo*, immediately before lysis of the cells. Alternatively, protein denatured in the course of the purification may become associated with GroEL. Much of this contaminating protein can be removed by passage of the standard GroEL preparation through a column of red agarose (see [8] in this volume[5a]), or, less expensively, by ion-exchange chromatography in the presence of 25% (v/v) methanol (see below).

[3] P. V. Viitanen, A. A. Gatenby, and G. H. Lorimer, *Protein Sci.* **1,** 363 (1992).

[4] S. M. Hemmingsen, C. Woolford, S. M. vanderVies, K. Tilly, D. T. Dennis, C. P. Georgopoulos, R. W. Hendrix, and R. J. Ellis, *Nature* (*London*) **333,** 330 (1988).

[5] T. E. Creighton, "Proteins: Structures and Molecular Principles." W. H. Freeman, New York, 1984.

[5a] A. C. Clark, R. Ramanathan, and C. Frieden, *Methods Enzymol.* **290,** Chap. 8, 1997 (this volume).

With these procedures, it is possible to obtain GroEL with less than 0.5 mol of tryptophan per mole of $GroEL_{14}$, which corresponds to about 0.1 mol of contaminating "average" protein per mole of $GroEL_{14}$. Urea-induced dissociation of $GroEL_{14}$ to monomers followed by reassembly of the oligomer has also been reported to remove contaminating proteins.[6]

Although standard preparations of GroEL may appear pure as judged by Coomassie-stained SDS–PAGE, silver staining of an overloaded (10 $\mu$g of protein) gel reveals a multiplicity of contaminating proteins (Fig. 1). Because of the nature of the silver-staining process, it is difficult to perform quantitatively. Nevertheless it is a valuable qualitative method and a reminder that absolute purity is a desirable but hard-to-achieve goal.

Some of the contaminating bands that show up following silver-stained SDS–PAGE of standard preparations of GroEL are in fact proteolytically damaged fragments of GroEL. This can be shown with a Western blot employing antibody raised against the intact molecule. These nicked subunits copurify with the intact 14-mer, and may compromise the functional properties of 14-mers in which they are harbored. If only 2.5% of all of the subunits are nicked, then on average, one in every three 14-mers will contain a nicked subunit. However, they can be removed by ion-exchange chromatography in 25% (v/v) methanol.

## Removing Contaminating Proteins from GroEL

To strip the contaminating proteins from a standard preparation of GroEL, ion-exchange chromatography in the presence of 25% (v/v) methanol has proved quite effective. Most chromatographic matrices bind denatured protein avidly. These denatured proteins are difficult to remove. However, when one passes ultrapure GroEL through a frequently used, but well-washed, analytical gel-filtration column, the protein has been observed to emerge from the column less pure than the starting material (M. J. Todd, unpublished data, 1997). For this reason the procedure described here is most reproducably and effectively performed with an unused chromatographic matrix. (The once-used matrix can be retained for other, less demanding separations.)

### *Protocol*

One hundred milligrams of a standard preparation of GroEL is brought to 25% (v/v) methanol by the addition of a one-third volume of methanol. This procedure and all subsequent steps are carried out at room tempera-

[6] J. Ybarra and P. M. Horowitz, *J. Biol. Chem.* **270,** 22962 (1995).

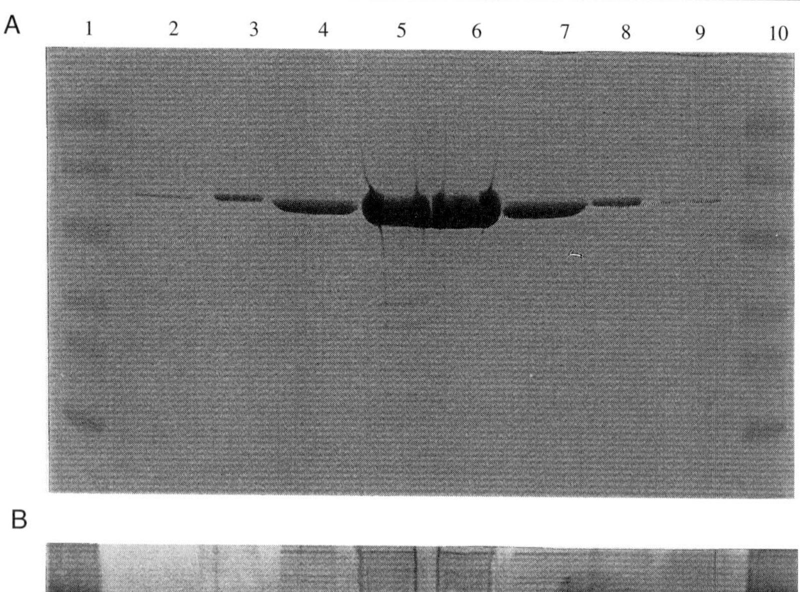


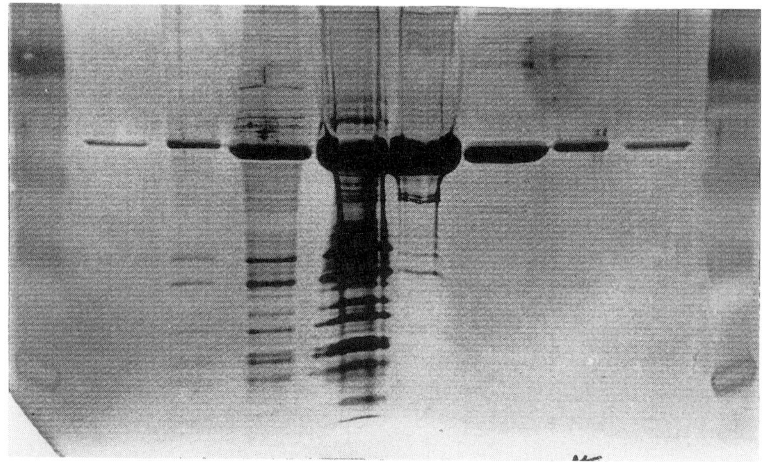


FIG. 1. SDS–polyacrylamide gels containing (lanes 2–5) 0.1, 0.5, 2.0, and 10 $\mu$g of a standard preparation of GroEL and (lanes 6–9) 0.1, 0.5, 2.0, and 10 $\mu$g of the same preparation after ion-exchange chromatography in 25% methanol. After SDS–PAGE the proteins were detected by (A) Coomassie blue staining or (B) silver staining. Lanes 1 and 10 contain molecular weight standards.

ture. Because GroEL has a limited solubility in 25% (v/v) methanol, the concentration of GroEL should be less than 1 mg/ml. The solution is next applied to a 1.6 × 16 cm column of Q-Sepharose Fast Flow (Pharmacia, Piscataway, NJ) equilibrated with 10 m*M* Tris-HCl (pH 7.6)–25% (v/v) methanol (buffer A). After washing the column with 300 ml of buffer A at 3 ml/min, a 300-ml gradient from 0.2 to 0.6 *M* NaCl in buffer A is applied. Fractions containing GroEL are analyzed fluorimetrically for tryptophan and by silver-stained SDS–PAGE (see below), and suitable fractions are pooled. Because this procedure typically removes about 80% of the contaminating proteins, occasionally it is desirable to repeat the procedure.

Ultrapure GroEL is unstable at 0° and dissociates slowly to monomers. (The presence of unfolded protein prevents this dissociation. The pure GroEL is also stabilized by 5 m*M* $Mg^{2+}$.) To remove the methanol, the pooled fractions are therefore dialyzed or washed and concentrated on a membrane filter versus 10 m*M* Tris-HCl (pH 7.6)–5 m*M* magnesium acetate. The purified GroEL is either flash frozen in small aliquots for long-term storage at −80° or sterile filtered and stored at 0°.

## Tryptophan Determination

This method of tryptophan determination is based on a modification of previously described methods[7,8] and employs bovine serum albumin (containing two tryptophan residues per protomer) in place of free tryptophan.

### *Stock Solutions*

Guanidine hydrochloride (ultrapure grade; Sigma Chemical Co., St. Louis, MO) is prepared at 8 *M* in 0.1 *M* sodium phosphate (pH 7.0)–1 m*M* EDTA–1 m*M* dithiothreitol, and filtered through a 0.2-$\mu$m pore size filter to remove dust and other light-scattering materials. After protein solutions have been added this solution, the guanidine hydrochloride concentration is adjusted to 6 *M* by addition of 0.1 *M* sodium phosphate, pH 7.0. Bovine serum albumin (BSA) is prepared in 0.1 *M* sodium phosphate (pH 7.0), at about 1.0 mg/ml. The exact molar concentration of this solution should be determined spectrophotometrically at 280 nm using $\varepsilon_{280\text{ nm}}$ 43,890 $M^{-1}$ cm$^{-1}$.[9] The concentration of tryptophan (twice the BSA concentration) of this solution should be about 30 $\mu M$.

[7] P. Payot, *Eur. J. Biochem.* **63,** 263 (1976).

[8] F. X. Schmid, *in* "Protein Structure" (T. E. Creighton, ed.), pp. 251–285. IRL Press, Oxford, 1989.

[9] C. N. Pace, F. Vajdos, L. Fee, G. Grimsley, and T. R. Gray, *Protein Sci.* **4,** 2411 (1995).

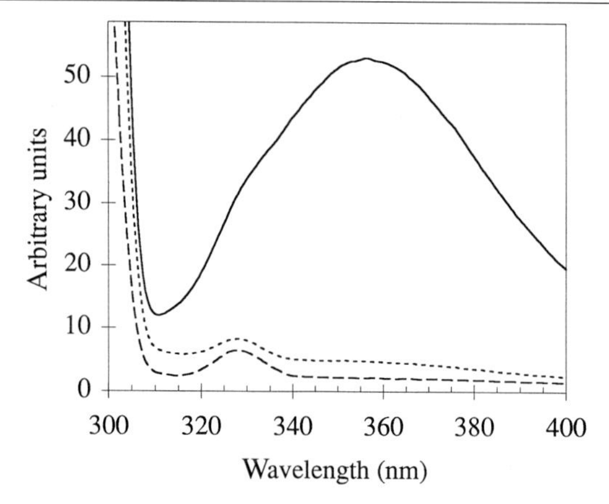


FIG. 2. Tryptophan fluorescence emission spectrum of GroEL before (solid line) and after (dotted line) ion-exchange chromatography in 25% (v/v) methanol. The baseline is shown as the broken line. The silver-stained SDS–PAGE analysis of these same samples is shown in Fig. 1. Excitation wavelength, 295 nm; bandwidth, 5 nm. The emission spectra (bandwidth, 10 nm) were recorded at 25°.

*Protocol*

The fluorescence emission spectrum of 6 *M* guanidine hydrochloride in a thermostatted (25°) 0.4 × 1.0 cm quartz cuvette is first recorded between 300 and 400 nm, with excitation at 295 nm, so as to establish a baseline. The spectrum of no more than 10 $\mu M$ GroEL subunits in 6 *M* guanidine hydrochloride is then recorded as described above. The fluorescence intensity at the maximum of 355 nm (after subtraction of the baseline) is measured. A standard curve using bovine serum albumin is likewise prepared. From the plot of fluorescence intensity at 355 nm versus micromolar tryptophan as bovine serum albumin, which should be linear, the tryptophan concentration in the original sample of GroEL can be determined. Standard preparations of GroEL typically contain 4–8 mol of tryptophan per mole of $GroEL_{14}$. However, using the purification procedures outlined above, this can be reduced to <0.5 mol of tryptophan per mole of $GroEL_{14}$ (Fig. 2).

## Silver Staining of Sodium Dodecyl Sulfate-Polyacrylamide Gels

This method of silver staining of SDS-polyacrylamide gels on a modification of the procedure described by Giulian *et al.*[10]

[10] G. G. Guilian, R. L. Moss, and M. Greaser, *Anal. Biochem.* **129,** 277 (1983).

*Stock Solutions*

Methanol (50%, v/v) in 10% (v/v) acetic acid

Silver stain solution (prepared fresh): Dissolve 0.8 g of $AgNO_3$ in 1 ml of water; dropwise to 21 ml of 0.1 *M* NaOH plus 1.4 ml of concentrated $NH_4OH$, being careful not to allow the silver to precipitate. This solution is diluted to 100 ml with water and used within 10 min

Developer: Citric acid (0.005%, w/v) in 0.05% (v/v) formaldehyde; prepare fresh

*Protocol*

Following electrophoresis, the proteins are fixed in the gel by soaking in 50% (v/v) methanol–10% (v/v) acetic acid for 30 min. The concentrations of residual gel buffers are lowered further by several rapid washes in water (3–5 min each). Protein is stained by incubating with $AgNO_3$ staining solution. The binding of silver is generally complete within 2–5 min, depending on the acrylamide concentration and the gel thickness. The gel is thoroughly washed with water to remove excess staining solution (three 5-min washes), and bands are developed by incubating with developer. To halt staining, the developer is removed by extensive washing with water.

# [11] Construction of Single-Ring and Two-Ring Hybrid Versions of Bacterial Chaperonin GroEL

*By* ARTHUR L. HORWICH, STEVEN G. BURSTON, HAYS S. RYE, JONATHAN S. WEISSMAN, and WAYNE A. FENTON

## Construction of Single-Ring Version of GroEL

A single-ring version of the bacterial chaperonin, GroEL, has been designed using information from the crystal structure of the native double-ring assembly.[1] This molecule, designated SR1, has proved to be functional as a "folding chamber," inside of which proteins such as rhodanese and green fluorescent protein (GFP), bound to the ring in nonnative form, can reach native form on addition of GroES and ATP.[2] In particular, the substrate protein folds to the native state within the central channel under-

[1] K. Braig, Z. Otwinowski, R. Hegde, D. C. Boisvert, A. Joachimiak, A. L. Horwich, and P. B. Sigler, *Nature (London)* **371,** 578 (1994).

[2] J. S. Weissman, H. S. Rye, W. A. Fenton, J. M. Beechem, and A. L. Horwich, *Cell* **84,** 481 (1996).





0076-6879/98 $25.00

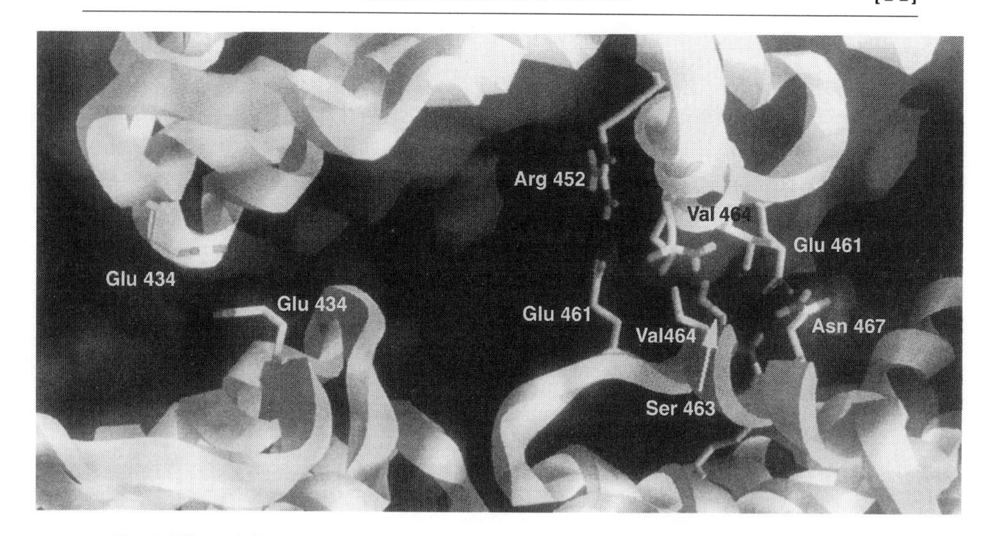


FIG. 1. View of the equatorial interface in wild-type GroEL. Part of the equatorial interface of wild-type GroEL is shown, along with residues making electrostatic and hydrophobic contact between the heptamer rings. The interface is approximately horizontal; the upper portion is derived from one subunit, whereas the lower comprises regions of two adjacent subunits.[1] The $\alpha$-carbon backbone is shown as a thick ribbon, and selected side chains are shown as tubes (drawn with the program GRASP). The residues changed in SR1 (R452, E461, S463, and V464) clearly form a major site of contact between the rings.

neath GroES and remains within this space because the absence of a second GroEL ring prevents the normal ATP mediated signaling that releases GroES.[3]

SR1 has been designed by examination of the model of unliganded GroEL at 2.8 Å, derived from the orthorhombic crystal.[1] This reveals a contact surface between the two rings that is composed of both electrostatic and hydrophobic interactions (see Fig. 1). As shown in Fig. 1, each subunit in the homotetradecameric complex forms contacts through the inferior aspect of its equatorial domain with that region of two neighboring subunits in the opposite ring. If a van der Waals surface is projected, it can be seen that the contacting surfaces are tightly apposed (see Ref. 1). The contact surface between rings totals ~2800 Å$^2$.[1] We have speculated that alteration of four major contacting residues contributed by each subunit to the interface would interfere with assembly of rings, resulting in a single ring structure. Codon R452 is changed to glutamate, and codons E461, S463, and V464 are altered to alanine in a plasmid bearing a cloned *groE* operon regulated by a *trc* promoter. This is accomplished by replacing a wild-type

[3] M. J. Todd, P. V. Viitanen, and G. H. Lorimer, *Science* **265,** 659 (1994).

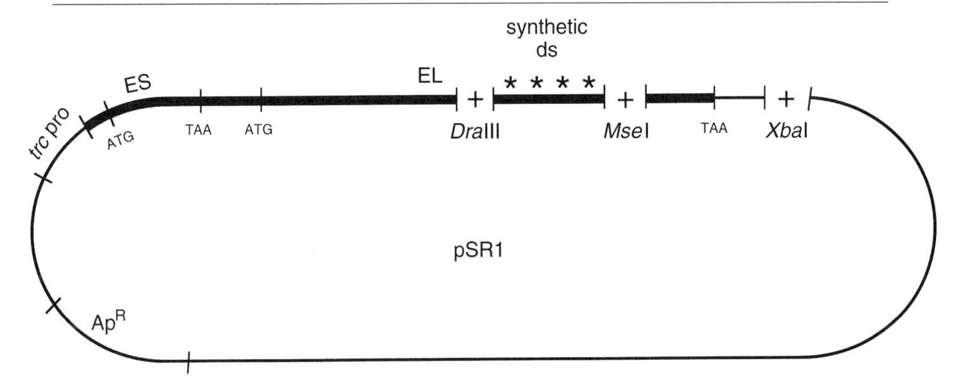


FIG. 2. Construction of pSR1. Because the residues changed were coded in a short DNA segment flanked by convenient restriction sites, a synthetic double-stranded fragment of 79 bp was made that programmed all of the necessary nucleotide changes simultaneously (represented by asterisks). This was used to replace the corresponding fragment in a wild-type *groE* operon expression plasmid in a three-part ligation, using the restriction enzymes noted. Clones were screened by restriction digestion, and the selected clones sequenced to confirm the introduction of the mutations. *trc* pro, the *trc* promoter, which replaces the natural promoter of the *groE* operon in this construct; ES and EL, the coding sequences for GroES and GroEL, respectively. Drawing is not to scale.

73-base pair (bp) *Dra*III–*Mse*I segment of the operon with a synthetic fragment bearing the mutational changes (Fig. 2).[4] The derived plasmid, called pSR1, is transformed into DH5$\alpha$ cells, and transformed cells are grown in liquid Luria broth (LB) medium containing ampicillin, 50 $\mu$g/ml, to an $OD_{650}$ 0.5 and then induced with isopropylthiogalactoside (IPTG, 1 m*M*). Cells are harvested 3 hr later, typically reaching an $OD_{650}$ no greater than 1.0. Direct solubilization of an aliquot of cells at this point and analysis by sodium dodecyl sulfate-polyacrylamide gel electrophoresis (SDS–PAGE) reveal a single major species of 57 kDa, the GroEL subunit, comprising >90% of total protein.

Cells from ~5 liters are collected in a Sorvall (Norwalk, CT) GS-3 rotor and resuspended at 4° in 50 ml of buffer containing 20 m*M* Tris (pH 7.4), 50 m*M* KCl, 1 m*M* ethylenediaminetetraacetic acid (EDTA), and the cells are broken by one pass in a microfluidizer (Microfluidics Corp., Newton, MA). Alternatively, breakage can be accomplished by sonication, adding first 2 ml of lysozyme, 10 mg/ml, and incubating for 5 min at 4°. The cell lysate is cleared by centrifugation in a Beckman (Fullerton, CA) Ti70.1 rotor at 55,000 rpm for 25 min at 4°. The supernatant fraction is then

[4] J. S. Weissman, C. M. Hohl, O. Kovalenko, Y. Kashi, S. Chen, K. Braig, H. R. Saibil, W. A. Fenton, and A. L. Horwich, *Cell* **83,** 577 (1995).

fractionated by chromatography on a Q-Sepharose Fast-Flow column (Pharmacia, Piscataway, NJ), eluting with a gradient from 0 to 1 *M* NaCl in 50 m*M* Tris (pH 7.4), 1 m*M* EDTA, 1 m*M* dithiothreitol (DTT). The major GroEL peak, eluting at ~400 m*M* NaCl, is further fractionated by high-performance liquid chromatography (HPLC) gel filtration on a TSK4000SW$_{xl}$ column (TosoHaas, Montgomeryville, PA) (applying no more than ~5 mg per HPLC run); this separates SR1 (400,000 Da) from wild-type GroEL (800,000 Da). SR1 characteristically elutes approximately 90 sec after double-ring wild-type GroEL (Fig. 3).

Studies with SR1 have shown that it stably binds GroES in the presence of hydrolyzable ATP.[2,4] This is associated with one round of hydrolysis of bound ATP, leaving ADP bound nonexchangeably. It should be noted that stability of GroES bound to SR1 is greatest with total salt concentrations held below 10–20 m*M*, assessed by off-rate measurements carried out with [$^{35}$S]methionine-radiolabeled GroES in the presence of an excess of nonlabeled GroES.[2] The $t_{1/2}$ for GroES release from SR1 under such conditions is at least a few hours. Conversely, as also observed for wild-type GroEL by Todd and co-workers,[3] GroES can be rapidly released from SR1 by

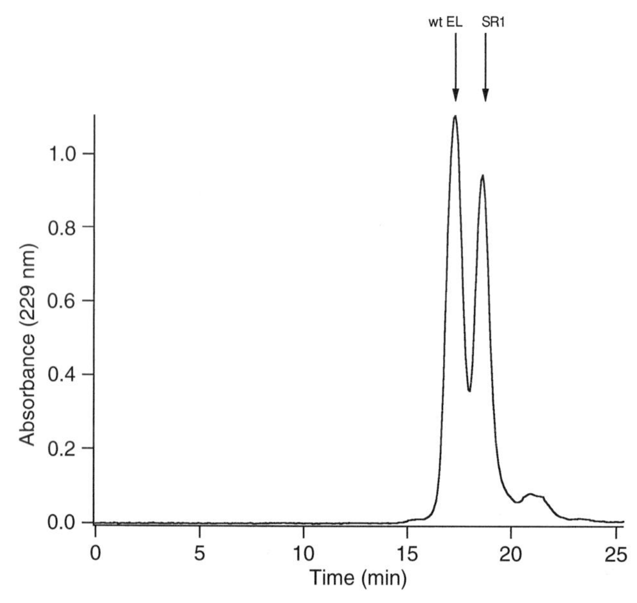


FIG. 3. Separation of wild-type and SR1 GroEL by HPLC gel filtration. Equal amounts of purified wild-type and SR1 mutant GroELs were mixed and injected onto a TSK4000SW$_{xl}$ HPLC column (7.8 × 300 mm), eluting at 0.4 ml/min. The labeled arrows indicate the elution positions of wild-type (wt EL) and SR1 when injected separately. After overexpression and ion-exchange purification of SR1, the wild-type peak is almost undetectable.

incubation at 4°.[5] This permits release, for example, of a substrate refolded inside the central channel underneath GroES. We presume that it is failure of normal release to take place, through absence of signaling from the second GroEL ring, that is the major defect of SR1, preventing, for example, subunits of oligomeric proteins, even though folded in the channel to a native conformation, from gaining access to partner subunits. As expected, expression of SR1 *in vivo* is unable to rescue growth of a GroEL-deficient *Escherichia coli* strain.[4]

## Formation of Mixed-Ring GroEL Complexes

The essential double-ring nature of GroEL, with critical signaling occurring between its rings during the folding reaction, encouraged the study of double-ring complexes containing defects in only one ring. For example, the ability of a GroEL ring to release both native and nonnative forms could be examined unambiguously using a mixed-ring complex in which

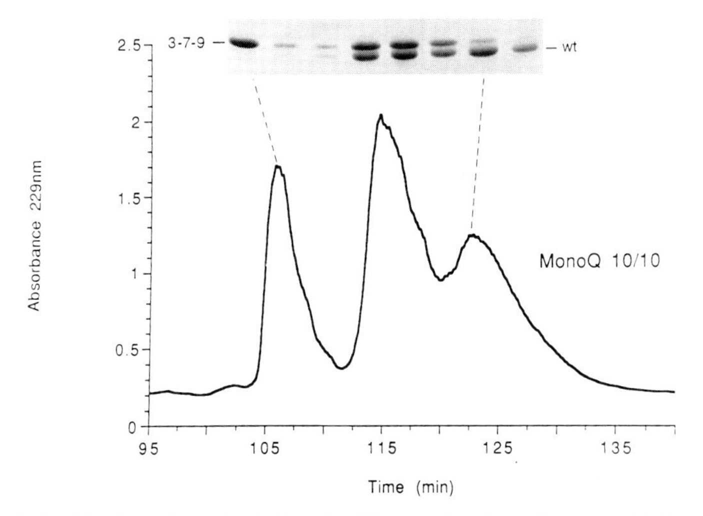


FIG. 4. Purification of a mixed-ring GroEL complex from the parental homotetradecameric complexes. Purified wild-type and mutant (Y203E/G337S/I349E; 3-7-9) GroELs were mixed and incubated as described in text, then subjected to anion-exchange chromatography on a Mono Q 10/10 column, eluted with an NaCl gradient. Aliquots of individual fractions were analyzed by SDS–PAGE, where the 3-7-9 and wild-type monomers separate (upper panel). The central peak of the mixed-ring complex, containing equal amounts of both monomers, was collected and an aliquot rechromatographed. It eluted as a single peak with no evidence of contaminating homotetradecamers. (From S. G. Burston, J. S. Weissman, G. W. Farr, W. A. Fenton, and A. L. Horwich, *Nature* (*London*) **383,** 96 (1996).)

[5] H. S. Rye, S. G. Burston, W. A. Fenton, J. M. Beechem, Z. Xu, P. B. Sigler, and A. L. Horwich, *Nature* (*London*) **388,** 792 (1997).

only one ring was capable of binding polypeptide.[6] The ability to produce and separate such mixed-ring complexes from parental assemblies was critical to performing any such studies. The availability of a mutant GroEL complex, G337D/I349E, that is physically separable from wild-type GroEL in anion-exchange chromatography,[7] prompted us to test whether a mixed-ring product, ostensibly migrating to an intermediate position, could be produced. Notably, both during coexpression of 337/349 and wild-type GroEL *in vivo* (A. Horwich and W. A. Fenton, unpublished, 1994) and in experiments of coincubation *in vitro* at room temperature with various salts and nucleotides (e.g., see Ref. 7), no such species has been observed during anion-exchange chromatography. However, when the wild-type and 337/349 assemblies are coincubated for 45 min at 42° in the presence of 5 m*M* $MgCl_2$, 5 m*M* Tris-HCl (pH 7.4), and 50 m*M* KCl, followed by addition of 10 m*M* *trans*-1,2-diaminocyclohexane-*N*,*N*,*N*′,*N*′-tetraacetic acid (CDTA) and return to 23°, a new species is observed by Mono Q (Pharmacia) chromatography (Fig. 4) and has been established through a number of analytical studies[6] to be a mixed-ring complex.

For production of several milligrams of mixed-ring species, chromatographic separation after incubation is carried out with a Mono Q 10/10 column (Pharmacia) at 23°, employing an initial wash at zero salt followed by a gradient of NaCl ranging from 0 to 500 m*M*, in 50 m*M* Tris-HCl, pH 7.4. The flow rate is set to 2 ml/min and the salt gradient is run over 160 min. The fractions are analyzed in 10% (w/v) SDS–PAGE, where 337/349 subunits migrate more slowly than wild type. The central protein peak (Fig. 4) is found to contain a 1:1 ratio of 337/349 and wild-type subunits and to chromatograph in gel filtration identically to wild-type GroEL.

The ability to separate a mixed-ring species can permit the testing of a variety of mutant mixed-ring structures. For example, the complex mentioned above, able to bind polypeptide on only one ring, is derived by mixing a wild-type ring with a ring containing the 337/349 mutations, along with the peptide-binding mutation F203E.[6,8] In designing such mixed-ring assemblies, it must be borne in mind that the 337/349 mutation, enabling separation, is not itself benign. The two apical domain mutations interfere with GroES binding. This behavior is associated in the 337/349 tetradecamer with the inability to release bound polypeptide, and this latter feature has been exploited in studying release of nonnative forms during GroEL-mediated folding.[7]

[6] S. G. Burston, J. S. Weissman, G. W. Farr, W. A. Fenton, and A. L. Horwich, *Nature (London)* **383,** 96 (1996).

[7] J. S. Weissman, Y. Kashi, W. A. Fenton, and A. L. Horwich, *Cell* **78,** 693 (1994).

[8] W. A. Fenton, Y. Kashi, K. Furtak, and A. L. Horwich, *Nature (London)* **371,** 614 (1994).

# [12] Chaperonin $60_{14}$ and Co-Chaperonin $10_7$ from *Chromatium vinosum*

*By* JOSE A. TORRES-RUIZ and BRUCE A. MCFADDEN

## Introduction

Understanding the mechanism by which molecular chaperones facilitate the process of protein folding in the cell constitutes one of the most challenging problems in biochemistry.

Molecular chaperones comprise a number of proteins that share the general property of interacting with other proteins in their nonnative conformations. Their basic function appears to be the prevention of incorrect associations within and between polypeptide chains during *de novo* protein folding and the protection of preexisting proteins under cellular stress.[1]

Among the most abundant molecular chaperones are the so-called "chaperonins," a specialized class of proteins that promote assembly, disassembly,[2] or translocation[3] of other proteins in an ATP-dependent reaction. The heat shock proteins, chaperonin 60 (Cpn60) and chaperonin 10 (Cpn10) from *Escherichia coli*, previously designated as GroEL and GroES, respectively, are classic representatives of the molecular chaperone family.[4] Chaperonins 60 and 10 are ubiquitous in nature as similar proteins have been identified in a wide variety of prokaryotes and organelles such as the peroxisome, chloroplast, mitochondrion, Golgi apparatus, and endoplasmic reticulum.[5,6]

In regard to their native structure, Cpn60 generally comprises 14 identical 60-kDa subunits arranged in a double-layered structure with a 7-fold symmetry. However, Cpn10 is a seven-subunit homooligomer with a subunit molecular mass of 10 kDa.[4]

The evidence that Cpn60 and Cpn10 interact functionally with each other has been strongly supported by both genetic and biochemical approaches. For instance, several studies of the properties of Cpn60 and

[1] U. Knauf, U. Jakob, K. Engel, J. Buchner, and M. Gaestel, *EMBO J.* **13,** 54 (1994).

[2] T. G. Chappell, W. J. Welch, D. M. Schlossman, K. B. Palter, M. J. Schlesinger, and J. E. Rothman, *Cell* **45,** 3C (1986).

[3] W. J. Chirico, M. G. Waters, and G. Blobel, *Nature* (*London*) **332,** 805 (1988).

[4] G. N. Chandrasekhan, K. Tilly, C. Woolford, R. Hendrix, and C. Georgopoulos, *J. Biol. Chem.* **261,** 12414 (1986).

[5] T. W. Mullin and R. L. Hallberg, *Mol. Cell Biol.* **8,** 371 (1988).

[6] S. M. Hemmingsen, C. Woolford, S. M. Vander Vies, K. Tilly, D. T. Dennis, C. Georgopoulos, R. W. Hendrix, and R. J. Ellis, *Nature* (*London*) **333,** 330 (1988).



0076-6879/98 $25.00

Cpn10 on the refolding of dimeric[7] and monomeric[8] enzymes have shown that Cpn60 binds, selectively, to unfolded proteins and eventually releases them in a folded form. This process appears to be coupled to the hydrolysis of ATP, $K^+$, and direct physical interaction with Cpn10. In this and other investigations, the functional stoichiometry is consistent with one 7-subunit Cpn10 oligomer per 14-subunit Cpn60 oligomer.

Along these lines, we have identified Cpn60 and Cpn10 homologs in *Chromatium vinosum*, an anoxygenic photosynthetic purple sulfur bacterium. In general terms, the data indicate that Cpn60 and Cpn10 from *C. vinosum* resemble, structurally and functionally, their counterparts from *E. coli.*[9,10]

## Methods

### *Bacterial Strain and Growth Conditions*

*Chromatium vinosum* (strain D), obtained from R. Chollet at the University of Nebraska (Lincoln, NE), is grown photoautotrophically at 32° using the $HCO_3^{1-}/S_2O_3^{2-}/Na_2S$ medium of Hurlbert and Lascelles.[11] *Chromatium vinosum* is also grown photoheterotrophically at 30° under anaerobic conditions on a minimal salt medium with 20 $\mu$g of vitamin $B_{12}$ per liter. The photoheterotrophic carbon source was 20 m*M* DL-malic acid, 20 m*M* succinic acid, 20 m*M* fumaric acid, or 20 m*M* pyruvic acid, each preadjusted to pH 6.8 with NaOH before addition to the medium.

### *Purification of Chaperonin 60 and Chaperonin 10 Homologs from Chromatium vinosum*

The purification of Cpn60 and Cpn10 from *C. vinosum* is accomplished by a protocol that combines sucrose density gradient centrifugation and Bio-Gel (Bio-Rad, Hercules, CA) A-1.5m gel-filtration chromatography.[9,10] All steps are conducted at 4°, unless otherwise indicated. *Chromatium vinosum* cells (60 g) are resuspended in 140 ml of MEMMB buffer [50 m*M* morpholine propanesulfonic acid (MOPS), 0.1 m*M* ethylenediaminetetraacetic acid (EDTA), 1 m*M* $MgCl_2$, 1 m*M* 2-mercaptoethanol, and 50 m*M*

[7] P. V. Viitaren, H. L. Thomas, J. Reed, P. Goloubinoff, D. P. O'Keefe, and G. H. Lorimer, *Biochemistry* **29,** 5665 (1990).

[8] J. Martin, T. Langer, R. Boteva, A. Schramel, A. L. Horwich, and F. U. Hartl, *Nature (London)* **352,** 36 (1991).

[9] J. A. Torres-Ruiz and B. A. McFadden, *Arch. Microbiol.* **142,** 55 (1985).

[10] J. A. Torres-Ruiz and B. A. McFadden, *Arch. Biochem. Biophys.* **295,** 172 (1992).

[11] R. E. Hulbert and J. Lascelles, *J. Gen. Microbiol.* **33,** 445 (1963).

$NaHCO_3$ (pH 7.3)] containing 1 m*M* *p*-tolylsulfonyl fluoride and 10 mg each of DNase I and RNase A. After disrupting the cells by sonic treatment, the cell debris is removed by centrifugation, and the resulting supernatant is diluted 1:1 with MEMMB buffer. The solution is adjusted to a final concentration (w/v) of 10% polyethylene glycol (PEG) 8000, and stirred for 45 min. Precipitated membranes and proteins are removed by centrifugation at 17,300 *g* for 30 min. Sufficient 1 *M* $MgCl_2$ is added to the supernatant fluid to yield a final concentration of 50 m*M*. After 30 min, the suspension is subjected to centrifugation at 17,300 *g* for 20 min. The resulting pellets are resuspended in 16 ml of MEMMB buffer (containing 10 m*M* KCl) and 20 ml is loaded on each of eight 40-ml 0.2 to 0.8 *M* linear sucrose density gradients in a VTi50 Beckman (Fullerton, CA) rotor. Because a physical interaction between Cpn60 and Cpn10 from *E. coli*, in the presence of ATP (and $Mg^{2+}$), has been well documented, 1 m*M* ATP is included at this sucrose density gradient centrifugation step to facilitate the identification of chaperonins from *C. vinosum*. After ultracentrifugation, at 175,000 *g* for 120 min, the gradients are fractionated and the chaperonins initially identified by Western immunoblotting analysis under polyclonal antibodies to Cpn60 from pea plants (1:3000 dilution) and to Cpn10 from *E. coli* (1:100 dilution).

Sucrose gradient fractions containing Cpn60 and Cpn10 are pooled, and the proteins are precipitated after adjusting the solution to 10% (w/v) PEG 8000 and 50 m*M* $MgCl_2$. After centrifugation, the resulting pellet is resuspended in MNM buffer [50 m*M* MOPS, 100 m*M* NaCl, 5 m*M* 2-mercaptoethanol (pH 7.5)], dialyzed against the same buffer system, and passed through a $1.6 \times 60$ cm column of Bio-Gel A-1.5m (Bio-Rad). Column fractions containing purified Cpn60 or Cpn10 are pooled and subjected to extensive dialysis. Protein concentrations are determined by the standard Bio-Rad dye-binding assay, using lysozyme as a protein standard. Polyclonal antibodies to purified Cpn60 or Cpn10 from *C. vinosum* are developed in rabbits and used to help in the identification of those proteins in subsequent experiments.

## Characterization of Chaperonin 60 and Chaperonin 10 from *Chromatium vinosum*

### *Molecular Weight and Quaternary Structure of Chaperonin 60 and Chaperonin 10*

The molecular weights for native Cpn60 and Cpn10 from *C. vinosum* are estimated by gel filtration using a calibrated Bio-Gel A-1.5m column.[10]

The elution volumes indicate molecular masses of 840 and 95 kDa for Cpn60 and Cpn10, respectively.

On the basis of the subunit molecular weight of 60,000, Cpn60 is composed of 14 closely similar or identical subunits. However, because in 15% (w/v) sodium dodecyl sulfate-polyacrylamide gel electrophoresis (SDS–PAGE) the purified Cpn10 migrates as a single 12.3-kDa subunit, the results suggest that Cpn10 from *C. vinosum* is an oligomer consisting of approximately seven subunits.[10]

*Amino Acid Composition and N-Terminal Amino Acid Sequence Analysis*

Aliquots containing purified Cpn60 and Cpn10 from *C. vinosum* (300–400 $\mu$g/ml) are transferred to hydrolysis vials for lyophilization. Hydrolysis is performed in 6 *N* HCl *in vacuo* under $N_2$ at 100 to 113° for 12, 48, and 62 hr.[10] All hydrolysates are analyzed with a Beckman 121 automatic amino acid analyzer by ion-exchange chromatography using the single-column method. Methionine is determined after performic acid oxidation of the protein for 3 hr at 0° followed by acid hydrolysis. For amino acid-sequencing analysis, preparations containing purified Cpn60 and Cpn10 are subjected to SDS–PAGE and polymerized from 15% (w/v) acrylamide, under conditions that prevent or reduce N-terminal blockage of proteins. The proteins from the gel are then electroblotted to polyvinylidenedifluoride (PVDF) membranes (Bio-Rad) for 12 hr at 30 V (25°) in a Tris–glycine–methanol transfer buffer. After transfer, the PVDF membrane is stained for 15 min in a mixture of water–methanol–acetic acid (5:5:1, v/v/v) containing 0.1% (w/v) Amido black and then destained with the mixture of water–methanol–acetic acid. The protein bands (~300 nmol) corresponding to the 60-kDa protein (Cpn60) and the 12.3-kDa protein (Cpn10) are excised from the Amido black-stained PVDF membrane, washed three times with doubly distilled $H_2O$, and then subjected to N-terminal amino acid sequence analysis using an Applied Biosystems (Foster City, CA) gas-phase sequenator.[10]

The amino acid compositions of purified Cpn60 and Cpn10 are comparable to those of their counterparts from *E. coli*. In that regard, the $S\Delta Q$ analysis by Marchelonis and Weltman,[12] which enables a hypothetical probe for sequence relatedness between proteins, reflects a considerable degree of homology between *C. vinosum* Cpn60 and Cpn10 and its homologs from other prokaryotes including *E. coli, Mycobacterium tuberculosis, Synechococcus* sp. strain PCC 7942, and *Coxiella brunetii*.[10] Moreover, *C. vinosum* Cpn60 exhibits a single N-terminal sequence, further suggesting that this

[12] J. J. Marchelonis and J. K. Weltman, *Comp. Biochem. Physiol.* **38B,** 609 (1971).

protein is a homooligomer. This sequence is similar to that of both the $\beta$ and $\alpha$ subunits of Cpn60 from pea plants.[13] The results also show that the N-terminal amino acid sequence for Cpn60 from *C. vinosum* is strikingly homologous (14 of the first 20 amino acids are identical) to its counterpart from *E. coli* and even more similar to that from *C. brunetii.*

*In Vitro Formation of Binary Complexes between Purified Chaperonin 60 and Chaperonin 10*

Several studies have been conducted to probe for interaction between the two chaperonins from *C. vinosum.*[10] In the first study, purified Cpn60 (5 mg/ml) and Cpn10 (1 mg/ml) are equilibrated in a reaction buffer [50 m*M* MOPS, 100 m*M* NaCl, 10 m*M* $MgCl_2$, 10 m*M* KCl, 5 m*M* 2-mercaptoethanol (pH 7.5)] with and without 1 m*M* ATP, or 1 m*M* 5′-adenylylimidodiphosphate (AMP-PNP) for 30 min at 4°. The three possible combinations are then subjected to gel-filtration chromatography on a Bio-Gel A-1.5m ($1.6 \times 60$ cm) column that has been previously equilibrated with the reaction buffer. Protein fractions eluting from the column are analyzed by 15% (w/v) SDS–PAGE. When ATP is excluded from the reaction mixture, three protein peaks emerge from the column. In order of decreasing molecular weight, the three peaks contain Cpn60 (oligomeric form), Cpn10 (oligomeric form), and free Cpn60 subunits, respectively. In marked contrast, under conditions in which ATP is included in the reaction mixture, Cpn60 and Cpn10 copurify as a single protein complex. Although no detectable uncomplexed oligomeric Cpn10 is detected under those conditions, a second peak reflects free 60-kDa subunits. On the other hand, the inclusion of the nonhydrolyzable ATP analog, AMP-PNP, does not support the formation of the Cpn60/10 complex, suggesting that for copurification to occur, hydrolysis of ATP is required.

In a second study,[10] the specific binding of Cpn10 to purified Cpn60 conjugated to an Affi-Gel 10 matrix column is also detected. Coupling of the Cpn60 protein from *C. vinosum* is carried out according to methods published elsewhere with several modifications. Approximately 10 ml of Affi-Gel 10 resin (Bio-Rad) is washed with 500 ml of phosphate buffer (10 m*M*, pH 4.5). Purified Cpn60 from *C. vinosum* is dialyzed against the coupling buffer [100 m*M* $NaHCO_3$, 0.5 *M* NaCl (pH 8.0)] and mixed with Affi-Gel 10 (5 mg of Cpn60/ml resin) for 36 hr at 2°. At the end of the incubation, the mixture is subjected to centrifugation to collect the resin, and the remaining reactive groups are blocked by incubation with 1 *M* glycine ethyl ester, pH 8.0, for 24 hr. The matrix is then washed with 1.5 *M* NaCl and then equilibrated with a column buffer [100 m*M* MOPS, 10 m*M* $MgCl_2$, 10 m*M* KCl, 1 m*M* ATP (pH 7.5)].

[13] J. A. Torres-Ruiz and B. A. McFadden, *Arch. Biochem. Biophys.* **261,** 196 (1988).

A crude extract from *C. vinosum* is obtained by disrupting the cells at 2° by sonic treatment in MEMMB buffer. The extract is diluted 1:1 with the column buffer (containing 1 m$M$ ATP) and it is then subjected to dialysis for 24 hr against the same buffer. After this, the *C. vinosum* crude extract is incubated with the Cpn60-coupled Affi-Gel 10 matrix (4.8 mg/ml resin) for 6 hr at 2°. The matrix, in a glass column, is then washed with the column buffer until the effluent gives a stable and low absorbance at 280 nm. The column is finally washed with the elution buffer [100 m$M$ MOPS, 100 m$M$ KCl, 10 m$M$ $MgCl_2$, 1 m$M$ 2-mercaptoethanol (pH 7.5)], and the eluting protein fractions are analyzed by SDS–PAGE.

The data illustrate that in the presence of 1 m$M$ ATP, under conditions similar to those used to demonstrate the formation of a Cpn60/10 complex by gel filtration, there is a specific retention of a 12.3-kDa protein by the affinity column. This 12.3-kDa protein comigrates, in 15% (w/v) SDS–PAGE, with previously purified Cpn10 from *C. vinosum*; also, antibodies to Cpn10 from *C. vinosum* cross-react with this protein, which confirms its identity.

The fact that the formation of binary complexes between the two chaperonins has been conserved in evolution, at least in mitochondria and bacterial systems, suggests that the establishment of this interaction is crucial for chaperonin function.

### *Assay of Chaperonin 60 ATPase Activity*

The release of radioactive inorganic phosphate from [$\gamma$-$^{32}$P]ATP (3000 Ci/nmol; New England Nuclear, Boston, MA) is monitored essentially as described elsewhere with various modifications.[10,14] Briefly, all reactions are conducted in 50 m$M$ MOPS (pH 7.5), 10 m$M$ $MgCl_2$, and Cpn60 (300 $\mu$g/ml). Some reactions are supplemented with Cpn10 (100 $\mu$g/ml) or with one of a variety of different chloride salts (KCl, $NH_4Cl$, NaCl, or CsCl) at 1 m$M$. Reactions are initiated by addition of [$\gamma$-$^{32}$P]ATP to 400 m$M$ (0.1–0.16 Ci/mmol) at 25°. At different time intervals, an aliquot of 25 $\mu$l is removed and added to a test tube containing 175 $\mu$l of 1 $M$ perchloric acid and 2 m$M$ sodium phosphate. The samples are maintained on ice. Ammonium molybdate (20 m$M$, 0.4 ml) and 0.4 ml of isopropyl acetate are then added and the solutions are vortexed vigorously. The phases are allowed to separate, and 100 $\mu$l of the upper organic phase is removed and assayed for radioactivity by liquid scintillation counting. In all instances, control reactions (without Cpn60) are conducted to correct for the low levels of nonenzymatic hydrolysis of [$\gamma$-$^{32}$P]ATP.

[14] R. Lill, K. Cunningham, L. A. Brundage, K. Ito, D. Oliver, and W. Wickner, *EMBO J.* **8,** 961 (1989).

This experiment clearly indicates that the *C. vinosum* Cpn60 contains a weak ATPase activity.[10] It catalyzes the hydrolysis of ATP, at 25°, at a rate corresponding to a specific activity of 0.059 $\mu$mol of ATP/min · mg of protein, and a $k_{cat}$ of 0.06 $sec^{-1}$ (based on protomers). When Cpn60 is incubated with a slight molar excess of Cpn10 (8 $\mu M$ Cpn10 subunits relative to 5 $\mu M$ Cpn60 subunits) or with 2 m$M$ AMP–PNP, the ATPase activity is more than 90% abolished. Similar results are obtained when Cpn60 from *C. vinosum* is incubated with Cpn10 purified from *E. coli* (our unpublished results, 1995). In the absence at Cpn10, the addition of $K^+$ stimulates the hydrolysis of ATP to a $k_{cat}$ of 0.14 $sec^{-1}$. That value is about twofold that observed in the absence of $K^+$. The effect of other monovalent cations on the initial rate for ATPase activity in Cpn60 has been explored. Although maximum initial activity for ATPase is achieved in the presence of $K^+$, other monovalent cations ($NH_4^+$, $Na^+$, $Cs^+$, and $Li^+$, each at 1 m$M$) support 87, 81, 74, and 65% of that velocity, respectively.

This is an interesting observation because, unlike other $K^+$-requiring enzymes including Cpn60 from *E. coli*, Cpn60 from *C. vinosum* displays considerable ATPase activity in the presence of monovalent cations such as $Li^+$, $Cs^+$, and $Na^+$. Further, even in the absence of any monovalent cation, *C. vinosum* Cpn60 supports more than 60% of the ATPase activity in the presence of added $K^+$. On the basis of these results, we suggest that, regardless of the structural and functional similarities between the two bacterial chaperonins from *E. coli* and *C. vinosum*, their active sites diverged during evolution (presumably owing to environmental adaptation), creating a *C. vinosum* Cpn60 less stringent in terms of monovalent cation requirements.

### *Comparison of Rubisco and Chaperonin 60 Levels in Chromatium vinosum*

The presence of Cpn60 and ribulose-bisphosphate carboxylase/oxygenase (Rubisco) in the crude extracts of *C. vinosum*, grown photoheterotrophically, may be quantified by enzyme-linked immunosorbent assay (ELISA), using the "sandwich method" according to established protocols.[15] Primary antibodies are either rabbit polyclonal antibodies to Rubisco or antibodies to Cpn60, from *C. vinosum*, at 1:3000 dilution. Goat polyclonal antibodies that have been raised against rabbit immunoglobulin G (IgG) and conjugated to alkaline phosphatase are used at 1:1000 dilution as the secondary antibodies according to the specifications of the commercial distributor (Sigma, St. Louis, MO). After addition of an alkaline phospha-

[15] S. Busby, A. Kumar, M. Joseph, and R. Woodbury, *Nature (London)* **316,** 271 (1985).

tase substrate, *p*-nitrophenyl phosphate (2 mg/ml) in 100 m*M* $NaHCO_3$, pH 9.5, and 10 m*M* $MgCl_2$, color development of positive wells is recorded at 415 nm using a microplate reader (model 700) from Cambridge Technology (Cambridge, MA).

Interestingly, measurements of Rubisco activity and immunological measurements of Rubisco and Cpn60 establish that levels of the two proteins vary together in *C. vinosum* grown on different carbon sources. In the limits, fumarate-grown cells are comparatively low in both Rubisco and Cpn60, and bicarbonate-grown cells contain the highest levels.

Along these lines, it appears that the expression of Rubisco and Cpn60, and presumably Cpn10, is coordinated under certain conditions.[12]

## Acknowledgments

We thank Dr. Gerhard Munske for the peptide sequence analysis, Dr. S. M. Gurusiddaiah (director of the Bioanalytical Center) for performing the amino acid composition analysis, Dr. C. Georgopoulos (at the University of Utah Medical Center) for providing antibodies to Cpn10 for *E. coli*, and Dr. Harry Roy (at the Rensselaer Polytechnic Institute at Troy, NY) for providing antibodies to Cpn60 from pea plants. This research was supported in part by Grant S06 GM-08239 (MBRS) to J. A. T. R. and by Grant GM-19,972 to B. A. M. from the National Institutes of Health (NIH).

# [13] Chaperonins of the Purple Nonsulfur Bacterium *Rhodobacter sphaeroides*

*By* W. Theodore Lee, Gregory W. M. F. Watson, and F. Robert Tabita

## Introduction

The chaperonins are highly conserved proteins that are involved in mediating the assembly of other proteins. Chaperonin 60 (cpn60, hsp60, GroEL) is a tetradecameric protein consisting of two stacked rings containing seven subunits ($M_r$ 60,000) in each ring. Chaperonin 10 (cpn10, hsp10, GroES) is a single ring of seven subunits ($M_r$ 10,000) that binds to cpn60 in the presence of MgATP. Chaperonin 60 possesses ATPase activity that is modulated by cpn10. The chaperonins are a class of ubiquitous molecular chaperones that are believed to play a major role in the pathway leading to the folding of proteins within all cells.

The purple nonsulfur bacterium *Rhodobacter sphaeroides* is a metabolically versatile organism capable of being cultured under widely different



0076-6879/98 $25.00

growth conditions.[1] Used as a model to study the regulation and biochemistry of carbon fixation in prokaryotes, *R. sphaeroides*, possesses two ribulosebisphosphate carboxylase/oxygenase (Rubisco, EC 4.1.1.39) enzymes. Rubisco, the key enzyme of the Calvin reductive pentose phosphate pathway, is a multimeric enzyme that has been demonstrated to require chaperonins for folding and subsequent assembly both *in vivo* and *in vitro*.[2,3] When grown photoautotrophically, and to a lesser extent photoheterotrophically, *R. sphaeroides* synthesizes large amounts of Rubisco.[1] The two Rubisco enzymes of *R. sphaeroides* are encoded by different operons and expression of the genes of each is differentially and independently regulated depending on the growth conditions.[1,4] Previously it was shown that cpn60 levels were elevated in cells grown photolithoautotrophically,[5] conditions under which Rubisco is synthesized at its highest level. The fact that cpn60 concentrations varied when *R. sphaeroides* was grown with different sources of carbon[5] suggests that there may be some specificity of the chaperonins with respect to substrate proteins. In addition, other bacteria in the alpha ($\alpha$) subdivision of the purple bacteria contain more than one *groE* operon. One of these bacteria, *Bradyrhizobium japonicum*, contains five *groE* operons, one of which is specifically expressed under conditions in which nitrogen fixation genes are required for growth.[6]

The *R. sphaeroides* cpn60/cpn10 proteins were purified and initially characterized in independent studies.[5,7] The heat-induced *groESL* operon from *R. sphaeroides* was cloned and characterized.[8] A second *groESL* operon was also cloned but a transcript has not yet been identified for this operon.

## Purification of Chaperonin 60/Chaperonin 10

Method 1 is from the protocol of Terlesky and Tabita.[5] The highest yield of the cpn60/cpn10 proteins is observed for cells growing under photo-

[1] F. R. Tabita, *in* "Anoxygenic Photosynthetic Bacteria" (R. E. Blankenship, M. T. Madigan, and C. E. Bauer, eds.), p. 885. Kluwer, Dordrecht, 1995.

[2] P. Goloubinoff, A. A. Gatenby, and G. F. Lorimer, *Nature (London)* **337,** 44 (1989).

[3] P. Goloubinoff, J. T. Christeller, A. A. Gatenby, and G. F. Lorimer, *Nature (London)* **342,** 884 (1989).

[4] J. L. Gibson, *in* "Anoxygenic Photosynthetic Bacteria" (R. E. Blankenship, M. T. Madigan, and C. E. Bauer, eds.), p. 1107. Kluwer, Dordrecht, 1995.

[5] K. C. Terlesky and F. R. Tabita, *Biochemistry* **30,** 3181 (1991).

[6] H. M. Fischer, M. Babst, T. Kaspar, G. Acuna, F. Arigoni, and H. Hennecke, *EMBO J.* **12,** 2901 (1993).

[7] G. M. F. Watson, N. H. Mann, G. A. MacDonald, and B. Dunbar, *FEMS Microbiol. Lett.* **72,** 349 (1990).

[8] W. T. Lee, K. C. Terlesky, and F. R. Tabita, *J. Bacteriol.* **179,** 487 (1997).

autotrophic conditions, although high levels of Cpn60 are also observed in cells grown photoheterotrophically and chemoheterotrophically (see below).

The *R. sphaeroides* cells are grown to midexponential phase. The cells are harvested by centrifugation (6000 *g*) for 15 min at 4° and resuspended in breakage buffer (50 m*M* *N*-[tris(hydroxymethyl)methyl]-2-aminoethane sulfonate buffer (TES) (pH 7.0) containing 10 m*M* $MgCl_2$ and 0.2 mg of DNase I per milliliter. Cells are lysed by passage through a French pressure cell twice at 20,000 lb/in$^2$. Unlysed cells and debris are removed by centrifugation at 6000 *g* for 10 min at 4°. Sucrose density gradient centrifugation is used in the initial step of the purification process. Sucrose solutions are prepared in breakage buffer containing 5 m*M* ATP. The gradients contain the following sucrose layers: 9.0 ml of 0.2 *M* sucrose, 8.5 ml of 0.4 *M* sucrose, 8.5 ml of 0.6 *M* sucrose, and 9.0 ml of 0.8 *M* sucrose. The extract is layered over the gradients and centrifuged in a Beckman (Fullerton, CA) SW28 swinging bucket rotor (20 hr, 25,000 rpm, 4°). Fractions are collected (1.0 ml) and the proteins in each fraction are analyzed by sodium dodecyl sulfate-polyacrylamide gel electrophoresis (SDS–PAGE) to visualize the high molecular weight chaperonin complex. Those fractions containing the cpn60–cpn10 complex are pooled. The pooled fraction is then chromatographed over a Green A–agarose (Amicon Division, W. R. Grace & Co., Danvers, MA) column to remove contaminating Rubisco. The column (10-ml bed volume) is equilibrated with buffer A [50 m*M* TES (pH 7.0), 10 m*M* $MgCl_2$, and 10% (v/v) ethylene glycol] and the sucrose density gradient pool is loaded onto the column. The column is washed with 80 ml of buffer A and the effluent collected. Any Rubisco present remains bound to the Green A matrix. The proteins in the Green A effluent are fractionated by fast protein liquid chromatography (FPLC) on a Mono Q 10/10 anion-exchange column (Pharmacia, Piscataway, NJ); chaperonin proteins are eluted after applying a linear gradient of 0.0–1.0 *M* KCl in buffer A. The chaperonin proteins are usually purified at this point, as analyzed by SDS–PAGE. If necessary, a final Superose-6 (Pharmacia, Piscataway, NJ) gel-filtration column purification step is employed.

Method 2 for the purification of Cpn60 is from the protocol of Watson *et al.*,[7] which is similar to the method described above. Here the cells are grown photoheterotrophically to an $A_{650}$ of 2.0. The cells are harvested and washed in TEMMB [20 m*M* Tris-HCl (pH 8.0), 10 m*M* $MgCl_2$, 50 m*M* $NaHCO_3$, 10 m*M* EDTA, 5 m*M* 2-mercaptoethanol]. The cells are resuspended in TEMMB and lysed by passage through a French pressure cell at 10,000 lb/in$^2$.

A sucrose gradient composed of 0.8, 0.7, 0.6, 0.5, 0.4, 0.3, and 0.2 *M* sucrose (2 ml each) in TEMMB is centrifuged at 240,000 *g* in an 8 × 25 ml angle-head rotor for 2.5 hr at 4°. Alternatively, an 8–28% (w/v) linear sucrose gradient is used to purify the cpn60. The 30-ml linear sucrose gradients in TEMMB are centrifuged at 242,000 *g* for 2.5 hr at 4°.[9] Fractions are collected (1.0 ml) and samples containing the chaperonins are pooled and applied to a 1-ml Pharmacia FPLC Mono Q column. The column is washed with 2 ml of 20 m*M* Tris-HCl, pH 7.5, and chaperonin proteins eluted after applying a linear 25-ml gradient of 0.0–1.0 *M* NaCl in 20 m*M* Tris-HCl, pH 7.5.

## Determination of ATPase Activity of Chaperonin 60

The ATPase activity[5] of the purified cpn60 is assayed by measuring the amount of radioactive inorganic phosphate released from [$\gamma$-$^{32}$P]ATP. The reaction contains 50 m*M* Tris-HCl (pH 7.5), 10 m*M* $MgCl_2$, 1 m*M* KCl, and 0.5 $\mu M$ cpn60. Reactions are initiated by the addition of the [$\gamma$-$^{32}$P]ATP (7000–9000 mCi/nmol) to a final concentration of 1.0 m*M*. Reaction vials are incubated at 37° and at given time intervals, samples (5 $\mu$l) are removed to a microcentrifuge tube containing 175 $\mu$l of 1 *M* perchloric acid and 1 m*M* sodium phosphate to quench the reaction. Next, 0.4 ml of 20 m*M* ammonium molybdate and 0.4 ml of isopropyl acetate are added. After mixing vigorously the phases are allowed to separate, and 100 $\mu$l of the top organic phase is removed to a scintillation vial. The amount of $^{32}$P is determined by counting in a Beckman LS 5000 TD liquid scintillation counter.

## Cloning of *Rhodobacter sphaeroides groESL* Operons and Expression in *Escherichia coli*

A heat-inducible *groESL* operon has been cloned using a heterologous probe.[8] The gene has been sequenced and transcripts characterized. Subclones have been constructed that express the *R. sphaeroides groES* and *groEL* genes in *Escherichia coli*. A second *groESL* operon has been cloned using a $groEL_1$ fragment as a probe. It appears that the $groES_2$ gene is a pseudogene.[8,10] It does not possess the usual codon bias of genes expressed in *R. sphaeroides* and appears to have a frameshift mutation. In addition, there is only 30% identity with the $cpn10_1$ sequence. The $groEL_2$ gene does not appear to have any frameshift mutations and has a typical codon bias. There is 70% identity between $cpn60_1$ and $cpn60_2$, based on their deduced amino acid sequences.

[9] G. M. F. Watson, unpublished observations (1990).

[10] W. T. Lee and F. R. Tabita, unpublished observations (1997).

## Properties of Purified *Rhodobacter sphaeroides* Chaperonins

The proteins from both purifications have been sequenced at the N terminus.[5,7] The sequences agree well with the deduced N-terminal sequence of the cloned heat-induced *groE* genes (Table I). On native PAGE the purified cpn60 has an apparent molecular weight of 889,200 in one study[5] and 670,000 in the other.[7] On denaturing SDS–PAGE the protein has a molecular weight of 60,000[5] or 58,000[7] in the different studies. The subunit molecular weight deduced from the nucleotide sequence is 57,946, with an isoelectric point (p*I*) of 4.81. Interestingly, the deduced amino acid sequence contains no cysteine residues. Gel filtration of cpn10 indicates a native molecular weight of 60,000 and SDS–PAGE shows a subunit molecular weight of 12,700.[5] The $groES_1$ gene encodes a protein of deduced molecular weight of 10,196 with a p*I* of 5.23. ATP is hydrolyzed by *R. sphaeroides* cpn60 at 37° with a specific activity of 134 nmol $min^{-1}$ $mg^{-1}$ and a $k_{cat}$ of 0.13 $sec^{-1}$. A preparation of *E. coli* cpn60 catalyzes ATP hydrolysis with a specific activity of 751 nmol $min^{-1}$ $mg^{-1}$ and a $k_{cat}$ of 0.75 $sec^{-1}$. The presence of cpn10 inhibits the ATPase activity of cpn60.[5] Using heterologous chaperonins it has been found that the ATPase activity of the *R. sphaeroides*

TABLE I
AMINO ACID SEQUENCES OF *Rhodobacter sphaeroides* CHAPERONINS

| Protein | Sequence |
|---|---|
| cpn60[a] | AKDVKFDTDAR??ML?GVSILADA |
| cpn60[b] | AAKDVKFDTDARD?MLRGV |
| $GroEL_1$[c] | MAAKDVKFDTDARDRMLRGVNILADA |
| $GroEL_2$[d] | MSAKEVRFGTDARGRMLKGINTLADT |
| Cpn10[a] | FKPLHDRVLVRRVQSDEKTKG |
| $GroES_1$[c] | MAFKPLHDRVLVRRVQSDEKTKG |

[a] N-terminal sequence of purified cpn60 or cpn10; K. C. Terlesky and F. R. Tabita, *Biochemistry* **30,** 3181 (1991).

[b] N-terminal sequence of purified cpn60; G. M. F. Watson, N. H. Mann, G. A. MacDonald, and B. Dunbar, *FEMS Microbiol. Lett.* **72,** 349 (1990).

[c] Deduced N-terminal sequences of $groEL_1$ or $groES_1$; W. T. Lee, K. C. Terlesky, and F. R. Tabita, *J. Bacteriol.* **179,** 487 (1997).

[d] Deduced N-terminal sequence of $groEL_2$; W. T. Lee, K. T. Terlesky, and F. R. Tabita, *J. Bacteriol.* **179,** 487 (1997).

TABLE II
LEVELS OF *Rhodobacter sphaeroides* Cpn60 AND *groESL*$_1$ TRANSCRIPT UNDER DIFFERENT GROWTH CONDITIONS

| Growth mode | Levels of Cpn60[a] (% soluble protein) | *groESL*[b] (cpm) | 16s rRNA[b] (cpm) |
|---|---|---|---|
| Photolithoautotrophic | 9.3 ± 2.0 | 913 | 21,100 |
| Photoheterotrophic | 6.7 ± 0.7 | 3,870 | 19,500 |
| Chemoheterotrophic | 3.5 ± 1.5 | 2,090 | 11,500 |

[a] Determined by rocket immunoassay.
[b] Determined by PhosphorImage analysis.

cpn60 is not inhibited by the *E. coli* cpn10. However, the *R. sphaeroides* cpn10 does inhibit the ATPase activity of the *E. coli* cpn60.

Electron microscopy of cpn60 reveals a distinctive four-layered density that possesses sevenfold symmetry.[11] Two major domains of approximately equal size give the chaperonin tetradecamer a cagelike structure. On incubation with MgATP there is a significant rearrangement of the staining pattern, which is most clearly seen on the side views. When apyrase is added to hydrolyze the ATP the subunit rotations are reversed. In the same study cpn60–cpn10 complexes have also been viewed by electron microscopy. The staining of *E. coli* cpn60 is similar to that of *R. sphaeroides* cpn60. Addition of *E. coli* cpn10 to the cpn60 complex reveals a small additional layer of density present at one end of the cpn60 molecule. In the complex, the cpn60 conformation resembles that which occurs on MgATP binding, suggesting that this may be a prerequisite for cpn10 binding.

Expression of cpn60 has been demonstrated to be regulated depending on the growth conditions in *R. sphaeroides*[5] (Table II). Levels of cpn60 are highest in cells grown photolithoautotrophically, as determined using antibodies raised against cpn60. However, Northern analysis of transcripts from the heat-induced *groESL*$_1$ operon indicates that levels of expression are fourfold higher in photoheterotrophically than in photolithoautotrophically grown cells.[8] This discrepancy in under investigation and may possibly be due to the presence of more than one *groE* operon in *R. sphaeroides*. However, as noted above, the N-terminal amino acid sequences obtained from cells grown photolithoautotrophically and photoheterotrophically are in agreement with the amino acid sequence deduced from the heat-inducible *groESL*$_1$ operon. In addition, no transcript has yet been identified for the

[11] H. R. Saibil, D. Zheng, A. M. Roseman, A. S. Hunter, G. M. F. Watson, S. Chen, A. auf der Mauer, B. P. O'Hara, S. P. Wood, N. H. Mann, L. K. Barnett, and R. J. Ellis, *Curr. Biol.* **3,** 265 (1993).

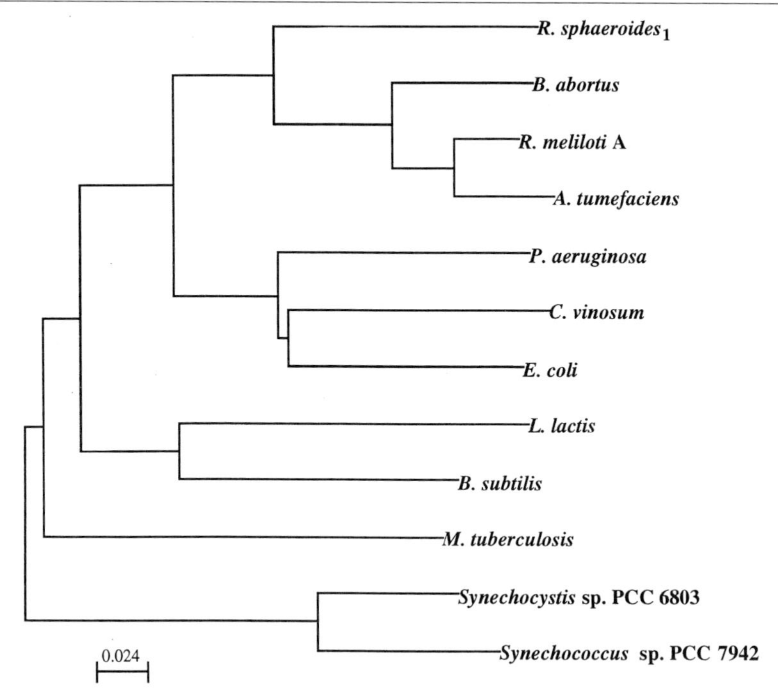


FIG. 1. Phylogenetic tree of cpn60 sequences from diverse eubacteria created using the Clustalw program [J. D. Thompson, D. G. Higgins, and T. J. Gibson, *Nucleic Acids Res.* **22,** 4673 (1994)]. *Rhodobacter sphaeroides* [W. T. Lee, K. C. Terlesky, and F. R. Tabita, *J. Bacteriol.* **179,** 487 (1997); *Brucella abortus* [D. Gor and J. E. Mayfield, *Biochim. Biophys. Acta* **1130,** 120 (1992)]; *Rhizobium meliloti* [E. Rusanganwa and R. S. Gupta, *Gene* **126,** 67 (1993)]; *Agrobacterium tumefaciens* [G. Segal and E. Z. Ron, *J. Bacteriol.* **175,** 3038 (1993)]; *Pseudomonas aeruginosa* [A. Sipos, M. Klocke, and M. Frosch, *Infect. Immun.* **59,** 3219 (1991)]; *Chromatium vinosum* [R. G. Ferreyra, F. C. Soncini, and A. M. Viale, *J. Bacteriol.* **175,** 1514 (1993)]; *E. coli* [S. M. Hemmingsen, C. Woolford, S. M. van der Vies, K. Tilly, D. T. Dennis, C. P. Georgopoulos, R. W. Hendrix, and R. J. Ellis, *Nature* (*London*) **333,** 330 (1988)]; *Lactococcus lactis* [S. G. Kim and C. A. Batt, *Gene* **127,** 121 (1993)]; *Bacillus subtilis* [M. Li and S.-L. Wong, *J. Bacteriol.* **174,** 3981 (1992); A. Schmidt, M. Schiesswohl, U. Volker, M. Hecker, and W. Schumann, *J. Bacteriol.* **174,** 3993 (1992)]; *Mycobacterium tuberculosis* [T. M. Shinnick, *J. Bacteriol.* **169,** 1080 (1987)]; *Synechocystis* sp. PCC 6803 [C. Lehel, D. Los, H. Wada, J. Gyorgyei, I. Horvath, E. Kovacs, N. Murata, and L. Vigh, *J. Biol. Chem.* **268,** 1799 (1993)]; and *Synechococcus* sp. PCC 7942 [R. Webb, K. J. Reedy, and L. A. Sherman, *J. Bacteriol.* **172,** 5079 (1990)].

*groESL*$_2$ operon. To investigate further the levels of expression and concentrations of the chaperonins, transcription, translation, and deletion constructs are being synthesized.

An association between *R. sphaeroides* form I Rubisco and the cpn60 chaperonin in the presence of ATP has been demonstrated.[12] When active Rubisco is incubated with *R. sphaeroides* cpn60 at a specific molar ratio of 18 (Rubisco to cpn60) in the presence of ATP, nearly all of the Rubisco and most of the cpn60 fail to enter a nondenaturing gel. At higher concentrations of cpn60 the proteins do enter the nondenaturing gel, but to a lesser extent than samples incubated in the absence of ATP.

The structure of the *E. coli* GroEL protein has been described.[13] The *E. coli* and *R. sphaeroides* cpn60s have 68% identity, and not surprisingly, several of the key amino acid residues identified in the *E. coli* protein are conserved in the *R. sphaeroides* cpn60. Of the 14 *E. coli* cpn60 amino acid residues identified by site-directed mutagenesis that are involved in ATP hydrolysis,[14] 12 are conserved in *R. sphaeroides* cpn60. Similarly, most of the resides involved in *E. coli* GroEL binding, polypeptide binding, and folding of heterologous proteins are conserved in the *R. sphaeroides* cpn60. The relationships between cpn60s of diverse eubacteria show that the *R. sphaeroides* cpn60 is most closely related to proteins from the $\alpha$ subdivision of the purple bacteria (Fig. 1).

[12] X. Wang and F. R. Tabita, *J. Bacteriol.* **174,** 3607 (1992).

[13] K. Braig, Z. Otwinowski, R. Hedge, D. C. Boisvert, A. Joachimiak, A. R. Norwich, and P. B. Sigler, *Nature (London)* **371,** 578 (1994).

[14] W. A. Fenton, Y. Kashl, K. Furtak, and A. L. Horwich, *Nature (London)* **371,** 614 (1994).

# [14] Chaperonins from *Thermoanaerobacter* Species

*By* ROBERT K. SCOPES and KAYE TRUSCOTT

Of the many heat shock proteins that have been detected, the most studied are those known as chaperonins, in particular the GroEL and GroES proteins from *Escherichia coli.* These are induced by exposure of an organism to temperatures several degrees higher than their normal growth temperature.[1,2] They have a protective role in heat shock, preventing aggregation of denatured polypeptides and allowing them to refold to active conformations. Unlike *E. coli*, thermophilic bacteria generally do not live in a controlled temperature environment except under laboratory conditions. In the hot springs where most are found, there are sharp temperature gradients, varying temporarily with local weather conditions, and individual

[1] M.-J. Gething and J. Sambrook, *Nature (London)* **355,** 33 (1992).

[2] H. Saibil and S. Wood, *Curr. Opin. Struct. Biol.* **3,** 207 (1993).



0076-6879/98 $25.00

organisms may experience a wide range of temperatures from well below their optimum to well above. Consequently, it is possible that the roles and activities of their heat shock proteins might differ from those of *E. coli.*

Although many chaperonin genes have been sequenced, relatively few of the proteins have been isolated and studied in any detail. The *E. coli* GroEL and GroES homologs (which are referred to as cpn60 and cpn10 in this chapter) from a few thermophiles have been isolated. The most studied are those from *Thermus* strains.[3–5] A notable difference between the *E. coli* GroE proteins and the *Thermus* chaperonins is that whereas the former interact only in the presence of ATP, the latter interact sufficiently strongly to be isolated together as a complex $(cpn60)_{14}$ $(cpn10)_7$.[3,6] This is generally presumed to be the composition of the main active form of the GroE complex in *E. coli.* Archaeal hyperthermophiles, such as *Sulfolobus* and *Pyrodictium*, have also been shown to contain heat shock proteins, but these are more homologous with the mammalian cytoplasmic TCP protein than with *E. coli* GroEL.[7–9] They are single or possibly double rings containing not seven but eight or even nine subunits per ring: a smaller co-chaperonin like GroES has not been conclusively identified in these hyperthermophiles.

Whereas *Thermus* species are aerobic bacteria, found free living in many hot water environments, there are many anaerobic thermophilic bacteria occurring in more sheltered unaerated parts of hot springs. Of these the most common are strains of *Thermoanaerobacter*, one of which, *Thermoanaerobacter thermohydrosulfuricus*, often makes up more than 90% of the microbial biomass in these environments.[10] Originally called *Clostridium thermohydrosulfuricum* on the basis of its ability to form spores, this has now been reclassified along with a number of other organisms in the *Thermoanaerobacter* genus.[11] In addition, the much studied and closely related *Thermoanaerobium brockii*[12] has been reclassified as *Thermoanaerobacter*

[3] H. Taguchi, J. Konishi, N. Ishii, and M. Yoshida, *J. Biol. Chem.* **266,** 22411 (1991).

[4] H. Tamada, T. Ohta, T. Hamamoto, Y. Otawara-Hamamoto, M. Yanagi, H. Hiraiwa, H. Hirata, and Y. Kagawa, *Biochem. Biophys. Res. Commun.* **179,** 565 (1991).

[5] K. Mikukik and O. Benada, *Biochem. Biophys. Res. Commun.* **197,** 716 (1993).

[6] N. Ishii, H. Taguchi, M. Sumi, and M. Yoshida, *FEBS Lett.* **299,** 169 (1992).

[7] J. D. Trent, E. Nimmesgern, J. S. Wall, F.-U. Hart, and A. L. Horwich, *Nature (London)* **354,** 490 (1991).

[8] B. M. Phipps, D. Typke, R. Hegerl, S. Volker, A. Hoffmann, K. O. Stetter, and W. Baumeister, *Nature (London)* **361,** 475 (1993).

[9] A. Guagliardi, L. Cerchia, S. Bartolucci, and M. Rossi, *Protein Sci.* **3,** 1436 (1994).

[10] J. Wiegel, L. G. Ljungdahl, and J. R. Rawson, *J. Bacteriol.* **139,** 800 (1979).

[11] Y. E. Lee, M. K. Jain, C. Y. Lee, S. E. Lowe, and J. G. Zeikus, *Int. J. Syst. Bacteriol.* **43,** 41 (1993).

[12] J. G. Zeikus, P. W. Hegge, and M. A. Anderson, *Arch. Microbiol.* **122,** 41 (1979).

*brockii*, and it is a strain of this species that forms the subject of the present chapter.

## Growth and Extraction of Cells

Several strains of *Thermoanaerobacter* species have been provided by H. Morgan (Thermophile Research Unit, University of Waikato, Hamilton, New Zealand). These have been isolated from hot spring environments at temperatures around 70° and neutral pH, and most have been classified as strains of *T. brockii* or of *T. thermohydrosulfuricus*. The principal strain used in these studies, which is referred to here simply as *T. brockii*, is code numbered Rt8.G4.

Cells are grown anaerobically between 68 and 70° in a Braun Biostat E fermentor, with pH controlled at pH 7.0, in a medium consisting of 0.2% (w/v) yeast extract, growth salts, 0.1% (w/v) cysteine hydrochloride, and glucose (20 g/liter) as fermentable substrate.[12] An anaerobic inoculum from a previous fermentation is used. From 10 liters of medium the cell yield is generally close to 100 g. Cells are harvested by centrifugation at 4000 *g* for 20 min at 4°, without precautions to maintain anaerobic conditions, and the cell paste stored frozen at −70°.

Cells are extracted by lysis. To each gram of cell paste, 5 ml of extraction buffer [30 m*M* $K_2HPO_4$ plus 0.1% (v/v) 2-mercaptoethanol] is added, together with lysozyme (0.2 mg/ml) and DNase (10 $\mu$g/ml). The suspension is stirred at room temperature for 1 hr, then centrifuged (20,000 *g* for 10 min at 4°) to remove cell debris. To the clear brown extract add enough fresh acetone solution of phenylmethylsulfonyl chloride to give a final concentration of 0.5 m*M*, and pepstatin to $10^{-7}$ *M*.

## Purification of Chaperonin 60

We have found that cpn60 is one of the most hydrophobic proteins in the extract, and substantial purification can be achieved using a hydrophobic adsorbent at the first step. The best adsorbent ($C_6$) is prepared by reacting epichlorohydrin-activated Sepharose CL-4B with hexane thiol overnight (*caution*—stench!), and the resulting adsorbent is used. It could be cleaned after use with a brief wash with 0.1 *M* NaOH. Approximately 1 ml of adsorbent is required to process the extract from each gram of cell paste. Although this can be done batchwise, it is more convenient to process the extract in a column. The purification procedure described here is a minor modification of the one already reported.[13]

[13] K. N. Truscott, P. B. Høj, and R. K. Scopes, *Eur. J. Biochem.* **222,** 277 (1994).

Extract (100 ml) is made to 0.5 *M* in $Na_2SO_4$, and passed through a 5 cm × 5 $cm^2$ $C_6$ column at 100 ml/hr. The column is preequilibrated with 20 m*M* Tris-HCl containing 0.5 *M* $Na_2SO_4$, and it is washed with a further 50 ml of this buffer after applying the extract. The column is then washed with another 50 ml of Tris-HCl buffer without $Na_2SO_4$. Chaperonin 60 remains on the column, and is eluted by reducing the hydrophobic interactions with organic solvent. Fifty milliliters of Tris-HCl buffer containing 2-propanol (20%, v/v) is used, and the eluted fraction collected. This is generally between 50 and 80% pure cpn60 as judged by sodium dodecyl sulfate-polyacrylamide gel electrophoresis (SDS–PAGE).

Further purification is by anion-exchange chromatography. The sample as eluted from the $C_6$ column can be applied directly to an anion-exchange column. We use a 2 $cm^2$ × 10 cm column of Q-Sepharose (Pharmacia, Piscataway, NJ) preequilibrated in 20 m*M* Tris-HCl buffer. After applying the sample, an elution scheme is commenced, which involves first a stepwise change to 0.18 *M* NaCl, followed by a linear gradient to 0.4 *M* NaCl. cpn60 elutes at 0.22 to 0.3 *M* NaCl. Monitoring at 280 nm is not entirely indicative, because the cpn60 has a low absorbance at this wavelength, and other ultraviolet (UV)-absorbing materials elute also. It is beneficial to confirm the major protein peak with a direct dye-binding assay of the fractions. Most experimental studies have been carried out on the major protein fraction from the Q-Sepharose column; the yield is 40 mg. Minor impurities still present can be removed by gel filtration (Superdex 200, Pharmacia), which also removes any monomer–dimer that can form on long-term storage of cell paste.

## Purification of Chaperonin 10

### *Purification Using Hydrophobic and Anion-Exchange Chromatography*

The flowthrough from the $C_6$ column, plus the washes both with and without $Na_2SO_4$, contain the bulk of the cpn10, as well as most of the other extracted proteins. Whereas cpn60 is one of the most hydrophobic proteins, cpn10 appears to be one of the least hydrophobic, and passes through many columns of greater hydrophobicity than $C_6$. We have found that one particular hydrophobic (thiophilic) adsorbent, consisting of mercaptobenzimidazole (Aldrich, Milwaukee, WI) linked to divinyl sulfone (DVS)-activated agarose,[13] is the most efficient in binding unwanted proteins, while allowing the cpn10 through.

Sufficient $Na_2SO_4$ is added to the combined wash fractions from the $C_6$ column to take its concentration to 0.5 *M*. In every 100 ml, 20 g of $(NH_4)_2SO_4$ is dissolved, and the precipitate removed. A further 6.4 g of $(NH_4)_2SO_4$

per 100 ml is dissolved, the precipitate collected, and dissolved in Tris-HCl buffer containing 0.5 *M* $Na_2SO_4$. This solution, at 10 mg of protein per milliliter, is passed through a 3 cm × 3 $cm^2$ column of DVS-mercaptobenzimidazole adsorbent, and washed with 20 ml of buffer. The nonadsorbed material is reprecipitated by dissolving 5 g of $(NH_4)_2SO_4$ per 10 ml. The precipitate is collected, redissolved in a small amount of Tris-HCl buffer, and dialyzed against 500 ml of Tris-HCl.

The dialyzed fraction is applied to a 10-ml Q-Sepharose column in Tris-HCl buffer, and elution is with a gradient to 0.5 *M* NaCl. Chaperonin 10 emerges in two main peaks at approximately 0.25 and 0.4 *M*, with more (as analyzed by SDS–PAGE) at an intermediate value (Fig. 1). It has been subsequently found that the 0.4 *M* peak is the most genuine cpn10, a 7-mer of molecular mass 70 kDa, whereas the other fractions contain aggregated cpn10 of higher molecular weight. The yield of cpn10 in the 0.4 *M* peak is 2 mg.

### *Piggy-Back Affinity Chromatography*

Because *E. coli* GroE proteins interact in the presence of ATP, but separate in its absence, it is possible to purify cpn10 by "piggy-back" affinity chromatography on the cpn60 adsorbed to the $C_6$ column. This is not satisfactory in the presence of 0.5 *M* $Na_2SO_4$, but in the absence of $Na_2SO_4$ the $C_6$ column still binds cpn60, although with a lower dynamic loading capacity. The 25-ml column used above is preequilibrated with Tris-HCl buffer with no $Na_2SO_4$, but with 1 m*M* ATP and 2 m*M* $MgCl_2$. Fifty

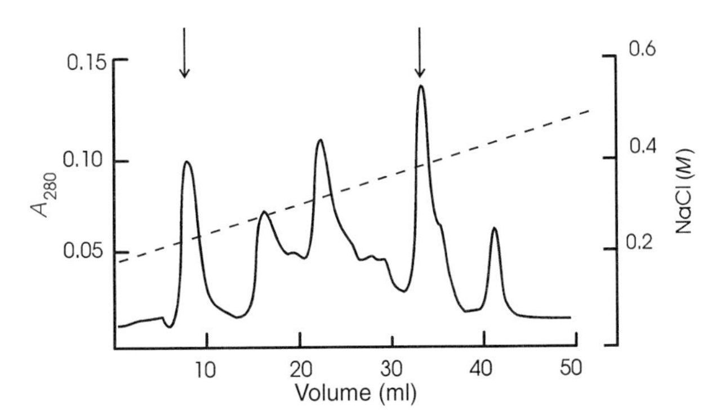


FIG. 1. Elution of *T. brockii* chaperonin 10 from a Q-Sepharose column. The sample that did not absorb to the mercaptobenzimidazole column, after concentration by ammonium sulfate precipitation and dialysis, was applied in a Tris buffer, and eluted with an NaCl gradient (dashed line). An aggregated form, which still possessed activity, was eluted in the first peak arrowed. The second arrow indicates the main 7-mer active component.

milliliters of extract to which 1 m$M$ ATP and 2 m$M$ $MgCl_2$ have been added is passed through the column. Most of the cpn60 binds, with at least some of the cpn10 attached to it. The column is washed with 50 ml of starting buffer, followed by 50 ml of Tris-HCl buffer without any additives. The latter wash fraction is concentrated by ultrafiltration, and contains cpn10. However, the yield is low, on the order of 200 $\mu$g from 50 ml of extract. The fraction is judged to be about 50% pure.

### *GroEL Affinity Chromatography*

A GroEL affinity column[14] is prepared by reacting recombinantly expressed *E. coli* GroEL (180 mg) with 15 ml of *N*-hydroxysuccinimide-activated agarose (Affi-Gel 15; Bio-Rad, Hercules, CA) in a 0.1 $M$ morpholinepropanesulfonic acid (MOPS) buffer, pH 7.5. GroEL is isolated from a strain of *E. coli* containing heat-inducible expression plasmid with the genes for both GroEL and GroES.[15]

The absorbent is equilibrated with a 25 m$M$ Tris-HCl buffer, pH 8, containing 7 m$M$ $MgCl_2$, 10 m$M$ KCl, and 1 m$M$ ATP. Eighty milliliters of a *T. brockii* extract is adjusted to pH 8 and $MgCl_2$, KCl, and ATP added to the same concentrations. This is passed through the adsorbent at 1 ml/min, and the column is then washed with the same buffer containing 1 $M$ NaCl. The column is then eluted with 25 m$M$ Tris-HCl, pH 8.0, and fractions analyzed by SDS–PAGE. Approximately 1 mg of cpn10 is obtained from 800 mg of protein in the original extract, and this is estimated to be at least 90% pure as judged by SDS–PAGE.

## Properties of *Thermoanaerobacter brockii* Chaperonins

The cpn60 of *T. brockii* differs from most other prokaryotic cpn60s in that as isolated, in the absence of ATP, it is a 7-mer, presumably a single ring of seven subunits. Of the other described cpn60s, that of *Neisseria gonorrhoeae*,[16] and of eukaryotic mitochondria,[17] are reported to be 7-mers, but others consist of double-ring 14-mers. Gel filtration (at 60°) of freshly prepared extracts of *T. brockii* indicated that the cpn60, as detected by SDS–PAGE, eluted at 400–450 kDa, not at around 900 kDa as would be

[14] G. N. Chandrasekhar, K. Tilly, C. Woolford, R. Hendrix, and C. Georgopoulos, *J. Biol. Chem.* **261,** 12414 (1986).

[15] D. J. Hartman, B. P. Surin, N. E. Dixon, N. J. Hoogenraad, and P. B. Høj, *Proc. Natl. Acad. Sci. U.S.A.* **90,** 2276 (1993).

[16] Y. Pannekoek, J. P. M. van Putten, and J. Dankert, *J. Bacteriol.* **174,** 6928 (1992).

[17] P. V. Viitanen, G. H. Lorimer, R. Seetharam, R. S. Gupta, J. Oppenheim, J. O. Thomas, and N. J. Cowan, *J. Biol. Chem.* **267,** 695 (1992).

expected if it existed as a 14-mer, or as a 14-mer in conjunction with cpn10. But if these procedures were carried out in the presence of MgATP in the buffer used for equilibration of the column, the molecular weight corresponded with that of a 14-mer. Native gradient electrophoresis also demonstrated that the cpn60 was a 7-mer if run under normal conditions, but a 14-mer if the gel was preelectrophoresed to introduce $MgATP^{2-}$ or $MgADP^{-}$ ions through the gel (Fig. 2).

The subunit size of cpn60 is 57,949 ± 10 Da (determined by electrospray mass spectrometry), close to that of GroEL, and there is substantial sequence homology with other cpn60s in the first 40 amino acids. UV absorption studies have demonstrated the lack of tryptophan residues; the absorption coefficient for a 1 mg/ml solution is low at 0.31. ATPase activity was measured at 50°; the value of $k_{cat}$ of 0.5 $sec^{-1}$ per 14-mer was in the range expected in comparison with GroEL ATPase. Some preparations have included significant amounts of a monomer or dimer form that has no ATPase activity, and that is inactive in protein-folding experiments (see below). This can be separated from the 7-mer by gel filtration. We have not been able to reconstitute the 7-mer from the monomer–dimer, but the 7-mer reassociates to form 14-mer complexes in the presence of ATP. Of the described chaperonin complexes, some have remained in the 21-mer complex on isolation (*Thermus thermophilus*), some have dissociated into 14-mer cpn60 and 7-mer cpn10 (e.g., *E. coli*), and the *T. brockii* cpn60

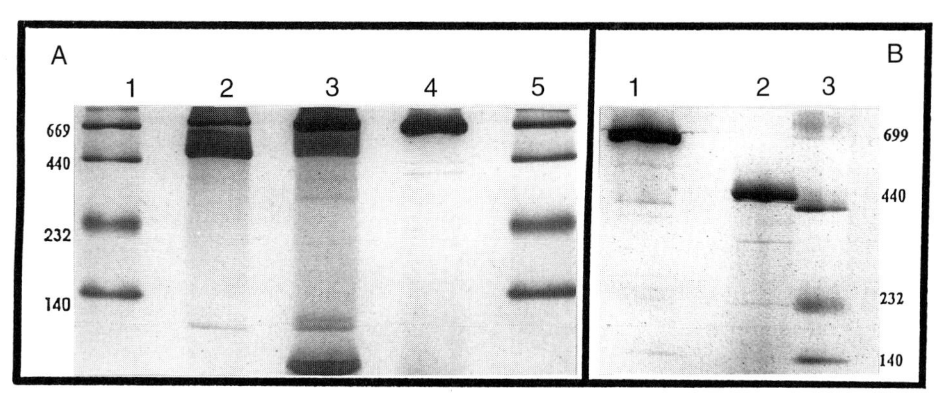


FIG. 2. Native gradient gel electrophoresis of *T. brockii* chaperonin 60. (A) The gel was preelectrophoresed with a buffer containing 7 m*M* $MgCl_2$ and 0.5 m*M* ATP for 1 hr. Lanes 1 and 5, native molecular weight standards; lane 2, cpn60; lane 3, cpn60 plus cpn10; lane 4, *E. coli* GroEL. (B) Gel run in the absence of ATP. Lane 1, *E. coli* GroEL; lane 2, *T. brockii* cpn60; lane 3, molecular weight standards. In the presence of ATP, a significant proportion of the cpn60 ran level with the GroEL protein, indicating 14-mer formation. In the presence of cpn10, this proportion increased.

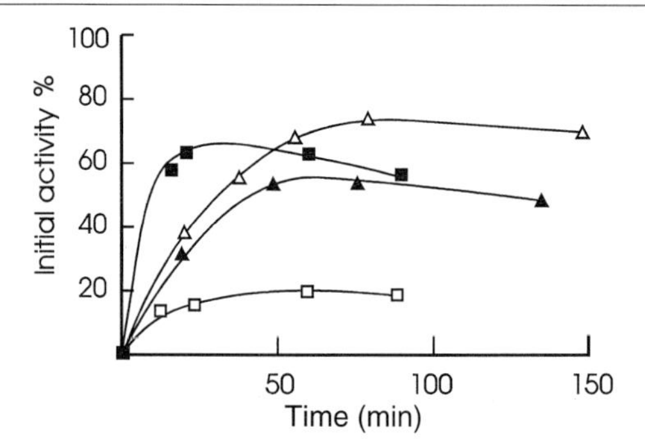


FIG. 3. Refolding of pig heart mitochondrial malate dehydrogenase using combinations of *T. brockii* chaperonins and *E. coli* GroE proteins. Malate dehydrogenase was denatured with guanidine hydrochloride, and diluted into the reaction mix; for details see Ref. 15. (■) GroEL plus GroES; (△) cpn60 plus GroES; (▲) cpn60 plus cpn10; (□) GroEL plus cpn10.

has further dissociated into 7-mers. In all cases, the 21-mer cpn60–cpn10 complex reforms on the addition of ATP.

*Thermoanaerobacter brockii* cpn10 is a 7-mer in its freshly purified form, and although there can be aggregation into high molecular weight complexes, this aggregated form is still active in protein folding, and presumably disaggregation occurs to allow the 7-mer to interact with 14-mer cpn60. The aggregation occurs mainly with the cpn10 prepared by method 1, and may be an artifact of the high salt concentrations used in the isolation. Its subunit molecular mass was determined to be 10,254 ± 0.4 Da by mass spectrometry. As with cpn60, the cpn10 protein contains no tryptophan and has a low absorption coefficient, 0.29 for a 1-mg/ml solution.

The optimum conditions for storage of the *T. brockii* chaperonin 60 is as an ammonium sulfate suspension at 4–25°. For chaperonin 10, however, it is preferable to store at −70° in a neutral buffer in the absence of salts.

## Protein Folding

The complex of *T. brockii* cpn60 and cpn10, together with ATP, is capable of reactivating denatured enzymes in much the same fashion as the *E. coli* GroE proteins and the *Thermus* chaperonins. At 40°, mitochondrial malate dehydrogenase could be renatured; it was found to be possible to carry out this process with any combination of GroEL or cpn60, with GroES or cpn10[13] (Fig. 3). At higher temperatures, however, the GroE proteins themselves denatured, so only the *T. brockii* chaperonins could be used. Secondary alcohol dehydrogenase[18] isolated from *T. brockii* was

[18] Y. Nagata, K. Maeda, and R. K. Scopes, *Bioseparation* **2,** 353 (1992).

reactivated at 65° to 80% of the original value after guanidine hydrochloride denaturation using cpn60, cpn10, and ATP.[13]

*Thermoanaerobacter brockii* cpn60, either in the 7-mer or the 14-mer form, was able to capture denatured malate dehydrogenase and preserve it for subsequent reactivation on adding cpn10 and ATP. Active enzyme has been refolded after a heat treatment of the cpn60–malate dehydrogenase complex at 60°.

## Acknowledgments

We thank Dr. P. B. Høj for suggestions and encouragement, and Mr. M. Ryan for providing the GroEL affinity column.

# [15] Chaperonin from Thermophile *Thermus thermophilus*

*By* HIDEKI TAGUCHI and MASASUKE YOSHIDA

## Introduction

Chaperonin, which is one of the well-studied molecular chaperones, is a ubiquitous, abundant, and indispensable protein that is considered to be involved in protein folding in the cell.[1-7] Although many studies on the chaperonin function have been carried out by using *Escherichia coli* chaperonins GroEL and GroES, we have developed a unique chaperonin system using the chaperonin from one of the thermophilic eubacteria, *Thermus thermophilus*.[8-10] Although amino acid sequences of the chaperonin from *T. thermophilus* (*Thermus* chaperonin) have a significant similarity to those of GroEL and GroES, there are some distinct features in the *Thermus* chaperonin. (1) Unlike GroEL and GroES, *Thermus* chaperonin consisting of homologs to GroEL (cpn60) and GroES (cpn10) is copurified as a large

[1] S. M. Hemmingsen, C. Woolford, S. M. van der Vies, K. Tilly, D. T. Dennis, C. P. Georgopoulos, R. W. Hendrix, and R. J. Ellis, *Nature (London)* **333,** 330 (1988).
[2] R. J. Ellis and S. M. van der Vies, *Annu. Rev. Biochem.* **60,** 321 (1991).
[3] M.-J. Gething and J. Sambrook, *Nature (London)* **355,** 33 (1992).
[4] C. Georgopoulos and W. J. Welch, *Annu. Rev. Cell Biol.* **9,** 601 (1993).
[5] A. L. Horwich and K. R. Willison, *Phil. Trans. R. Soc. Lond. B* **339,** 313 (1993).
[6] J. P. Hendrick and F. U. Hartl, *Annu. Rev. Biochem.* **62,** 349 (1993).
[7] A. A. Gatenby and P. V. Viitanen, *Annu. Rev. Plant Physiol. Plant Mol. Biol.* **45,** 469 (1994).
[8] H. Taguchi, J. Konishi, N. Ishii, and M. Yoshida, *J. Biol. Chem.* **266,** 22411 (1991).
[9] H. Taguchi and M. Yoshida, *J. Biol. Chem.* **268,** 5371 (1993).
[10] H. Taguchi, Y. Makino, and M. Yoshida, *J. Biol. Chem.* **269,** 8529 (1994).



0076-6879/98 $25.00

complex containing endogenous ADP (holo-cpn)[11–13]; (2) because *Thermus* chaperonin is thermostable up to around 80°, it provides a suitable assay system for studying the chaperonin in protein folding at high temperature and in heat denaturation of relatively heat-labile enzymes.[8,9] This chapter describes the methods for manipulating the *Thermus* chaperonin.

## Purification of Chaperonin from *Thermus thermophilus*

### *Detection of Thermus Chaperonin*

Because chaperonin 60's from various sources have common features, such as their abundance in cells, their subunit molecular mass (around 60 kDa), their existence as large tetradecameric complexes, and their weak ATPase activities,[14–16] we have isolated a chaperonin from *T. thermophilus* by searching for a protein complex with the preceding features.[8] In fact, an analysis of a crude extract from *T. thermophilus* using polyacrylamide gel electrophoresis in the presence of sodium dodecyl sulfate (SDS–PAGE) shows the existence of an abundant protein with a molecular mass of 58 kDa in *T. thermophilus*. Staining of a polyacrylamide gel under nondenaturing conditions (native PAGE) and subsequent two-dimensional PAGE show that this protein exists as large complex with weak ATPase activity. Therefore, throughout the purification steps described as follows, chaperonin from *T. thermophilus* is monitored by SDS–PAGE or native PAGE or both. We recommend the use of ATPase staining of the native gels or immunoblotting by some anti-chaperonin antibody (e.g., anti-GroEL antibody) to detect the chaperonin precisely for the initial purification.

### *General Procedures*

Protein concentrations are determined by the method of Bradford [Bio-Rad (Hercules, CA) protein assay] with bovine serum albumin (BSA) as the standard.[17] Spectrophotomeric determination at 280 nm, which is routinely used in the GroE system,[18] is not appropriate because endoge-

[11] N. Ishii, H. Taguchi, M. Sumi, and M. Yoshida, *FEBS Lett.* **299,** 169 (1992).

[12] M. Yoshida, N. Ishii, E. Muneyuki, and H. Taguchi, *Phil. Trans. R. Soc. Lond. B* **339,** 305 (1993).

[13] N. Ishii, H. Taguchi, H. Sasabe, and M. Yoshida, *J. Mol. Biol.* **236,** 691 (1994).

[14] A. Ishihama, T. Ikeuchi, A. Matsumoto, and S. Yamamoto, *J. Biochem.* **79,** 927 (1976).

[15] T. Hohn, B. Hohn, A. Engel, M. Wurtz, and P. R. Smith, *J. Mol. Biol.* **129,** 359 (1979).

[16] R. W. Hendrix, *J. Mol. Biol.* **129,** 375 (1979).

[17] M. M. Bradford, *Anal. Biochem.* **72,** 248 (1976).

[18] P. V. Viitanen, T. H. Lubben, J. Reed, P. Goloubinoff, D. P. O'Keefe, and G. H. Lorimer, *Biochemistry* **29,** 5665 (1990).

nously bound ADP in the *Thermus* chaperonin[12] affects the profile of the ultraviolet (UV) absorption spectrum. Proteins are analyzed by polyacrylamide gel electrophoresis either on 15% (w/v) polyacrylamide gels in the presence of SDS (SDS–PAGE) or on 7.5% (w/v) polyacrylamide gels without SDS (native PAGE).[19] It is not necessary to treat the protein solution containing *Thermus* chaperonin with SH reagents such as 2-mercaptoethanol because *Thermus* chaperonin does not contain any cysteine in the deduced amino acid sequences.[20] Gels are stained routinely with Coomassie Brilliant Blue R-250 (Sigma, St. Louis, MO). If necessary, ATPase activity is located by incubating the native polyacrylamide gel in a solution containing 5 m*M* MgATP, 0.02% (w/v) lead acetate, and 25 m*M* Tris-HCl, pH 7.5, at 60° until white bands appear. The stained gel is then treated with ammonium sulfide solution, and washed with distilled water.

*Step 1: Preparation of cytosol. Thermus thermophilus* strain HB8 (ATCC 27634)[21] is relatively easy to cultivate. A wide range of growth media used for *E. coli* are suitable for growth of *T. thermophilus*. For routine growth of cultures we usually use 2×YT medium. To avoid contamination by other thermophiles, temperature control is important. The culture is grown at 75–80° with vigorous aeration. Cells are harvested at the late log phase and stored at −80° until use.

Frozen cells (100 g, wet weight) are thawed in warm (30–40°) water, suspended in 300 ml of buffer A [25 m*M* Tris-HCl (pH 7.5), 5 m*M* $MgCl_2$] containing a trace amount of DNase I (Boehringer Mannheim, Indianapolis, IN) and disrupted by French press at 1200 kg/cm$^2$ in a cold room. The lysate is centrifuged at 40,000 rpm for 60 min at 4°. The supernatant thus obtained contains typically 10 g of soluble protein.

*Step 2: DEAE-Cellulose Chromatography.* The supernatant fraction from step 1 is passed through a column (7.4 × 24 cm) of DE52 cellulose (Whatman, Clifton, NJ) equilibrated with buffer A at room temperature. The column is washed with 1 liter of buffer A and 1 liter of buffer A containing 0.03 *M* NaCl, and *Thermus* chaperonin is then eluted with 3 liters of a linear gradient of 0.03 to 0.15 *M* NaCl in buffer A. The elution is done at about 5 ml/min, and 20-ml fractions are collected. Each fraction (10 $\mu$l) is analyzed by SDS–PAGE to detect *Thermus* chaperonin. Typically, *Thermus* chaperonin is eluted at around 0.12 *M* NaCl. The fractions containing *Thermus* chaperonin are pooled, and the protein is precipitated

[19] U. K. Laemmli, *Nature (London)* **227,** 680 (1970).

[20] K. Amada, M. Yohda, M. Odaka, I. Endo, N. Ishii, H. Taguchi, and M. Yoshida, *J. Biochem.* **118,** 347 (1995).

[21] T. Oshima and K. Imahori, *Int. J. Syst. Bacteriol.* **24,** 102 (1974).

with the addition of solid ammonium sulfate to 65% (w/v) saturation (0.43 g of ammonium sulfate/ml).

*Step 3: Sepharose CL-4B Chromatography.* The precipitate from step 2 is pelleted at 8000 rpm for 20 min at 4°, dissolved in a minimum volume of buffer A containing 100 m*M* $Na_2SO_4$. An aliquot of the concentrated solution (up to 50 ml) is applied to a column (4 × 110 cm) of Sepharose CL-4B (Pharmacia, Piscataway, NJ) at room temperature, which has been equilibrated with buffer A containing 100 m*M* $Na_2SO_4$. The column is eluted at about 3.5 ml/min, and 10-ml fractions are collected. Fractions eluted slightly after the void volume (approximately 700 ml) are pooled and analyzed by SDS–PAGE. Fractions containing *Thermus* chaperonin are collected and stored as a suspension in 65% (w/v) saturated ammonium sulfate solution at 4°.

### *Comments on Purification*

Fractions containing *Thermus* chaperonin along with some contaminating proteins from the Sepharose CL-4B column can be rechromatographed using the same gel-filtration column. An alternative method of purification is hydrophobic interaction chromatography using Butyl-Toyopearl (Tosoh, Tokyo, Japan). Separation using Butyl-Toyopearl is effective, but the yield is poor, probably owing to the tendency of *Thermus* chaperonin to stick to the resin.

We routinely store purified *Thermus* chaperonin as a suspension in ammonium sulfate. The chaperonin can also be stored at −80° after adding glycerol to a concentration of about 10% (v/v). Note that lyophilization of *Thermus* chaperonin after replacement of buffer by distilled water leads to the partial dissociation of the chaperonin 60 oligomer to the monomeric form.

## *In Vitro* Folding Assay Using Purified *Thermus* Chaperonin

An *in vitro* assay system of chaperonin-dependent protein folding for the *E. coli* chaperonin (GroEL and GroES)[18,22] was modified for the *in vitro* folding assay system for *Thermus* chaperonin.[8] Unlike the folding assay using the GroE chaperonin, use of the *Thermus* chaperonin enables us to examine the folding assay at high temperature, up to approximately 75°.

Basically, the *in vitro* folding assay is composed of three steps.

*Step 1: Denaturation of Substrate Protein.* The substrate protein is denatured, usually with a high concentration of denaturants such as guanidine

[22] P. Goloubinoff, J. T. Christeller, A. A. Gatenby, and G. H. Lorimer, *Nature (London)* **342,** 884 (1989).

hydrochloride or urea. In the case of proteins containing cysteine or cystine residues, appropriate reducing agents such as dithiothreitol should be added.

*Step 2: Dilution of Denatured Protein into Buffer Containing Chaperonin.* Folding is initiated by diluting the denatured protein solution 20- to 100-fold into a dilution buffer containing various additions. Typical conditions for the dilution buffer are as follows: (1) Spontaneous folding—the dilution buffer contains nothing. As a substitute for chaperonin, proteins such as bovine serum albumin or ribonuclease A may be added; (2) chaperonin-dependent arrest of folding—under conditions in which the substrate protein can refold spontaneously, chaperonin in the dilution buffer can arrest the folding by capturing a folding intermediate; (3) chaperonin- and MgATP-dependent folding—dilution buffers in the presence of both chaperonin and MgATP could lead to the productive folding of substrate proteins. Addition of MgATP after the arrest of folding by chaperonin without the nucleotide is also effective.

*Step 3: Estimation of Folding Yield by Measuring Enzymatic Activity of Substrate Protein.* An aliquot of the dilution buffer is withdrawn at the desired times, and the enzyme activity recovered after the dilution is then measured by specific methods.

### *Search for Optimum Conditions*

Although the conditions for the folding assay depend on the features of each protein of interest, several parameters should be considered for the chaperonin-dependent folding assay.

*Temperature and Protein Concentration.* According to our experiments, spontaneous folding depends strongly on both temperature and protein concentration. This dependency can be interpreted on the basis of the principle that kinetic competition between correct folding and the nonproductive aggregation process occurs during protein folding.[23,24] The rate of the latter aggregation process is accelerated when the concentration of the folding intermediate increases or when the temperature is high. For example, isopropylmalate dehydrogenase (IPMDH) from *T. thermophilus* (described in procedures 1 and 2) cannot refold spontaneously above 60°, although native IPMDH is stable at that temperature.[8] Therefore, to examine the arrest of spontaneous folding by chaperonin, the experiment using IPMDH would have to be done below 60°.

*Chaperonin Concentration.* Chaperonin interacts with the substrate protein stoichiometrically but not catalytically.[7] Therefore, the molar ratio of

[23] A. Mitraki and J. King, *Bio/Technology* **7,** 690 (1989).

[24] T. Kiefhaber, R. Rudolph, H.-H. Kohler, and J. Buchner, *Bio/Technology* **9,** 825 (1991).

chaperonin to the substrate protein would have to be greater than 1:1, when the molecular masses of chaperonin and the substrate protein are calculated as the oligomer and monomer, respectively. Practically, the molar ratio of *Thermus* chaperonin to IPMDH monomer from *T. thermophilus* is estimated to be about 4:1 to achieve optimum chaperonin-dependent folding at 68°.[8] Note that an excess amount of chaperonin causes an inhibition of chaperonin-dependent folding.[8,22,25,26]

*Optimum Temperature Ranges for Chaperonin.* An advantage of using *Thermus* chaperonin is that the chaperonin can promote folding over wide temperature ranges[8]—from 30 to 75°.[8] Although ATPase activity of *Thermus* chaperonin is greatest at 80°, optimum temperature for the chaperonin-dependent folding depends on the substrate protein. Even if the chaperonin could bind and then release the substrate protein in an ATP-dependent manner, the released substrate protein might not be stable at that temperature.

*Procedure 1: Thermus Chaperonin-Dependent Folding of IPMDH from Thermus thermophilus at Various Temperatures*

To date, we have used several enzymes as substrate proteins for *Thermus* chaperonin-dependent folding: isopropylmalate dehydrogenase and isocitrate dehydrogenase from *T. thermophilus,* lactate dehydrogenase and glucose-6-phosphate dehydrogenase from *Bacillus stearothermophilus,*[8] and bovine mitochondrial rhodanese.[9,10] Here we describe a folding assay system using IPMDH (a dimer of 37-kDa subunits) from *T. thermophilus.* This enzyme is stable up to 80°. However, as described above, no or very little spontaneous folding occurs above 60°.

*Materials and Reagents*

*Thermus* chaperonin: 5–10 mg/ml in 50 m*M* potassium phosphate, pH 7.8
Isopropylmalate dehydrogenase (IPMDH) from *T. thermophilus*[27]: 1.5 mg/ml in 50 m*M* potassium phosphate, pH 7.8
Guanidine hydrochloride solution: 8.0 *M* in 20 m*M* potassium phosphate, pH 7.8
MgATP solution: 50 m*M* (equimolar Mg and ATP), adjusted to pH 7 with NaOH
Dilution buffer: 50 m*M* potassium phosphate, pH 7.8

[25] J. A. Mendoza, G. H. Lorimer, and P. M. Horowitz, *J. Biol. Chem.* **266,** 16973 (1991).
[26] W. Zhi, S. J. Landry, L. M. Gierasch, and P. A. Srere, *Protein Sci.* **1,** 522 (1992).
[27] T. Yamada, N. Akutsu, K. Miyazaki, K. Kakinuma, M. Yoshida, and T. Oshima, *J. Biochem.* **108,** 449 (1990).

IPMDH assay mixture: 0.4 m*M* (*2R*,3S**)-3-isopropylmalic acid (Wako Pure Chemical, Osaka, Japan), 0.8 m*M* $NAD^+$ (Boehringer Mannheim), 100 m*M* potassium phosphate (pH 7.8), 1.0 *M* KCl, and 10 m*M* $MgCl_2$

*Method*

1. Denaturation: IPMDH (20 μl) is mixed with 180 μl of 8 *M* guanidine hydrochloride solution. The denatured IPMDH (7.2 *M* guanidine hydrochloride) is incubated for at least 30 min at room temperature.

2. Dilution of the denatured IPMDH: The denatured IPMDH (2.5 μl) is diluted into 97.5 μl of dilution buffer containing 40 μg of *Thermus* chaperonin with or without 0.5 m*M* MgATP, preincubated at the desired temperature (18–80°). To examine the temperature dependence of spontaneous folding of IPMDH, dilution buffers without the chaperonin are used. As a control, native IPMDH is treated with the same procedures, except that guanidine hydrochloride is contained not in the enzyme solution but in the dilution buffer.

3. Assay of enzyme activities: After 10 min of incubation at constant temperature, recovered IPMDH activities are measured. The reactions are started by addition of 50 μl of the dilution buffer prepared as described above into 1.2 ml of assay mixture preincubated in a cuvette at 68°. The initial rate of increase of the absorbance at 340 nm (generated NADH) is measured using a spectrophotometer. An activity of the same amount of native IPMDH is taken as 100%.

*Procedure 2: Time Course of Chaperonin-Dependent Folding of Isopropylmalate Dehydrogenase Monitored with Continuous Assay System*

To monitor the chaperonin-dependent reactivation of IPMDH continually, we have developed a continuous folding assay.[8] In the continuous assay, dilution buffer contains components for the assay of IPMDH activity, as well as chaperonin. The denatured IPMDH is then added directly to the dilution buffer. The time course of the folding can be monitored in real time. By changing the temperature, we can control the mode of spontaneous folding of IPMDH. At 68°, a temperature at which the native IPMDH is stable but the spontaneous folding on dilution of guanidine hydrochloride fails, *Thermus* chaperonin induces productive folding. We can observe the arrest of IPMDH folding, however, by the chaperonin at 50°, at which a significant amount of the IPMDH can refold spontaneously.

*Materials and Reagents.* The same materials and reagents as described in procedure 1 are used.

*Method*

1. Denaturation: The same procedure as described in procedure 1 (step 1) is used.

2. Dilution of the denatured IPMDH: The dilution buffer (190 $\mu$l) containing both *Thermus* chaperonin (160 $\mu$g/ml) and the components for the assays is poured into a cuvette in a spectrophotometer, and preincubated at 68 or 50°. Ten microliters of the denatured IPMDH solution is then injected into the cuvette. Folding of IPMDH can be directly monitored by the rate of increasing absorbance at 340 nm.

## Protection from Irreversible Heat Denaturation by *Thermus* Chaperonin

It has been demonstrated that chaperonins from various sources are heat shock proteins.[28] To understand better the function of the chaperonins in heat-stressed cells, the effect of chaperonin on heat inactivation of proteins has been studied in *in vitro* assay systems.[9,29–32] As mentioned previously, chaperonin isolated from *T. thermophilus* is stable up to 80°. Taking advantage of this heat stability, we have developed a system to investigate the effect of chaperonin on heat denaturation of several relatively heat-labile enzymes.[9] The experiment to assess the effect consists of three steps.

*Step 1: First Incubation of Native Protein at Denaturing Temperatures.* In the absence of chaperonin, the enzyme heat denatures irreversibly, whereas the enzyme is protected from irreversible inactivation in the presence of chaperonin.

*Step 2: Second Incubation at Moderate Temperatures.* The enzyme solution is transferred to a temperature at which the enzyme is stable, with concomitant addition of MgATP. Chaperonin releases the captured enzyme, and the released enzyme can then refold.

*Step 3: Assay of Recovered Enzyme Activity.* An aliquot of the enzyme solution is withdrawn at desired times and the enzyme activity is measured. This activity is the sum of remaining activity after step 1 and recovered activity after step 2.

[28] C. Georgopoulos, K. Liberek, M. Zylicz, and D. Ang, *in* "The Biology of Heat Shock Proteins and Molecular Chaperones" (R. I. Morimoto, A. Tissieres, and C. Georgopoulos, eds.), p. 209. Cold Spring Harbor Laboratory Press, Cold Spring Harbor, New York, 1994.

[29] B. Höll-Neugebauer, R. Rudolph, M. Schmidt, and J. Buchner, *Biochemistry* **30,** 11609 (1991).

[30] J. A. Mendoza, G. H. Lorimer, and P. M. Horowitz, *J. Biol. Chem.* **267,** 17631 (1992).

[31] J. Martin, A. L. Horwich, and F.-U. Hartl, *Science* **258,** 995 (1992).

[32] D. J. Hartman, B. P. Surin, N. E. Dixon, N. J. Hoogenraad, and P. B. Høj, *Proc. Natl. Acad. Sci. U.S.A.* **90,** 2276 (1993).

*Procedure 3: Effect of Chaperonin on Heat Denaturation of Lactate Dehydrogenase from Bacillus stearothermophilus*

Lactate dehydrogenase (LDH) from *B. stearothermophilus* (a tetramer of 35-kDa subunits) is not stable above 60°[8] and is inactivated irreversibly after a 30-min incubation at 72°.[9] The presence of *Thermus* chaperonin protects the LDH from irreversible heat denaturation, because LDH activity can be recovered after the protein solution is transferred to 55° with concomitant addition of MgATP.[9] This result indicates that during heat denaturation chaperonin recognizes and binds to a denaturation intermediate, and thus the irreversible heat inactivation of LDH is protected.

*Materials and Reagents*

LDH from *B. stearothermophilus* (Seikagaku Corp., Tokyo, Japan): 0.1 mg/ml in 50 m*M* potassium phosphate, pH 7.8

MgATP solution: 50 m*M* (equimolar Mg and ATP), adjusted to pH 7 with NaOH

Dilution buffer: 50 m*M* potassium phosphate, pH 7.8, 5 m*M* dithiothreitol, and *Thermus* chaperonin (0.46 mg/ml)

LDH assay mixture: 100 m*M* potassium phosphate (pH 6.0), 10 m*M* sodium pyruvate, and 0.25 m*M* NADH (Boehringer Mannheim)

*Method*

1. Heat inactivation at 72°: Ten microliters of "native" LDH (1 $\mu$g) is diluted into 490 $\mu$l of the dilution buffer preincubated at 72°. After the desired time of incubation, an aliquot (50 $\mu$l) is removed and the remaining LDH activity is measured as described below.
2. Transfer to 55°: Following incubation at 72° for 30 min, the solution is moved to a water bath equilibrated at 55° and the recovered LDH activity is measured at the desired time. MgATP at a final concentration of 0.5 m*M* is added at the time of the temperature shift.
3. Assay of enzyme activities: The aliquot of enzyme solution withdrawn (50 $\mu$l) is added into 1.2 ml of assay mixture preincubated at 55°, and the initial rate of decrease of the absorbance at 340 nm is measured.

*Procedure 4: Heat Stability of Folding Intermediate Captured by Thermus Chaperonin*

Chaperonin binds folding intermediates or denaturation intermediates and the complex formed between chaperonin and the intermediates is stable to some extent in the absence of MgATP. It is therefore of interest to investigate the heat stability of the binary complex. The heat stability of proteins captured by *Thermus* chaperonin, assessed as the ability to resume productive folding under optimal conditions for the enzyme, is measured using chemically produced folding intermediate–chaperonin complexes.[9] This ability is lost at about 78° irrespective of the variable heat stability of individual enzymes. Once the folding intermediate is captured by chaperonin it is protected from irreversible denaturation until the chaperonin itself is collapsed by heat, because the denaturation temperature of *Thermus* chaperonin itself is 80–85° under *in vitro* conditions.

Here we describe the assay system using LDH from *B. stearothermophilus* as a substrate protein.

*Materials and Reagents*

Denatured LDH from *B. stearothermophilus:* 0.1 mg/ml in 20 m*M* potassium phosphate (pH 7.8), 6.4 *M* guanidine hydrochloride, and 5 m*M* dithiothreitol

MgATP solution: 50 m*M* (equimolar Mg and ATP), adjusted to pH 7 with NaOH

Dilution buffer: 50 m*M* potassium phosphate (pH 7.8), 5 m*M* dithiothreitol, and *Thermus* chaperonin (0.46 mg/ml)

LDH assay mixture: See procedure 3

*Method*

1. Complex formation between chaperonin and folding intermediate at 37°: Ten microliters of LDH (1 $\mu$g) is added to 490 $\mu$l of the dilution buffer preincubated at 37°. The diluted solution is incubated for 5 min at 37°.
2. Exposure to various high temperatures: The solution containing the complex is shifted to various high temperatures (55–85°), and incubated for 30 min.
3. Shift to a moderate temperature to permit folding with concomitant addition of MgATP: The solution is adjusted to 55° with concomitant addition of 0.5 m*M* MgATP (final concentration). After the desired time of incubation, an aliquot (50 $\mu$l) is removed and the recovered LDH activity is measured as described.
4. Enzyme activities are determined as described in procedure 3.

## Activity of Monomeric Chaperonin 60 as Molecular Chaperone

Chaperonin 60 (cpn60) is typically isolated as a unique tetradecameric form arranged as two stacked seven-member rings,[15,16,33] and this structure had been considered to be required for functioning as a molecular chaper-

[33] K. Braig, Z. Otwinowski, R. Hegde, D. C. Boisvert, A. Joachimiak, A. L. Horwich, and P. B. Sigler, *Nature (London)* **371,** 578 (1994).

one, i.e., promoting protein folding and protecting protein from irreversible heat denaturation. However, we discovered that the monomeric form of chaperonin 60 ($cpn60_m$) possesses the ability to interact with nonnative proteins, to suppress aggregation, and to promote protein folding.[10] In contrast to tetradecameric chaperonin 60, folding promoted by the chaperonin 60 monomer does not require ATP or chaperonin 10 (cpn10).

In addition to $cpn60_m$, we found that the proteolytic fragment of $cpn60_m$ (termed the 50-kDa fragment) is as effective as $cpn60_m$ in preventing aggregation and promoting protein folding.[10] The 50-kDa fragment that starts from Thr-79 in the whole sequence of *Thermus* cpn60 cannot assemble to the oligomeric structure even in the presence of MgATP and cpn10.

*Isolation of Chaperonin 60 Monomer and Proteolytic 50-kDa Fragment*

The purified *Thermus* chaperonins are denatured in 0.1% (v/v) trifluoroacetic acid, applied on a reversed-phase high-performance liquid chromatography (HPLC) column (Hi-Pore RP304, Bio-Rad, Richmond, CA), and eluted with a linear gradient of acetonitrile. The peak fractions containing cpn60 polypeptides and cpn10 polypeptides are dried *in vacuo*, solubilized with a buffer containing 6 *M* guanidine hydrochloride, and diluted 40-fold in 50 m*M* Tris-HCl buffer, pH 7.5, at 60°. The purified cpn60 and cpn10 spontaneously refold when guanidine hydrochloride is diluted. The refolded cpn60 exists as monomers and does not assemble into oligomers. The refolded cpn10 assembles into oligomers, probably heptamers.

The 50-kDa fragment of $cpn60_m$ is prepared as follows. Purified $cpn60_m$ (15 mg in 3 ml of 50 m*M* Tris-HCl buffer, pH 7.5) is treated with 2.8 units of immobilized TPCK–trypsin (Pierce, Rockford, IL) for 3 hr at 25°. Judging from gel electrophoresis, $cpn60_m$ is mostly converted into the 50-kDa fragment after this procedure, and the fragment is further purified with gel-permeation HPLC (Tosoh, $G3000SW_{XL}$).

*Procedure 5: Dependence of Reactivation of Rhodanese on Concentrations of Chaperonin 60 Monomer and 50-kDa Fragment*

Although chaperone activity of $cpn60_m$ has been detected, it should be noted that $cpn60_m$-promoted folding does not require ATP or chaperonin 10, is more rapid, and produces lower yield of reactivated enzymes. Therefore, to detect clearly the chaperone activity, folding experiments should be done under "nonpermissive"[34] conditions in which spontaneous folding cannot occur. In the case of rhodanese, a nonpermissive condition is set

[34] M. Schmidt, J. Buchner, M. J. Todd, G. H. Lorimer, and P. V. Viitanen, *J. Biol. Chem.* **269,** 10304 (1994).

by increasing the denatured rhodanese concentration (final, ~1.5 $\mu M$) at 37°.[10,35]

*Materials and Reagents*

$cpn60_m$ and the 50-kDa fragment: 10 mg/ml in 50 m*M* potassium phosphate, pH 7.8

Bovine mitochondrial rhodanese (thiosulfate sulfurtransferase; EC 2.8.1.1; type II, Sigma): 5 mg/ml in 50 m*M* potassium phosphate, pH 7.8, containing 10% (v/v) glycerol

Guanidine hydrochloride solution: 8.0 *M* in 20 m*M* potassium phosphate, pH 7.8, containing 6.7 m*M* dithiothreitol

Dilution buffer: 50 m*M* potassium phosphate, pH 7.8, containing 5 m*M* dithiothreitol and 50 m*M* $Na_2S_2O_3$

Rhodanese assay mixture: 62.5 m*M* $Na_2S_2O_3$, 50 m*M* potassium phosphate, and 62.5 m*M* KCN

Formaldehyde (38%, v/v)

$Fe^{3+}$ solution: 50 g of $Fe(NO_3)_3 \cdot 9H_2O$, 100 ml of $HNO_3$ (specific gravity, 1.40) in 500 ml distilled water

*Method*

1. Denaturation: Ten microliters of rhodanese is mixed with 30 $\mu$l of 8 *M* guanidine hydrochloride solution containing 6.7 m*M* dithiothreitol. The denatured rhodanese (6 *M* guanidine hydrochloride) is incubated for at least 30 min at room temperature.

2. Dilution of the denatured rhodanese: One microliter of the denatured rhodanese (1.25 $\mu$g) is diluted into 24 $\mu$l of dilution buffer containing various amounts of $cpn60_m$ or the 50-kDa fragment (12.5–125 $\mu$g) preincubated at 37°.

3. Assay of enzyme activities[36]: After 30 min of incubation at 37°, recovered rhodanese activities are measured. The reactions are started by addition of 25 $\mu$l of the dilution buffer into the 1 ml of assay mixture preincubated at 20°. The reaction is terminated by adding 200 $\mu$l of formaldehyde (38%, v/v) after 15 min. One milliliter of $Fe^{3+}$ solution is then added to the solution, and the absorbance at 460 nm is measured using a spectrophotometer. An activity of the same amount of native rhodanese is taken as 100%.

[35] J. A. Mendoza, E. Rogers, G. H. Lorimer, and P. M. Horowitz, *J. Biol. Chem.* **266,** 13587 (1991).

[36] B. H. Sörbo, *Acta Chem. Scand.* **7,** 1129 (1953).

# [16] Insect Chaperonin 60: Symbionin

*By* MIZUE MORIOKA and HAJIME ISHIKAWA

## Introduction

The study of insect chaperonins (Cpn) has been advanced by using the intracellular symbiotic bacterium (endosymbiont) of aphids (Homoptera; Insecta). Endosymbiont, when harbored by the aphid bacteriocyte, a specialized fat body cell, selectively synthesizes only one protein, symbionin.[1] Symbionin has been purified to homogeneity and its gene has already been cloned.[2,3] Nucleotide sequence analysis of the symbionin region of the endosymbiont genome has revealed that it contains the two-cistronic operon structure, which is referred to as the *sym* operon.[3] The symbionin gene, designated *symL,* encodes a polypeptide of 548 amino acids that belongs to the Cpn60 family. An additional coding region, designated *symS,* for a polypeptide of 96 amino acids that belongs to the Cpn10 family, is found immediately upstream from *symL.* Expression of *symS* is somehow repressed in the aphid endosymbiont. According to the nucleotide sequence, *symL* is 85.8% identical with *groEL* at the amino acid sequence level. The oligomeric structure of symbionin is also similar to that of GroEL.[3] As expected from the structural similarity of symbionin to GroEL, symbionin is functional as a molecular chaperone *in vivo* and *in vitro.*[3,4]

Further study of the biochemical properties of symbionin demonstrated that it is autocatalytically phosphorylated *in vitro* in response to temperature increases.[5] This phosphorylated form of symbionin contains the high-energy phosphate bond that is transferable to other components such as ADP to produce ATP.[6] Also, symbionin catalyzes the phosphoryl group transfer from ATP to GDP to produce GTP.[6] All these results suggested that symbionin functions dually in the transphosphorylation reaction, serving as an "intermediary substrate" on the one hand and an "enzyme" on the other hand. These characteristics of symbionin remind us of the two-component pathway of signal transduction for histidine protein kinase.

[1] H. Ishikawa, *Insect Biochem.* **14,** 417 (1984).
[2] E. Hara and H. Ishikawa, *Insect Biochem.* **20,** 421 (1990).
[3] C. Ohtaka, H. Nakamura, and H. Ishikawa, *J. Bacteriol.* **174,** 1869 (1992).
[4] K. Kakeda and H. Ishikawa, *J. Biochem.* **110,** 583 (1991).
[5] M. Morioka and H. Ishikawa, *J. Biochem.* **111,** 431 (1992).
[6] M. Morioka, H. Muraoka, and H. Ishikawa, *J. Biochem.* **114,** 246 (1993).



0076-6879/98 $25.00

Thus, it is likely that endosymbiont chaperonin 60 functions not only as a molecular chaperone but also as an energy-coupling protein.[7]

In this chapter, we describe assay methods determining symbionin as an energy-coupling protein.

## Purification of Symbionin from Pea Aphid

Symbionin is routinely purified from a long-established parthenogenic clone of pea aphids, *Acyrthosiphon pisum* (Harris).[4] Insects are maintained on young broad bean plants, *Vicia faba* (L.), at 15° in a long-day regime with 18 hr of light and 6 hr of dark, and stored at −80° until used for purification. All purification steps are carried out at 0–4°, except for those involving high-performance liquid chromatography (HPLC), which are performed at room temperature.

### *Step 1: Homogenization and Preparation of Cytoplasm*

In a typical purification procedure, frozen tissues of the pea aphid (about 30 g) are sterilized in 70% (v/v) ethanol, quickly washed twice with water, crushed in a glass mortar and pestle, and then homogenized in 240 ml of buffer A [0.15 *M* Tris-HCl (pH 6.4) containing 0.3 *M* KCl, 1 m*M* EDTA, and 0.5 m*M* phenylmethylsulfonyl fluoride (PMSF)] using a motor-driven Potter–Elvehjem type homogenizer. The homogenate is filtered through a nylon mesh (with a pore size of 90 $\mu$m) to remove unbroken tissue, and the highly viscous filtrate is centrifuged at 15,000 *g* for 20 min at 4°. The pellet containing DNA and cell debris and a small amount of viscous materials overlaying the pellet are discarded. The nonviscous, dark green postmitochondrial supernatant is pooled, filtered through a cotton layer in a glass funnel to remove a yellow film of fat particles, and then centrifuged at 105,000 *g* for 60 min at 4° in a Hitachi RP-30 rotor to obtain the clear, light green cytoplasm (fraction I). A protein profile of this fraction is shown in Fig. 1 (lane 2).

### *Step 2: Ammonium Sulfate Fractionation*

To fraction I, solid ammonium sulfate is added to give 40% saturation with stirring over the course of 10–15 min. After stirring for 20 min more, the suspension is centrifuged 15,000 *g* for 20 min at 4° and the supernatant is recovered. Ammonium sulfate is then added to give 50% saturation. After stirring for 30 min more, the suspension is centrifuged at 15,000 *g* for 20 min at 4°. A crude fraction of symbionin recovered by the second

[7] M. Morioka, H. Muraoka, K. Yamamoto, and H. Ishikawa, *J. Biochem.* **116,** 1075 (1994).

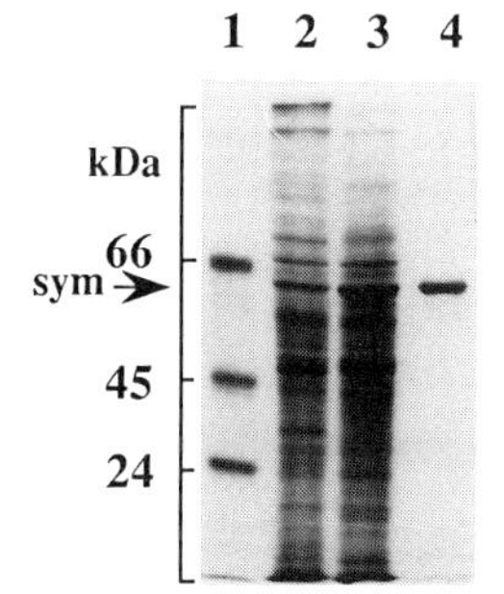


FIG. 1. Purification of symbionin. The proteins resolved in a 10% SDS–polyacrylamide gel are stained with Coomassie Brilliant Blue R-250; lane 1, molecular weight marker proteins; lane 2, cytoplasm (fraction I); lane 3, proteins precipitated between 40 and 50% saturation (fraction II); lane 4, symbionin fraction after sucrose density gradient centrifugation (fraction III). An arrow labeled "sym" indicates the position of symbionin detected by immunoblotting using anti-symbionin antiserum.

ammonium sulfate precipitation is dissolved in a minimum volume (about 3 ml) of buffer B [50 m*M* Tris-HCl (pH 7.6), 35 m*M* KCl, 10 m*M* $Mg(CH_3COO)_2$, 25 m*M* $NH_4Cl$, and 1 m*M* dithiothreitol (DTT)] and dialyzed against an excess volume of buffer B overnight with three changes to give fraction II (Fig. 1, lane 3).

### *Step 3: Sucrose Gradient Centrifugation*

Dialyzed fraction II (about 4 ml) is centrifuged at 15,000 *g* for 20 min at 4° to remove insoluble materials and 1-ml aliquots are carefully layered over four tubes of linear gradient from 10 to 30% (w/v) sucrose made in 36 ml of buffer B. After centrifugation in a Beckman (Fullerton, CA) SW-28 swinging bucket-type rotor at 25,000 rpm for 20 hr at 4°, gradients are fractionated into 1-ml fractions from the bottom of the gradient. Symbionin is identified by sodium dodecyl sulfate-polyacrylamide gel electrophoresis (SDS–PAGE) using a 10% (w/v) polyacrylamide gel followed by immunoblotting with polyclonal antibody raised against symbionin. Anti-GroEL antibody is also useful for detection of symbionin. The fractions containing symbionin (about 12 ml) are pooled and concentrated to a volume of about 1 ml using an ultrafiltration membrane (Mol-Cut UFP2 TMK-24; Millipore, Bedford, MA) to obtain fraction III. At this stage of purification the symbionin is usually more than 99% pure as judged by SDS–PAGE and Coomassie blue staining (Fig. 1, lane 4), and it is pure enough to determine its chaperonin activity. Using this purification procedure, the yield of symbionin typically ranges from 1 to 2 mg from 10 g of frozen tissues of pea aphid.

### *Step 4 (Optional Step): Gel-Permeation High-Performance Liquid Chromatography*

An additional optional step is described here, but this is generally omitted unless highly purified protein is desired for purposes such as determinations of *in vitro* autophosphorylating activity and phosphotransferase activity. An appropriate amount of fraction III is loaded onto a Shim-Pack DIOL-300 gel-permeation HPLC column (7.9-mm i.d. × 50 cm; Shimadzu) equilibrated with 10 m*M* potassium phosphate buffer, pH 6.9, containing 200 m*M* $Na_2SO_4$. Symbionin is eluted from the column at 12.6 min with the same buffer (fraction IV). This column removes a higher molecular weight contaminant present in minute amounts.

### *Storage*

Purified symbionin (fractions III and IV) appears to be stable and it can be stored for long periods of time (about 1 year) at −80° in 50 m*M* Tris-HCl buffer, pH 6.5 to 7.5, in the presence of 10% (v/v) glycerol. Repeated freezing and thawing and relatively short exposure to temperatures above 10° lead to rapid degradation of protein.

### *General Properties*

Purified symbionin is a protein with a molecular mass of 63 kDa when determined by SDS–PAGE under reducing conditions and has an apparent native molecular mass of about 810 kDa by gel filtration on a Sephacryl S-300 column.

The electron microscopic examination of symbionin negatively stained with uranyl acetate also reveals that symbionin is a 14-subunit homooligomer composed of two stacked rings of 7 subunits each, similar to the *Escherichia coli* GroEL rather than mitochondrial chaperonin 60.

## Chaperonin Activity of Symbionin

### *Reconstitution of Rubisco Mediated by Symbionin*

To investigate the chaperonin activity of symbionin,[4] the reconstitution of the active dimeric ribulose-bisphosphate carboxylase (Rubisco, EC 4.1.1.39) from its unfolded subunits is examined, *in vitro*.

The substrate for the reconstitution experiment is prepared by incubating native Rubisco purified from *Rhodospirillum rubrum*[8] in denaturation

[8] F. R. Tobita and B. A. McFadden, *J. Biol. Chem.* **249,** 3453 (1974).

buffer [100 m$M$ Tris-HCl (pH 8.0), 4 m$M$ EDTA, and 7.6 $M$ urea] at 25° for 2 hr.[9] Reconstitution is initiated by diluting a 1-$\mu$l aliquot of the denatured Rubisco (8 $\mu M$ as a protomer concentration) into 100 $\mu$l of reconstitution buffer [50 m$M$ Tris-HCl (pH 7.8), 10 m$M$ KCl, 7 m$M$ $MgCl_2$, and 3 m$M$ ATP supplemented with 8.0 $\mu M$ (as a protomer) GroES] containing either 7.5 $\mu M$ (as a protomer) symbionin (fraction III, see above) or GroEL.[10] After incubation at 18° for 2 hr, the degree of reconstitution is estimated by determining the $CO_2$-fixing activity of Rubisco.[11]

On denaturation with 7.6 $M$ urea, the $CO_2$-fixing activity of Rubisco is completely abolished, indicating that the enzyme protein is converted into unfolded subunits. When denatured Rubisco is incubated with GroEL in the presence of GroES and MgATP, about 50% of the enzyme activity is restored. In contrast, when GroEL is replaced by symbionin, about 15% of the $CO_2$-fixing activity is restored. The reason why symbionin is only 30% as effective as GroEL in reconstituting active Rubisco may be that aphid symbionin is combined with the Cpn10 from a different source, GroES from *E. coli.*

### *Self-Reconstitution of Symbionin*

The substrate for the self-reconstitution experiment[12] is prepared by denaturing symbionin in urea. Seven micrograms of purified native symbionin (fraction III, see above) is incubated at 23° for at least 30 min in 10 $\mu$l of denaturation buffer [50 m$M$ potassium phosphate buffer (pH 6.9) containing 5 $M$ urea]. Reconstitution is initiated by diluting a 4-$\mu$l aliquot of denatured symbionin into 400 $\mu$l of reconstitution buffer [20 m$M$ triethylammonium acetate (pH 7.5), 100 m$M$ potassium acetate, 1 m$M$ DTT, and 0.1 m$M$ EDTA] supplemented with 1.7 m$M$ ATP and 5 m$M$ $MgCl_2$. After incubation at 23°, a 200-$\mu$l aliquot is subjected to a DIOL-300 gel-permeation HPLC column (7.9-mm i.d. × 50 cm; Shimadzu). Proteins are eluted with 10 m$M$ potassium phosphate buffer (pH 6.9) containing 200 m$M$ $Na_2SO_4$ at a flow rate of 1.0 ml/min at room temperature, and scanned for ultraviolet (UV) absorption at 230 nm.

As a result, a protein with a retention time of 12.6 min, which coincides with that of native tetradecameric symbionin, is yielded with high efficiency. Self-reconstitution of symbionin proceeds in an MgATP-dependent fashion

[9] P. Goloubinoff, J. T. Christeller, A. A. Gatenby, and G. H. Lorimer, *Nature (London)* **342,** 884 (1989).

[10] P. V. Viitanen, T. H. Lubben, J. Reed, P. Goloubinoff, D. P. O'Keefe, and G. H. Lorimer, *Biochemistry* **29,** 5665 (1990).

[11] A. E. Grebanier, D. Champagne, and H. Roy, *Biochemistry* **17,** 5150 (1978).

[12] M. Morioka and H. Ishikawa, *J. Biochem.* **114,** 468 (1993).

and hyperbolically with time. The yield of self-reconstituted symbionin increases up to 70% in response to an increase in the initial concentration of denatured symbionin. Self-reconstituted symbionin not only has the same molecular mass but also exhibits the same ATPase activity and phosphotransferase activity as native symbionin.

## Symbionin Synthesis by Isolated Endosymbionts

### *Isolation of Endosymbiont from Pea Aphid*

Living pea aphids (parthenogenic females, 14 to 15 days old)[5] are sterilized in 70% (v/v) ethanol and quickly washed twice with water. One gram of insect is lightly crushed in 50 ml of cold buffer C [35 m*M* Tris-HCl (pH 7.6), 250 m*M* sucrose, 25 m*M* KCl, 10 m*M* $MgCl_2$, 5 m*M* PMSF, and 1 m*M* DTT] using a Dounce-type homogenizer with a loosely fitting Teflon pestle. The homogenate is filtered through a nylon mesh with a pore size of 50 $\mu$m to remove the unbroken tissue and cuticle. The crude endosymbiont is sedimentated by centrifuging the filtrate at 1700 *g* for 10 min at 4°. The dark green pellet is gently suspended in 10 ml of buffer C by using a Dounce-type homogenizer and 1-ml aliquots of the suspension are layered over 10 ml of discontinuous Percoll gradients [9, 18, 27, 36, and 45% (v/v) Percoll, 2 ml each] made in Wasserman test tubes. Percoll gradient is prepared by diluting 100% (v/v) Percoll, which contains 5% (w/v) polyethylene glycol 6000, 1% (w/v) bovine serum albumin (BSA), 1% (w/v) Ficoll 400, and 8.6% (w/v) sucrose, with buffer C. Ten tubes of gradient are centrifuged at 1000 *g* for 20 min at 4° in a swinging-bucket type rotor by accelerating and decelerating slowly to avoid mixing of each layer of the gradients. Because endosymbionts are concentrated at the boundary between 36 and 45% (v/v) Percoll, upper layers are discarded by aspiration and the boundary fraction is recovered with a capillary pipette. The endosymbionts are washed in buffer C by a light and brief homogenization and centrifugation at 1700 *g* for 5 min at 4°.

*Comments.* Special care is taken not to damage the endosymbionts mechanically during homogenization of the insect materials. Experience in our laboratory has indicated that high yield cannot be expected if the tight-fitting homogenizer is used. The homogenization is performed by three or four up-and-down manual strokes.

The final preparation of endosymbiont is pale green and homogeneously round-shaped as observed by light microscopy. Although an essentially quantitative recovery of endosymbionts is not achieved, the fraction contains no residual subcellular components such as mitochondria except for a small portion of nuclei. Unbroken bacteriocytes are at the bottom of the gradient and membrane fragments are found in the area above the upper band of endosymbiont in the gradient.

*Determination of Symbionin Synthetic Activity*

The endosymbionts isolated from 0.1 g of insect tissue are suspended in 0.5 ml of Grace's medium (without methionine and serum free) and incubated with 0.75 MBq of L-[$^{35}$S]methionine (38.44 TBq/mmol) at various temperatures for 20 min with gentle shaking.[5] After the incubation period the $^{35}$S-labeled endosymbionts are collected by centrifugation and then homogenized in 1 ml of ice-cold 10% (w/v) trichloroacetic acid (TCA). The acid-insoluble precipitate is washed with ice-cold 10% (w/v) TCA four times and with cold acetone twice. The final precipitate is boiled for 3 min in 50 $\mu$l of SDS–PAGE lysis buffer containing 3% (v/v) 2-mercaptoethanol and subjected to electrophoresis using a slab separating gel of 10% (w/v) polyacrylamide. The $^{35}$S-labeled proteins are detected by fluorography following Enlightning (New England Nuclear Research Products, Boston, MA) enhancement.

When two-dimensional electrophoretic analysis is desired, $^{35}$S-labeled endosymbionts are directly lysed in 50 $\mu$l of lysis buffer[13] and loaded on an isoelectric focusing (IEF) gel (3-mm i.d. $\times$ 15 cm) containing a mixture of pH 3.5–10 ampholines and pH 5–7 ampholines (Pharmacia-LKB Biotechnology AB, Uppsala, Sweden) (1 : 4, v/v) at a final concentration of 2% (w/v).

The syntheses of three species of protein with molecular masses of 73, 63, and 45 kDa are selectively augmented in response to the temperature increase over their constitutive levels at 20° (Fig. 2A). Other stress-causing agents such as $Cd^{2+}$, $As^{2+}$, and ethanol also induce the syntheses of these proteins at 20°, whereas nalidixic acid, 2,4-dinitrophenol, 5-azacytidine, sodium azide, and hydroxyurea do not induce synthesis. Only the 63-kDa protein has the same subunit molecular mass and native molecular mass on a sucrose density gradient as symbionin (see above). The 63-kDa protein also shares immunogenicity with symbionin, whereas the isoelectric point of the 63-kDa protein is more acidic than that of authentic symbionin when analyzed by two-dimensional PAGE (Fig. 2B).

*In Vivo* Phosphorylation of Symbionin by Heat Shock

The isolated endosymbionts are incubated with ortho[$^{32}$P]phosphoric acid (carrier free, 10.5 TBq/mg phosphorus) in Grace's medium[14] having

[13] P. H. O'Farrell, *J. Biol. Chem.* **250,** 4007 (1975).

[14] T. D. C. Grace, *Nature* (*London*) **195,** 788 (1962).

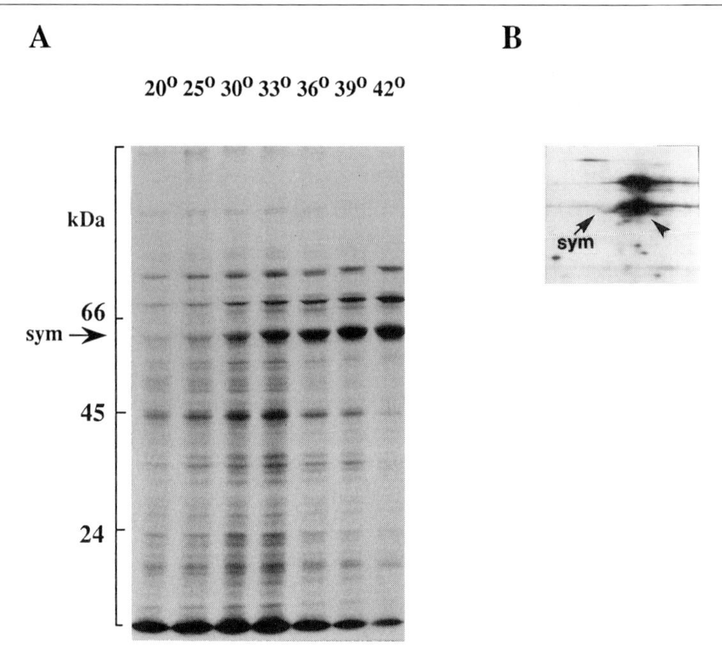


FIG. 2. Synthesis of 63-kDa protein by isolated endosymbionts at various temperatures. (A) SDS–PAGE analysis of proteins synthesized by isolated endosymbionts at various temperatures. An arrow labeled "sym" indicates the position of symbionin detected by immunoblotting using polyclonal antibody raised against purified symbionin. (B) Two-dimensional analysis of 63-kDa protein synthesized at 39°. An arrow labeled "sym" and an arrowhead indicate the positions of symbionin and 63-kDa protein, respectively.

a low phosphate concentration (0.05 m*M*) at 39° for 30 min and then lysed by sonication in 1 ml of immunoprecipitation buffer [50 m*M* Tris-HCl (pH 7.5), 100 m*M* NaCl, 2 m*M* EDTA, 0.4% (w/v) SDS, 1% (w/v) Triton X-100, and 2 m*M* PMSF]. The lysate is boiled for 3 min and the clear supernatant is incubated with 20 $\mu$l of anti-symbionin antiserum with gentle shaking for 120 min at 0°. The $^{32}$P-labeled immunoprecipitate is further incubated with 5 mg of protein A–Sepharose CL-4B beads {Sigma (St. Louis, MO), 5 mg/35 $\mu$l of washing buffer [50 m*M* Tris-HCl buffer (pH 7.5), 1% (w/v) sodium deoxycholate, 1% (w/v) Triton X-100, 0.1% (w/v) SDS, 150 m*M* NaCl, 1 m*M* EDTA, and 1 m*M* PMSF]} for 60 min at 0°. The $^{32}$P-labeled immunoprecipitate–resin complex is subjected to two-dimensional PAGE followed by autoradiography.

As a result, both symbionin and the 63-kDa protein are found to be labeled, although the radioactive signal of the 63-kDa protein is stronger than that of symbionin. This indicates that the 63-kDa protein, the major protein synthesized by isolated endosymbiont in response to the temperature increase, is a phosphorylated form of symbionin.

## *In Vitro* Autophosphorylation of Symbionin by Heat Shock

### *Determination of Autophosphorylating Activity*

To study the mechanism underlying the phosphorylation of symbionin,[5] freshly purified symbionin (fraction IV, see above) is incubated *in vitro* with 2.25 MBq of [$\gamma$-$^{32}$P]ATP (14.8 TBq/mmol, 10 $\mu M$) in a phosphorylation buffer [50 m$M$ 2-($N$-morpholino)ethanesulfonic acid–NaOH buffer (pH 6.5), 5 m$M$ 2-mercaptoethanol, 5 m$M$ $MgCl_2$, and 10% (v/v) glycerol] at either 20 or 39° for 30 min. The reaction is terminated by adding sodium deoxycholate and TCA to final concentrations of 0.2 and 10% (w/v), respectively. The acid-insoluble materials are washed three times with 10% (w/v) TCA and twice with cold acetone, and then subjected to SDS–PAGE followed by autoradiography.

The radioactivity is found to be associated with symbionin that is incubated at 20° (Fig. 3), but symbionin is labeled more strongly when incubated at 39° (Fig. 3). Addition of $ZnCl_2$ instead of $MgCl_2$ results in a remarkable stimulation of the autophosphorylation, whereas $CaCl_2$ and $MnCl_2$ show a slightly inhibitory effect. The maximum level of autophosphorylation is achieved at around pH 6.5. The number of subunits phosphorylated during a 60-min incubation under the optimized conditions is 1 of 14, which is equivalent to 1 per native symbionin molecule. The *in vitro* phosphorylation of symbionin is not inhibited to a large extent by prior incubation at 95° for 10 min, and the phosphorylating activity survives in the immunocomplex of symbionin and its antiserum. Symbionin is not labeled with [$\alpha$-$^{32}$P]ATP at either 20 or 39°.

### *Identification of Autophosphorylation Site*

To identify the phosphorylated residue,[7] symbionin (about 200 $\mu$g, fraction III) is phosphorylated with [$\gamma$-$^{32}$P]ATP at 39° for 60 min in the phos-

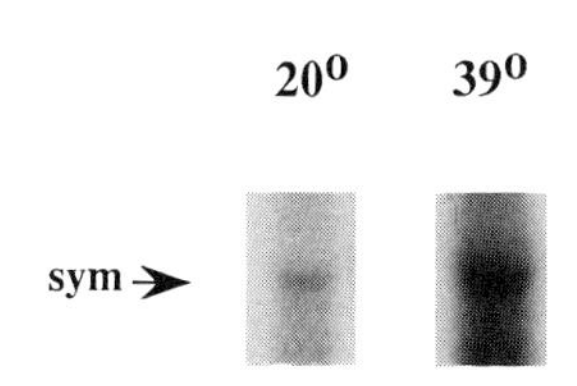


FIG. 3. *In vitro* phosphorylation of symbionin. Symbionin is labeled with [$\gamma$-$^{32}$P]ATP at 20 and 39°. An arrow labeled "sym" indicates the position of symbionin.

phorylation buffer containing 5 m$M$ $ZnCl_2$ instead of $MgCl_2$, and then digested with 10 $\mu$g of TPCK-treated trypsin (Sigma) in 50 m$M$ Tris-HCl buffer (pH 8.9) containing 5 m$M$ $CaCl_2$. After incubation at 37° for 60 min, a 1/6 volume of acetic acid is added to terminate the digestion and the whole digest is subjected to reversed-phase HPLC. Peptides are fractionated on a Shim-Pack ODS-H column (4.0-mm i.d. × 25 cm, 5-$\mu$m pore size; Shimadzu) using a linear gradient from 10 to 60% (v/v) acetonitrile made in 0.1% (v/v) trifluoroacetic acid (TFA) over 60 min at a flow rate of 0.8 ml/min at 38° with monitoring at $A_{214\,nm}$. The eluate is automatically fractionated every 15 sec for Cerenkov $^{32}P$ counting. The $^{32}P$-labeled symbionin fragment [retention time ($R_t$), 14.95 min] is immediately evaporated to dryness under reduced pressure. The dried material is dissolved in a small volume (about 10 $\mu$l) of 3 $N$ KOH and hydrolyzed at 110° for 3 hr. The resulting hydrolysate is applied to a silica gel PK5F thin-layer plate (Whatman, Clifton, NJ) together with the internal marker phospho amino acids. After developing in ascending order with chloroform–methanol–12.7% (w/v) ammonium hydroxide (1:3:1, v/v/v) at room temperature, the plate is extensively air dried and subjected to ninhydrin staining followed by autoradiography.

As a result, the radioactivity is detected on the ninhydrin-stained spot with an $R_f$ of 0.82 that corresponds to 1-phosphohistidine.

To identify the phosphorylation site of symbionin, the above-described radioactive fragment with a retention time of 14.95 min is analyzed on a PSQ-1 gas-phase peptide sequencer (Shimadzu). The result reveals that the radioactive fragment is a decapeptide, beginning at His-133, with the sequence His-Leu-Ser-Val-Pro-Cys-Ser-Asp-Ser-Lys. Thus, it is concluded that symbionin is autocatalytically phosphorylated at the histidine residue at position 133.

## Special Activities of Symbionin

### *Assay of Phosphotransferase Activity*

To investigate a possibility that symbionin functions as a histidine protein kinase of the two-component pathway in the aphid endosymbiont, the phosphotransferase activity of symbionin is examined.[6] Fifteen micrograms of purified symbionin (fraction IV, see above) is incubated with 0.37 MBq of [$\gamma$-$^{32}P$]ATP (10 $\mu M$) in 50 $\mu$l of reaction mixture [50 m$M$ Tris-HCl (pH 6.5), 5 m$M$ 2-mercaptoethanol, 5 m$M$ $MgCl_2$, and 10% (v/v) glycerol] containing 10 $\mu M$ unlabeled GDP. After incubation at 37° for 20 min, the protein is removed by passing it through an ultrafiltration membrane (Mol-Cut UFP2 TMK-24; Millipore). A 5-$\mu$l aliquot of the filtrate is applied to a WAX-1 ion-exchange HPLC column (4.0-mm i.d. × 5 cm, 3-$\mu$m particle size; Shimadzu) equilibrated with 20 m$M$ potassium phosphate, pH 7.1. The nucleotides are eluted with a linear gradient from 20 m$M$ (pH 7.1) to 480 m$M$ (pH 6.8) potassium phosphate at a flow rate of 0.8 ml/min at 35°. Fractions are collected every 1.0 min and counted directly for $^{32}$P (Cerenkov, cpm) (Fig. 4).

A large amount of radioactivity is found in the GTP fraction with a retention time of 26.0 min, suggesting that symbionin mediates the phosphoryl-group transfer from ATP to GDP. The reaction is time dependent during the first 20 min and equilibrates when the amount of GTP reaches that of ATP. When symbionin is treated with anti-symbionin antiserum, its phosphotransferase activity is entirely abolished.

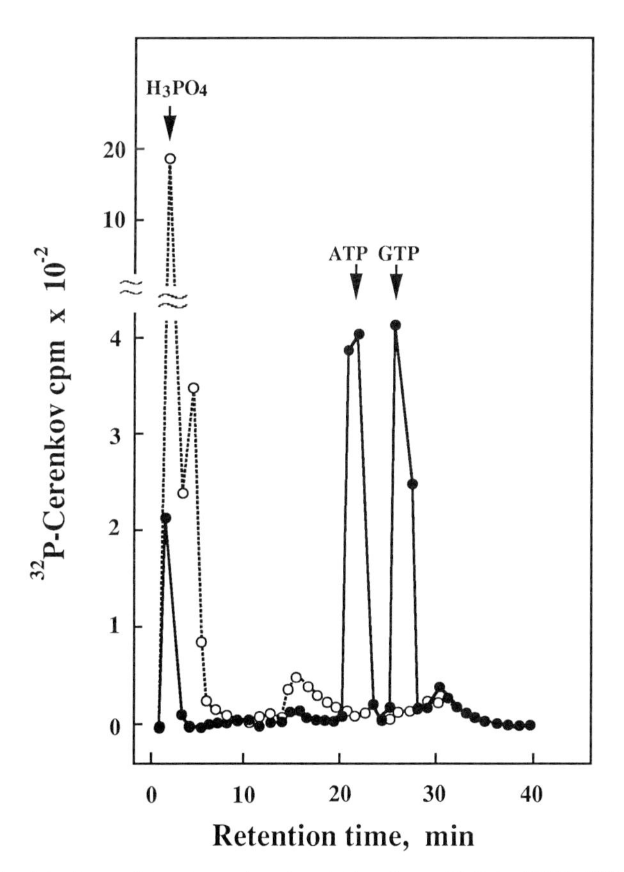


FIG. 4. Symbionin mediates transphosphorylation from ATP to GDP. (●) Symbionin; (○) GroEL.

*Transphosphorylation from Symbionin to Other Endosymbiotic Proteins*

As described above, symbionin is autocatalytically phosphorylated at a histidine residue and catalyzes the phosphoryl-group transfer from ATP to GDP, suggesting that it functions as a histidine protein kinase of the two-component pathway.[7] To identify the response regulator, a partner molecule of the two-component pathway that accepts the phosphoryl group from histidine protein kinase, the following experiment is carried out.

The isolated endosymbionts are lysed for SDS–PAGE. Following electrophoresis using a 10% (w/v) polyacrylamide separating gel, proteins are electrophoretically transferred to Immobilon-P [polyvinylidene difluoride (PVDF) membrane; Millipore] in transfer buffer (100 m*M* glycine and 25 m*M* Tris base). Transfer is carried out using a constant current at 160 m*M* for 90 min with a wet-type Trans-Blot apparatus (Marisol). Before transfer, the Immobilon-P membrane must be wetted in methanol for 30 sec, washed with water for 2 min, and then soaked in transfer buffer. The proteins transferred and immobilized onto a membrane are incubated in 7 *M* guanidine hydrochloride, pH 8.3, at room temperature for 60 min, and then allowed to renature in 50 m*M* Tris-HCl (pH 7.5) containing 100 m*M* NaCl, 2 m*M* DTT, 2 m*M* EDTA, 1% (w/v) bovine serum albumin (BSA), and 0.1% (v/v) Nonidet P-40 at 4° for 18 hr. After renaturation of the immobi-

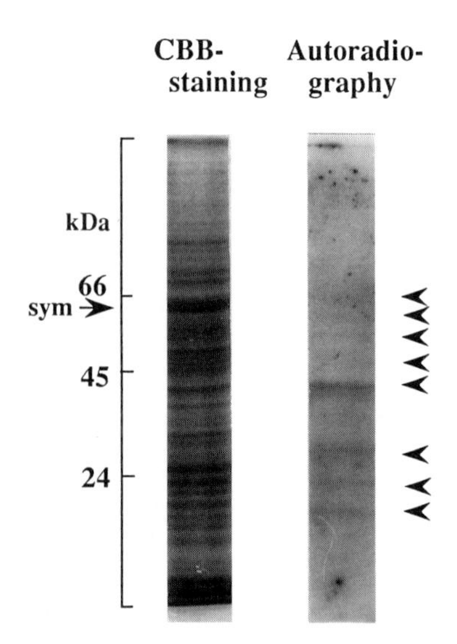


FIG. 5. Transphosphorylation from $^{32}P$-labeled symbionin to other endosymbiotic proteins. Arrowheads indicate the proteins, candidates for response regulator of the two-component pathway, which accept the phosphoryl group from $^{32}P$-labeled symbionin. CBB, Coomassie Brilliant Blue.

lized proteins, the blot is blocked with 5% (w/v) BSA in 30 m*M* Tris-HCl (pH 7.5) by incubating at room temperature for 1 hr. The BSA-blocked blot is overlaid with $^{32}$P-labeled symbionin (100 $\mu$g of protein, about 3 $\times$ $10^4$ cpm) in 30 m*M* Tris-HCl (pH 7.5) containing 5 m*M* $MgCl_2$ at 37° for 1 hr. During this incubation, the phosphoryl-group transfer from $^{32}$P-labeled autocatalytically phosphorylated symbionin to other endosymbiotic proteins immobilized onto a membrane may occur. After terminating the reaction by dousing the blot into an excess volume of 30 m*M* Tris-HCl (pH 7.5), the blot is washed five times with 30 m*M* Tris-HCl (pH 7.5), once with the same buffer containing 0.05% (v/v) Nonidet P-40, once with buffer, once with the buffer containing 1 *M* NaCl, and finally once with buffer, successively, at room temperature for 10 min each. Even after extensive washing the background radioactivity remaining on the blot is fairly high. This background can be reduced by incubating the blot with 1 *N* KOH for 10 min. The alkali-treated blot is soaked several times with water and 10% (v/v) acetic acid, dried, and subjected to autoradiography.

As a result, radioactive signals are detected on at least eight electrophoretically distinct polypeptides as shown in Fig. 5. The radioactive bands survived successive treatments with detergent, 1 *M* NaCl, 1 *N* KOH, and 10% (v/v) acetic acid as described previously, suggesting that the phosphoryl group is covalently linked to immobilized symbiotic polypeptides. It is possible that at least some of the eight polypeptides are response regulators and/or those chaperoned by symbionin in the endosymbiont.

# [17] Purification of Yeast Mitochondrial Chaperonin 60 and Co-Chaperonin 10

*By* YVES DUBAQUIÉ, GOTTFRIED SCHATZ, and SABINE ROSPERT

## Introduction

The chaperonin (cpn) system of the yeast *Saccharomyces cerevisiae* is located in the mitochondrial matrix, where it promotes the refolding of particular newly imported precursor proteins[1] and prevents protein denaturation under heat stress conditions.[2] Yeast heat shock protein 60 (hsp60) was identified as a protein that cross-reacts with an antiserum against a

[1] S. Rospert, R. Looser, Y. Dubaquié, A. Matouschek, B. S. Glick, and G. Schatz, *EMBO J.* **15,** 764 (1996).

[2] J. Martin, A. L. Horwich, and F. U. Hartl, *Science* **258,** 995 (1992).





0076-6879/98 $25.00

heat shock protein from *Tetrahymena thermophilus.*[3] Electron microscopy of the oligomeric complex revealed a double-toroidal structure with a sevenfold rotational axis of symmetry, similar to bacterial GroEL.[3] Subsequent cloning of the *HSP60* gene revealed a 54% amino acid identity to *Escherichia coli* GroEL.[4]

The structure and function of GroEL have been studied extensively *in vitro.*[5–8] However, more recent studies have not been performed with purified yeast hsp60, because large amounts of this protein cannot easily be isolated from yeast. We describe here a large-scale purification procedure of *S. cerevisiae* hsp60 expressed in *Pichia pastoris.*

Yeast cpn10 was isolated by taking advantage of the fact that bacterial GroEL forms a functional complex with co-chaperonins from mitochondria or chloroplasts.[9,10] In a biochemical approach, yeast mitochondrial fractions were assayed for their ability to promote refolding of dimeric ribulose-1,5-bisphosphate carboxylase in the presence of GroEL.[11] In a different approach, GroEL was added to yeast mitochondrial extracts in the presence of MgATP. Under these conditions, cpn10 and GroEL form a complex that can be recovered in the high molecular weight fraction of a gel-filtration column.[12] The *S. cerevisiae CPN10* gene encodes a polypeptide of about 12 kDa, which is 36.5% identical to *E. coli* GroES.[12,13] Electron micrographs show seven subunits arranged in a single ring (A. Engel, personal communication, 1997) similar to the heptameric toroidal structure of bacterial GroES.[14] The N-terminal portion of yeast cpn10 resembles a typical matrix-targeting signal that, unlike that of most other matrix-targeted proteins, is not cleaved on import.[13]

[3] T. W. McMullin and R. L. Hallberg, *Mol. Cell. Biol.* **8,** 371 (1988).

[4] D. S. Reading, R. L. Hallberg, and A. M. Myers, *Nature* (*London*) **337,** 655 (1989).

[5] K. Braig, Z. Otwinowski, R. Hedge, D. C. Boisvert, A. Joachimiak, A. L. Horwich, and P. B. Sigler, *Nature* (*London*) **371,** 578 (1994).

[6] W. A. Fenton, Y. Kashi, K. Furtak, and A. L. Horwich, *Nature* (*London*) **371,** 614 (1994).

[7] M. J. Todd, P. V. Viitanen, and G. H. Lorimer, *Science* **265,** 659 (1994).

[8] M. K. Hayer-Hartl, J. Martin, and F. U. Hartl, *Science* **269,** 836 (1995).

[9] U. Bertsch, J. Soll, R. Seetharam, and P. V. Viitanen, *Proc. Natl. Acad. Sci. U.S.A.* **90,** 10967 (1992).

[10] T. H. Lubben, A. A. Gatenby, K. D. Gail, G. H. Lorimer, and P. V. Viitanen, *Proc. Natl. Acad. Sci. U.S.A.* **87,** 7683 (1990).

[11] S. Rospert, B. S. Glick, P. Jenö, G. Schatz, M. J. Todd, G. H. Lorimer, and P. V. Viitanen, *Proc. Natl. Acad. Sci. U.S.A.* **90,** 10967 (1993).

[12] J. Höhfeld and F.-U. Hartl, *J. Cell Biol.* **126,** 305 (1994).

[13] S. Rospert, T. Junne, B. S. Glick, and G. Schatz, *FEBS Lett.* **335,** 358 (1993).

[14] G. N. Chandrasekhar, K. Tilly, C. Woodford, R. Hendrix, and C. Georgopoulos, *J. Biol. Chem.* **261,** (1986).

Although the concentration of both hsp60 and cpn10 in the mitochondrial matrix is increased two- to threefold after heat shock, hsp60 and cpn10 are essential for viability at all temperatures.[3,12,13] This indicates that mitochondrial polypeptides are unable to fold without the aid of the chaperonin system *in vivo*. However, more recent studies suggest that the chaperonin system is not involved in the folding of all mitochondrial proteins.[1] Assaying the interaction of purified hsp60 and cpn10 with authentic substrate proteins should give insight into the specificity of the mitochondrial chaperonin system. In addition, a comparison of the bacterial GroEL/ES system with the eukaryotic hsp60/cpn10 system may help to elucidate the chaperonin reaction cycle.

## Method

### *General Procedures*

Growth media, unless specified otherwise, are prepared according to standard procedures using products from Difco Laboratories (Detroit, MI).

Protein concentrations are determined with the Pierce (Rockford, IL) bicinchoninic acid (BCA) protein assay kit using bovine serum albumin (BSA) as the standard. Values for cpn10 are corrected by a factor of 1.41. This factor was deduced from amino acid analysis of cpn10. hsp60 and cpn10 are detected by immunoblotting with the corresponding rabbit antisera followed by $^{125}$I-labeled protein A decoration.[15] All FPLC (fast protein liquid chromatography) purification steps are carried out at 4°. A cocktail containing the following protease inhibitors (Sigma, St. Louis, MO) at the indicated final concentrations is used throughout the purification procedures: leupeptin (1.25 $\mu$g/ml), antipain (0.75 $\mu$g/ml), chymostatin (0.25 $\mu$g/ml), elastinal (0.25 $\mu$g/ml), pepstatin (5 $\mu$g/ml), phenylmethylsulfonyl fluoride (PMSF, 1 m*M*), and ethylenediaminetetraacetic acid (EDTA, 1 m*M*; only in the absence of MgATP). Gel electrophoresis for hsp60 is on sodium dodecyl sulfate (SDS)–10% (w/v) polyacrylamide gels,[16] that for cpn10 on SDS–10% (w/v) polyacrylamide gels using Tris–Tricine buffers.[17]

The yield of hsp60 and cpn10 during purification is quantified by densitometric analysis [Molecular Dynamics (Sunnyvale, CA) 300A] of immunoblots.

[15] A. Haid and M. Suissa, *Methods Enzymol.* **96,** 192 (1983).

[16] U. K. Laemmli, *Nature* (*London*) **227,** 680 (1970).

[17] H. Schägger and G. von Jagow, *Anal. Biochem.* **166,** 368 (1987).

## Purification of Recombinant Heat Shock Protein 60

### *Generation of a Pichia pastoris Strain Expressing Saccharomyces cerevisiae Heat Shock Protein 60*

*Saccharomyces cerevisiae* hsp60 and *E. coli* GroEL are highly homologous proteins[4] with similar calculated isoelectric points of 5.06 and 4.67 for hsp60 and GroEL, respectively. It is therefore not surprising that *S. cerevisiae* hsp60 expressed in *E. coli* is not separated completely from GroEL by standard chromatographic methods (Y. Dubaquié, unpublished, 1997). Therefore, mature *S. cerevisiae* hsp60 is expressed in the cytosol of *P. pastoris.*

The first amino acid of mature hsp60 isolated from *S. cerevisiae* mitochondria is Ser-22 (K. Suda, unpublished, 1997). *Pichia pastoris* strain YJP1 is constructed by integrating the DNA sequence of mature *S. cerevisiae* hsp60 into the *AOX1* locus (encoding alcohol oxidase) of strain GS115 (Invitrogen, San Diego, CA). hsp60 expression in YJP1 is under control of the *AOX1* promoter. Shifting YJP1 from glycerol- to methanol-containing medium therefore induces hsp60 expression. Disruption of the *AOX1* gene abolishes growth on methanol as a sole carbon source. Cells therefore are grown to stationary phase before hsp60 induction (for detailed information, refer to Ref. 17a).

### *Preparation of Pichia pastoris Cytosol Containing Mature Heat Shock Protein 60*

For a typical purification procedure, 10 liters of minimal glycerol medium [MGY; yeast nitrogen base (13.4 g/liter), biotin (0.4 mg/liter), 1% (v/v) glycerol] is inoculated with 100 ml of preculture of *P. pastoris* strain YJP1. Cells are grown at 30° for 3 days to stationary phase and are collected by centrifugation in a Sorvall rotor HG-4L at room temperature (4500 rpm, 10 min). hsp60 expression is induced by resuspending the cell pellet in 6 liters of minimal methanol medium [MM; yeast nitrogen base (13.4 g/liter), biotin (0.4 mg/liter), 0.5% (v/v) methanol] followed by incubation for 3 days at 30°. Two days after the start of induction, methanol in the medium is replenished to 0.5% (v/v) to compensate for evaporative losses. hsp60 expression is monitored by immunoblotting of total protein extracts.[18] After maximal induction is achieved, cells are harvested by centrifugation. Typical yields range from 10 to 14 g wet cell paste per liter.

[17a] Invitrogen, "*Pichia* Expression Manual." Invitrogen, San Diego, California.
[18] M. P. Yaffe and G. Schatz, *Proc. Natl. Acad. Sci. U.S.A.* **81,** 4819 (1984).

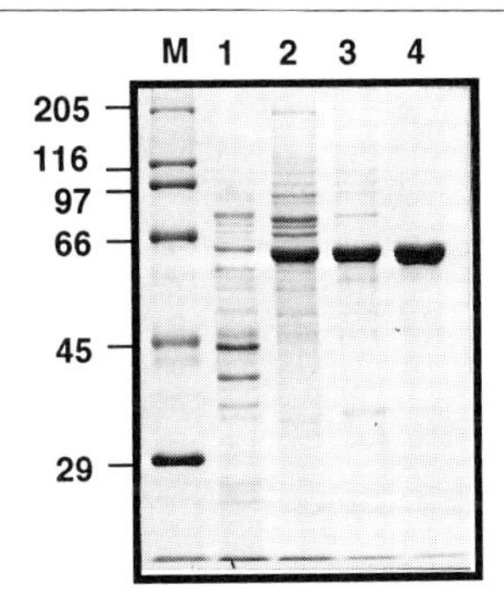


FIG. 1. Purification of *S. cerevisiae* hsp60. Aliquots of 10 $\mu$g of total protein per lane were separated on an SDS–10% (w/v) polyacrylamide gel and stained with Coomassie blue. Lane 1, cytosolic fraction from *P. pastoris;* lane 2, TMAE pool; lane 3, Mono Q pool; lane 4; Superose 6 eluate. Lane M, molecular weight markers; values ($\times$ $10^{-3}$) are indicated on the left-hand side.

To avoid contamination of the recombinant *S. cerevisiae* hsp60 with hsp60 from *P. pastoris* mitochondria, we make use of the different intracellular locations of the two proteins. Cell extract is prepared according to Glick and Pon.[19] The cytosol containing *S. cerevisiae* hsp60 is first cleared of mitochondria by centrifugation for 15 min at 10,000 *g*,[19] and is then clarified by ultracentrifugation [Beckman (Fullerton, CA) rotor TFT45.94, 40 krpm = 125,000 *g*] for 30 min at 2° (Fig. 1, lane 1).

*Trimethylaminoethyl Chromatography.* Half of the clarified extract (60–70 ml) is loaded at a rate of 1 ml/min onto a trimethylaminoethyl (TMAE) Superformance anion-exchange column (150 $\times$ 10 mm; Merck, Darmstadt, Germany) equilibrated with 50 m*M* *N*-2-hydroxyethylpiperazine-*N'*-2-ethanesulfonic acid (HEPES)–NaOH, pH 7.4. Fractionation is achieved in 50 m*M* HEPES–NaOH at pH 7.4 by applying the following gradient: 0–140 ml, no NaCl; 140–260 ml, 0.2 *M* NaCl; followed by a linear increase to 0.5 *M* NaCl from 260 to 310 ml and a plateau at 0.5 *M* NaCl from 310 to 360 ml. hsp60-containing fractions (2-ml fraction size) elute between 0.45 and 0.5 *M* NaCl. The collected material, termed the *TMAE pool* (Fig. 1, lane 2), is supplemented with 20 m*M* KCl, 20 m*M* $MgCl_2$, and 1 m*M* MgATP, and incubated on ice for 1 hr to dissociate chaperonin–substrate complexes formed in the *P. pastoris* cytosol or during the purification procedure.

*Mono Q Chromatography.* The TMAE pool (14 ml) is diluted twofold with 50 m*M* HEPES–NaOH, pH 7.4, and loaded at a rate of 1 ml/min onto

[19] B. S. Glick and L. A. Pon, *Methods Enzymol.* **260,** 213 (1995).

TABLE I
PURIFICATION OF hsp60

| Material | Total protein (mg) | Total hsp60 (mg) | Purification (-fold) | Yield (%) |
|---|---|---|---|---|
| *Pichia pastoris* cytosol | 712 | 36 | 1 | 100 |
| TMAE pool | 48 | 29 | 12 | 81 |
| Mono Q pool | 31 | 26 | 16 | 72 |
| Superose 6 pool | 8 | 8 | 20 | 22 |

a Mono Q column (HR 10/10; Pharmacia, Uppsala, Sweden) equilibrated with 50 m*M* HEPES–NaOH (pH 7.4), 200 m*M* NaCl. The protein is eluted with a 50-ml gradient from 200 to 800 m*M* NaCl in 50 m*M* HEPES–NaOH, pH 7.4. The peak fractions containing hsp60 are pooled; hsp60 elutes between 600 and 700 m*M* NaCl. The collected material (Mono Q pool; Fig. 1, lane 3) is concentrated to 0.5–1 ml with a Centricon 100 microconcentrator (Amicon, Beverly, MA) in the presence of 20 m*M* KCl, 20 m*M* $MgCl_2$, and 1 m*M* MgATP.

*Superose 6 Chromatography.* The Mono Q concentrate is subjected to gel filtration on a Superose 6 column (HR 10/30; Pharmacia) equilibrated with 50 m*M* HEPES–NaOH (pH 7.4), 50 m*M* NaCl. Fractions (0.5 ml) are collected at a flow rate of 0.1 ml/min and analyzed by sodium dodecyl sulfate-polyacrylamide gel electrophoresis (SDS–PAGE) and Coomassie blue staining. hsp60-containing fractions (Fig. 1, lane 4) are pooled, adjusted to 10% (v/v) glycerol, and frozen in liquid nitrogen. At −80°, the protein is stable for at least 2 months as judged from ATPase assays. The purification procedure is summarized in Table I.

## Purification of Recombinant Authentic and Hexahistidine-Tagged Chaperonin 10

### *Cloning of Saccharomyces cerevisiae Chaperonin 10 into Escherichia coli Expression Vector*

Eukaryotic cpn10 is more basic than GroES from *E. coli* and most other bacteria.[11,20] The predicted p*I* of *S. cerevisiae* cpn10 is 9.8, as compared with p*I* 5.2 for *E. coli* GroES. Therefore, despite the sequence similarity between *S. cerevisiae* cpn10 and GroES, the two proteins can be separated

[20] R. Dickson, B. Larson, P. V. Viitanen, M. B. Tormey, J. Geske, R. Strange, and L. T. Bemis, *J. Biol. Chem.* **269,** 26858 (1994).

by ion-exchange chromatography. Thus, *E. coli* is a suitable host for expression of cpn10 from eukaryotes.

The DNA fragment encoding the 106 amino acids of yeast cpn10 is cloned into expression vector pQE60 (QIAexpress system; Qiagen, Chatsworth, CA) to yield either the authentic cpn10 or a C-terminally hexahistidine-tagged version (cpn10–$H_6$). The resulting plasmids are termed pYD3 and pRL1, respectively. Protein expression is controlled by the *lac* operator and is efficiently prevented by high levels of *lac* repressor, encoded on the additional multicopy plasmid pREP4.

## *Purification of Authentic Chaperonin 10*

*Preparation of Escherichia coli Crude Extract Containing Saccharomyces cerevisiae Chaperonin 10.* Typically, 1 liter of LB [1% (w/v) Bacto-tryptone, 0.5% (w/v) yeast extract, 0.5% (w/v) NaCl] medium containing ampicillin (100 $\mu$g/ml) and kanamycin (50 $\mu$g/ml) is inoculated with 10 ml of preculture (same medium) of *E. coli* strain M15[pYD3 pREP4]. To minimize formation of inclusion bodies, cells are grown under vigorous shaking to an $OD_{600}$ of 0.5 at 30°. Expression of cpn10 is induced by adding IPTG (isopropyl-$\beta$-D-thiogalactopyranoside; Sigma) to a final concentration of 0.3 m*M*. After 3 to 4 hr of induction, cells are harvested and either frozen in liquid nitrogen and kept at −80°, or processed directly as described below. The level of induction can be assessed by comparing whole cell extracts on a 10% (v/v) Tris–Tricine gel (Fig. 2A, lanes 1 and 2). A 1-liter culture typically yields 2–3 g of wet cell paste.

Cells are resuspended in 10 ml of 50 m*M* morpholineethanesulfonic acid (MES)–NaOH, pH 6.0, supplemented with protease inhibitors as described in General Procedures, and submitted to sonication (sonicator XL 2015; Misonix, Farmingdale, NY). Lysozyme is not used for cell wall lysis because of the similarities with cpn10 in apparent size and basic p*I*. The resulting suspension is cleared by ultracentrifugation for 20 min at 2°, using a Beckman TFT 65.13 rotor (50 krpm = 160,900 *g*). SDS–PAGE analysis of crude extracts before and after ultracentrifugation (Fig. 2A, lanes 3 and 4) shows decreased cpn10 levels after ultracentrifugation, indicating the removal of an insoluble cpn10 fraction. Addition of 0.1% (w/v) Triton X-100 during the sonication procedure results in release of 90% of the total cpn10 into the supernatant. This suggests that cpn10 may stay attached to membranes or other sedimentable complexes, rather than forming inclusion bodies. However, Triton X-100 disrupts the oligomeric structure of cpn10 and is therefore not used during the purification procedure (S. Rospert and A. Lustig, unpublished, 1997).

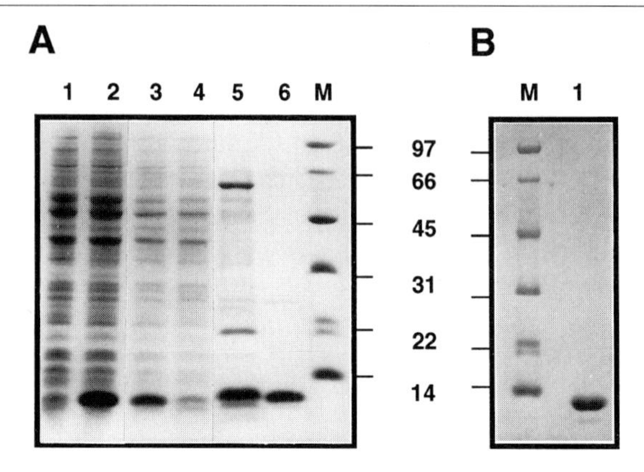


FIG. 2. Purification of authentic cpn10 (A) and cpn10–$H_6$ (B). (A) Protein profile of fractions in the purification procedure for authentic cpn10. Lanes 1 and 2, total *E. coli* cells before and after IPTG induction, respectively; lanes 3 and 4, 10 $\mu$g of crude cell extract before and after ultracentrifugation, respectively; lane 5, Sepharose S pool (10 $\mu$g); lane 6, cpn10 fraction after butyl–Sepharose chromatography (10 $\mu$g). (B) One-step purification of cpn10–$H_6$ on Ni–agarose. A 10-$\mu$g aliquot of proteins eluted from Ni–agarose beads is run on a 10% (v/v) Tris–Tricine gel and stained with Coomassie blue [as for (A)]. Lane M, molecular weight markers; values ($\times$ $10^{-3}$) are indicated on the left-hand side.

*Ion-Exchange Chromatography.* The clear *E. coli* extract is loaded at a rate of 1 ml/min on a trimethylaminoethyl (TMAE) Superformance anion-exchange column (150 × 10 mm; Merck) that is connected directly to a column of Fast Flow Sepharose S (150 × 10 mm; Pharmacia). Both columns have been previously equilibrated with 50 m*M* MES–NaOH, pH 6.0. When no more protein is detected in the flowthrough, the TMAE column is disconnected, and cpn10 is eluted from the Sepharose S column with 60 ml of a linear 0 to 600 m*M* NaCl gradient. The cpn10-containing fractions (2 ml), which elute at 200 to 375 m*M* NaCl, are pooled and concentrated in Centricon-30 microconcentrators (Amicon). Simultaneously, the buffer is exchanged for 50 m*M* HEPES–NaOH (pH 7.4), 1 *M* NaCl. The resulting fraction, termed the *Sepharose S pool* (Fig. 2A, lane 5), contains cpn10 and a major contaminant with an apparent size of 56–60 kDa. This contaminant cannot be removed by gel filtration on a Superose 12 column (HR 10/30; Pharmacia) or by using a different ion-exchange column (Y. Dubaquié, unpublished, 1997).

*Hydrophobic Interaction Chromatography.* The Sepharose S pool is loaded at a rate of 0.5 ml/min onto a Fractogel TSK butyl–Sepharose column (150 × 10 mm; Merck) equilibrated with 50 m*M* HEPES–NaOH, 1 *M* NaCl. The major contaminant still present after the Sepharose S column treatment is eluted in the flowthrough, whereas cpn10 binds to the column.

cpn10 is eluted from the butyl–Sepharose with 50 m$M$ HEPES–NaOH lacking NaCl (Fig. 2A, lane 6). Fractions (1 ml) containing cpn10 are pooled, concentrated to 1–2 mg/ml in Centricon-30 microconcentrators, adjusted to 10% (v/v) glycerol, 50 m$M$ NaCl, and immediately frozen in liquid nitrogen. At −80°, cpn10 remains stable for at least 2 months, as judged from malate dehydrogenase refolding assays. The purification procedure is summarized in Table II.

*Purification of Chaperonin 10–Hexahistidine*

Adsorption of hexahistidine-tagged proteins on nickel–nitrilotriacetic acid (Ni–NTA) agarose beads is a widely used purification procedure (QIA-express system; Qiagen). The following adaptations of standard buffers are used for purification of cpn10–$H_6$: Buffer A containing 300 m$M$ NaCl, 20 m$M$ imidazole, and 5% (v/v) glycerol in 20 m$M$ HEPES–NaOH, pH 7.4; buffer B is the same as buffer A, but contains 600 m$M$ NaCl; buffer C is the same as buffer A, but contains 300 m$M$ imidazole. In addition, all buffers contain the protease inhibitor cocktail described in General Procedures. *Escherichia coli* strain M15[pREP4] containing the plasmid pRL1, which encodes a histidine-tagged version of cpn10, is grown as described previously for the authentic cpn10 purification procedure. Crude cytosolic extract is prepared by sonication of a cell suspension in buffer A (20 ml for 2.0–3.0 g wet cell paste), cleared by centrifugation for 20 min in a Beckman SS-34 rotor (15,000 rpm = 26,900 $g$) at 4°, diluted by addition of buffer A to a final volume of 80 ml, and mixed with 5 ml of a 1 : 1 Ni–bead suspension in buffer A. cpn10–$H_6$ is allowed to adsorb for 1–2 hr at 4° on a rocking shaker. Beads are spun down at 4000 rpm (3000 $g$) in a Heraeus Minifuge RF (Osterode, Germany) and washed twice with 30 ml of buffer B (5 min on shaker at 4°). Buffer B (6 ml) is added, and cpn10–$H_6$ is eluted

TABLE II
PURIFICATION OF cpn10 AND cpn10–$H_6$

| Material | Total protein (mg) | Total cpn10 (mg) | Purification (-fold) | Yield (%) |
|---|---|---|---|---|
| cpn10 | | | | |
| Cell extract | 52 | 3 | 1 | 100 |
| Sepharose S pool | 3 | 2 | 6 | 67 |
| Butyl–Sepharose pool | 0.75 | 0.7 | 8 | 23 |
| cpn10–$H_6$ | | | | |
| Cell extract | 54 | 5 | 1 | 100 |
| Ni–agarose eluate | 4 | 3.7 | 10 | 74 |

for 45 min on a rocking shaker at 4°. The supernatant is collected at 4000 rpm, and the elution procedure is repeated once with 6 ml of buffer. The cpn10–$H_6$ eluted from the beads is 90–95% pure (Fig. 2B, lane 1), and can be directly used if imidazole does not interfere with further experiments. If imidazole needs to be removed, the buffer can be exchanged using Centricon 30 microconcentrators.

An additional FPLC purification step yields highly purified cpn10–$H_6$. The pooled eluates are loaded onto a Mono Q column (HR 5/5; Pharmacia) equilibrated with 50 m*M* MES–NaOH, pH 6.0. cpn10–$H_6$ is recovered in the flowthrough, whereas most of the contaminating proteins bind to the column. The recovered material is concentrated, adjusted to 10% (v/v) glycerol and 50 m*M* NaCl, frozen in liquid nitrogen, and kept at −80°. One liter of *E. coli* culture (2.0–3.0 g wet cell paste) yields about 4 mg of purified cpn10–$H_6$ (Table II).

## Characterization of Purified Yeast Chaperonins

Ultracentrifugation analysis of purified chaperonins yields a molecular mass of 797 ± 10 kDa for hsp60, 75 ± 5 kDa for cpn10, and 80 ± 10 kDa for cpn10–$H_6$ (A. Lustig and Y. Dubaquié, unpublished, 1997). These results indicate a correct assembly of the oligomeric proteins, being tetradecameric for hsp60 and heptameric for cpn10 and cpn10–$H_6$, respectively. Recombinant hsp60 has an ATPase activity comparable to that of hsp60 purified from *S. cerevisiae* mitochondria.[11] In the presence of cpn10 or cpn10–$H_6$, purified hsp60 promotes the ATP-dependent refolding of denatured malate dehydrogenase. Both co-chaperonins are also functional in combination with bacterial GroEL. The purified recombinant chaperonins are thus fully functional.[21]

Electron micrographs of cpn10–$H_6$ reveal a torus-like structure with sevenfold symmetry (A. Engel, personal communication, 1997). In addition, cpn10–$H_6$ can fully complement a *CPN10* disruption in *S. cerevisiae.*

[21] Y. Dubaquié, R. Looser, and S. Rospert, *Proc. Natl. Acad. Sci. USA* **94,** 9011 (1997).

# [18] Purification of Mammalian Mitochondrial Chaperonin 60 through *in Vitro* Reconstitution of Active Oligomers

*By* PAUL V. VIITANEN, GEORGE LORIMER, WOLFGANG BERGMEIER, CELESTE WEISS, MARTIN KESSEL, and PIERRE GOLOUBINOFF

The GroEL homolog of mammalian mitochondria (e.g., mt-cpn60[1]) is nuclear encoded and posttranslationally imported into mitochondria,[2-4] where it assists in the folding of other mitochondrial proteins.[5] The gene for the mammalian mt-cpn60 precursor has been cloned from multiple species,[2,3] and recombinant versions of the protein have been overexpressed in *Escherichia coli* and purified to homogeneity.[1,6] The mammalian mt-cpn60 has also been purified from natural sources[2,7]; however, the quantities obtained are insufficient for detailed biochemical or structural analysis.

Even though mammalian mt-cpn60[1,6] and mt-cpn10[8-11] are both available as recombinant proteins (including those of humans), little is known about how they interact with one another during protein folding. Owing to technical difficulties associated with the mammalian mt-cpn60 (see below), functional characterization of the mammalian mt-cpn10 has largely been restricted to *in vitro* interactions with GroEL.[10-15] In this heterologous test

[1] P. V. Viitanen, G. H. Lorimer, R. Seetharam, R. S. Gupta, J. Oppenheim, J. O. Thomas, and N. J. Cowan, *J. Biol. Chem.* **267,** 695 (1992).

[2] D. J. Picketts, C. S. K. Mayanil, and R. S. Gupta, *J. Biol. Chem.* **264,** 12001 (1989).

[3] S. Jindal, A. K. Dudani, B. Singh, C. B. Harley, and R. S. Gupta, *Mol. Cell. Biol.* **9,** 2279 (1989).

[4] R. S. Gupta and R. A. Austin, *Eur. J. Cell Biol.* **45,** 170 (1987).

[5] M. Y. Cheng, F.-U. Hartl, J. Martin, R. A. Pollock, F. Kalousek, W. Neupert, E. M. Hallberg, R. L. Hallberg, and A. L. Horwich, *Nature (London)* **337,** 620 (1989).

[6] S. Jindal, E. Burke, B. Bettencourt, S. Khandekar, R. A. Young, and D. S. Dwyer, *Immunol. Lett.* **39,** 127 (1994).

[7] L. A. Mizzen, C. Chang, J. I. Garrels, and W. J. Welch, *J. Biol. Chem.* **264,** 20664 (1989).

[8] S. J. Pilkington and J. E. Walker, *DNA Sequencing* **3,** 291 (1993).

[9] M. T. Ryan, N. J. Hoogenraad, and P. B. Høj, *FEBS Lett.* **337,** 152 (1994).

[10] R. Dickson, B. Larsen, P. V. Viitanen, M. B. Tormey, J. Geske, R. Strange, and L. T. Bemis, *J. Biol. Chem.* **269,** 26858 (1994).

[11] G. Legname, T. Fossati, G. Gromo, N. Monzini, F. Marcucci, and D. Modena, *FEBS Lett.* **361,** 211 (1995).

[12] T. H. Lubben, A. A. Gatenby, G. K. Donaldson, G. H. Lorimer, and P. V. Viitanen, *Proc. Natl. Acad. Sci. U.S.A.* **87,** 7683 (1990).

[13] D. J. Hartman, N. J. Hoogenraad, R. Condron, and P. B. Høj, *Proc. Natl. Acad. Sci. U.S.A.* **89,** 3394 (1992).





0076-6879/98 $25.00

system, GroES and the mammalian mt-cpn10 are functionally interchangeable by a number of criteria, including the ability to assist GroEL in the facilitation of protein folding. The resounding conclusion from these experiments is that the fundamental mechanism of the GroE-related chaperonins has been highly conserved from bacteria to mitochondria.

However, only a limited number of experiments have been performed with the purified mammalian mt-cpn60.[1,10] For example, it has been shown that the latter can form a stable complex with nonnative ribulose-bisphosphate carboxylase (Rubisco, EC 4.1.1.39), and that the recovery of active Rubisco from the binary complex requires ATP hydrolysis and mammalian mt-cpn10. Neither bacterial GroES nor the chloroplast cpn10 (ch-cpn10) can replace the mammalian mt-cpn10 in this reaction. Thus, the mammalian mt-cpn60 is far more discriminating than GroEL (see [19] in this volume[16]) or the chloroplast cpn60[17] in its preference for a co-chaperonin.

Other significant differences also exist between the mammalian mt-cpn60 and other GroEL homologs. Unlike the double-ringed GroEL, the recombinant mammalian mt-cpn60 used for the *in vitro* studies cited above, as well as the authentic protein purified from mitochondria,[2] consists of a single seven-member ring.[1,10] Because this preparation supported the facilitated folding of Rubisco, when augmented with ATP and mammalian mt-cpn10, it appeared as if a single toroid were the minimal unit capable of chaperonin function. However, the transient formation of a double-ringed intermediate during this reaction could not be excluded.

The major obstacle that has limited *in vitro* studies with the mammalian mt-cpn60 is its notorious instability. In contrast to the cpn60 homologs of bacteria, chloroplasts, and yeast mitochondria, all of which survive purification as stable 14-mers, the mammalian mt-cpn60 readily dissociates into inactive monomeric subunits. This occurs at each step of its purification, resulting in low yields of the desired product. To date, none of our attempts to stabilize the native oligomer during purification have been successful. To circumvent this problem, we have devised an alternative strategy that actually exploits the instability of the protein. Here we describe the purification of monomeric mammalian mt-cpn60 and its subsequent reassembly

---

[14] M. T. Ryan, D. J. Naylor, N. J. Hoogenraad, and P. B. Høj, *J. Biol. Chem.* **270,** 22037 (1995).

[15] A. C. Cavanagh and H. Morton, *Eur. J. Biochem.* **222,** 551 (1994).

[16] P. V. Viitanen, K. Bacot, R. Dickson, and T. Webb, *Methods Enzymol.* **290,** Chap. 19 (1998) (this volume).

[17] P. V. Viitanen, M. Schmidt, J. Buchner, T. Suzuki, E. Vierling, R. Dickson, G. H. Lorimer, A. Gatenby, and J. Soll, *J. Biol. Chem.* **270,** 18158 (1995).

into functional oligomers using a general approach that has worked well with other GroEL homologs.[18–21]

*Materials.* $NaH^{14}CO_3$ is from Du Pont-New England Nuclear (Boston, MA). MgATP, *N*-2-hydroxyethylpiperazine-*N'*-2-ethanesulfonic acid (HEPES), bovine serum album (BSA), isopropyl-$\beta$-D-thiogalactopyranoside (IPTG), dithiothreitol (DTT), phenylmethylsulfonyl fluoride (PMSF), deoxyribonuclease I (DNase I), and hexokinase (type F-300) are purchased from Sigma (St. Louis, MO). Ribulose 1,5-bisphosphate is prepared as before.[22] The TSK gel-filtration columns are obtained from Toso Haas (Montgomeryville, PA). GroEL, GroES, and the dimeric Rubisco from *Rhodospirillum rubrum* are purified and quantitated as previously described.[23] The purification of mouse mt-cpn10 and spinach ch-cpn10 is described in [19] in this volume.[16] For all purified proteins, the concentrations given in text refer to subunit molecular masses.

## Expression of Mammalian Mitochondrial Chaperonin 60

The T7 promoter-driven plasmid construct that is used for the expression of the "mature" Chinese hamster mt-cpn60 has been described.[1] As determined by N-terminal amino acid sequence analysis, the purified recombinant protein is identical to the authentic protein purified from mitochondria. *Escherichia coli* BL21(DE3) cells, harboring the mt-cpn60 expression plasmid, are grown at 37° in LB medium containing ampicillin (150 $\mu$g/ml). At an $A_{550}$ of ~1.0, the cells are induced for expression with 0.5 m*M* IPTG, and are harvested 12 hr later by centrifugation (10,000 *g*, 20 min, 4°). Cell pellets are stored at −80° for subsequent use.

## Preparation of *Escherichia coli* Cell-Free Extracts

All steps are performed at 0–4°. The frozen cell pellets are rapidly thawed and resuspended (3 ml/g wet weight) by manual homogenization in 100 m*M* Tris-HCl (pH 7.7), 10 m*M* $MgSO_4$, 1 m*M* DTT, DNase I (30 $\mu$g/

[18] N. M. Lissin, S. Yu Venyaminov, and A. S. Girshovich, *Nature (London)* **348,** 339 (1990).
[19] N. M. Lissin, *FEBS Lett.* **361,** 55 (1995).
[20] M. Morioka and H. Ishikawa, *J. Biochem.* **114,** 468 (1993).
[21] R. S. Boston, P. V. Viitanen, and E. Vierling, *Plant Molecular Biology* **32,** 191 (1996).
[22] S. Gutteridge, G. S. Reddy, and G. H. Lorimer, *Biochem. J.* **260,** 711 (1989).
[23] M. Schmidt, J. Buchner, M. J. Todd, G. H. Lorimer, and P. V. Viitanen, *J. Biol. Chem.* **269,** 10304 (1994).

ml), and 0.5 m*M* PMSF.[24] Cell disruption is achieved by two passages through a French pressure cell at 20,000 psi. Cell debris is removed by centrifugation (60,000 *g*, 1 hr), and the resulting supernatant is supplemented with 5% (v/v) glycerol and stored at −80° for subsequent use. The protein concentration of the cell-free extract (Fig. 2, lane 1) is ~20 mg/ml.

## Determination of Protein Concentrations and Assessment of Purity

Throughout the purification procedure, protein concentrations are determined by the method of Lowry *et al.*,[25] using BSA as a standard. Chromatographic fractions are analyzed for purity and recovery of the recombinant protein by sodium dodecyl sulfate-polyacrylamide gel electrophoresis (SDS–PAGE),[26] with Coomassie blue staining. Appropriate column fractions are pooled and subjected to further manipulation. Because of factors that are frequently difficult to control (e.g., slight variations in column specifications, flow rates, and temperature) the investigator should not rely exclusively on values cited in text for retention times or gradient positions where proteins elute.

## Purification Strategy

Several technical difficulties were encountered during our initial attempts to purify the mammalian mt-cpn60.[1] In particular, the oligomeric protein proved to be highly unstable and readily dissociated into subunits. Because dissociation was enhanced at lower temperatures (cf. Fig. 2 of Ref. 1), the entire purification procedure had to be carried out at 25°.[27] Although this manipulation stabilized the oligomers to some extent, the recombinant protein still behaved as if it were multiple species. For example, when *E. coli* cell-free extracts, prepared from the recombinant strain, were fractionated on a Mono Q (Pharmacia, Piscataway, NJ) column, two peaks of mt-cpn60 were detected, which eluted between 0.09–0.14 *M* NaCl and 0.27–0.35 *M* NaCl.[1] The material in the first Mono Q peak (at least half of the total recombinant protein) was unable to bind to ATP–agarose, did not support facilitated protein folding, and was not further characterized. In contrast, the mt-cpn60 in the later eluting peak bound to ATP–agarose, was functional in folding assays, and proved to be a heptamer, following its

[24] All pH values reported in text were measured at 25°.

[25] O. H. Lowry, N. J. Rosebrough, A. L. Farr, and R. J. Randall, *J. Biol. Chem.* **193,** 265 (1951).

[26] U. K. Laemmli, *Nature* (*London*) **227,** 680 (1970).

[27] Although not to the same extent that is observed with the mammalian mt-cpn60, lower temperatures also destabilize other GroEL homologs (cf. Ref. 19).

purification by affinity chromatography. However, with time the heptameric material dissociated into inactive monomers.

We know now that the mt-cpn60 eluting in the first Mono Q peak is monomeric (see below). Given this observation and the problematic instability of the native heptamer, it was reasoned that it might be easier to purify the monomeric subunits and reassemble the desired oligomer *in vitro*. With this goal in mind, conditions were sought to maximize the dissociation of the recombinant protein before its attempted purification. As shown in Fig. 1C, when *E. coli* cell-free extracts are incubated at 0° with EDTA, and subsequently analyzed by gel filtration, virtually all of the mt-cpn60 elutes as a monomer (fractions 9–11). In contrast, little monomer is observed when the same cell-free extract is preincubated at 25° with MgATP and the mammalian mt-cpn10 (Fig. 1B). Under these conditions, the majority of the mt-cpn60 exhibits an average molecular mass of ~350 kDa, consistent with its previously reported heptameric structure.[1,2] Shifting the equilibrium of the recombinant protein in the cell-free extract toward the monomeric state is an essential component of the purification procedure described as follows.

## Purification of Monomeric Mammalian Mitochondrial Chaperonin 60

### *Ion-Exchange Chromatography*

The first step in the purification of monomeric mt-cpn60 entails anion-exchange chromatography. The scale described below is for a single run

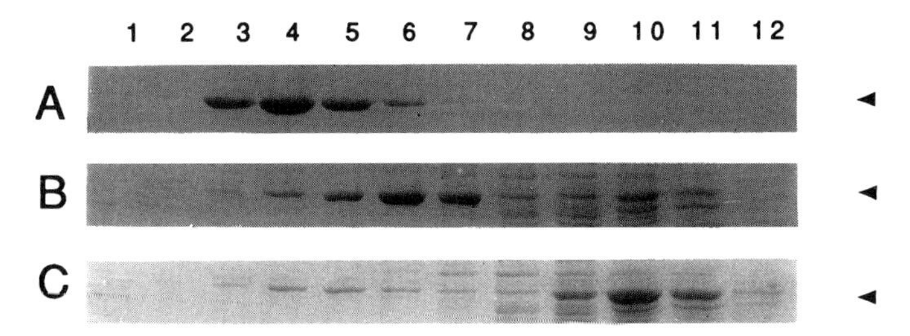


FIG. 1. Size distribution of recombinant mt-cpn60. *Escherichia coli* cell-free extract (~20 mg of protein/ml) was incubated for 2 hr at 0° with 20 m*M* EDTA (C). Alternatively, the extract was supplemented with ATP (4 m*M*), KCl (16 m*M*), and mt-cpn10 (25 $\mu M$), and incubated for 40 min at 25° (B). Aliquots (100 $\mu$l) of these mixtures were then applied to a TSK G4000SW gel-filtration column (7.5 × 300 mm), which was developed at 1.0 ml/min (25°) with buffer B (see text). Starting at 6.5 min, fractions (0.4 ml each) were collected and analyzed by SDS–PAGE and Coomassie blue staining using 15% (w/v) gels. Lanes 1–12, successive fractions from the gel-filtration columns. The profile obtained with purified GroEL is shown in (A). On the basis of protein standards, the mt-cpn60 eluting in fractions 5–7 (B) and fractions 9–11 (C) yield average molecular masses of ~350 and ~60 kDa, respectively.

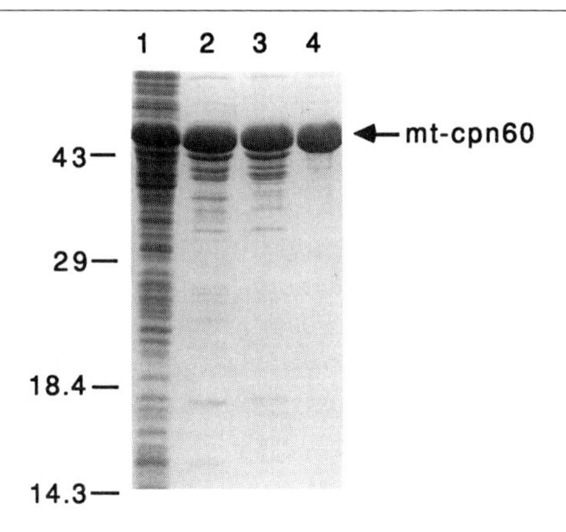


FIG. 2. Purification of recombinant mt-cpn60. Lane 1, cell-free extract; lane 2, material after Mono Q; lane 3, partially purified mt-cpn60 monomers after the preparative TSK G3000SW gel filtration; lane 4, purified mt-cpn60 oligomers after *in vitro* reconstitution and a second round of gel filtration. Lanes 1–4 contain ~30, 15, 11, and 6 $\mu$g of total protein, respectively. Analysis was by SDS–PAGE using 15% (w/v) gels and Coomassie blue staining.

on a Mono Q HR 10/10 column (Pharmacia). Because the entire chromatographic procedure (including column regeneration) requires less than 1 hr, up to eight columns can be run in a single day. An aliquot (7.5 ml) of the *E. coli* cell-free extract (~0.15 g of total protein) is rapidly thawed and filtered through a 0.2-$\mu$m pore size Acrodisc filter (Gelman Sciences, Ann Arbor, MI). The filtrate is supplemented with 20 m*M* EDTA (pH 8) and incubated for 2 hr at 0°, to induce the nearly complete dissociation of mt-cpn60 oligomers. Following buffer exchange[28] into 50 m*M* Tris-HCl (pH 7.7), 1 m*M* EDTA, 10 m*M* sodium sulfite (buffer A), the entire sample is applied to the Mono Q column. The latter is washed with 20 ml of buffer A at 4 ml/min (25°), and is then developed with a linear gradient (60 ml) of 0–0.25 *M* NaCl (in buffer A). Fractions eluting between ~35–70 m*M* NaCl are pooled,[29] supplemented with 2.5 m*M* DTT and 5% (v/v) glycerol, concentrated to ~12 mg of protein/ml,[30] and stored at −80° for subsequent use. Before processing additional aliquots of the cell-free extract, the column is washed with 1 *M* NaCl (in buffer A) and reequilibrated to starting conditions. The recovery of total protein from a single Mono Q column is ~12 mg (Fig. 2, lane 2).

[28] All buffer exchange steps are carried out at 4° using PD-10 gel-filtration columns (Pharmacia, Piscataway, NJ).

[29] The NaCl concentrations cited have been corrected for the 8-ml transit through the column.

[30] Unless indicated, protein samples are concentrated at 4° using Centriprep-30 and Centricon-30 concentrators; both are available from Amicon (Danvers, MA).

### *Gel-Filtration Chromatography*

Two milliliters of the concentrated Mono Q pool (~24 mg of total protein) is applied to a preparative TSK G3000SW gel-filtration column (21.5 × 600 mm). The column is developed at 4 ml/min (25°) with 50 m*M* Tris-HCl (pH 7.7), 0.3 *M* NaCl, and the material eluting between ~30 and 33 min is collected (e.g., monomeric mt-cpn60). The sample is supplemented with glycerol (5%, v/v) and DTT (2.5 m*M*), concentrated to ~75 mg of protein/ml, and stored at −80° for use in the large-scale reconstitution procedure described below. The yield of monomeric mt-cpn60 from a single preparative gel filtration column is ~16 mg, corresponding to ~5% of the total protein present in the cell-free extract.

## Reconstitution of Oligomeric Mitochondrial Chaperonin 60

### *General Considerations*

Judging from Coomassie-stained gels the monomeric mt-cpn60 recovered from the preparative gel-filtration column is only 80–90% pure (Fig. 2, lane 3). However, this material is suitable for *in vitro* reconstitution experiments, which result in nearly homogeneous preparations of oligomeric mt-cpn60. The reconstituted particles are readily purified by a second round of gel-filtration chromatography, because the only protein contaminants remaining in the monomeric starting material are of relatively low molecular weight (e.g., <100 kDa). There are two attractive features of the *in vitro* reconstitution procedure. First, intact chaperonin rings, such as GroEL, bind unfolded protein substrates cooperatively and avidly. Consequently, certain protein contaminants copurify with the oligomer. However, this problem is circumvented by initially purifying the monomer. Second, the *in vitro* reassembly process is selective for molecules that are functional. Invariably, a portion of the monomeric mt-cpn60 is not incorporated into oligomers (~20–30%), and much of this material has been nicked by proteases. The remaining assembly-incompetent molecules are likely denatured or damaged in some other way.

As reported for other GroEL homologs, the *in vitro* assembly of oligomeric mt-cpn60 is strictly dependent on adenine nucleotides (Fig. 3A), and ATP is most effective in this regard. At 30°, reconstitution is essentially complete by 90 min,[31] and the resulting oligomers are easily resolved from protein contaminants and unincorporated monomers by gel filtration. In accordance with analogous studies,[18–21] the percentage yield of oligomeric

[31] P. V. Viitanen and W. Bergmeier, unpublished observations (1996).

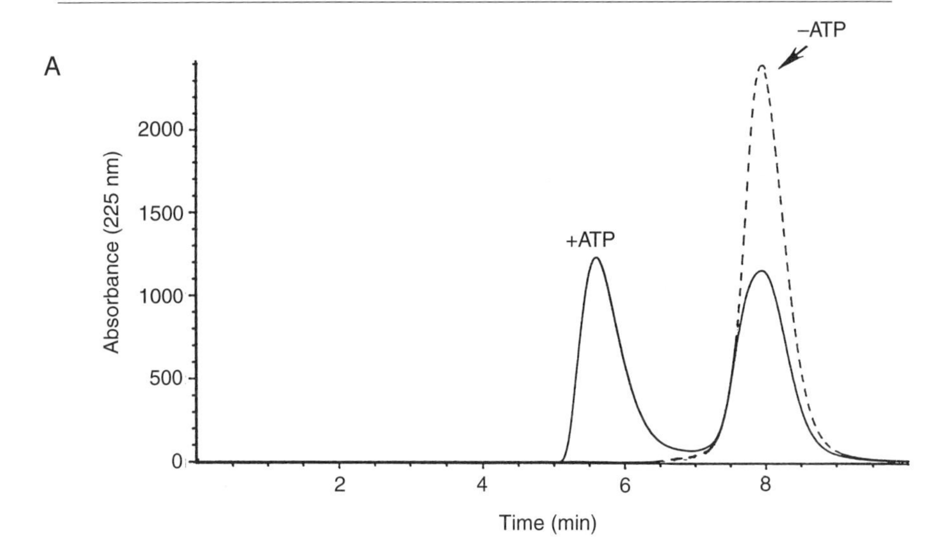


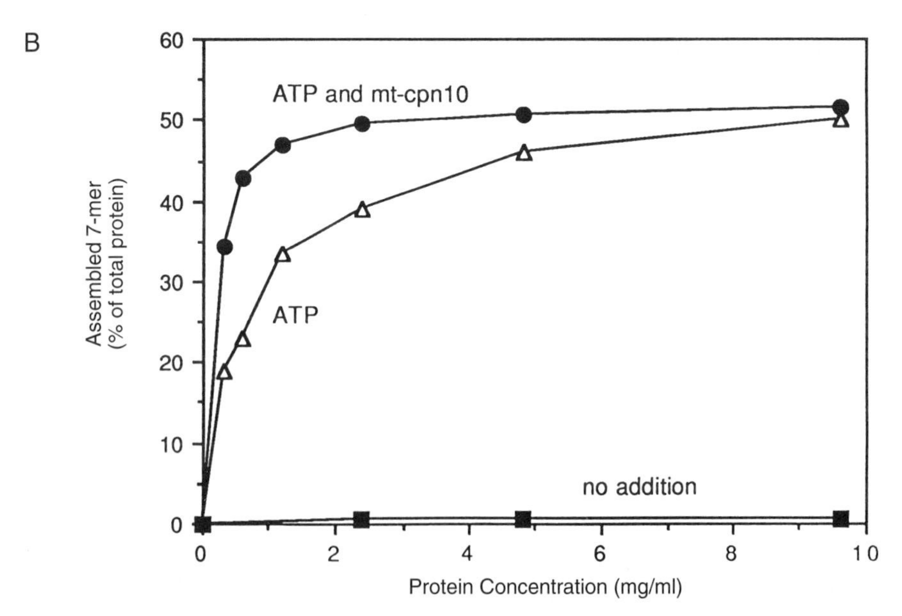


FIG. 3. Reconstitution of mt-cpn60 oligomers. (A) Monomeric mt-cpn60 (4.7 mg of protein/ml) was incubated in 50 m*M* Tris-HCl (pH 7.5), 0.3 *M* NaCl, 20 m*M* magnesium acetate, and 20 m*M* KCl, in the presence (solid trace) and absence (broken trace) of 4 m*M* ATP. After a 90-min incubation period at 30°, aliquots (100 $\mu$l) were injected onto a TSK G3000SW gel-filtration column (7.5 × 300 mm). The latter was developed at 1.0 ml/min (25°) with buffer

mt-cpn60 is highly dependent on the total protomer concentration (Fig. 3B), suggesting that oligomerization is cooperative. In the presence of ATP alone, half-maximal assembly is observed at a protein concentration of ~0.7 mg/ml. Assuming that the monomeric precursor is only 80% pure, this translates to ~10 $\mu M$ protomer. Maximal assembly is observed at protomer concentrations in excess of ~100 $\mu M$, with more than 50% of the monomers being incorporated into oligomers. Although it is not required for the formation of oligomers per se, the mammalian mt-cpn10 potentiates the ATP-dependent assembly reaction at lower protomer concentrations (Fig. 3B). However, the mitochondrial co-chaperonin has no effect on the maximum yield of reconstituted particles. As will become apparent, many of these observations are exploited in the reconstitution procedure described in the next section.

### *Large-Scale Reconstitution and Purification of Oligomeric Mitochondrial Chaperonin 60*

Owing to the instability of the oligomeric mt-cpn60 (see below), it is imperative that all of the following manipulations be carried out at temperatures of at least 25°. The *in vitro* assembly reaction mixture is constructed by mixing the following components together: 223 $\mu$l of monomeric mt-cpn60 (75 mg of protein/ml), 13 $\mu$l of 1 $M$ KCl, 13 $\mu$l of 1 $M$ magnesium acetate, 350 $\mu$l of buffer B [50 m$M$ Tris-HCl (pH 7.7), 0.3 $M$ NaCl, 10 m$M$ $MgCl_2$], and 52 $\mu$l of 0.05 $M$ ATP (pH 7.5). Following a 90-min incubation period at 30°, the first 100-$\mu$l aliquot of six to be processed is removed from the reaction vessel, and injected onto a TSK G3000SW gel-filtration column (7.5 × 300 mm); the remaining sample is returned to 30° for subsequent processing. The column is developed at 1.0 ml/min with buffer B (25°), and oligomeric mt-cpn60, eluting between ~5.1 and 6.5 min (Fig. 2, lane 4), is collected. The purified oligomers are kept at 25° while the rest of the reaction mixture is subjected to gel filtration.

---

B, and $A_{225}$ was monitored. On the basis of retention time, the molecular mass of monomeric mt-cpn60 is ~65 kDa. In contrast, the reconstituted 7-mers elute between 5.1 and 6.5 min, just after the void volume (4.7 ml) of the column. (B) ATP-dependent reconstitution reactions (△) were performed with indicated concentrations of monomeric mt-cpn60, and analyzed by gel filtration. The percentage of monomeric starting material that was reconstituted into heptamers at each concentration is plotted. These values were calculated from the individual chromatographic tracings, by integrating the area under the peak where mt-cpn60 heptamers elute (5.1–6.5 min) and expressing it as a percentage of the total area. (●) Identical to (△), but also containing an equivalent amount of mouse mt-cpn10. (■) Identical to (△), but ATP was omitted.

The pooled oligomers are concentrated to 15–20 mg of protein/ml, supplemented with glycerol (5%, v/v), and divided into small aliquots. The latter are rapidly frozen on dry ice and stored at −80°. On the basis of protein determination,[25] the yield of purified reconstituted particles is ~10 mg, or roughly 60% of the total monomeric starting material. However, this value underestimates the efficiency of the *in vitro* reconstitution procedure, because the monomeric starting material is not pure. Indeed, it would appear that virtually all of the assembly-competent monomers are incorporated into oligomers.

## Properties of *in Vitro*-Reconstituted Particles

### *Stability*

Provided that the following precautions are taken, the concentrated stock solution of oligomeric mt-cpn60 is relatively stable to multiple freeze–thaw cycles: (1) The preparation should be rapidly thawed and maintained at 25° when not in the frozen state; (2) after removing the required amount of material, the stock solution should be rapidly refrozen and returned to −80°. Dilute solutions of the reconstituted particles are more problematic, the degree of dissociation depending on the protein concentration and experimental conditions.

For example (Fig. 4), when the concentrated stock solution is diluted nearly 100-fold with buffer B and incubated for 6 hr at 25°, more than 90% of the mt-cpn60 oligomers are recovered by gel filtration (Fig. 4, ▲). However, at 0° (Fig. 4, △), the reconstituted particles dissociate into monomers with a $t_{1/2}$ of ~30 min. The destabilizing effect of lower temperature is greatly potentiated by ATP, which further reduces the $t_{1/2}$ for dissociation to <3 min (Fig. 4, ○). Although the effect is not nearly as dramatic as at 0°, in dilute solutions ATP also causes substantial dissociation of oligomers at 25° (Fig. 4, ●).

Thus, the effect of ATP on the oligomeric state of the mammalian mt-cpn60 is seemingly paradoxical. It is absolutely required for the assembly of oligomers at higher protein concentrations (Fig. 3), and induces oligomer dissociation at lower protein concentrations (Fig. 4). However, these observations are consistent with the behavior that is observed with other GroEL homologs. According to Lissin,[19] in the presence of ATP, cpn60 monomers (mt-cpn60) are converted to an assembly-competent state (mt-cpn60*), which in turn is in rapid equilibrium with the oligomeric state (Scheme I). In the absence of ATP, the energy barrier separating monomers from oligomers is so great that equilibrium is either never attained or is only slowly attained. Consequently, whether ATP causes oligomerization or

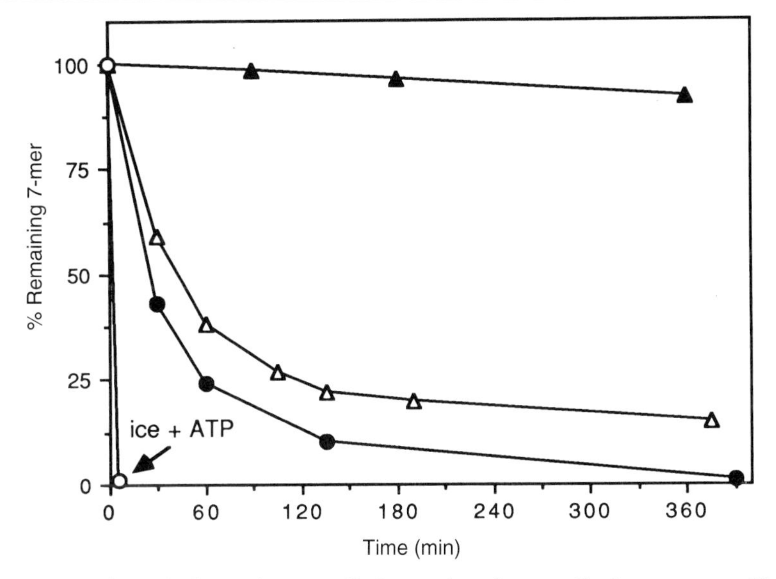


FIG. 4. Stability of oligomeric mt-cpn60. Reconstituted mt-cpn60 oligomers were diluted with buffer B at 25° to a protomer concentration of ~3.3 $\mu M$. Two portions of this mixture were supplemented with 3.8 m$M$ ATP (●,○), while the other two received no further addition (▲, △). The reactions were then incubated at 0° (open symbols) or 25° (filled symbols). At the indicated times, aliquots (100 $\mu$l) were removed and analyzed by gel filtration as described in the caption to Fig. 3. Plotted is the percentage of oligomeric mt-cpn60 remaining as a function of time.

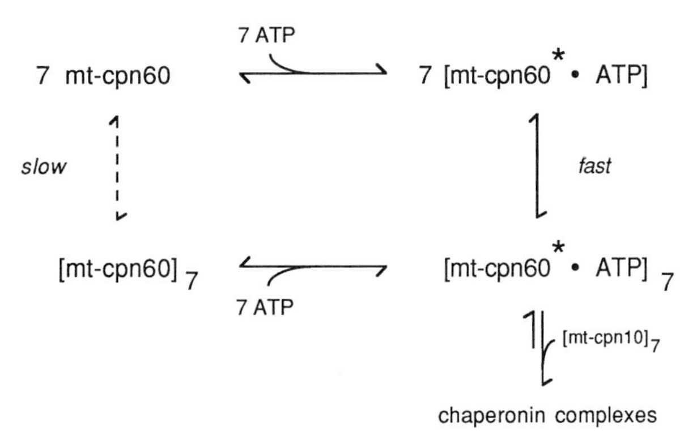


SCHEME I

dissociation depends only on the prevailing equilibrium constant and the total concentration of mt-cpn60. Obviously, various factors can influence $K_{eq}$ including temperature, pH, and ionic strength, and complex formation with mt-cpn10 would favor oligomerization.

*Assistance of Protein Folding*

The facilitated folding reaction mediated by the mammalian mt-cpn60 requires both ATP hydrolysis and the participation of cpn10.[1,10] Consequently, $K^+$ and $Mg^{2+}$ ions are also required, because the mammalian mt-cpn60 cannot hydrolyze ATP in their absence. These are the expected properties of cpn60. However, in contrast with other GroEL homologs,[16,17] the mammalian mt-cpn60 is only able to interact with its natural partner, the mammalian mt-cpn10. Although the basis of this extreme functional specificity is not understood, the reassembled mt-cpn60 oligomers are equally discriminating (Fig. 5).

When stable binary complexes formed between acid-denatured Rubisco and the reconstituted particles are incubated in a nonpermissive environment where unassisted spontaneous folding does not occur,[23] the recovery of active Rubisco is observed only in the presence of ATP and the mammalian mt-cpn10 (Fig. 5, ●). In the experiment shown, the $t_{1/2}$ for facilitated folding was ~1 min and the yield of native Rubisco was nearly 60%. In contrast, neither GroES (Fig. 5, □) nor chloroplast cpn10 (ch-cpn10; Fig. 5, ■) is able to discharge active Rubisco from the binary complex. Thus, the reconstituted mt-cpn60 particles are functional in *in vitro* protein-folding assays and exhibit the same general characteristics, including co-chaperonin specificity, that have previously been described.[1,10]

*Oligomeric Structure*

As determined by gel filtration, the average molecular mass of the reconstituted oligomeric mt-cpn60 is ~400 kDa (Fig. 1B), a value that is closer to a heptamer than a tetradecamer. However, the peak is broad and trace amounts of 14-mers are likely included in the leading edge. Consistent with previous observations,[1] electron micrographs of the reconstituted particles yield predominantly top views, toroidal structures that consist of seven subunits, arranged with sevenfold symmetry (Fig. 6A and C, row 1). Whereas 7-mers and 14-mers are indistinguishable in this orientation, two populations of reconstituted particles are evident in side views. One population consists of 7 × 13-nm rectangular structures with two striations running parallel to the long axis,[32] and these correspond to the single-ring heptamers

[32] M. Kessel and P. V. Viitanen, unpublished observations (1996).

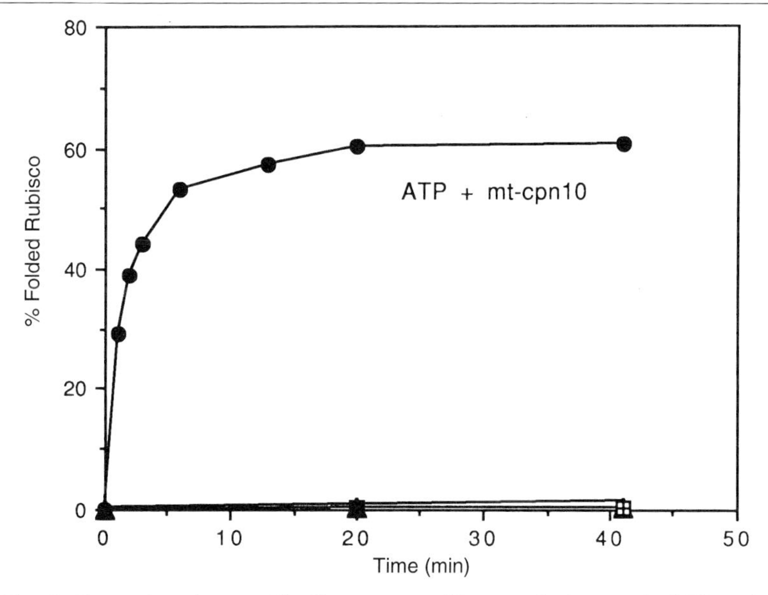


FIG. 5. Reconstituted mt-cpn60 oligomers are able to assist in protein folding. Acid-denatured Rubisco (see [19] in this volume[16]) was rapidly diluted to 85 n*M* into a solution (at 25°) containing 94 m*M* HEPES–KOH (pH 7.6), 5 $\mu M$ BSA, 2.3 m*M* DTT, and reconstituted mt-cpn60 oligomers (2.5 $\mu M$, based on protomer). After adding magnesium acetate to 10 m*M*, aliquots of this mixture (100 $\mu$l) were incubated at 25° with the following additions: (●) mouse mt-cpn10 plus ATP; (□) GroES plus ATP; (■) spinach ch-cpn10 plus ATP. The concentration of ATP was 2 m*M* and the molar ratio of cpn10 to mt-cpn60 was ~3:1. At the indicated times, folding reactions were terminated by the addition of 4 $\mu$l of a solution containing 0.35 *M* glucose and hexokinase (1250 U/ml), and Rubisco activity was determined. *Note:* Active Rubisco was not recovered in the absence of ATP.

that were previously described.[1] However, larger rectangles (14 × 13-nm) with four striations are also present, and these resemble the two stacked rings of GroEL (Fig. 6C, row 2). That 14-mers were not seen in the earlier study is probably a reflection of their *in vitro* instability and subtle differences in experimental conditions. Regardless, even in the present study, heptamers were still the predominant species, as evidenced by the paucity of side views that were observed with the reconstituted particles (<5%). Whereas double toroids such as GroEL show no strongly preferred orientation on the carbon grid (side views and top views are both well represented), single toroids yield almost exclusively top views.[1,33]

Interestingly, when the reconstituted particles are incubated with ATP and mt-cpn10, the proportion of side views increases dramatically. More-

[33] N. Ishii, H. Taguchi, H. Sasabe, and M. Yoshida, *FEBS Lett.* **362,** 121 (1995).

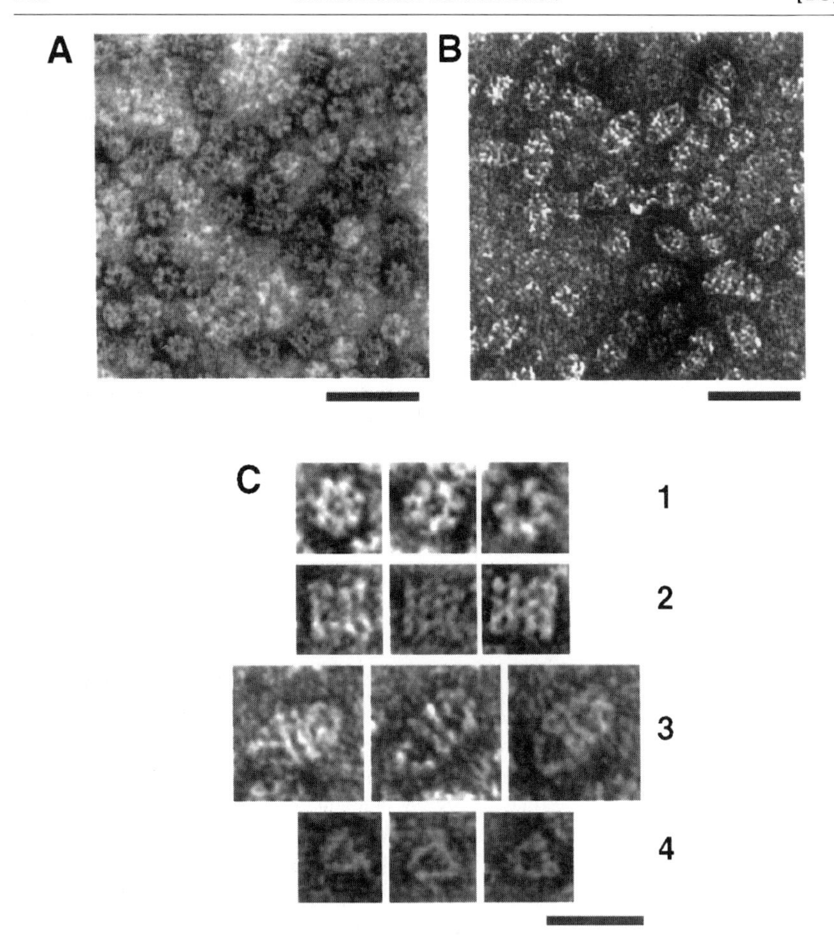


FIG. 6. Electron micrographs of reconstituted mt-cpn60 oligomers, alone and in combination with ATP and mt-cpn10. Reconstituted mt-cpn60 oligomers (5 $\mu M$, based on protomer) were incubated in 90 m$M$ HEPES–KOH (pH 7.5), 30 m$M$ magnesium acetate in the presence (B) and absence (A) of 3.1 m$M$ ATP and a twofold molar excess of mouse mt-cpn10. After a 15-min incubation period at 25°, the samples were negatively stained with 1% uranyl acetate, and examined in a Zeiss 902 transmission electron microscope, operating at 80 kV. Micrographs were recorded at a nominal magnification of 50,000. Bars: 50 nm. (C) Enlargements of selected particles: 1, top views of reconstituted mt-cpn60 oligomers; 2, side views of mt-cpn60 14-mers; 3, side views of "football-shaped" chaperonin complexes (e.g., $[\text{mt-cpn60}]_{14} \cdot 2[\text{mt-cpn10}]_7$); 4, side views of "half-footballs" (e.g., $[\text{mt-cpn60}]_7 \cdot [\text{mt-cpn10}]_7$). Bar: 20 nm.

over, the predominant species are now symmetrical football-shaped particles (Fig. 6B and C, row 3), virtually identical to those obtained with the *E. coli* chaperonins in the presence of ATP.[34] Apparently, mt-cpn60 14-mers are more stable in the presence of their ligands. Through analogy to GroEL and GroES,[35–37] we conclude that the football-shaped particles consist of 1 mol of mt-$cpn60_{14}$ and 2 mol of mt-$cpn10_7$, and further suggest that they are important intermediates in the mitochondrial chaperonin reaction cycle. It was therefore anticipated that some bullet-shaped particles would also be observed in the presence of ATP and mt-cpn10, but this was not the case. Apparently, asymmetric complexes consisting of 1 mol of mt-$cpn60_{14}$ and 1 mol of mt-$cpn10_7$ are not stable under the experimental conditions. Nevertheless, they likely exist *in vivo,* because in the case of the *E. coli* chaperonins, "footballs" and "bullets" are both intermediates in a cycle in which each species begets the other.[37]

Smaller, triangular-shaped particles were also observed when the two mitochondrial chaperonins were incubated with ATP (Fig. 6C, row 4). These resemble "half-footballs" and presumably consist of one ring each of mt-cpn60 and mt-cpn10. Similar observations have been made with the holochaperonin of *Thermus thermophilus,* which undergoes an equatorial split in the presence of ATP.[33] Although the mechanistic significance of half-footballs requires further investigation, the preceding results leave little doubt that the mammalian mt-cpn60 functions as a tetradecamer *in vivo.*

[34] ATP hydrolysis is not required since "footballs" are also observed with AMP-PNP, but not ADP.

[35] M. Schmidt, K. Rutkat, R. Rachel, G. Pfeifer, R. Jaenicke, P. Viitanen, G. Lorimer, and J. Buchner, *Science* **265,** 656 (1994).

[36] A. Azem, M. Kessel, and P. Goloubinoff, *Science* **265,** 653 (1994).

[37] M. J. Todd, P. V. Viitanen, and G. H. Lorimer, *Science* **265,** 659 (1994).

# [19] Purification of Recombinant Plant and Animal GroES Homologs: Chloroplast and Mitochondrial Chaperonin 10

*By* PAUL V. VIITANEN, KAREN BACOT, RAMONA DICKSON, and TOM WEBB

The chaperonins[1] of chloroplasts and mitochondria consist of two distinct family members, generically referred to as chaperonin 60 (cpn60) and chaperonin 10 (cpn10). Together these proteins facilitate the folding of numerous other proteins through a complicated reaction cycle that requires ATP hydrolysis.[2–4] The best studied of the chaperonins are the bacterial prototypes, GroEL (cpn60) and GroES (cpn10). GroEL oscillates between states of high and low affinity for nonnative proteins.[5,6] In the absence of adenine nucleotides it forms a stable complex with the unfolded protein substrate,[7] suppressing both aggregation[8] and productive[9] folding. Although the bound protein substrate can be released from GroEL with ATP alone, this is not always sufficient to ensure proper folding. Depending on the stringency of the experimental conditions, the chaperonin-assisted folding reaction also requires GroES.[10,11] It is believed that the latter coordinates the hydrolysis of ATP, such that all of the subunits in one GroEL ring convert to their low-affinity state in synchrony.[6] This ensures the transient,

[1] S. M. Hemmingsen, C. Woolford, S. M. van der Vies, K. Tilly, D. T. Dennis, C. P. Georgopoulos, R. W. Hendrix, and R. J. Ellis, *Nature* (*London*) **333,** 330 (1988).

[2] P. Goloubinoff, J. T. Christeller, A. A. Gatenby, and G. H. Lorimer, *Nature* (*London*) **342,** 884 (1989).

[3] A. A. Gatenby and P. V. Viitanen, *Annu. Rev. Plant Physiol. Plant Mol. Biol.* **45,** 469 (1994).

[4] J. P. Hendrick and F.-U. Hartl, *Annu. Rev. Biochem.* **62,** 349 (1993).

[5] G. S. Jackson, R. A. Staniforth, D. J. Halsall, T. Atkinson, J. J. Holbrook, A. R. Clarke, and S. G. Burston, *Biochemistry* **32,** 2554 (1993).

[6] M. J. Todd, P. V. Viitanen, and G. H. Lorimer, *Science* **265,** 659 (1994).

[7] P. V. Viitanen, A. A. Gatenby, and G. H. Lorimer, *Protein Sci.* **1,** 363 (1992).

[8] J. Buchner, M. Schmidt, M. Fuchs, R. Jaenicke, R. Rudolph, F. X. Schmid, and T. Kiefhaber, *Biochemistry* **30,** 1586 (1991).

[9] A. A. Laminet, T. Ziegelhoffer, C. Georgopoulos, and A. Plückthun, *EMBO J.* **9,** 2315 (1990).

[10] M. Schmidt, J. Buchner, M. J. Todd, G. H. Lorimer, and P. V. Viitanen, *J. Biol. Chem.* **269,** 10304 (1994).

[11] J. Martin, T. Langer, R. Boteva, A. Schramel, A. L. Horwich, and F.-U. Hartl, *Nature* (*London*) **352,** 36 (1991).



0076-6879/98 $25.00

but complete, release of the substrate protein, affording it an opportunity to fold unhindered in free solution.[5,6,12] Those molecules that fail to achieve their native conformation are efficiently recaptured by GroEL, and eventually partition to the folded state through iterative cycles of release and rebinding.

Although only a limited number of studies have been performed with the purified chaperonins of mitochondria (see Refs. 13–18, and see [18] in their volume[19]) and chloroplasts,[20–26] it has already been demonstrated that cpn60 and cpn10 are also both required for protein folding in these organelles. Here we describe detailed large-scale purification schemes for spinach chloroplast cpn10 (ch-cpn10) and mouse mitochondrial cpn10 (mt-cpn10), two eukaryotic GroES homologs that are available as functional recombinant proteins.[16,24,25] Although both proteins can be purified from their natural sources,[14,15,22,27] the yields obtained are insufficient for biochemical and structural studies. In contrast, the two recombinant proteins are readily abundant and easily purified to homogeneity. It should be noted that purification schemes for recombinant human[17] and rat[18] mt-cpn10 are also available in the literature, and an affinity purification procedure, potentially applicable to all cpn10 homologs, has been published.[18]

[12] J. S. Weissman, Y. Kashi, W. A. Fenton, and A. L. Horwich, *Cell* **78,** 693 (1994).

[13] T. H. Lubben, A. A. Gatenby, G. K. Donaldson, G. H. Lorimer, and P. V. Viitanen, *Proc. Natl. Acad. Sci. U.S.A.* **87,** 7683 (1990).

[14] D. J. Hartman, N. J. Hoogenraad, R. Condron, and P. B. Høj, *Proc. Natl. Acad. Sci. U.S.A.* **89,** 3394 (1992).

[15] P. V. Viitanen, G. H. Lorimer, R. Seetharam, R. S. Gupta, J. Oppenheim, J. O. Thomas, and N. J. Cowan, *J. Biol. Chem.* **267,** 695 (1992).

[16] R. Dickson, B. Larsen, P. V. Viitanen, M. B. Tormey, J. Geske, R. Strange, and L. T. Bemis, *J. Biol. Chem.* **269,** 26858 (1994).

[17] G. Legname, T. Fossati, G. Gromo, N. Monzini, F. Marcucci, and D. Modena, *FEBS Lett.* **361,** 211 (1995).

[18] M. T. Ryan, D. J. Naylor, N. J. Hoogenraad, and P. B. Høj, *J. Biol. Chem.* **270,** 22037 (1995).

[19] P. V. Viitanen, G. Lorimer, W. Bergmeier, C. Weiss, M. Kessel, and P. Goloubinoff, *Methods Enzymol.* **290,** Chap. 18 (1998) (this volume).

[20] S. M. Hemmingsen and R. J. Ellis, *Plant Physiol.* **80,** 269 (1986).

[21] J. E. Musgrove, R. A. Johnson, and R. J. Ellis, *Eur. J. Biochem.* **163,** 529 (1987).

[22] U. Bertsch, J. Soll, R. Seetharam, and P. V. Viitanen, *Proc. Natl. Acad. Sci. U.S.A.* **89,** 8696 (1992).

[23] N. M. Lissin, *FEBS Lett.* **361,** 55 (1995).

[24] F. Baneyx, U. Bertsch, C. E. Kalbach, S. M. van der Vies, J. Soll, and A. A. Gatenby, *J. Biol. Chem.* **270,** 10695 (1995).

[25] P. V. Viitanen, M. Schmidt, J. Buchner, T. Suzuki, E. Vierling, R. Dickson, G. H. Lorimer, A. Gatenby, and J. Soll, *J. Biol. Chem.* **270,** 18158 (1995).

[26] R. S. Boston, P. V. Viitanen, and E. Vierling, *Plant Molecular Biology* **32,** 191 (1996).

[27] A. C. Cavanagh and H. Morton, *Eur. J. Biochem.* **222,** 551 (1994).

## Unexpected Properties of Eukaryotic GroES Homologs

The higher plant ch-cpn10[22] and mammalian mt-cpn10[13] were initially identified through their abilities to form stable, isolatable complexes with GroEL. Previous studies, demonstrating that the two *Escherichia coli* chaperonins physically interact with each other in the presence of ATP,[28,29] suggested that it might be possible to use the bacterial GroEL protein to "go fishing" for unknown GroES homologs in crude chloroplast and mitochondrial extracts. The two eukaryotic proteins that were identified in this manner were interchangeable with GroES by a number of criteria, including the ability to assist GroEL in the folding of prokaryotic Rubisco (ribulose-bisphosphate carboxylase, EC 4.1.1.39).[13,22] However, certain structural and functional properties clearly distinguish them from each other and their bacterial counterparts.

For example, the molecular mass of the ch-cpn10 promoter (~21 kDa) is nearly twice that of bacterial or mitochondrial cpn10,[22] and actually consists of two complete GroES-like sequences that are fused "head-to-tail" to form a single protein. Remarkably, both "halves" of this unusual protein can function autonomously in certain GroES-deficient mutants,[24] and are highly conserved at amino acid residues that are believed to be important for cpn10 function (Fig. 1). In addition, the two tandemly linked domains each contain a polypeptide segment analogous to the so-called "mobile loop" region of GroES,[30] which is critical for the interaction between the two bacterial chaperonins.[31] The binary organization of the ch-cpn10 presumably reflects a special adaptation of plants that has occurred in response to their possession of two divergent cpn60 isoforms.[32]

The mammalian mt-cpn10 also exhibits substantial sequence homology to GroES (Fig. 1), but it consists of the more conventional 10-kDa subunits. Although this protein can effectively interact with GroEL,[13,14] the reciprocal heterologous combination is not functional. Thus, during *in vitro* protein folding experiments, only the mammalian mt-cpn10 can satisfy the stringent requirements of the mammalian mt-cpn60 (see Refs. 15 and 16, and see [18] in this volume[19]), and GroES cannot substitute. More intriguing is a report that the mammalian mt-cpn10 is also an extracellular growth factor,

[28] G. N. Chandrasekhar, K. Tilly, C. Woolford, R. Hendrix, and C. Georgopoulos, *J. Biol. Chem.* **261,** 12414 (1986).

[29] P. V. Viitanen, T. H. Lubben, J. Reed, P. Goloubinoff, D. P. O'Keefe, and G. H. Lorimer, *Biochemistry* **29,** 5665 (1990).

[30] E. V. Koonin and S. M. van der Vies, *Trends Biochem. Sci.* **20,** 14 (1995).

[31] S. J. Landry, J. Zeilstra-Ryalls, O. Fayet, C. Georgopoulos, and L. M. Gierasch, *Nature* (*London*) **364,** 255 (1993).

[32] R. Martel, L. P. Cloney, L. E. Pelcher, and S. M. Hemmingsen, *Gene* **94,** 181 (1990).

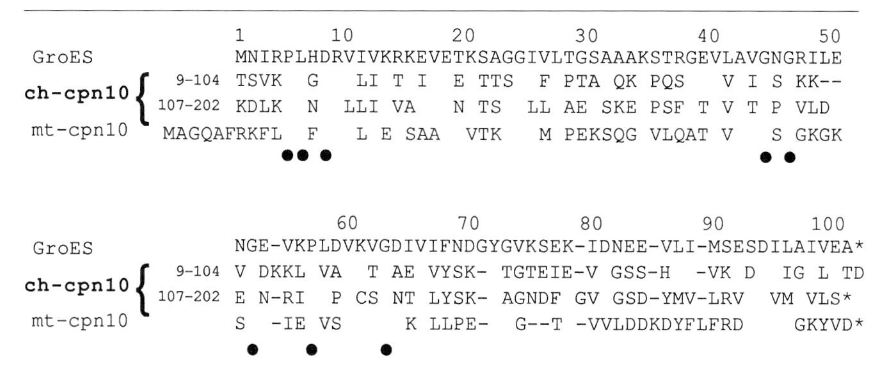


FIG. 1. Mouse mt-cpn10 and both halves of the mature spinach ch-cpn10 exhibit strong homology to GroES. Two tandem amino acid sequences, present in the open reading frame of the spinach ch-cpn10 (residues 9–104 and residues 107–202, according to Fig. 4 of Bertsch *et al.*[22]), and the sequence of mouse mt-cpn10 (Dickson *et al.*[16]) are aligned with GroES (Chandrasekhar *et al.*[28]). Single amino acid gaps (–) have been introduced for optimal alignment. Blank spaces denote residues that are identical to those of GroES. Filled circles indicate residues that are present in most cpn10 sequences, and that are also present in both "halves" of the spinach ch-cpn10.

originally described as early pregnancy factor, that possesses potent immunosuppressive properties.[27] This unexpected dual role could explain the high degree of sequence conservation that is observed among all the mammalian mt-cpn10s that have been cloned (e.g., the proteins from humans and cows are identical).[16,33–35]

*Materials.* $NaH^{14}CO_3$ is from Du Pont-New England Nuclear, Boston, MA). MgATP, ADP, *N*-2-hydroxyethylpiperazine-*N'*-2-ethanesulfonic acid (HEPES), 4-morpholineethanesulfonic acid (MES), bovine serum album (BSA), isopropyl-β-D-thiogalactopyranoside (IPTG), dithiothreitol (DTT), phenylmethylsulfonyl fluoride (PMSF), deoxyribonuclease I (DNase I), and hexokinase (type F-300) are purchased from Sigma (St. Louis, MO). Ribulose 1,5-bisphosphate is prepared as before.[36] GroEL, GroES, and the dimeric Rubisco from *Rhodospirillum rubrum* are purified and quantitated as previously described.[10] For all proteins, the concentrations given in text refer to subunit molecular masses.

[33] S. J. Pilkington and J. E. Walker, *DNA Sequencing* **3,** 291 (1993).

[34] M. T. Ryan, N. J. Hoogenraad, and P. B. Høj, *FEBS Lett.* **337,** 152 (1994).

[35] N. Monzini, G. Legname, F. Marcucci, G. Gromo, and D. Modena, *Biochim. Biophys. Acta* **1218,** 478 (1994).

[36] S. Gutteridge, G. S. Reddy, and G. H. Lorimer, *Biochem. J.* **260,** 711 (1989).

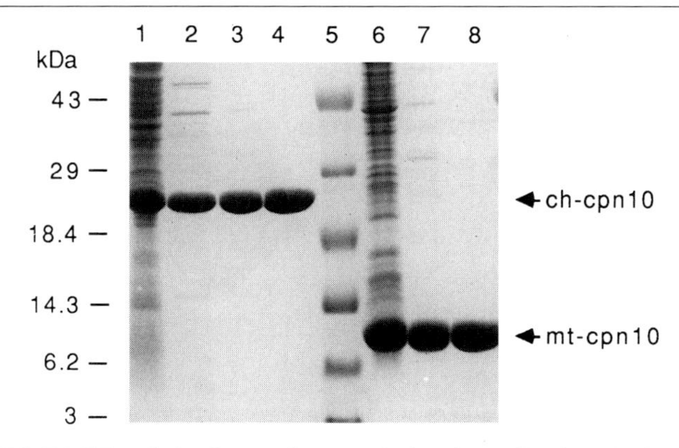


FIG. 2. SDS–PAGE analysis of successive steps during the purification of eukaryotic GroES homologs. Purification of spinach ch-cpn10: lane 1, cell-free extract; lane 2, material after treatment with S-Sepharose Fast Flow; lane 3, material after hydroxylapatite column treatment; lane 4, material after Mono Q column treatment. Lanes 1–4 contain ~28, 6, 6, and 10 $\mu$g of total protein, respectively. Purification of mouse mt-cpn10: lane 6, cell-free extract; lane 7, material after Mono S column treatment; lane 8, material after the preparative gel-filtration column treatment. Lanes 6–8 contain ~28, 10, and 15 $\mu$g of total protein, respectively. Lane 5, protein molecular mass standards. Analysis was by SDS–PAGE using 15% (w/v) gels and Coomassie blue staining.

## General Considerations

The recombinant spinach ch-cpn10 and mouse mt-cpn10 are both well expressed in *E. coli,* at levels exceeding 10% of the total soluble protein (Fig. 2, lanes 1 and 6). Thus, a mere 10-fold purification theoretically results in homogeneous preparations. However, as with the purification of any recombinant protein, the level of expression in *E. coli* is extremely important. Poor inductions translate to lower yields and additional purification steps. The investigator is therefore encouraged to optimize protein expression before attempting a purification. If shaker flasks are employed for bacterial growth, adequate aeration is essential. The cells should be vigorously shaken and the growth medium should not exceed 20% of the total volume of the flask. Because the time of induction is also critical for success, cell growth should be carefully monitored spectrophotometrically.

Throughout both purification procedures, the purity and approximate recovery of the recombinant proteins are monitored by sodium dodecyl sulfate-polyacrylamide gel electrophoresis (SDS–PAGE), using 15% (w/v) gels[37] and Coomassie blue staining. For mouse mt-cpn10, which migrates

[37] U. K. Laemmli, *Nature* (*London*) **227,** 680 (1970).

as an extremely diffuse band in freshly poured gels, the resolving gel should be allowed to polymerize at least 24 hr before use. Depending on the desired level of purity at each step, appropriate column fractions are pooled and subjected to further manipulation. It is neither practical nor informative to monitor cpn10 activity profiles. Such assays are difficult to perform on multiple samples and are subject to interference by various contaminants and monovalent ions that are present in crude *E. coli* extracts and column buffers.[10,29]

The most efficient step in the purification of both recombinant proteins exploits their abilities to bind to cation-exchange resins under conditions in which most other proteins do not. For mammalian mt-cpn10, this useful property is adequately explained by its abnormally high isoelectric point. Indeed, mouse,[16] rat,[34] cow,[33] and human[35] mt-cpn10 all have predicted p*I* values that are >9.0. In contrast, the predicted p*I* of GroES and most other procaryotic cpn10s is <6.0. The ionic behavior of spinach ch-cpn10 is not as well understood. With a predicted p*I* of ~5.5, it should bear a net negative charge under the conditions employed. However, the recombinant plant protein binds to cation exchangers at pH values as high as pH 6.5,[24,25] presumably through interaction of its positively charged N-terminal domain.

## Plasmid Constructs

The expression plasmid pCK25, which encodes the spinach ch-cpn10 without its transit peptide, has previously been described.[24] As documented elsewhere,[25] the initiator methionine residue of the recombinant protein—which was added for expression in *E. coli*—is cleaved by the bacterial host to yield the authentic mature chloroplast protein.

The cloning, expression, and purification of mouse mt-cpn10 has been reported.[16] However, the recombinant protein used in that study was modified at its C terminus by the addition of a polyhistidine tail. Here we describe the expression and purification of the authentic mouse mt-cpn10. The polymerase chain reaction (PCR) product described in Ref. 16 is digested with *Nco*I and *Pst*I, and ligated into pGEM5Zf(+) (Promega, Madison, WI), cut with the same enzymes. This step provides a unique *Sac*I site just downstream from the termination codon of the protein. The resulting plasmid is digested with *Nco*I and *Sac*I, and the small fragment containing mt-cpn10 is isolated and ligated into *Nco*I/*Sac*I-digested pET24d(+) (Novagen). Although the final expression construct, pMES, encodes the full-length mouse mt-cpn10, Edman degradation reveals that the initiator methionine residue of this protein is also removed in *E. coli.* The recombinant mouse mt-cpn10 is therefore identical to the authentic mitochondrial protein, but is not acetylated at its N terminus.[16,18,38]

## Expression of Spinach Chloroplast Chaperonin 10 and Mouse Mitochondrial Chaperonin 10

*Escherichia coli* JM105, harboring the spinach ch-cpn10 expression construct, pCK25, is grown at 37° in LB medium containing 0.2% (w/v) glucose and ampicillin (50 $\mu$g/ml). The cells are induced for expression with 1 m*M* IPTG at an $A_{600}$ of ~1.0, and are harvested 6 hr later by centrifugation (10,000 *g*, 20 min, 4°). The cells are rapidly frozen at −80° for subsequent use. Mouse mt-cpn10 is expressed in *E. coli* BL21(DE3), transformed with the plasmid pMES. Growth is at 37° in LB medium containing kanamycin (30 $\mu$g/ml). At an $A_{600}$ of ~0.80, the cells are induced for expression with 1 m*M* IPTG, and after 4 hr are harvested as described previously.

## Determination of Protein

Throughout both purification procedures, total protein is determined by the method of Lowry *et al.*,[39] using BSA as a standard. Molar extinction coefficients, calculated from the translated DNA sequences, are used to quantitate the purified recombinant proteins. At 280 nm, these values are 3840 and 7740 $M^{-1}$ for mt-cpn10 and ch-cpn10, respectively. The molar concentrations given in text are based on protomer molecular masses of 10,831 Da (mt-cpn10) and 21,385 Da (ch-cpn10).

## Preparation of *Escherichia coli* Cell-Free Extracts

All steps are performed at 0–4°. The frozen cell pellets are rapidly thawed and resuspended (2.5 ml/g wet weight) by manual homogenization in 100 m*M* Tris-HCl (pH 7.5), 5 m*M* $MgSO_4$, 1 m*M* DTT, DNase I (30 $\mu$g/ml), and 0.5 m*M* PMSF.[40] Cell disruption is achieved by subjecting the suspension to two passages through a motor-driven French pressure cell at 20,000 psi. Cell debris is removed by ultracentrifugation (100,000 *g*, 1 hr), and the supernatant is supplemented with 5% (v/v) glycerol and stored at −80° for subsequent use. The protein concentration of the cell-free extract is ~35–50 mg/ml.

[38] D. J. Hartman, N. J. Hoogenraad, R. Condron, and P. B. Høj, *Biochim. Biophys. Acta* **1164,** 219 (1993).

[39] O. H. Lowry, N. J. Rosebrough, A. L. Farr, and R. J. Randall, *J. Biol. Chem.* **193,** 265 (1951).

[40] All pH values reported in text are measured at room temperature.

### Purification of Spinach Chloroplast Chaperonin 10

The first step in the purification of the spinach ch-cpn10 uses the cation-exchange resin S-Sepharose Fast Flow (Pharmacia, Piscataway, NJ).[25] All steps are performed at 0–4°. Thirty-five milliliters of the cell-free extract (~1.3 g of total protein) is rapidly thawed and buffer exchanged[41] into 20 m*M* MES–NaOH (pH 6.2), 0.5 m*M* DTT (buffer A). The sample is equally distributed to seven disposable columns (Bio-Rad, Hercules, CA) that each contain 4 ml (settled bed volume) of S-Sepharose Fast Flow, preequilibrated with same buffer. The columns are washed three times by gravity with 8 ml of buffer A, allowing the resin to drain completely in between washes. Spinach ch-cpn10 is then eluted from the resin by adding 5 ml of 0.3 *M* NaCl (in buffer A) to each of the columns. This step is repeated and the combined eluents are supplemented with 5% (v/v) glycerol and concentrated to ~15 mg of protein/ml.[42] Occasionally, a flocculent precipitate forms during sample concentration. This material—which does not contain ch-cpn10—is removed by centrifugation (15,000 *g*, 10 min), and the resulting supernatant is stored at −80°. The recovery of total protein is ~250 mg, and the spinach ch-cpn10 is at least 70% pure (Fig. 2, lane 2).

Next, the entire sample is exchanged into 10 m*M* sodium phosphate (pH 6.8), 0.01 m*M* $CaCl_2$, 0.5 m*M* DTT (buffer B), and applied to a preparative high-performance liquid chromatography (HPLC)-hydroxylapatite column (Bio-Rad) that is preequilibrated with the same buffer. The column is developed at 25° with a linear gradient (150 ml) of 10–500 m*M* sodium phosphate, pH 6.8 (in buffer B), at a flow rate of 3 ml/min. Fractions eluting between ~195 and 250 m*M* sodium phosphate are pooled, supplemented with 5% (v/v) glycerol, concentrated to ~35 mg of protein/ml, and stored at −80°. The recovery of total protein is ~115 mg, and the spinach ch-cpn10 is nearly homogeneous (Fig. 2, lane 3).

In the last step, the entire sample is exchanged into 50 m*M* Tris-HCl (pH 7.7), 0.5 m*M* DTT (buffer C), and applied to a Mono Q HR 10/10 column (Pharmacia). The latter is developed with a linear gradient (125 ml) of 0–0.5 *M* NaCl (in buffer C), at a flow rate of 3.5 ml/min (25°). Fractions eluting between ~180 and 210 m*M* NaCl are pooled and supple-

[41] Buffer exchange steps are carried out at 4° using PD-10 gel-filtration columns (Pharmacia, Piscataway, NJ), according to manufacturer instructions.

[42] Samples are concentrated at 4° using Centriprep-30 and Centricon-30 concentrators (Amicon, Danvers, MA). It should be emphasized that the spinach ch-cpn10 exhibits a strong propensity to adsorb to various surfaces, including the walls of microcentrifuge tubes. This can result in extremely low yields of purified protein if certain precautions are not taken. The greatest losses occur at low ionic strength and pH values greater than pH 7. However, we have found that the inclusion of glycerol (5%, v/v) and NaCl (0.3 *M*), during sample concentration and storage, greatly enhances recovery.

mented with 5% (v/v) glycerol and additional NaCl to a final concentration of ~0.3 *M*. The material is then concentrated to ~15 mg of protein/ml, and stored at −80° for subsequent use. The final yield of purified spinach ch-cpn10 (Fig. 2, lane 4) is ~40 mg, corresponding to about 3% of the total protein present in the cell-free extract.

## Purification of Mouse Mitochondrial Chaperonin 10

Twelve milliliters of the cell-free extract (~0.6 g of total protein) is buffer exchanged[41] into 50 m*M* Tris-HCl (pH 7.2), 0.1 m*M* EDTA, 1.0 m*M* DTT (buffer D), and passed through a 0.2-$\mu$m pore size Acrodisc filter (Gelman Sciences, Ann Arbor, MI). The sample is applied to a Mono Q HR 16/10 column (Pharmacia) that is preequilibrated with buffer D at 25°; the flow rate is 4 ml/min. Under these conditions, mouse mt-cpn10 elutes isocratically. Appropriate fractions are pooled, supplemented with 5% (v/v) glycerol, and concentrated[42] to ~40 mg of protein/ml. This material is exchanged into 20 m*M* MES–NaOH (pH 6.0), 0.1 m*M* EDTA (buffer E), and applied to a Mono S HR 10/10 column (Pharmacia) that is preequilibrated with the same buffer at 25°. The column is washed with 25 ml of buffer E, and is then developed with a linear gradient (160 ml) of 0–0.5 *M* NaCl (in buffer E) at a flow rate of 2.5 ml/min. Fractions eluting between ~20 and 100 m*M* NaCl are pooled, supplemented with 5% (v/v) glycerol, concentrated to ~40 mg of protein/ml, and stored at −80°. The recovery of total protein is ~120 mg, and mouse mt-cpn10 is substantially pure (Fig. 2, lane 7).

Next, the entire sample is buffer exchanged into 10 m*M* sodium phosphate (pH 6.8), 0.01 m*M* $CaCl_2$ (buffer F), and applied to the same preparative hydroxylapatite column that was used above for the spinach ch-cpn10. The column is preequilibrated with buffer F, and developed at 25° (3 ml/min) with a linear gradient (150 ml) of 10–350 m*M* sodium phosphate, pH 6.8 (in buffer F). Fractions eluting between ~200 and 240 m*M* sodium phosphate are pooled, supplemented with 5% (v/v) glycerol, concentrated to ~30 mg of protein/ml, and stored at −80°. The recovery of total protein is ~70 mg.

Finally, half of the above material is applied to a 21.5 × 600 mm TSK G3000SW gel-filtration column (Toso Haas, Montgomeryville, PA) that is preequilibrated with 50 m*M* Tris-HCl (pH 7.7), 0.3 *M* NaCl. The column is developed at 3 ml/min (25°) with the same buffer. Highly purified mouse mt-cpn10, eluting between ~40 and 43 min, is collected and kept on ice, while the remaining half of the sample is processed. The pooled fractions from both runs are supplemented with 5% (v/v) glycerol, concentrated to

~80 mg of protein/ml, and stored at −80° for subsequent use. The final yield of purified mouse mt-cpn10 (Fig. 2, lane 8) is ~60 mg, or ~10% of the total protein present in the cell-free extract.

## Assay for Chaperonin 10 Function

A sensitive and reliable assay for the purified, recombinant, eukaryotic GroES homologs exploits their ability to assist GroEL during *in vitro* protein folding. Previous studies have shown that under nonpermissive conditions,[10] the recovery of active Rubisco from the GroEL–Rubisco binary complex strictly requires MgATP, $K^+$ ions, and the participation of GroES.[2,10,29] In this reaction, however, ch-cpn10[22,24] and mammalian mt-cpn10[13–17] can both effectively substitute for the normal partner of GroEL.

Chaperonin assisted-folding reactions are performed essentially as described,[10] taking precautions to minimize the final concentration of $Cl^-$ ions.[43] The dimeric Rubisco of *R. rubrum* is denatured for 15 min at 25° in 10 m*M* HCl at a final concentration of 10 μ*M*. Acid-denatured Rubisco is rapidly diluted to 80 n*M* in a solution (at 0°) containing 94 m*M* HEPES–KOH (pH 7.6), 5 μ*M* BSA, 2.7 m*M* DTT, and 2.0 μ*M* GroEL. The temperature is immediately brought to 25° and magnesium acetate is added to 10 m*M*. Aliquots (100 μl) of this mixture are supplemented with GroES or GroES homologs (0 to 10 μ*M*), and folding reactions are then initiated with ATP (2 m*M*). The final reaction volume is 125 μl. It is essential to include control reactions in which ATP and/or the GroES homologs are omitted. Folding is terminated at various times by the addition of 4 μl of an ATP-scavaging solution that contains 0.35 *M* glucose and hexokinase (1250 U/ml), and Rubisco activity is then determined.[2] To quantitate the extent of refolding, identical experiments are performed with native Rubisco, substituting KCl for HCl during the initial denaturation step.

## Properties of Purified Eukaryotic GroES Homologs

### *Oligomeric Structure*

In accordance with earlier expectations,[28] X-ray crystallography has now proved that native GroES is a heptameric toroid.[44] A similar structure is likely shared by the mammalian mt-cpn10, because this protein also

[43] As documented in Schmidt *et al.*,[10] millimolar concentrations of $Cl^-$ ions promote the spontaneous folding of Rubisco, thus obviating the need for GroES.

[44] J. F. Hunt, A. J. Weaver, S. J. Landry, L. Gierasch, and J. Deisenhofer, *Nature* (*London*) **379,** 37 (1996).

consists of seven identical ~10.8-kDa subunits,[18] and forms stable complexes with the sevenfold symmetrical mammalian mt-cpn60 (see [18] in this volume[19]). Consistent with this notion, the purified recombinant mouse mt-cpn10 exhibits an apparent molecular mass of ~67 kDa during gel filtration,[45] and superficially resembles the toroidal GroES in electron micrographs.[46]

In contrast, the spinach ch-cpn10 exhibits peculiar behavior during gel filtration. As a heptamer of ~21-kDa subunits, its native molecular mass should be approximately 150 kDa. However, the retention time of the recombinant protein strictly depends on its molarity. In dilute solutions, it dissociates into low molecular weight species that range in size from dimers to tetramers,[18,22] whereas at higher protein concentrations its apparent molecular mass can greatly exceed that of a heptamer.[24,45] Indeed, electron micrographs of the purified recombinant protein reveal a heterogeneous population of chainlike molecules that resemble individual rings stacked on top of one another.[24] As with other GroES homologs, the fundamental unit of these peculiar structures is probably a heptameric toroid. Although it is extremely doubtful that the long "chains" that are observed *in vitro* have any physiological significance, the propensity of the recombinant spinach protein to self-aggregate has already imposed severe limitations on structural studies. Consequently, efforts are underway to determine whether other ch-cpn10s might exhibit more desirable properties.

### *In Vitro Protein Folding*

The purified eukaryotic GroES homologs are both able to assist GroEL in the *in vitro* refolding of prokaryotic Rubisco (Fig. 3). However, they do so with different efficiencies. Maximal rates of folding with mt-cpn10 are observed when it and GroEL are present in roughly equal amounts (based on protomer).[45] In contrast, a nearly fivefold molar excess of the purified spinach homolog is required to achieve saturation with GroEL.[45] Although this could simply be a manifestation of the heterologous system, it more likely reflects the tendency of the ch-cpn10 to interact with itself *in vitro*. Such behavior would obviously limit the amount of co-chaperonin that is available for interaction with GroEL.

In any case, one study has shown that the purified recombinant spinach ch-cpn10 is also functionally active with its cognate partner.[25] In these experiments, the two homologous chloroplast chaperonins were able to assist in the *in vitro* refolding of prokaryotic Rubisco and mitochondrial malate dehydrogenase under nonpermissive conditions. Although ATP hy-

[45] P. V. Viitanen, unpublished observations (1995).

[46] M. Kessel and P. V. Viitanen, unpublished observations (1996).

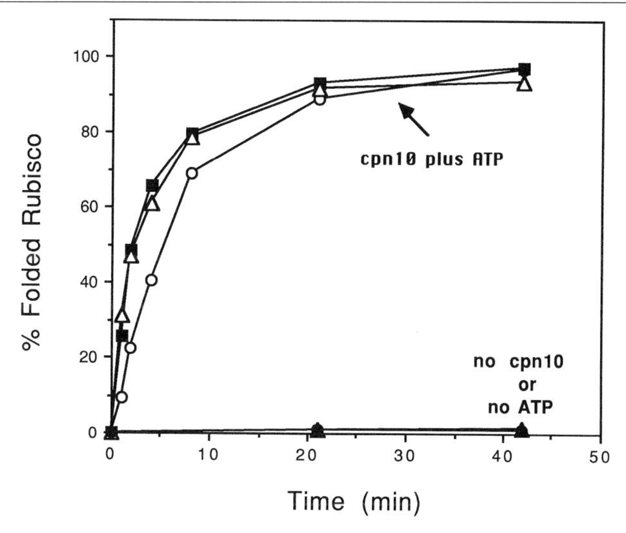


FIG. 3. The purified spinach ch-cpn10 and mouse mt-cpn10 are both functional in *in vitro* protein-folding assays. Preformed GroEL–Rubisco binary complexes were incubated at 25° with the following additions: (△) ATP and GroES; (■) ATP and mouse mt-cpn10; (○) ATP and spinach ch-cpn10. The molar ratio of cpn10 to cpn60 was ~2:1 (based on protomer). At the indicated times, reactions were terminated with glucose/hexokinase and Rubisco activity was determined. (▲) Received ATP, but no cpn10; (●) received cpn10, but no ATP. See text for additional details.

drolysis was required for both of these reactions, the unique binary ch-cpn10 was not an obligate co-chaperonin for the ch-cpn60. Despite the unusual nature of its subunits, the former could effectively be replaced by either GroES or the mammalian mt-cpn10. The extent to which this surprising observation pertains to other unfolded protein substrates remains to be determined.

In contrast, the mammalian mt-cpn60 is extremely selective about its co-chaperonin, and can assist protein folding only in the presence of the mammalian mt-cpn10. As already noted, the only difference between recombinant and authentic mammalian mt-cpn10 is that the latter is acetylated at its N terminus.[38] Indeed, it was initially speculated that this posttranslational modification might be required for functional compatibility with its natural partner. However, we have shown that the recombinant protein is also able to satisfy the stringent requirements of the mammalian mt-cpn60, at least with regard to Rubisco folding (see Ref. 16 and [18] in this volume[19]). Although N-terminal acetylation might be required for some other important physiological function,[18] it does not appear to be essential for the facilitated folding reaction mediated by the mitochondrial chaperonins. Thus, only additional studies with the purified mammalian mt-cpn60 and mt-cpn10 will reveal the unique structural features that enable these two proteins to interact productively with each other.

# [20] Mammalian Cytosolic Chaperonin

*By* NICHOLAS J. COWAN

## Introduction

The homolog of GroEL in the cytosol of eukaryotes is the cytosolic chaperonin, c-cpn[1] (also referred to by some laboratories as TRiC[2] or TCP-1 complex[3]). In common with all known chaperonins, c-cpn is a toroidal structure; however, c-cpn is unique in that it is assembled from eight distinct but related subunits, each in the range of 57–62 kDa, one of which is t-complex polypeptide 1.[4,5] Both genetic[6–8] and biochemical evidence[1,9–11] implicate c-cpn in the facilitated folding of actins and tubulins, the two most abundant proteins in the eukaryotic cytosol. An actin-related protein (actin-RPV) is also folded via interaction with c-cpn,[11] but the role of c-cpn in mediating the folding of other cytosolic proteins has not been firmly established.

Although c-cpn is both more complex than its prokaryotic, mitochon-

[1] Y. Gao, J. O. Thomas, R. L. Chow, G.-H. Lee, and N. J. Cowan, *Cell* **60,** 1043 (1992).
[2] J. Frydman, E. Nimmesgern, H. Erdjument-Bromage, J. S. Wall, P. Tempst, and F.-U. Hartl, *EMBO J.* **11,** 4767 (1992).
[3] V. A. Lewis, G. M. Hynes, D. Zheng, H. Saibil, and K. Willison, *Nature (London)* **352,** 36 (1992).
[4] H. Rommelaere, M. Van Troys, Y. Gao, R. Melki, N. J. Cowan, J. Vandekerckhove, and C. Ampe, *Proc. Natl. Acad. Sci. U.S.A.* **90,** 11975 (1993).
[5] H. Kubota, G. Hynes, A. Carne, A. Ashworth, and K. Willison, *Curr. Biol.* **4,** 89 (1994).
[6] D. Ursic and M. R. Culbertson, *Mol. Cell Biol.* **11,** 2629 (1991).
[7] X. Chen, D. S. Sullivan, and T. Huffaker, *Proc. Natl. Acad. Sci. U.S.A.* **91,** 9111 (1994).
[8] D. B.-N. Vinh and D. Drubin, *Proc. Natl. Acad. Sci. U.S.A.* **91,** 9116 (1994).
[9] M. B. Yaffe, G. W. Farr, D. Miklos, A. L. Horwich, M. L. Sternlicht, and H. Sternlicht, *Nature (London)* **358,** 245 (1992).
[10] Y. Gao, I. E. Vainberg, R. L. Chow, and N. J. Cowan, *Mol. Cell Biol.* **13,** 2478 (1993).
[11] R. Melki and N. J. Cowan, *J. Cell Biol.* **122,** 1301 (1993).



0076-6879/98 $25.00

drial, or chloroplastic homologs and more expensive to prepare, studies using c-cpn have yielded unique insights into chaperonin function. For example, when unfolded actin is presented to GroEL or its mitochondrial homolog, a binary complex is formed and, on incubation with ATP, the target protein undergoes cycles of binding and release; however, no native product is formed. Only on inclusion of c-cpn in the folding reaction does the target protein transfer to its cognate chaperonin, with the resulting production of native actin. These data demonstrate that whereas GroEL or mitochondrial chaperonin is capable of cycling actin intermediates in an ATP-dependent manner, only c-cpn can produce intermediates that will partition to the native state. The same results are obtained in parallel experiments done with tubulin target proteins. It follows that different chaperonins produce distinctive spectra of folding intermediates.[12]

A second feature that makes c-cpn useful for studying chaperonin-mediated folding rests on the observation that the c-cpn-mediated folding of $\alpha$- and $\beta$-tubulin requires not only the participation of c-cpn itself, but also a number of additional protein cofactors.[10] The role of these cofactors is not yet certain, although four that participate in the $\beta$-tubulin-folding pathway have now been purified to homogeneity and cloned.[13] Nonetheless, they are likely to prove useful in dissecting the c-cpn-mediated folding reaction, because their action is ATP independent. For example, in the case of $\alpha$- and $\beta$-tubulin folding, it is possible to uncouple the ATP-dependent cycling of folding intermediates by c-cpn from the action of cofactors that are required for the production of native protein. In these circumstances, $\alpha$- or $\beta$-tubulin-folding intermediates accumulate.

Because c-cpn is a large heterooligomeric complex, the prospect of engineering its expression in a prokaryotic host is a daunting one, and, in any event, there is no guarantee that functional c-cpn would be assembled under the intracellular conditions that prevail in bacteria. Mammalian c-cpn must therefore be purified from a tissue source. We have found rabbit reticulocyte lysate to be the starting material of choice. Although an alternative procedure has been described for the purification of c-cpn from bovine testis,[2] we have not found this method to yield functionally active chaperonin reproducibly. We describe here our method for purifying c-cpn from rabbit reticulocyte lysate, and the procedure that we use to assay its function.

[12] G. Tian, I. E. Vainberg, W. D. Tap, S. A. Lewis, and N. J. Cowan, *Nature* (*London*) **375,** 250 (1995).

[13] G. Tian, Y. Huang, H. Rommelaere, J. Vandekerckhove, C. Ampe, and N. J. Cowan, *Cell* **86,** 287 (1996).

## Purification of Cytosolic Chaperonin from Rabbit Reticulocyte Lysate

### *Materials*

Rabbit reticulocyte lysate is available from a number of vendors, but not all preparations yield active c-cpn. We have found lysate from Promega (Madison, WI) to be the most reliable—if expensive—source of biologically active c-cpn. Because the lysate is prepared as a pool from several animals, there is inevitably some batch-to-batch variation both in the quality and yield of material, but the protocol described here (which is for 10 ml of starting material) almost always yields useful c-cpn, typically in the range 0.4–0.5 mg (enough for about 100 *in vitro* folding reactions; see below). The lysate can be stored at −70° for several months without compromising yield or activity of c-cpn.

### *Procedures*

#### *Lysate Preparation*

1. Thaw 10 ml of lysate (Promega) in a water bath at 37°.
2. Transfer to four 3-ml polycarbonate 1/2 × 2 in. ultracentrifuge tubes (Beckman, Fullerton, CA) and centrifuge at 100,000 rpm (350,000 *g*) for 30 min at 4° using the TL-100 rotor in a Beckman Optima TL ultracentrifuge (or equivalent). This step removes polyribosomes and other particulate material.
3. Pool the supernatants in a 15-ml plastic centrifuge tube on ice.
4. Add 0.2 ml of a 1 *M* solution of $MgCl_2$ and mix by swirling.
5. Filter the dark red solution by passage through a disposable 0.2-$\mu$m pore size filtration unit (e.g., Acrodisc; Gelman, Ann Arbor, MI). The filtrate is now ready for application to an ion-exchange column.

#### *Ion-Exchange Chromatography*

1. Charge a Pharmacia (Piscataway, NJ) 10/10 Mono Q FPLC (fast protein liquid chromatography) ion-exchange column by successive passage of 30 ml of buffers A, B, and A (Table I).
2. Inject 10 ml of rabbit reticulocyte lysate cleared of particulate material (as described in the preceding section), using the 10-ml FPLC superloop.
3. Run the FPLC using the program parameters listed in Table II, collecting 1-ml fractions. Because the entire column run takes less than 30 min, we routinely perform this fractionation step at room temperature, although it is advisable to maintain the emerging fractions in the cold, e.g., by putting some crushed ice in the fraction collector bowl.

TABLE I
COMPOSITION OF BUFFERS USED IN PURIFICATION OF CYTOSOLIC CHAPERONIN FROM RABBIT RETICULOCYTE LYSATE (A–E) AND IN *in Vitro* FOLDING ASSAYS (F)

| Buffer | Composition[a] | pH |
|---|---|---|
| A | 20 m*M* $MgCl_2$, 20 m*M* KCl, 1 m*M* EGTA, 1 m*M* DTT, 20 m*M* Tris-HCl | 7.2 |
| B | 0.5 *M* $MgCl_2$, 20 m*M* KCl, 1 m*M* EGTA, 1m*M* DTT, 20 m*M* Tris-HCl | 7.2 |
| C | 20 m*M* KCl, 1 m*M* EGTA, 1 m*M* DTT, 20 m*M* Tris-HCl | 7.2 |
| D | 5 m*M* $MgCl_2$, 20 m*M* KCl, 1 m*M* EGTA, 1 m*M* DTT, 20 m*M* Tris-HCl | 7.2 |
| E | 5 m*M* $MgCl_2$, 5 m*M* ATP, 20 m*M* KCl, 1 m*M* EGTA, 1 m*M* DTT, 20 m*M* Tris-HCl | 7.2 |
| F | 1 m*M* $MgCl_2$, 20 m*M* KCl, 1 m*M* EGTA, 1 m*M* DTT, 20 m*M* MES | 6.8 |

[a] EGTA, Ethylene glycol-bis($\beta$-aminoethyl ether)-*N,N,N',N'*-tetraacetic acid; DTT, dithiothreitol; MES, morpholineethanesulfonic acid.

4. Fractions containing chaperonin emerge from the Mono Q column at 80–110 m*M* $MgCl_2$. As a guide, we make use of the observation that a distinctly pink fraction (of unknown identity) is eluted at about 85 m*M* $MgCl_2$; this colored fraction serves as a convenient internal marker. Generally we pool the three fractions following this pink material; alternatively,

TABLE II
PARAMETERS FOR ANION-EXCHANGE CHROMATOGRAPHY OF RABBIT RETICULOCYTE LYSATE[a]

| Parameter | Setting |
|---|---|
| 0.00 CONC %B | 0.0 |
| 0.00 ML/MIN | 2.00 |
| 0.00 CM/ML | 0.20 |
| 0.00 PORT.SET | 6.0 |
| 0.00 VALVE.POS | 1.2 |
| 20.00 CONC %B | 0.0 |
| 20.00 VALVE.POS | 1.1 |
| 30.00 PORT.SET | 6.1 |
| 50.00 PORT.SET | 6.0 |
| 65.00 CONC %B | 50.0 |
| 75.00 CONC %B | 100 |
| 100.00 CONC %B | 100 |
| 100.00 CONC %B | 0.0 |
| 120.00 CONC %B | 0.0 |

[a] On Pharmacia FPLC system using a Mono Q 10/10 column.

the fractions can be assayed for c-cpn activity using the folding assay described below before proceeding to the affinity chromatography step.

*Affinity Chromatography*

1. Dilute the fractions containing c-cpn emerging from the Mono Q column with an equal volume of buffer C (Table I). This step lowers the concentration of $MgCl_2$ sufficiently so as to allow efficient binding of c-cpn to the affinity column.
2. Apply this material to a column (1–2 ml) of ATP–agarose ($C_8$ linked; Sigma, St. Louis, MO) equilibrated at 4° with buffer D (Table I); the flow rate should not exceed one drop per 5 sec.
3. Wash the column with four bed volumes of buffer D (Table I).
4. Elute the column with four bed volumes of buffer E (Table I); measure the protein yield using, for example, the Bio-Rad (Hercules, CA) protein assay reagent; this should be in the range 0.4–0.6 mg.
5. Concentrate the product on a Centricon-30 ultrafiltration unit (Amicon, Danvers, MA) until the final volume is 0.2 ml.

*Gel Filtration*

1. Apply the c-cpn preparation purified by ion-exchange and affinity chromatography to a Pharmacia 10/30 Superose 6 gel-filtration column equilibrated in buffer E (Table I) at room temperature and run at 0.3 ml/min. Cytosolic chaperonin elutes as a symmetrical peak at about 10.5 ml, with an apparent molecular mass (relative to standard molecular size markers) of about 600 kDa, although the actual molecular mass is probably closer to 900 kDa.
2. Concentrate the c-cpn peak on a Centricon-30 ultrafiltration unit to give a final protein concentration of about 5 mg/ml. Flash-freeze in small aliquots and store at −70°. Cytosolic chaperonin can be stored under these conditions for several months without loss of activity.
3. Check the purity of the preparation by running a few micrograms on an 8% sodium dodecyl sulfate (SDS) polyacrylamide gel. Cytosolic chaperonin runs as a closely spaced cluster of polypeptides in the range 57–62 kDa (Fig. 1).

## Assay of Folding Activity

*Principle*

The principle of the assay depends on presentation of an unfolded labeled target protein to c-cpn via dilution from urea. The assay can readily

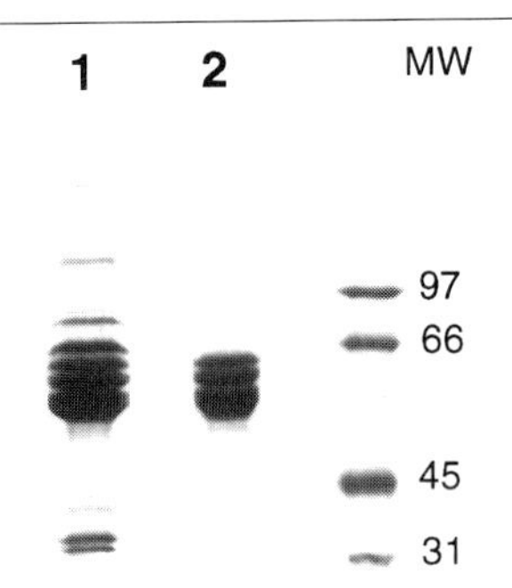


FIG. 1. Analysis on an 8.5% (w/v) SDS–polyacrylamide gel of c-cpn purified from rabbit reticulocyte lysate after the affinity chromatography step (lane 1) and after gel filtration on Superose 6 (lane 2). Molecular weight markers ($\times 10^{-3}$) are shown at the right.

be adapted for measuring the biological activity of chaperonins other than c-cpn, as well as other chaperones. On dilution, an unfolded target protein may either fold to the native state in solution, or become kinetically trapped as a misfolded intermediate. In the case of actins and tubulins, it has never proved possible to obtain detectable yields of native protein via spontaneous refolding from guanidine or urea. In the presence of chaperonin, however, the opportunity exists for the formation of a binary complex. Once formed, such binary complexes appear to be quite stable.

Binary complex formation following presentation from urea of unfolded actin or tubulin to c-cpn is fairly rapid (with a $t_{1/2}$ of about 4 min at 30°), although much slower than the corresponding rate for GroEL or its mitochondrial homolog.[14] The reason for this difference is not clear. In any event, the ATP-dependent cycling of target proteins by c-cpn is a relatively slow reaction compared with GroEL, with a half-time for the production of native product of 15–30 min at 30°.[12] Folding of $\alpha$- or $\beta$-tubulin, both of which are GTP-binding proteins, also requires the presence of GTP plus additional protein cofactors and native carrier tubulin.[10,13] Native actin or tubulin can be conveniently detected following either resolution on a nondenaturing gel or fractionation by gel-filtration chromatography.

The sensitivity of the assay depends critically on the specific radioactivity of the unfolded target protein. In general, it is not convenient to label purified protein (for example, by iodination), in part because the specific activities obtained are too low to allow analysis of reaction products within a reasonable time frame, but also because the procedure itself involves the covalent modification of the target protein, and this could interfere with

[14] R. Melki and N. J. Cowan, *Mol. Cell Biol.* **14,** 2895 (1994).

the folding reaction. It is therefore preferable to obtain labeled target protein via expression in a bacterial host. This procedure offers several advantages. First, the protein is expressed in unmodified form. Second, the expression of actin or tubulin (and, indeed, many other target proteins) in *Escherichia coli* results in the production of inclusion bodies, which are relatively easy to purify by virtue of their extreme insolubility. Third, *in vivo* labeling of target proteins via expression in *E. coli* can yield material of high ($10^6$–$10^7$ cpm/$\mu$g) specific radioactivity. Finally, the labeling procedure can conveniently be done on a small scale, and takes less than 1 day.

*Cloned Target Proteins*

Full-length cDNA clones encoding $\beta$-actin and $\alpha$-, $\beta$-, or $\gamma$-tubulin inserted into pET vectors are available from the author. pET vectors are designed for the T7-driven expression of cloned inserts, and are fully described in Ref. 15. Any potential target proteins may be expressed using these vectors following insertion of the appropriate coding sequence by standard cloning procedures; the only requirement for successful labeling is that the expressed amino acid sequence should contain at least one methionine or cysteine residue.

*Labeling Procedure*

1. Using a standard transformation protocol, transform the pET plasmid encoding the target protein of interest into host *E. coli* BL21(DE3) cells.[15]
2. Inoculate 1 ml of Luria broth containing ampicillin (0.1 mg/ml) with freshly transformed bacteria and grow with vigorous shaking at 37° to $A_{550}$ 1.2.
3. Induce the production of T7 polymerase by addition of isopropyl-$\beta$-D-thiogalactopyranoside (IPTG) to 0.5 m$M$; continue incubating at 37° for a further 30 min.
4. Harvest the cells by centrifugation at room temperature and discard the supernatant.
5. Resuspend the bacterial pellet in 0.7 ml of labeling medium [minimal medium supplemented with IPTG (0.5 m$M$), ampicillin (20 $\mu$g/ml), rifampicin (200 $\mu$g/ml), and a mixture of 18 amino acids (i.e., all except methionine and cysteine) at a concentration of 0.1 m$M$ each].
6. Continue incubating at 37° for a further 10 min.

[15] F. W. Studier, A. H. Rosenberg, J. J. Dunn, and J. W. Dubendorff, *Methods Enzymol.* **185,** 60 (1990).

7. Add 1 mCi of $^{35}$S-labeling mix ($^{35}$S-Express; New England Nuclear, Boston, MA); this relatively inexpensive product contains a mixture of [$^{35}$S]methionine and [$^{35}$S]cysteine.

8. Continue incubating at 37° for a further 1.5 hr.

9. Carefully transfer the culture to an Eppendorf tube using a pulled Pasteur pipette and harvest the cells by centrifugation. Discard the supernatant in the radioactive waste.

10. Resuspend the bacterial pellet in 50 $\mu$l of 10 m*M* Tris-HCl (pH 7.5), 1 m*M* EDTA, by vortexing.

11. Freeze the bacterial pellet on dry ice and thaw by brief incubation at 37°. This weakens the bacterial cell walls and facilitates subsequent lysis.

12. Add 1 $\mu$l of freshly prepared lysozyme (10 mg/ml in 10 m*M* Tris-HCl, pH 8.0).

13. Tap the tube so as to mix the contents. Within a few minutes or less, the culture should lyse, yielding a viscous mass.

14. Freeze and thaw the lysate twice by placing the Eppendorf tube on dry ice; this helps to complete cellular lysis.

15. Add 1 $\mu$l of 1.0 *M* $MgCl_2$ and 0.5 $\mu$l of DNase I [0.5 mg/ml, reconstituted in water (code DPRF; Worthington, Freehold, NJ). Note that many commercial DNase preparations from other sources are contaminated with proteases that will result in degradation of the expressed protein]. Incubate for 1 min on ice; note the resulting loss of viscosity.

16. Add 6 $\mu$l of 10% (v/v) Triton X-100 and mix by tapping.

17. Centrifuge the lysate in an Eppendorf centrifuge for 15 min at 4° to recover the insoluble inclusion bodies.

18. Remove and discard the supernatant in the radioactive waste, using a pulled Pasteur pipette.

19. Add 50 $\mu$l of freshly prepared 8 *M* guanidine hydrochloride in 20 m*M* Tris-HCl (pH 7.2), 10 m*M* dithiothreitol (DTT), 2 m*M* ethylenediaminetetraacetic acid (EDTA).

20. Vortex vigorously and persistently until the pellet is no longer visible. This may take several minutes. (The solubilization can be accelerated by adding a few extra crystals of guanidine hydrochloride to the tube; this saturates the solution and provides internal abrasion of the pellet during the vortexing.)

21. Centrifuge the solubilized inclusion bodies in an Eppendorf centrifuge for 5 min at room temperature to remove any remaining insoluble material.

22. Apply the supernatant to a 0.5 × 5.0 cm column of Sephadex G-50 (medium grade) (this may conveniently be constructed using a plastic 1-ml disposable syringe plugged with glass wool) equilibrated in 7.5 *M* urea

(freshly prepared and deionized by stirring with Amberlite mixed-bed resin (Bio-Rad) containing 20 m*M* Tris-HCl (pH 7.2), 10 m*M* DTT, 2 m*M* EDTA.

23. Collect single drops emerging from the column and monitor the radioactivity. This can be done rapidly and conveniently by spotting out 1-$\mu$l aliquots onto paper, and estimating the radioactivity content using a hand-held Geiger counter placed directly over the spot.

24. Pool the two most highly labeled fractions, divide into 10- or 20-$\mu$l aliquots, and store at $-70°$.

25. Determine the specific radioactivity by scintillation counting and measuring total protein (e.g., using the Bio-Rad protein assay reagent). Typical values are in the range of $10^6$–$10^7$ cpm/$\mu$g. The actual specific activity will depend on the target protein in question, i.e., its level of expression in the host *E. coli* cells, and its content of methionine and cysteine residues.

26. Determine the biochemical and radiochemical purity of the probe by running an SDS–polyacrylamide gel, staining with Coomassie blue (to visualize the target protein and any contaminants), and autoradiography of the dried gel (to assess radiochemical purity). Typically target proteins prepared by the preceding protocol are 70–80% biochemically pure and about 90% radiochemically pure. Contaminating radiolabeled species tend to be either prematurely terminated or internally initiated polypeptide chains generated in the bacterial host.

Unfortunately, the shelf-life of probes made by this protocol is relatively short, and rarely exceeds 2 weeks, even when stored at $-70°$. Probes that are more than 2 weeks old do not lose their ability to efficiently form a binary complex with c-cpn, but the efficiency of native product formation declines rapidly. The reason for this is not entirely clear, but presumably is at least in part due to radiochemical damage.

*Folding Assays and Analysis of Folding Reaction Products*

The example given here is for actin folding.

1. Dilute 4 $\mu$g (5 pmol) of c-cpn to 20 $\mu$l with folding buffer (Table I, buffer F) in an Eppendorf tube.

2. Add 1 $\mu$l of 20 m*M* MgATP (pH 7).

3. Add 0.2 $\mu$l of $^{35}$S-labeled urea-unfolded actin target protein, mix immediately by tapping the tube to disperse the urea rapidly, and incubate at 30° for 45 min.

4. Terminate the reaction by adding 5 $\mu$l of a solution of 80% (v/v) glycerol, 80 m*M* MES (pH 6.8), 0.2% (v/v) bromphenol blue, and freezing on dry ice.

*Product Analysis on Nondenaturing Polyacrylamide Gels.* The following protocol is for nondenaturing gels cast and run in the Bio-Rad Mini-PROTEAN II system using 1-mm-thick spacers and combs, but can be adapted to any similar vertical gel apparatus. The buffer conditions described are suitable for the analysis of the products of actin- or tubulin-folding reactions; obviously, it may be necessary to change the pH of the gel and running buffer to accommodate proteins having different isoelectric points.

1. Mix 0.4 ml of 1 *M* MES, pH 6.8 (adjusted with KOH), 0.75 ml of [30% (w/v) acrylamide, 0.8% (w/v) bisacrylamide], 3.75 ml of $H_2O$, 5 $\mu$l of 1.0 *M* $MgCl_2$, and 10 $\mu$l of 1.0 *M* EGTA. In some cases [for example, in experiments involving the generation of native actin (which binds ATP) or tubulin (which binds GTP)] it is necessary to include 5 $\mu$l of 1 *M* ATP or 5 $\mu$l each of 1 *M* ATP and 1 *M* GTP.
2. Start the polymerization reaction by adding 67 $\mu$l of 10% (w/v) ammonium persulfate (freshly prepared) and 4 $\mu$l of TEMED (*N,N,N',N'*-tetramethylethylenediamine). Pour the mixture between the gel plates and insert the comb (typically with 15 slots).
3. Following polymerization (which takes 15–30 min), remove the comb, insert the gel into the electrophoresis apparatus and load the samples (which should contain enough glycerol to ensure that they remain in the slot on addition of the running buffer, and enough bromphenol blue to serve as a visible tracking dye).
4. Add running buffer [80 m*M* MES (pH 6.8), 1 m*M* $MgCl_2$, 1 m*M* EGTA, 0.2 m*M* ATP, 0.1 m*M* GTP—the latter two components being optional, depending on whether the proteins of interest require the presence of these nucleotides to maintain their native state].
5. Run the gel at 5-W constant power at room temperature until the tracking dye has completely eluted from the bottom of the gel (i.e., 1–1.5 hr). It is not advisable to run two gels in the same electrophoresis unit, because 10 W may generate enough heat to cause protein denaturation during the run. At half-hour intervals, stop the run and mix and replace the anode and cathode buffers to avoid extreme pH changes generated during electrophoresis. Alternatively, use a buffer-circulating pump.
6. Remove the gel from the apparatus, fix and stain with Coomassie blue, and destain. Cytosolic chaperonin appears as a sharp stained band that migrates approximately 8 mm below the gel origin.
7. Prepare the gel for fluorography, dry on a gel drier, and expose to film at $-70°$. Correctly folded labeled target protein should comigrate with unlabeled, control native material run in a parallel track. Assays done with

probes in the range of $10^6$–$10^7$ cpm/$\mu$g typically require exposure times in the range of 4–24 hr (Fig. 2).

Other functional assays (such as ability to copolymerize with native material in the case of actin or tubulin) can be done with folding reaction products isolated on a gel-filtration column such as Superose 6 (Pharmacia, Piscataway, NJ).

## Concluding Remarks

The protocols described provide a relatively simple means of assaying the interaction of an unfolded target protein with c-cpn and identifying the potential production of native protein. Given the high specific activities of labeled target proteins that can be generated via expression of recombinant sequences in *E. coli*, the assay is sensitive; for example, it is possible to detect the generation of 10 ng or less of native actin folded via interaction with c-cpn. However, it should be noted that it is not practical to use the folding assays described here for the production of biochemical quantities of c-cpn-folded material. For example, if one were to commit 1 mg of c-cpn, the theoretical yield of correctly folded product, presented as an equimolar amount of unfolded actin and assuming 100% efficiency of folding to the native state, would be approximately $(4.5 \times 10^4)/(9 \times 10^5) = 0.05$ mg. Notwithstanding the cost of producing c-cpn in milligram quantities, such theoretical yields are never obtained in practice; indeed, the efficiency of c-cpn-mediated *in vitro*-folding reactions is typically in the range of 30–40%, calculated as a fraction of the total counts recovered as either binary complex or native product. Even this figure is potentially misleading, as the efficiency with which unfolded input target protein forms a binary complex with c-cpn is itself never more than 50%; the remaining radiolabeled material simply "disappears," presumably because it either aggregates and/or sticks to the walls of the reaction vessel. In spite of these limitations,

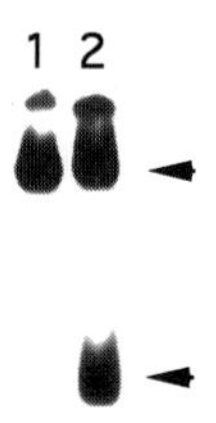


FIG. 2. Analysis on a 4.5% (w/v) nondenaturing polyacrylamide gel of the products of $\beta$-actin folding reactions done in the absence (lane 1) or presence (lane 2) of ATP. Upper and lower arrows show the bands corresponding to c-cpn–$\beta$-actin binary complex and native $\beta$-actin, respectively.

however, there is much that can be learned concerning the mechanism of c-cpn action using the plentiful radiochemical quantities of product that can be generated *in vitro*.

Although the reasons for the relatively poor efficiency of c-cpn-mediated folding reactions are not entirely clear, several possible explanations can be ruled out. The stability of c-cpn itself is not a contributing factor, because preincubation of our c-cpn preparations at 30° for several hours before the presentation of target protein has no influence on the yield of native product. Nor is there reason to believe that additional factors might be required for efficient folding: folding reactions supplemented with crude extracts (e.g., from reticulocyte lysate, the natural source of our c-cpn) do not improve the overall yield. Finally, the reaction itself does not become poisoned by a buildup of ADP as a result of c-cpn-mediated ATP hydrolysis, because inclusion of an ATP-generating system in the folding reaction also has no detectable effect.

A plausible explanation for the relative inefficiency of c-cpn-mediated folding in *in vitro* reactions lies in the mode of presentation of target protein. In contrast to reactions done via dilution of the target protein from denaturant, the efficiency of folding reactions done in a translational context in rabbit reticulocyte lysate is relatively high: for example, after 30 min at 30°, only about 10% of recovered label remains as binary complex in an actin *in vitro* translation reaction, with the remaining 90% appearing as native product. It seems probable, therefore, that the vectorial extrusion of newly synthesized polypeptides from ribosomes plays an important role in generating intermediates that can be efficiently recognized and cycled by c-cpn.[16]

## Summary

Cytosolic chaperonin, the eukaryotic cytosolic homolog of GroEL, has certain unusual features that make it uniquely useful for studying the mechanism of chaperonin action. It is of particular interest as an essential component in the generation of native actin and tubulin *in vivo*. We describe a method for the purification of mammalian c-cpn from rabbit reticulocyte lysate via a three-step procedure involving ion-exchange chromatography, affinity selection on ATP–agarose, and gel filtration. We also describe a sensitive *in vitro*-folding assay for the activity of c-cpn and other chaperone proteins, and a simple nondenaturing gel assay for the analysis of folding reaction products.

[16] J. Frydman, E. Nimmesgern, K. Ohtsuka, and F.-U. Hartl, *Nature* (*London*) **370,** 111 (1994).

# [21] Electron Microscopy of Chaperonins

*By* S. Chen, A. M. Roseman, and H. R. Saibil

## Introduction: Contribution of Electron Microscopy to Chaperonin Studies

Electron microscopy (EM) plays an important role in understanding chaperonin structure and function. The large size and high symmetry of chaperonin oligomers make it possible to deduce detailed two-dimensional (2D) and 3D structural information, and the time resolution of cryo-EM in particular makes it a powerful technique for dynamic studies of chaperonin-assisted protein folding. The double-ring, two-domain structure of chaperonin 60 (cpn60) with its central cavity was originally revealed by negative-stain EM.[1,2] The large conformational changes caused by nucleotide and cpn10 binding were defined by cryo-EM experiments, as were the sites and density of substrate binding.[3] The preservation of native structure in frozen-hydrated specimens, and developments in image processing of single particles in solution, now make it possible to discriminate different orientations of the complexes and different species in a reaction mixture. This allows 3D structure determination from a set of randomly oriented views, and, uniquely, correlation of kinetic and structural information. In this chapter we discuss several EM and image-processing methods that are useful for chaperonin studies and the different kinds of information that they provide.

## Negative-Stain Electron Microscopy

Electron microscopy can provide a level of resolution to 0.1–0.2 nm in materials science. However, a resolution better than 1 nm is not easily obtained with protein samples, which are limited by low contrast and radiation sensitivity. High contrast can be generated by use of heavy metal staining or shadowing, at the expense of loss of native structure by dehydration, and loss of internal detail in the imaging. Negative staining resolves the surface shape to around 2 nm under favorable conditions. The specimen is embedded in an electron-dense stain, on the surface of a carbon-coated

[1] R. W. Hendrix, *J. Mol. Biol.* **129,** 375 (1979).

[2] E. G. Hutchinson, W. Tichelaar, G. Hofhaus, H. Weiss, and K. R. Leonard, *EMBO J.* **8,** 1485 (1989).

[3] S. Chen, A. M. Roseman, A. S. Hunter, S. P. Wood, S. G. Burston, N. A. Ranson, A. R. Clarke, and H. R. Saibil, *Nature* (*London*) **371,** 261 (1994).





0076-6879/98 $25.00

EM grid. On drying, the heavy metal stain envelopes the specimens, which appear light in the EM, surrounded by an electron-dense halo of dried stain. The advantages of negative staining are its speed, ease of use, high contrast, and relatively low sensitivity to the electron beam. It is therefore useful as a routine technique, for assessing the homogeneity of specimens, estimating concentration, for determining symmetry, and for low-resolution structure analysis. Its main limitation is that only the imprint left by proteins is visualized. Factors such as partial staining, positive staining, and stain migration in the beam affect the accuracy of the interpretation. Dehydration, essential for observation in the high-vacuum environment of the EM, decreases the intrinsic structural resolution dramatically and causes variable flattening of the 3D structure. Structure distortion may also occur owing to unfavorable pH of the stain or to interaction with the carbon film.

*Methods*

There are three commonly used negative stains: uranyl acetate (UA), phosphotungstic acid (PTA), and ammonium molybdate (AM), normally used at 2% (w/v). Among them UA gives the best contrast and shows good results with many, but not all, protein specimens. It has a fixative effect and favors the adhesion of protein particles to the supporting film. But UA has a low pH (3.7–4.5). Phosphotungstic acid and AM can be used at around neutral pH but show lower contrast than UA. Phosphotungstic acid (or sodium silicotungstate) is a good stain for many viruses and bacteriophages. In some cases, AM staining shows more structural detail, perhaps owing to better penetration with some protein complexes and membrane proteins.

Carbon-coated 300–400 mesh grids (thickness of the carbon film is about 10 nm) are glow discharged before use.[4] Three to five microliters of protein solution (0.1–0.2 mg/ml) is applied to a grid, allowing several seconds to several minutes for absorption, depending on the adhesion of the specimen. A large drop (50 $\mu$l) of negative stain is added to the grid. The grids can be rinsed with a few drops of distilled water before staining. After staining for 10–20 sec, the excess solution is blotted by a piece of filter paper from the edge of the grid. The grid is then left to dry thoroughly in air.

Grids are imaged in a transmission EM, using a low-dose system for best results. Regions in which the stain is well spread, giving a diffuse halo around the negatively contrasted molecules, should be chosen. One problem with cpn60 is that the end views (molecules oriented with their cylinder axis perpendicular to the plane of the carbon film) tend to predominate.

[4] M. A. Hayat, Negative staining and support films. *In* "Principles and Techniques of Electron Microscopy," 3rd Ed., pp. 328–376, Macmillan Press Ltd., London, 1994.

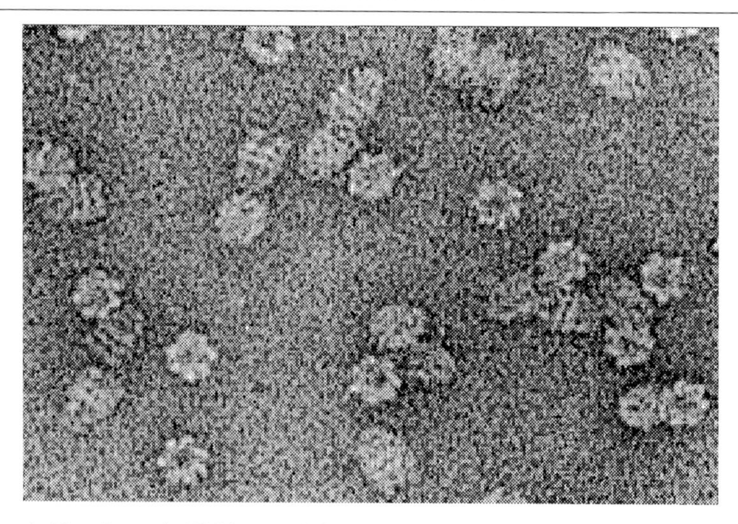

FIG. 1. Negative-stain EM image of GroEL–GroES complexes showing end views, with sevenfold symmetry, and bullet-shaped side views, with stripes of density. Protein is light and stain is dark. The short diameter of the complex is about 140 Å.

These are useful for determining the rotational symmetry, but once this is known, side views are preferable because they contain much more structural information. We are not aware of any procedure that gives control over the resulting orientations, aside from structural modification. For example, football-shaped particles ($cpn10_7$–$cpn60_{14}$–$cpn10_7$) orient in side view.[5]

*Antibody Labeling.* Chemically defined sites can be localized on the structure by combining EM with antibody labeling. Specific, high-affinity antibodies can be complexed with the chaperonin sample in solution, or after deposition on the carbon film, taking care not to let the sample dry out until labeling and staining are complete. For divalent antibodies, it may be necessary to do the labeling on molecules immobilized by prior adsorption to the carbon film, to avoid cross-linking and formation of dense aggregates. If Fab fragments are used, it should be possible to obtain heavier labeling. No further labeling with secondary antibodies is necessary, as the primary antibodies are large enough to see directly by negative staining, once they are bound to the chaperonin complexes (e.g., see Ref. 6). Fairly pure antibody preparations are necessary, to avoid background material obscuring the image.

[5] A. Engel, M. K. Hayer-Hartl, K. N. Goldie, G. Pfeifer, R. Hegerl, S. Müller, A. C. R. de Silva, W. Baumeister, and F. U. Hartl, *Science* **269,** 832 (1995).

[6] J. Martin, K. N. Goldie, A. Engel, and F. U. Hartl, *Biol. Chem. Hoppe-Seyler* **375,** 635 (1994).

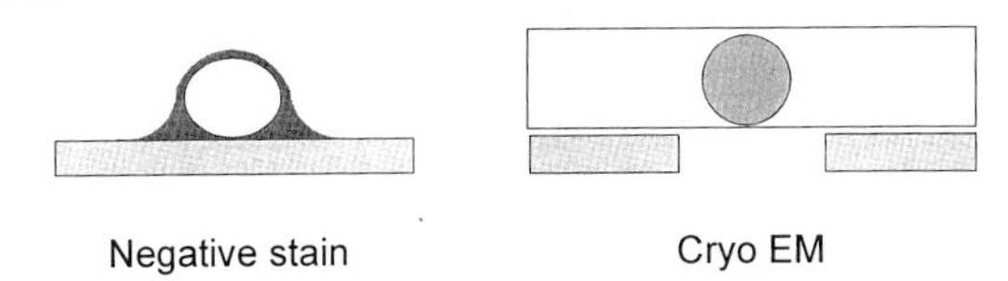


FIG. 2. Schematic side views of negative-stain and cryo-EM grids. The negatively stained particle is enveloped in electron-dense stain and flattened onto the plane of the support film. In the cryo-EM specimen, the particle is held in a layer of vitreous ice and viewed over a hole in the support film to maximize the obtainable contrast.

*Scanning Transmission Electron Microscopy and Gold Labeling.* Scanning transmission electron microscopy (STEM) uses a finely focused beam to scan over the sample, and can separately detect elastically and inelastically scattered electrons. The elastically scattered beam is proportional to the mass thickness of thin specimens. Thus it can be used for mass determination of single particles, and for localization of small clusters of heavy atoms.[7] A polypeptide (dihydrofolate reductase) was tagged with a gold cluster (1.4-nm diameter), unfolded, and then bound to GroEL. The unstained specimens were freeze-dried and examined by STEM. The gold clusters were observed within the cavity of GroEL.[8] STEM mass measurement was used to determine whether end views of cpn60 contained none, one, or two molecules of bound cpn10.[5] These forms cannot be distinguished from each other in transmission EM (TEM) end views.

*Structural Information from Negative Staining.* A negative-stain image of cpn60–cpn10 complexes is shown in Fig. 1, and a schematic diagram of negative stain and cryo-EM grids in Fig. 2. Because of its sevenfold symmetry, and the precise arrangement of its two major subunit domains in the four layers of its two, back-to-back rings, its cagelike 3D structure can be deduced from negative-stain EM images, as well as the large conformational change on GroES binding.[9] The location of unfolded protein substrates within the central cavity was indicated by negative stain,[10] and by gold and antibody labeling.[8,11]

[7] R. Freeman and K. R. Leonard, *J. Microsc.* **122,** 275 (1981).

[8] K. Braig, M. Simon, F. Furuya, J. F. Hainfeld, and A. L. Horwich, *Proc. Natl. Acad. Sci. U.S.A.* **90,** 3978 (1993).

[9] H. R. Saibil, D. Zheng, A. Roseman, A. S. Hunter, G. M. F. Watson, S. Chen, A. auf der Mauer, B. P. O'Hara, S. P. Wood, L. Barnett, N. H. Mann, and R. J. Ellis, *Curr. Biol.* **3,** 265 (1993).

[10] T. Langer, G. Pfeifer, J. Martin, W. Baumeister, and F.-U. Hartl, *EMBO J.* **11,** 4757 (1992).

[11] N. Ishii, H. Taguchi, H. Sasabe, and M. Yoshida, *J. Mol. Biol.* **236,** 691 (1994).

## Cryoelectron Microscopy

To preserve 3D structures of proteins in the hydrated state, cryo-EM was invented. It provides a direct visualization of native protein structure in solution or in two-dimensional crystals. The specimen is rapidly cooled to the vitreous state, and imaged in this frozen–hydrated state, which is stable in the microscope vacuum at temperatures around −160°.[12]

Electron microscopy is particularly suitable for characterizing large molecules or macromolecular assemblies in solution as single particles. Individual molecules or complexes can be inspected and measured, a useful characteristic for specimens with molecular flexibility or heterogeneity. In principle, cryo-EM images contain information to atomic resolution. This information can be extracted from images of 2D crystals, and in theory from single particle images, if enough molecules are processed ($10^4$–$10^5$ molecules; see Ref. 13), and if the position and orientation of each molecule can be determined accurately. By contrast, atomic force microscopy provides high-resolution surface topography, but not full 3D information. The measured electron-scattering density from cryo-EM is closely comparable to the electron density determined by X-ray crystallography.

### *Methods*

The main requirement of cryo-EM is to immobilize water molecules by rapidly cooling the protein solution to the vitreous state and to maintain this state through all the procedures of observation and image recording in the microscope.[12] The vitreous state is a solid state of water without the ice crystal formation that would damage protein structures. A relatively simple grid plunger is sufficient to vitrify a thin TEM specimen. A cryo-transfer stage is used to transfer the sample into the microscope vacuum while avoiding surface contamination, and to maintain a specimen temperature of −160 to −170°. At this temperature water sublimation in the column vacuum becomes negligible. The low contrast of unstained, frozen–hydrated specimens is enhanced by the use of phase contrast, obtained by defocusing the objective lens. Structural information from the resulting low signal-to-noise (*S/N*) images is extracted by averaging and other image-processing procedures (see below).

*Grid preparation: Holey Carbon Films.* For single particle EM, perforated carbon support films are used, so that the molecules can be observed in the holes. To make a holey carbon grid, 60 μl of 50% (v/v) glycerol in

[12] J. Dubochet, M. Adrian, J.-J. Chang, J.-C. Homo, J. Lepault, A. W. McDowell, and P. Schultz, *Q. Rev. Biophys.* **21,** 129 (1988).

[13] R. Henderson, *Q. Rev. Biophys.* **28,** 171 (1995).

water is added to 50 ml of 0.5% (w/v) Formvar in chloroform solution. The solution is vigorously sonicated to disperse the glycerol into micron-sized droplets. A precleaned glass slide is dipped into the Formvar solution immediately after sonication and then allowed to drain dry. The dried Formvar film is floated off the slide onto a clean surface of distilled water. The EM grids are put on the Formvar film with the dull side downward and then picked up with a piece of filter paper. The dried filter paper is then soaked in methanol for 20 min to remove the glycerol, i.e., to form perforated Formvar-coated grids. After drying, the Formvar grids can be stored in a dry place for months. Before freezing specimens, the Formvar grids on the filter paper are coated with a thin layer (about 30 nm) of carbon, then soaked in chloroform for 15–30 min to dissolve the Formvar film, leaving the holey carbon. Carbon film normally becomes hydrophobic with storage, which is unfavorable for spreading of the specimen. Glow discharge in a weak vacuum increases the hydrophilicity of carbon film and should be done one to several hours before freezing. Glow discharge in amylamine vapor results in moderate hydrophilicity on the carbon film, with altered charge, which is more favorable for some specimens. The type of surface preparation must be determined empirically.

*Specimen Preparation for Single Particles.* The protein concentration should be about 1 mg/ml, in a low salt buffer (e.g., 10–20 m*M* Tris-HCl, 10–40 m*M* ammonium acetate, 5 m*M* $MgCl_2$, and 5 m*M* KCl, pH 7.6), to avoid loss of contrast and concentration effects during evaporation from the thin film. One to five microliters of protein solution is added to a holey carbon-coated grid. If necessary, the specimen-absorbed grid can be rinsed by floating it (solution side down) on a drop of distilled water or low salt buffer for 5–10 sec. Two steps of washing can be used to remove unwanted solutes while retaining a proportion of the macromolecules absorbed at the air–water interface. The grid is mounted on a pair of forceps in a Guillotine-type plunger (Fig. 3). Excess water is removed by blotting a piece of filter paper against the surface of the grid. When a thin film is left over the surface (about 1–3 sec of blotting), the filter paper is withdrawn and the plunger is released, immersing the grid into liquid ethane (−160°) cooled by liquid nitrogen (−190°). The frozen grids can be transferred to the cryospecimen holder and observed in the microscope, or stored in liquid nitrogen.

*Time-Resolved Cryo Electron Microscopy.* The shortest time from mixing two components to plunging into cryogen is 5–10 sec in routine practice. To observe transient states with shorter lifetimes, the sample is prepared on an EM grid and the dynamic process is induced *in situ,* for example, by flash photolysis of caged reactants. The reaction is allowed to run for a set time, and terminated by vitrification. When specimens are left on the grids

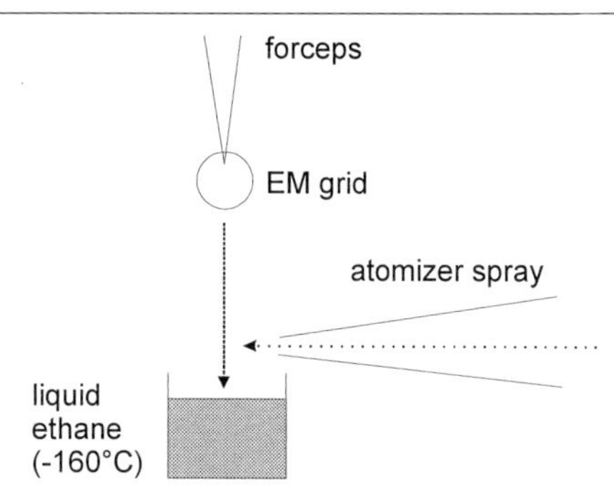


FIG. 3. Diagram of cryo-EM plunger, with atomizer spray for rapid mixing experiments, after Berriman and Unwin.[14] A reaction can be initiated milliseconds before vitrification, by spraying components onto the specimen as it plunges into the ethane.

for a period of time, although it may be as short as a few seconds, evaporation of the water may cause concentration changes in the solutions. A humid chamber can help to minimize this effect.

A procedure that can be used to mix components and trap reactions in as short a time as 1 msec has been introduced by Berriman and Unwin.[14] This is done by combining an atomizer spray with a standard freezing plunger (Fig. 3). The atomizer is used to spray the solution containing reactant and ferritin as a marker in fine droplets onto the blotted grid, during the operation of the plunger. The time between mixing and freezing depends on the distance between the height of the atomizer nozzle and the ethane surface, for a given speed of the plunger.

*Observation and Recording.* Once the frozen grid is in the cryostage of the microscope, the problems that need to be considered are contamination, radiation sensitivity, and phase contrast. An efficient anticontaminator, normally double blades close to the specimen, is important. When the grid is transferred into the microscope, it is left to stabilize for 10–15 min before opening the shield covering the specimen, to allow the water vapor pressure around the grid to decrease.

There are two major points to note about cryo-TEM image formation: First, the depth of focus is much greater than the specimen thickness, so that the recorded image is a 2D projection of the 3D structure; second, the amplitude contrast of proteins in ice is extremely low, and phase contrast must be used to enhance the contrast. Phase contrast is induced by defocusing the objective lens, which also has the effect of allowing particular ranges of spatial frequencies to appear in the image. This optical effect is described

[14] J. Berriman and N. Unwin, *Ultramicroscopy* **56,** 241 (1994).

by the contrast transfer function, an oscillatory function that modulates the frequency information. The defocus level is usually chosen as a compromise between loss of contrast and optical distortion. In high-resolution work, corrections are made for the effect of the transfer function.

## Image Processing

Photographic film is still the most practical medium for recording low electron dose images, at relatively low magnification, with high spatial resolution over a large area. CCD and/or image detectors may eventually replace film but at present they are extremely expensive and cannot deliver high resolution over large areas. Images on film must be digitized with at least 8-bit gray scale resolution (256 density levels), and 10 to 20-$\mu$m spatial resolution, equivalent to a spacing on the structure about one-fourth the desired resolution. This oversampling (beyond the theoretical requirement of half the resolution) is necessary to avoid interpolation errors.

Because of the high information content but low signal-to-noise ratio in cryomicrographs, computer image processing is an essential part of the structure analysis.[15] Averaging aligned similar images improves the signal by a factor of $n^{1/2}$, where $n$ is the number of images. The TEM images are projections of 3D density, and the 3D structure can be determined from a set of projected views at different orientations, by tomographic reconstruction, once the original orientations are determined.[16] The main task of 3D reconstruction is to generate and determine a sufficient set of independent views. Statistical analysis of large data sets is a powerful tool for identification of different orientations or of subclasses,[17] e.g., components in a reaction mixture. For higher resolution, the effects of the contrast transfer function must be corrected.[18,19] This is straightforward for 2D crystal data, less so for single particle images.

### *Image Alignment and Averaging*

In the case of chaperonins, both on carbon film and in the vitrified water layer over holes in the support film, two orientations of the cylinder predominate: end views, in which the cylinder axis is perpendicular to the plane of the film, and side views, with the axis lying in the plane. This

[15] M. F. Moody, Image analysis of electron micrographs. *In* "Biophysical Electron Microscopy" (P. W. Hawkes and V. Valdre, eds.), pp. 145–287. Academic Press, London, 1990.

[16] R. A. Crowther, D. J. DeRosier, and A. Klug, *Proc. R. Soc. London A* **317,** 319 (1970).

[17] J. Frank, *Q. Rev. Biophys.* **23,** 281 (1990).

[18] L. A. Amos, R. Henderson, and P. N. T. Unwin, *Prog. Biophys. Mol. Biol.* **39,** 183 (1982).

[19] R. H. Wade, *Ultramicroscopy* **46,** 145 (1992).

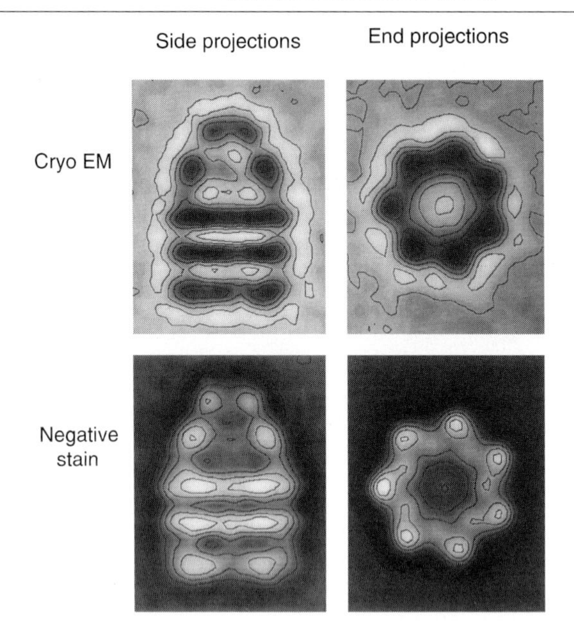


FIG. 4. Averaged images of GroEL–GroES complexes from negative stain and cryo-EM. Each view is the average of several hundred images and is a 2D projection of the 3D density. The cryo-EM images are from the work of Chen *et al.*[3] [The negative-stain images were processed as part of the M.Sc. project of P. Mginah.]

must be attributed to the interaction of the molecules with the air–water interface, as there is nothing else to align them.[20] All usable particle images are selected manually from large scanned areas, and subsequent steps are run automatically, according to selected criteria for cross-correlation coefficients. The first step in processing is to align the set of extracted molecule views, in translation and rotation. This is done by iterative cross-correlation.[21] The signal is maximized by band pass filtering to select the spatial frequency range that best represents the structure, excluding low-frequency background variations and high frequency noise. In Fig. 4, averages of several hundred images of GroEL–GroES complexes, both in negative stain and cryo-EM, are compared.

[20] M. Cyrklaff, N. Roos, and J. Dubochet, *J. Microsc.* **175,** 135 (1994).

[21] J. Frank, Correlation techniques in image processing. *In* "Computer Processing of Electron Microscope Images" (P. W. Hawkes, ed.), pp. 187–222. Springer-Verlag, New York, 1980.

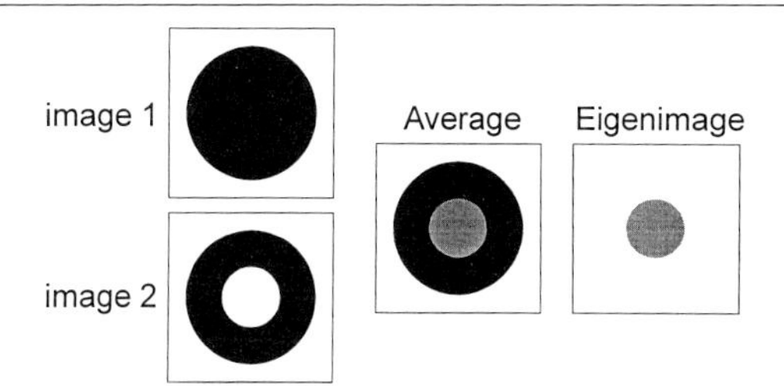


FIG. 5. Schematic diagram of correspondence analysis. A set of two images is shown as one full (containing bound substrate) and the other empty. Each image can be represented as the average plus or minus the eigenimage, or difference image. Eigenimages from a large data set show the principal components of variation in the data.

### *Correspondence Analysis and Image Classification*

Correspondence analysis, with the associated automated image classification, is an important development in single particle image processing.[17,22] These methods aim at detecting significant variations between particles within an image set and subsequently grouping them into consistent subsets. A set of several thousand aligned images is arranged so that the pixels of each image form a vector in the matrix of all images. Eigenvector analysis of the $\chi^2$ distances between all the image vectors yields a set of eigenimages, or difference images, that reveal the principal components of variation in the data set. This is shown schematically in Fig. 5. This analysis provides a classification of particles in different orientations, information necessary for 3D reconstruction of randomly oriented views (see the next section). Useful information that could also be provided by this analysis includes the identification of substrate-bound (full) and empty chaperonin complexes on a data set taken over the course of an assisted folding reaction.

### *Three-Dimensional Reconstruction from Tilted Images*

The main task in 3D reconstruction of single particles is to obtain and/or identify the orientations in a set of independent views. There are two main approaches suitable for low-dose, cryo-EM conditions. The random conical tilt method[23] makes use of a preferred orientation of molecules. A high-tilt image (45–60° tilt) is recorded, followed by a second image of the same area at 0° tilt. If the molecular orientation can be recognized in the untilted image, then the orientation of the same molecule in the tilted image is deduced. Because of the physical limitation on tilt angle (thin EM

[22] M. van Heel and J. Frank, *Ultramicroscopy* **6,** 187 (1981).
[23] M. Radermacher, T. Wagenknecht, A. Verschoor, and J. Frank, *J. Microsc.* **146,** 113 (1987).

specimens cannot be viewed at tilt angles higher than 60–70°), there is always a missing cone of space that is not sampled, resulting in poorer resolution in the $z$ direction. Highly tilted images are also difficult to record and are of poorer quality than untilted images. The missing data can be filled in by making use of symmetry averaging, depending on the orientation of the symmetry axis. The high rotational symmetries of the chaperonins provide a great advantage for 3D reconstruction.

### *Three-Dimensional Reconstruction from Untilted Images Containing Randomly Oriented Views*

An elegant procedure for 3D reconstruction from randomly oriented views has been developed by van Heel and colleagues.[24] Several thousand views from an untilted image of randomly oriented molecules are classified into subclasses of different orientation, using multivariate statistical analysis, similar to correspondence analysis. The relative orientations of the class averages are determined by comparing their 1D line projections. With these starting assignments, an initial 3D reconstruction is calculated, which is then used to generate a set of model projections. Refinement proceeds by correlating the original views with the model projections, reclassifying and assigning improved orientations according to the best correlation coefficients. This is an effective procedure that, in principle, avoids the missing cone problem and makes use of only high-quality, untilted images. In the case of highly preferred orientations, large data sets are necessary to provide enough views of rare orientations. Refinement by cross-correlation with projections of a starting model[25,26] has worked well with side views of cpn60 and cpn60–cpn10 complexes, giving good angular resolution.[27]

## Conclusion

Electron microscopy studies have made significant contributions to our understanding of chaperonin structure and function, aided by the high symmetry and the highly cooperative conformational changes within the rings of subunits. Negative cooperativity between the two rings of GroEL is demonstrated by its preference for single-sided binding of cpn10 and

[24] I. I. Serysheva, E. V. Orlova, W. Chiu, M. B. Sherman, S. L. Hamilton, and M. van Heel, *Nature Struct. Biol.* **2,** 18 (1995).

[25] P. A. Penczek, R. A. Grassucci, and J. Frank, *Ultramicroscopy* **53,** 251 (1994).

[26] T. S. Baker and R. H. Cheng, *J. Struct. Biol.* in press (1997).

[27] A. M. Roseman, S. Chen, H. E. White, K. Braig, and H. R. Saibil, *Cell* **87,** 241–251 (1996).

substrates. Even with crystal structures available for *E. coli* GroEL,[28] GroES[29] and their complex,[30] cryo-EM reconstructions can be combined with atomic structure information to study transient conformational changes and characterize complexes with folding substrates.

[28] K. Braig, Z. Otwinowski, R. Hegde, D. Boisvert, A. Joachimiak, A. L. Horwich, and P. B. Sigler, *Nature* (*London*) **371,** 578 (1994).

[29] J. F. Hunt, A. J. Weaver, S. J. Landry, L. Gierasch, and J. Deisenhofer, *Nature* (*London*) **379,** 37–45 (1996).

[30] Z. Xu, A. L. Horwich, and P. B. Sigler, *Nature* **388,** 741–750 (1997).

# [22] Structural Analysis of GroE Chaperonin Complexes Using Chemical Cross-Linking

*By* ABDUSSALAM AZEM, CELESTE WEISS, and PIERRE GOLOUBINOFF

## Introduction

The chaperonins GroEL and GroES from *Escherichia coli* belong to a large family of chaperone proteins that assist in the folding of nascent peptides, the translocation of proteins across membranes, and the protection of proteins from stress-induced denaturation.[1] Chaperone proteins may also be involved in pathways of protein degradation.[2] Whereas GroEL binds partially folded proteins and prevents their aggregation, GroES is involved in the ATP-dependent release of the bound protein from GroEL, in a folding-competent conformation.[3,4]

Electron microscopy (EM) showed that $GroEL_{14}$ is a cylindrical oligomer of 14 identical 58-kDa subunits organized in 2 stacked $GroEL_7$ toroids facing each other, with a large central cavity spanning the 7-fold axis of symmetry.[5–7] GroES is a heptameric ring of 10-kDa sub-

[1] C. Georgopoulos and W. Welch, *Annu. Rev. Cell Biol.* **9,** 601 (1993).

[2] M. Sherman and A. L. Goldberg, *EMBO J.* **11,** 71 (1992).

[3] P. Goloubinoff, J. T. Christeller, A. A. Gatenby, and G. H. Lorimer, *Nature* (*London*) **342,** 884 (1989).

[4] J. Buchner, M. Schmidt, M. Fuchs, R. Jaenicke, R. Rudolph, F. X. Schmid, and T. Kiefhaber, *Biochemistry* **30,** 1586 (1991).

[5] R. W. Hendrix, *J. Mol. Biol.* **129,** 375 (1979).

[6] T. Hohn, B. Hohn, A. Engel, and M. Wurts, *J. Mol. Biol.* **129,** 359 (1979).

[7] A. M. Roseman, S. Chen, H. White, K. Braig and H. R. Saibil, *Cell* **87,** 241 (1996).



METHODS IN ENZYMOLOGY, VOL. 290
0076-6879/98 $25.00

units.[8,9] X-Ray crystallography showed that each GroEL subunit is composed of an apical domain that interacts with $GroES_7$, an equatorial domain that contains the ATP-binding site, and an intermediate domain that serves as a hinge between the two major domains.[10] Analysis of GroEL mutants.[11] EM,[12–14] and chemical cross-linking,[15] suggested that the binding site of the unfolded protein is on the inner surface of the apical domain, at the entrance of the central cavity, near or overlapping the binding site for $GroES_7$.

Early studies using EM and gel filtration showed that $GroES_7$ binds asymmetrically to a single $GroEL_{14}$ core particle.[12,16,17] However, chemical cross-linking revealed that two $GroES_7$ co-chaperonins can successively bind $GroEL_{14}$.[18] This was confirmed by the observation of symmetrical $GroEL_{14}(GroES_7)_2$ particles by negative-stain EM.[9,19,20] The symmetrical $GroEL_{14}(GroES_7)_2$ particle was suggested to be an intermediate of the chaperonin protein-folding cycle on the basis of a correlation between the rate of protein folding and amount of symmetrical particles in an active chaperonin solution.[21]

Cross-linking of homooligomeric proteins with bifunctional reagents has been extensively used to obtain important information about the oligomeric state and symmetry of protein subunits within complexes.[22–29] We applied

[8] G. N. Chandrasekhar, K. Tilly, C. Woolford, R. Hendrix, and C. Georgopoulos, *J. Biol. Chem.* **261,** 12414 (1986).

[9] J. R. Harris, A. Pluckthun, and R. Zahn, *J. Struct. Biol.* **112,** 216 (1994).

[10] K. Braig, Z. Otwinowski, R. Hegde, D. Boivert, A. Joachimiak, A. L. Horwich, and P. B. Sigler, *Nature (London)* **371,** 578 (1994).

[11] W. A. Fenton, Y. Kashi, K. Furtak, and A. L. Horwich, *Nature (London)* **371,** 614 (1994).

[12] T. Langer, G. Pfeifer, J. Martin, W. Baumeister, and F. U. Hartl, *EMBO J.* **11,** 4757 (1992).

[13] N. Ishii, H. Taguchi, H. Sasabe, and M. Yoshida, *J. Mol. Biol.* **236,** 691 (1994).

[14] S. Chen, A. M. Roseman, A. S. Hunter, S. P. Wood, S. G. Burston, N. A. Ranson, A. R. Clarke, and H. R. Saibil, *Nature (London)* **371,** 261 (1994).

[15] E. S. Bochkareva and A. S. Girshovich, *J. Biol. Chem.* **267,** 25672 (1992).

[16] H. Saibil, Z. Dong, S. Wood, and A. auf der Mauer, *Nature (London)* **353,** 25 (1991).

[17] E. S. Bochkareva, N. M. Lissin, G. C. Flynn, J. E. Rothman, and A. S. Girshovich, *J. Biol. Chem.* **267,** 6796 (1992).

[18] A. Azem, M. Kessel, and P. Goloubinoff, *Science* **265,** 653 (1994).

[19] M. Schmidt, K. Rutkat, R. Rachel, G. Pfeifer, R. Jaenicke, P. Viitanen, G. Lorimer, and J. Buchner, *Science* **265,** 656 (1994b).

[20] O. Llorca, S. Marco, J. L. Carrascosa, and J. M. Valpuesta, *FEBS Lett.* **345,** 181 (1994).

[21] A. Azem, S. Diamant, M. Kessel, C. Weiss, and P. Goloubinoff, *Proc. Natl. Acad. Sci. U.S.A.* **92,** 12021 (1995).

[22] G. E. Davies and G. R. Stark, *Proc. Natl. Acad. Sci. U.S.A.* **66,** 651 (1970).

[23] F. H. Carpenter and K. T. Harrington, *J. Biol. Chem.* **247,** 5580 (1972).

[24] F. Hucho and M. Janda, *Biochem. Biophys. Res. Commun.* **57,** 1080 (1974).

[25] J. Hajdu, F. Bartha, and P. Friedrich, *Eur. J. Biochem.* **68,** 373 (1976).

[26] F. Hucho, H. Mullner, and H. Sund, *Eur. J. Biochem.* **59,** 79 (1975).

this method to $GroEL_{14}$ and $GroES_7$ oligomers, and were able to obtain important structural information following a relatively simple procedure.[30] First, native $GroEL_{14}$ or $GroES_7$ was exposed to a bifunctional cross-linking reagent, glutaraldehyde (GA), for various periods of time. The reaction was terminated by the addition of glycine, which inactivates the free cross-linker. The cross-linking products were then separated by denaturing sodium dodecyl sulfate (SDS) gel electrophoresis and identified in terms of the discrete numbers of cross-linked protein subunits in each species. The amount of each species was proportional to the intensity of the Coomassie stain, as measured by densitometric scanning. The kinetic analysis of the appearance and disappearance of transient species during the cross-linking reaction provided important information about the number and arrangement of subunits within the $GroEL_{14}$ or $GroES_7$ complexes.[30,31]

This chapter addresses the use of chemical cross-linking as a method to analyze the structure of the various GroE oligomers in solution. In the first part, we discuss parameters of the cross-linking reaction for GroE proteins, such as the protein concentration, the choice of cross-linking reagents and buffers, gel electrophoresis, and EM of cross-linked chaperonins, and finally, the identification of the GroE cross-linked products. In the second part, we discuss methods of quantitation by cross-linking of the various GroEL–GroES chaperonin heterooligomers in solution. In the third part, we discuss two control experiments that demonstrate that the cross-linking pattern of chaperonin heterooligomers reflects the true equilibrium between the various chaperonin species that existed in the solution before cross-linking was carried out.

## Cross-Linking Method

### *Protein Purity and Concentration*

Because the identification of protein cross-linking products is usually carried out in SDS–polyacrylamide gels stained with Coomassie Brilliant Blue, it is important that the protein be as pure as possible. The presence of contaminating proteins can confuse the analysis of the cross-linking pattern. Typically, when examined by Coomassie blue stain, each cross-linked electrophoretic species must contain 0.5–1 $\mu$g of protein to be clearly

[27] M. J. Sculley, G. B. Treacy, and P. D. Jeffrey, *Biophys. Chem.* **19,** 39 (1984).
[28] S. Darawshe and E. Daniel, *Eur. J. Biochem.* **201,** 169 (1991).
[29] Y. Tsfadia, I. Shaked, and E. Daniel, *Eur. J. Biochem.* **193,** 25 (1990).
[30] A. Azem, S. Diamant, and P. Goloubinoff, *Biochemistry* **33,** 6671 (1994).
[31] A. Azem, unpublished results.

visible in the gel. In the case of $GroEL_{14}$, each sample must contain at least 7.5–15 $\mu g$ of protein to obtain a clear signal for each of the 14 potential cross-linking intermediates.

The optimal protein concentration in the cross-linking reaction must be carefully calibrated, and varies from protein to protein. When the protein concentration is too high, there are more intermolecular collisions, which increase the risk of intermolecular cross-linking events between two individual particles. For $GroEL_{14}$, a protein concentration of up to 0.4 mg/ml (7 $\mu M$ protomer) was found to be optimal for intra-, as opposed to inter-, molecular cross-linking with 0.22% (w/v) glutaraldehyde at 37°. In the case of catalase, significant intermolecular cross-linking was already observed at 0.4 mg/ml (1.67 $\mu M$ oligomer).[32,33] It is possible to prevent artifactual intermolecular cross-linking by lowering the protein concentration. However, some protein oligomers may dissociate upon extensive dilution.[33]

### *Choice of Cross-Linking Reagent*

The choice of the cross-linking reagent depends on many parameters such as the specificity, homo- or heterofunctionality, stability, solubility, and cost of the reagent. One of the most stable and inexpensive cross-linkers in aqueous solution is glutaraldehyde (GA), which reacts mainly with ε-amino groups, and to a lesser extent with other groups such as thiols, alcohols, and guanidino groups.[34] Glutaraldehyde is commonly stored in a 25% (w/v) aqueous solution, which, at such a high concentration, contains a high percentage of polymeric species, together with the hemiacetal form. Therefore, at high concentrations, GA cannot be used for cross-linking. However, when diluted to ≤2.5% (w/v), between pH 3 and 8, most of the GA assumes a monomeric cyclic hemiacetal structure, and is in equilibrium with the noncyclic monomer.[35] Glutaraldehyde does not react with water in a nonreversible manner and is thus relatively stable in aqueous solution, at room temperature. For example, at 25°, we found that 0.1% (w/v) GA was active even after 24 hr, in the cross-linking of $GroEL_{14}$.[30] This high stability ensures the completion of the cross-linking reaction for proteins that undergo slow cross-linking, for example, when only a limited number of reactive groups is available at the interface of the subunits.

Depending on the length of the cross-linking arm, different kinds of

[32] J. Hajdu, R. S. Wyss, and H. Aebi, *Eur. J. Biochem.* **80,** 199 (1977).

[33] A. Azem, Ph.D. thesis. Tel Aviv University, Tel Aviv, Israel, 1991.

[34] J. Hajdu and P. Friedrick, *Anal. Biochem.* **65,** 273 (1975).

[35] J.-I. Kawahara, T. Ohmori, T. Ohkubo, S. Hattori, and M. Kawamura, *Anal. Biochem.* **201,** 94 (1992).

information can be extracted from the cross-linking pattern. Using GA, for example, cross-linking within the $GroEL_7$ toroids occurred much faster than between the two $GroEL_7$ toroids of the $GroEL_{14}$ particle. Consequently, the cross-linking pattern was biphasic, and indicated that the $GroEL_{14}$ oligomer was composed of two heptamers of seven subunits each[30,31] (Fig. 1A). In contrast, when a longer cross-linker, sulfo-SANPAH

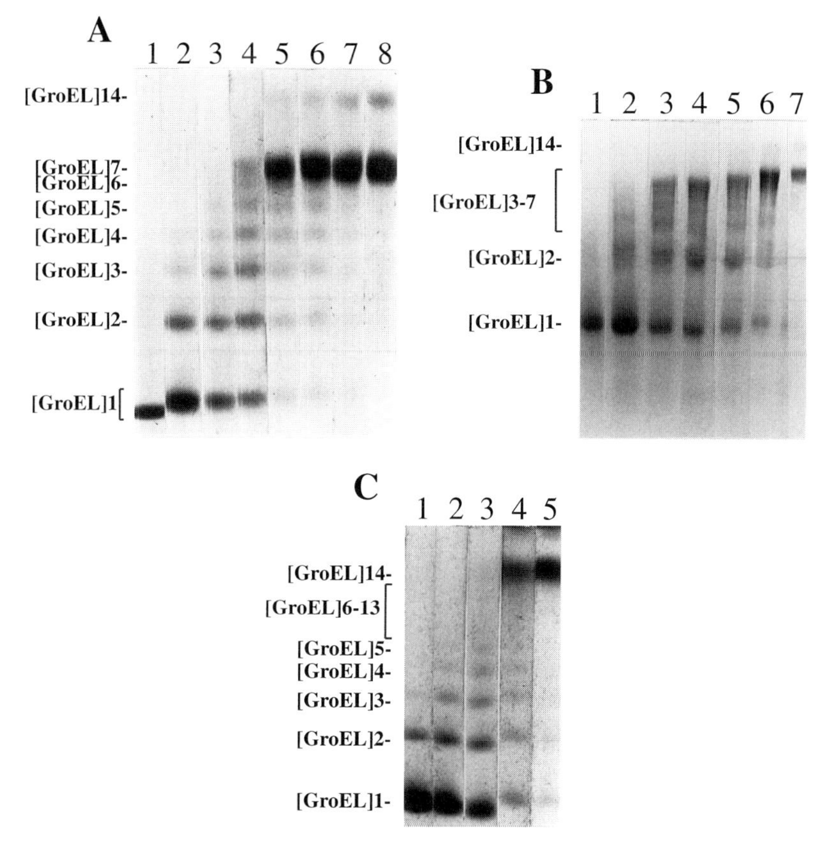


FIG. 1. Cross-linking of $GroEL_{14}$. Native $GroEL_{14}$ was incubated either in the absence of cross-linker (lane 1) or in the presence of 0.22% (w/v) GA at 37° for 10, 20, 50, 60, 70, 90, and 105 sec (lanes 2–8, respectively) as in Azem *et al.*[18] Cross-linking products were separated by denaturing electrophoresis on either (A) 3.3% (w/v) uniform polyacrylamide tube gels in phosphate buffer or (B) in a 3–25% (w/v) polyacrylamide gradient slab gel in Laemmli buffer. In (A) and (B) 30 and 15 $\mu$g of protein were loaded, respectively. (C) $GroEL_{14}$ was incubated as in (A) and (B), but in the presence of 0.6 m*M* SANPAH for 10 min, instead of GA, and then exposed to UV light (320 nm) for 0, 5, 10, 20, or 30 min (lanes 1–5, respectively). Samples were separated as in (A).

[sulfosuccinimidyl 6-(4′-azido-2′-nitrophenylamino)hexanoate], was reacted with $GroEL_{14}$, the kinetics of cross-linking within the $GroEL_7$ toroids was similar to that between the two $GroEL_7$ toroids of the $GroEL_{14}$ particle (Fig. 1C), and the structural arrangement of $GroEL_{14}$ subunits could not be determined.

Unlike GA, many cross-linking agents undergo rapid hydrolysis in aqueous solution at neutral pH. For example, diimido ester derivatives such as dimethyl suberimidate (DMS) are rapidly hydrolyzed in aqueous solution, some of them with a half-life of 20 min.[36–38] Thus, cross-linking of $GroEL_{14}$ with DMS at pH 7.5 proceeds less than 20% the extent of the cross-linking reaction obtained with the same concentration of GA (not shown).

*Choice of Cross-Linking Buffers*

Some of the buffers that are commonly used in protein biochemistry can react with the cross-linking reagents and should therefore be avoided. Reactivity of the buffer with the cross-linker not only reduces the effective concentration of the cross-linker in the reaction, but also changes the properties of the buffer. Thus, when $GroES_7$ and $GroEL_{14}$ were cross-linked for 15 min at 25° with 10 m*M* (0.1%, w/v) GA in 50 m*M* Tris, pH 7.5, as in Engel *et al.*,[39] the GA was inactivated and less than 5% of the GroES was found by SDS gel analysis to be cross-linked to GroEL. We found that under these conditions, owing to the reaction of the Tris buffer with GA, the pH of the sample decreased from pH 8.0 to less than pH 6.0 within 15 min.[31]

Buffers such as triethanolamine (TEA), 4-morpholinopropanesulfonic acid (MOPS), and phosphate are suitable for cross-linking of proteins with amino-reactive reagents.

The pH of the solution is also an important variable. For example, above neutral pH, the dialdehyde undergoes self-condensation. Polymerization of the GA increases the cross-linking arm and consequently increases the chances of nonspecific and artifactual inter-molecular cross-linking events. Therefore, cross-linking with GA must be carried out at a pH not higher than pH 8.0. In contrast, cross-linkers such as DMS are rapidly hydrolyzed above neutral pH.[22] For each cross-linker, the optimal pH must be determined empirically.

[36] K. Peters and M. Richards, *Annu. Rev. Biochem.* **46,** 523 (1977).

[37] F. Wold, *Methods Enzymol.* **25,** 6223 (1972).

[38] D. T. Browne and S. B. H. Kent, *Biochem. Biophys. Res. Commun.* **67,** 126 (1975).

[39] A. Engel, M. K. Hayer-Hartl, K. N. Goldie, G. Pfeifer, R. Hegerl, S. Muller, A. C. R. da Silva, W. Baumeister, and F. U. Hartl, *Science* **269,** 832 (1995).

The temperature of the cross-linking reaction must also be carefully calibrated. At high temperatures, the risk of intermolecular cross-linking increases. However, at low temperatures, the cross-linking reation is slow. We found that whereas at 25° nearly complete cross-linking of $GroEL_{14}$ with 0.22% (w/v) GA requires 24 hr, at 37° the same extent of cross-linking was obtained in about 1 hr.

It is noteworthy that, as the cross-linking reaction proceeds, proteins become saturated with intra-subunit cross-links, which reduce mobility of the oligomer in SDS gels (compare lane 1 with lane 2 in Fig. 1A). Identification of cross-linked species may therefore be difficult when comparing cross-linking patterns from different times of exposure.

*Controlling Time of Cross-Linking Reaction*

It is convenient to control the cross-linking reaction by inactivation of the free reagent after a given period of time. This is particularly important when quantitation of transient cross-linking species is desired for kinetic analysis. In the case of amino-reactive cross-linkers, such as GA, this can be achieved by the addition of moderately reactive compounds such as Tris buffer, amino acids such as lysine and glycine, or by the addition of highly reactive reagents like hydroxylamine and hydrazine. We arrest the cross-linking reaction before denaturing electrophoresis by adding sample buffer for final concentrations of 400 m*M* glycine, 50 m*M* Tris, 3% (w/v) SDS, 3% (v/v) 2-mercaptoethanol, and 10% (v/v) glycerol.

*Gel Electrophoresis*

To obtain maximal information from protein cross-linking, there is a need for clearly resolved and quantifiable protein signals on SDS gels. The best resolution of $GroEL_{14}$ intermediate cross-linked species can be observed in uniform tube gels containing 2.8–3.3% (w/v) acrylamide in phosphate buffer. Slab gels in Laemmli buffer[40] are not recommended, because protein signals often appear as poorly resolved smears.[30] Figure 1A shows the cross-linking products of $GroEL_{14}$ separated on SDS–polyacrylamide tube gels in the presence of phosphate buffer.[41] Protein signals ranging from the 57.3-kDa GroEL monomer to the 945-kDa $GroEL_{14}(GroES_7)_2$ heterooligomer can be detected on the same uniform polyacrylamide tube gel.[18] Figure 1B shows the same samples as in Fig. 1A, separated on a 3–25% (w/v) SDS–polyacrylamide slab gel in Laemmli buffer. Another advantage of the tube gel system is that the protein signals

[40] U. K. Laemmli, *Nature* (*London*) **227,** 680 (1970).

[41] K. Weber, J. R. Pringle, and M. Osborn, *Methods Enzymol.* **26,** 3 (1972).

are symmetric and can be resolved by fitting the density scan to a Gaussian function.[28] This is of importance when the protein signals are poorly separated.

For $GroEL_{14}$ and $GroES_7$, electrophoresis was carried out in $10 \times 0.6$ cm tube gels of 2.8–3.3% (w/v) acrylamide (acrylamide : bisacrylamide, 29 : 1, v/v), run in 0.1 *M* sodium phosphate buffer, pH 7.2. The current used was 4 mA/tube (~20 V) during the first 30 min, then 8 mA/tube (~40 V) during the remainder of the electrophoresis (5–7 hr). To visualize GroEL or GroES monomers, electrophoresis was stopped when the bromophenol blue was 1 cm from the end of the gel. However, to obtain clear resolution of cross-linked GroEL–GroES heterooligomers, electrophoresis was continued for 2 hr after the stain left the gel.

### *Electron Microscopy of Cross-Linked Chaperonins*

The structure of the asymmetric $GroEL_{14}GroES_7$ and symmetric $GroEL_{14}(GroES_7)_2$ particles as viewed by negative-stain electron microscopy was indistinguishable whether or not the molecules were treated with GA before EM.[18,21] This was true despite the fact that GA-cross-linked GroEL and GroES species were much more diffuse than uncross-linked species when separated on SDS-polyacrylamide tube gels. It should be noted that exposure to GA resulted in a higher proportion of lateral views, as opposed to top views, of the chaperonin particles.[21,39] Because the GA-induced bias in favor of lateral views was seen in the absence of bound $GroES_7$,[31] the GA treatment most likely results in an increase in the affinity of the $GroEL_{14}$ lateral surface for the carbon surface.

For EM preparation of cross-linked chaperonins, the cross-linking reaction was terminated by the addition of a one-third volume of 1.0 *M* glycine buffered with 50 m*M* Tris, pH 7.5. Clarity of the chaperonin particles in the micrographs decreased as the time between the GA quench and the fixation for EM increased. Therefore, chaperonin samples were fixed on EM grids within less than 5 min of terminating the cross-linking reaction.

### *Identification and Quantitation of Cross-Linked Homooligomeric Species*

The molecular weight of each intermediate cross-linking product of a homooligomer such as $GroEL_{14}$ is a multiple of the molecular weight of the monomeric subunit polypeptide chain. SDS gels can be used to separate cross-linked species of proteins composed of a maximum of about 10 polypeptide chains, regardless of the subunit size. Above 10 chains, however,

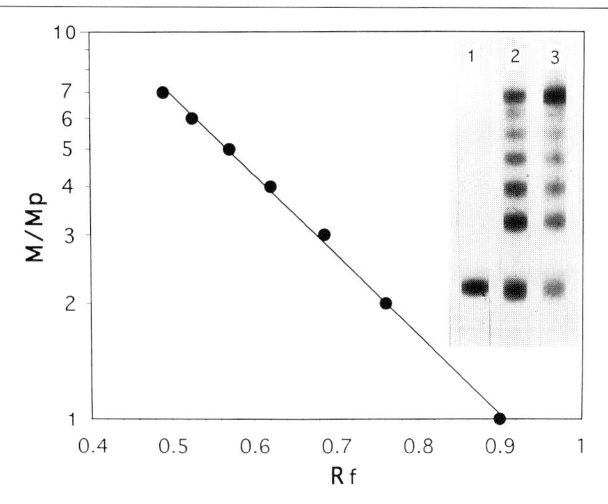


FIG. 2. Cross-linking pattern of the $GroES_7$ oligomer. $GroES_7$ (50 $\mu M$ protomer) was exposed to GA (0.1%, w/v) at 25° for 0, 2, or 24 hr (*inset,* lanes 1–3, respectively). Protein (2 $\mu$g) was loaded in lane 1, whereas 15 $\mu$g was loaded in lanes 2 and 3. The cross-linked species were separated on 5% (w/v) polyacrylamide tube gels as in Fig. 1A. The data from lane 2 were plotted and gave a linear relationship between the number of chains ($M/M_p$: 1, 2, 3, 4, 5, 6, and 7) and the relative migration distance ($R_f$) ($r^2$ = 0.998). All the protein accumulated in an electrophoretic species corresponding to an $M/M_p$ 7, indicating that $GroES_7$ is a heptamer.

the electrophoretic species migrate too close to one another to be well resolved.[42]

Cross-linking products can be characterized in terms of the number of polypeptide chains cross-linked together, $M/M_p$, where $M$ equals the relative molecular weight of the cross-linked species and $M_p$ is the molecular weight of the protomer. In the case of uniform tube gels, a linear correlation is obtained by plotting $\log(M/M_p)$ of each species against the distance that species has migrated in the gel divided by the distance that the bromphenol blue stain has migrated ($R_f$).[43] When a gradient gel is used, the molecular weight of the species can be determined from a plot of $\log(M/M_p)$ against $\log T\%$, where $T$ is the calculated concentration of acrylamide at which the electrophoretic species is found.[30] Figure 2 shows the molecular weight determination of each intermediate species from partial cross-linking of

[42] A. Azem, A. Pinhasy, and E. Daniel, *in* "Structure and Function of Invertebrate Dioxygen Carriers" (S. Vinogradov and O. Kapp, eds.), pp. 49–57. Springer-Verlag, New York, 1991.

[43] S. Darawshe, Y. Tsfadyah, and E. Daniel, *Biochem. J.* **242,** 689 (1987).

$GroES_7$ oligomers. For identification of lower molecular weight species such as GroES, higher concentrations of acrylamide (5%, w/v) were used.

The amount of protein in each electrophoretic species is proportional to the intensity of the Coomassie signal. Quantitation of cross-linked species was carried out by scanning the gels using an Ultrascan XL scanner (LKB) as in Azem *et al.*[30]

## Identification and Quantitation of Chaperonin Heterooligomers

### *Quantitation Using $GroEL_{14}GroES_7$ and $GroEL_{14}(GroES_7)_2$ Species*

When $GroEL_{14}$ and $GroES_7$ were incubated in the presence of increasing amounts of ATP and then cross-linked for 1 hr at 37° in the presence of 0.22% (w/v) GA, the major products were the final cross-linking products of the reaction: $GroES_7$, $GroEL_{14}$, $GroEL_{14}GroES_7$, and $GroEL_{14}(GroES_7)_2$.[18] Quantitation of each species was carried out by scanning and Gaussian fitting the protein signals.[18] The ratio of various heterooligomers was determined using the difference in relative mobilities ($R_f$) after long (1-hr) exposure times to cross-linker and long separation times: $GroEL_{14}$ ($R_f$ 0.19), $GroEL_{14}GroES_7$ ($R_f$ 0.14), and $GroEL_{14}(GroES_7)_2$ ($R_f$ 0.125).[18] However, separation of the $GroEL_{14}GroES_7$ and $GroEL_{14}(GroES_7)_2$ cross-linked species was at the limit of resolution of the tube gel system. For this reason, an alternative method for the quantitation of the $GroEL_{14}GroES_7$ and $GroEL_{14}(GroES_7)_2$ particles in solution was developed on the basis of the ratio between the intermediate cross-linking species, $GroEL_7GroES_7$ and $GroEL_7$ (see the next section).

### *Quantitation Using $GroEL_7GroES_7$ and $GroEL_7$ Intermediates*

When $GroEL_{14}$ and $GroES_7$ were incubated in the presence of ADP, only asymmetric $GroEL_{14}GroES_7$ particles were observed by EM and cross-linking, whereas in the presence of high concentrations of ATP, a majority of symmetric $GroEL_7(GroES_7)_2$ particles were observed.[18,21] We examined the time-dependent cross-linking patterns of GroELS heterooligomers generated either in the presence of excess ADP or ATP (Fig. 3A and B, respectively). In the presence of ADP, two intermediate-mobility species were observed in equal amounts, which were identified as $GroEL_7$ and $GroEL_7GroES_7$ (Fig. 3A, lanes 2 and 3, respectively). As the cross-linking time increased, a low-mobility, fully cross-linked $GroEL_{14}GroES_7$ species accumulated, with the simultaneous disappearance of $GroEL_7$ and $GroEL_7GroES_7$ (Fig. 3A, lane 4). Therefore, equal amounts of $GroEL_7$ and $GroEL_7GroES_7$ reflect a solution saturated with asymmetric $GroEL_{14}GroES_7$ chaperonins (Fig. 4).

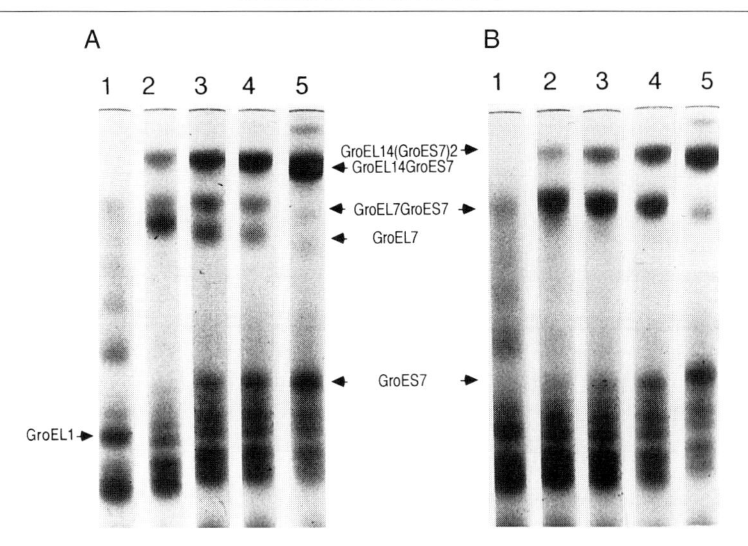


FIG. 3. Cross-linking of GroEL–GroES heterooligomers. $GroEL_{14}$ (3.5 $\mu M$ protomer) and $GroES_7$ (14 $\mu M$ protomer) in TEA (50 m$M$), magnesium acetate (20 m$M$) were preincubated for 2 min in the presence of 1.5 m$M$ ADP (A) or ATP (B). Cross-linking with GA (0.22%, w/v) was carried out for 0.5, 2.5, 5, 10, or 60 min at 37° (lanes 1–5, respectively). The samples were separated by SDS–PAGE, and allowed to migrate until the stain reached the end of the gel. *Note:* At cross-linking times >2.5 min, the high-mobility electrophoretic species represent the $GroES_7$ cross-linked species. The monomeric form of GroES has migrated out of the gel.

In contrast, in the presence of saturating amounts $GroES_7$ and ATP, a single $GroEL_7GroES_7$ intermediate species was observed with an intensity twice that of either the $GroEL_7$ or $GroEL_7GroES_7$ intermediate species seen in the ADP pattern (Fig. 3B, lanes 2 and 3). As the cross-linking time increased, a low-mobility fully cross-linked $GroEL_{14}(GroES_7)_2$ species accumulated, at the expense of the disappearing $GroEL_7GroES_7$ species (Fig. 3B, lane 4). Thus, a single $GroEL_7GroES_7$ intermediate species after short cross-linking is the signature of a chaperonin solution saturated with symmetrical $GroEL_{14}(GroES_7)_2$ particles (Fig. 4; Azem *et al.*[21]) This method allowed measurement of the relative amounts of $GroEL_{14}(GroES_7)_2$ and $GroEL_{14}GroES_7$ particles in chaperonin solutions containing various limiting amounts of $GroES_7$, $Mg^{2+}$ and $Mn^{2+}$ ions, ATP, ADP, and the analog AMP-PNP and correlation of the distribution of chaperonin heterooligomers with chaperonin functions, such as ATPase activity and protein folding. Thus, Azem *et al.*[21] showed that a correlation exists between the amount of symmetrical $GroEL_{14}(GroES_7)_2$ particles in a chaperonin solution and the rate of protein folding of that chaperonin solution. However, the validity

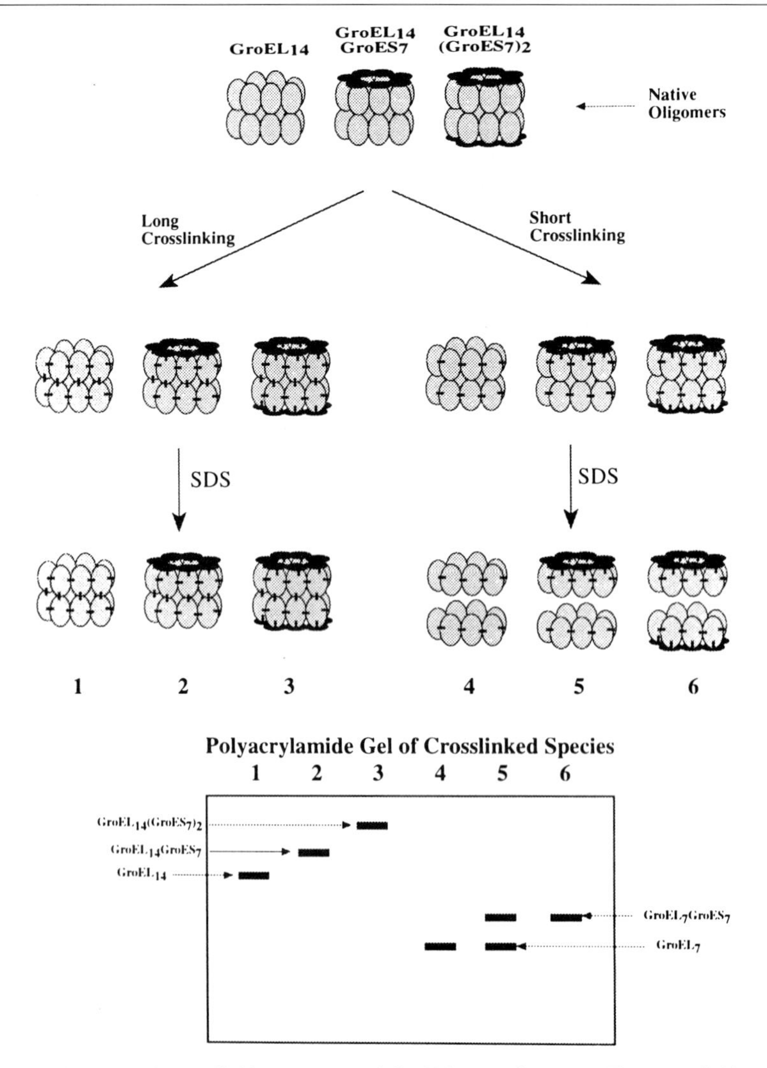


FIG. 4. Scheme of cross-linking patterns of GroE heterooligomers. The cross-linking patterns of chaperonin particles on denaturing gels vary with the time of exposure to GA. Whereas a long exposure (60 min at 37°) of $GroEL_{14}$ to GA produces a single $GroEL_{14}$ signal (lane 1), a short exposure to GA (6 min at 37°) produces a majority of $GroEL_7$ intermediates (lane 4). This is due to the faster cross-linking reaction within the $GroEL_7$ toroids, than between the two $GroEL_7$ toroids of the $GroEL_{14}$ particle (Azem *et al.*).[18] For the same reason, a long exposure of an asymmetric $GroEL_{14}GroES_7$ particle to GA produces a single $GroEL_{14}GroES_7$ signal (lane 2), whereas a short exposure produces a two intermediate signals in equal amounts: $GroEL_7GroES_7$ and $GroEL_7$ (lane 5). Likewise, a long exposure of the symmetric $GroEL_{14}(GroES_7)_2$ particle to GA produces a single $GroEL_{14}(GroES_7)_2$ signal (lane 3), but a short exposure produces a single intermediate signal $GroEL_7GroES_7$ as the major species (lane 6).

of such a correlation is dependent on the assumption that the ratios obtained after cross-linking the chaperonin heterooligomers with GA indeed reflect the true chaperonin equilibrium during protein folding.

## Cross-Linking Controls

Does cross-linking of chaperonin molecules with GA result in the recruitment of $GroES_7$ by $GroEL_{14}GroES_7$ during the reaction? Or, alternatively, does cross-linking with GA cause the dissociation of $GroES_7$ from $GroEL_{14}GroES_7$ particles? These questions must be addressed if a correlation is to be established between the function and true structure of chaperonin heterooligomers in a solution at equilibrium.

### *Lack of Binding to $GroEL_{14}GroES_7$ by $GroES_7$ during Glutaraldehyde Cross-Linking*

To verify whether $GroES_7$ can bind $GroEL_{14}$ in the presence of GA, a solution of unbound $GroEL_{14}$ and $GroES_7$ was exposed to GA for different periods of time, before, during, or after the addition of a saturating amount of ATP (Fig. 5). Within 2 seconds, GA was seen to have prevented the binding of $GroES_7$ to $GroEL_{14}GroES_7$ particles and only the high affinity binding of one $GroES_7$ ring to the $GroEL_{14}$ core particle could take place. Within 4 seconds, GA was seen to have prevented the ATP-dependent binding of $GroES_7$ co-chaperonins to both $GroEL_{14}GroES_7$ and $GroEL_{14}$ core particles.

Since there cannot be any recruitment of a second $GroES_7$ by $GroEL_{14}GroES_7$ particles beyond the first 2 seconds of the addition of GA, and since 2 seconds represents 1/12 of the time necessary for a full turnover of the $GroEL_{14}$ ATPase cycle,[44,45] we concluded that cross-linking with GA is unlikely to have caused an artifactual overestimation of the amounts of symmetrical $GroEL_{14}(GroES_7)_2$ in the chaperonin solution.

### *Lack of Dissociation of $GroES_7$ from $GroEL_{14}GroES_7$ during Glutaraldehyde Cross-Linking*

To verify whether bound $GroES_7$ can dissociate from the chaperonin during GA cross-linking, a solution of preformed asymmetrical $GroEL_{14}$ $[^3H]GroES_7$ particles was exposed to GA for periods of time before, during, or after the addition of a fourfold excess of unlabeled $GroES_7$ (Fig. 6). A 1-min $GroES_7$ chase, in the absence of GA, sufficed for the labeled and unlabeled $GroES_7$ to reach equilibrium. In contrast, when the unlabeled

[44] M. J. Todd, P. V. Viitanen, and G. H. Lorimer, *Biochemistry* **32,** 8560 (1993).
[45] M. J. Todd, P. V. Viitanen, and G. H. Lorimer, *Science* **265,** 659 (1994).

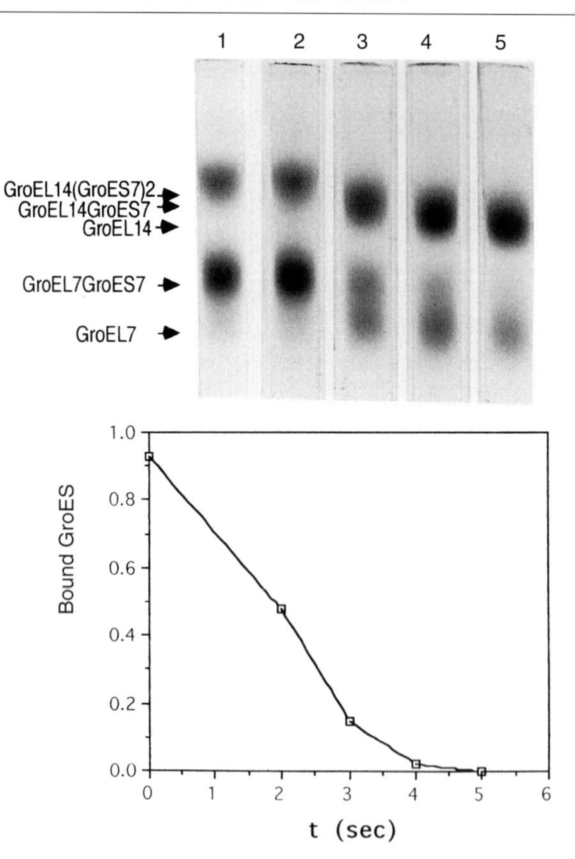


FIG. 5. Effect of GA on the equilibrium between chaperonin heterooligomers in solution. $GroEL_{14}$ (3.5 $\mu M$) and $GroES_7$ (10 $\mu M$) were incubated in the presence of 1.6 $\mu M$ ATP for 10 sec before (lane 1), during (lane 2), or 2, 3, and 4 sec (lanes 3–5, respectively) after the addition of 0.22% GA. The cross-linking reaction was allowed to proceed for 7.5 min at 37° and the products were separated by SDS–PAGE as in Fig. 2A. The gels were scanned and the relative intensity of $GroEL_7$ and $GroEL_7GroES_7$ signals was determined. The amount of $GroEL_{14}$, $GroEL_{14}GroES_7$, and $GroEL_{14}(GroES_7)_2$ particles in solution was calculated from the ratio of $GroEL_7$ to $GroEL_7GroES_7$ species as described in Azem *et al.*[21] The gels were scanned and bound GroES was calculated from the molar ratio of $GroEL_7GroES_7$, assuming two binding sites for GroES on the $GroEL_{14}$ molecule. In footballs, bullets, and bricks there are 2, 1 and 0 $GroES_7$ molecules bound to each core oligomer, respectively.

$GroES_7$ chase was applied 5 sec after the GA addition, about 80% of the bound $^3$H-labeled $GroES_7$ remained associated with the chaperonin. It can be concluded that active $GroES_7$ binding to and release from the chaperonin particles is prevented within the first few seconds of GA addition. Consequently, the distribution of chaperonin heterooligomers, as revealed by

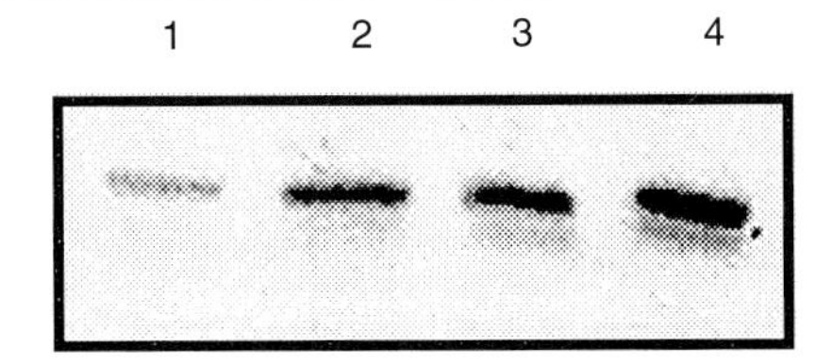


FIG. 6. Prevention of ATP-dependent $GroES_7$ binding/release turnover on $GroEL_{14}$ by GA cross-linking. GroEL (3.3 $\mu M$) and [$^3$H]GroES (1.35 $\mu M$, 4.3 $\times$ $10^5$ cpm/nmol) were incubated as in Fig. 1B with 0.22% (w/v) GA for various times before or after adding a four fold molar excess of unlabeled GroES (13.2 $\mu M$). GroES was added 1 min before GA (lane 1), simultaneously with GA (lane 2), 1 min after GA (lane 3), or 10 min after GA (lane 4). Following 1 hr of cross-linking at 37°, samples were separated on a 3–25% (w/v) SDS–acrylamide gradient gel. [$^3$H]GroES was detected by autoradiography.

SDS gels of cross-linked species, reflects the true equilibrium that existed in the chaperonin solution before GA was added.[21]

Moreover, this conclusion was independently confirmed by EM. Chaperonin solutions that were treated with GA before EM analysis displayed a similar ATP-dependent distribution of $GroEL_{14}GroES_7$ and $GroEL_{14}(GroES_7)_2$ particles as native, untreated heterooligomers under the same incubation conditions.[21]

*Importance of Cross-Linking for Accurate Determination of Equilibrium State*

The relatively low affinity of the second $GroES_7$ as compared with the first $GroES_7$ for the chaperonin[18,21] and the high sensitivity of the $GroEL_{14}(GroES_7)_2$ particle to ADP[21] and low $Mg^{2+}$[39] suggest that the bound second $GroES_7$ is highly labile. This stresses the importance of cross-linking as a method that can prevent the displacement of subunit equilibria in solution.

Several methods have been used for the structural analysis of chaperonin heterooligomers. Electron microscopy,[12,14,39] native gel electrophoresis,[46] gel filtration,[17] and dialysis[39] are all likely to displace the biochemical equilibrium and therefore result in a severe underestimation of the true amount of the labile $GroEL_{14}(GroES_7)_2$ particle. It is therefore important to cross-link chaperonin particles before their analysis by the previously mentioned methods or to use alternative methods, such as analytical ultracentrifugation, light scattering, or fluorescence, which preserve the integrity of the equilibrium state during analysis.

[46] A. Horovitz, E. S. Bochkareva, O. Kovalenko, and A. S. Girshovich, *J. Mol. Biol.* **231,** 58 (1993).

## Summary

In this chapter, we have shown how chemical cross-linking with a bifunctional reagent, GA, can be used to investigate the structure of large oligomeric complexes such as $GroEL_{14}GroES_7$ and $GroEL_{14}(GroES_7)_2$. Cross-linking, followed by denaturing electrophoresis, confirmed the number and arrangement of GroEL and GroES subunits within each individual oligomer, which was previously known from EM analysis.[5,9] Furthermore, cross-linking permitted a close examination of the effect of regulatory factors, such as nucleotides and free divalent cations, on the molecular structure of $GroEL_{14}$, $GroEL_{14}GroES_7$, and $GroEL_{14}GroES_7$.[30,47,48] Finally, cross-linking analysis permitted characterization and quantitation of various chaperonin heterooligomeric complexes, $GroEL_{14}$, $GroEL_{14}GroES_7$, and $GroEL_{14}GroES_7$ in solution, under conditions that also supported protein folding and ATP hydrolysis.[21]

It was shown that GA does not induce the artifactual association or the dissociation of $GroES_7$ from the chaperonin. On the contrary, chemical cross-linking is an obligatory procedure when the subsequent analysis is carried out using methods that can displace the equilibrium.

## Note Added in Proof

This approach of structural analysis of the GroE chaperonin complexes by chemical cross-linking has been validated and confirmed by independent methods that do not displace the biochemical equilibrium. Experiments using analytical ultra-centrifugation[49] and protein fluorescence[50] not only confirmed the existence of a major faction of $GroEL_{14}(GroES_7)_2$ complexes in optimally active protein-folding solutions at equilibrium, but showed identical nucleotide- and GroES-dependent distributions of the various chaperonin oligomers as initially described using cross-linking.[18,21]

## Acknowledgments

We thank Dr. Sophie Diamant for critical review of this manuscript. This work was supported by Grant 00015/1 from the United States–Israel Binational Science Fund; Grant 1180 from the Joint German–Israeli Research Program; Grant 158-93 from the Israeli Science Foundation; Grant 512 from the Levi Eshkol Fund of the Israeli Ministry of Science and Art to C. W.; and a grant from the Golda Meir Foundation to A.A.

[47] S. Diamant, A. Azem, C. Weiss, and P. Goloubinoff, *Biochemistry* **34,** 273 (1995).
[48] S. Diamant, A. Azem, C. Weiss, and P. Goloubinoff, *J. Biol. Chem.* **270,** 28387 (1995).
[49] J. Behlke, O. Ristau, and H.-J. Schönfeld, *Biochemistry* **36,** 5149 (1997).
[50] S. Török, L. Vigh, and P. Goloubinoff, *J. Biol. Chem.* **271,** 16180 (1996).

# [23] Molecular Chaperones and Their Interactions Investigated by Analytical Ultracentrifugation and Other Methodologies

*By* HANS-JOACHIM SCHÖNFELD and JOACHIM BEHLKE

Most chaperones form oligomers and the interaction of different chaperones often results in complex heterooligomeric structures. In many cases these homo- and heterooligomeric structures are highly conserved between different species and even throughout different biological kingdoms (prokaryotes, eukaryotes, plants, and mammals), indicating that the quaternary structures of chaperones are important for their functions.

The oligomeric properties of different chaperones and their complexes formed in solution have been studied by various equilibrium and nonequilibrium techniques such as chemical cross-linking, size-exclusion chromatography, Ferguson plot analysis, density gradient sedimentation, electron microscopy, equilibrium dialysis, fluorescence spectroscopy, quasi-elastic light scattering, and analytical ultracentrifugation. Some of these data are summarized in Table I.[1-19] Note that some results are characterized in their

[1] J. B. Weiss, P. H. Ray, and P. J. J. Bassford, *Proc. Natl. Acad. Sci. U.S.A.* **85,** 8978 (1988).
[2] M. Watanabe and G. Blobel, *Proc. Natl. Acad. Sci. U.S.A.* **86,** 2728 (1989).
[3] M. Zylicz and C. Georgopoulos, *J. Biol. Chem.* **259,** 8820 (1984).
[4] D. R. Palleros, K. L. Reid, L. Shi, and A. L. Fink, *FEBS Lett.* **336,** 124 (1993).
[5] H.-J. Schönfeld, D. Schmidt, H. Schröder, and B. Bukau, *J. Biol. Chem.* **270,** 2183 (1995).
[6] S. Blond Elguindi, A. M. Fourie, J. F. Sambrook, and M. J. Gething, *J. Biol. Chem.* **268,** 12730 (1993).
[7] N. Benaroudj, G. Batelier, F. Triniolles, and M. M. Ladjimi, *Biochemistry* **34,** 15282 (1995).
[8] M. Zylicz, D. Ang, and C. Georgopoulos, *J. Biol. Chem.* **262,** 17437 (1987).
[9] J. Osipiuk, C. Georgopoulos, and M. Zylicz, *J. Biol. Chem.* **268,** 4821 (1993).
[10] K. L. Reid and A. L. Fink, *Cell Stress Chaperones* **1,** 127 (1996).
[11] B. Wu, A. Wawrzynow, M. Zylicz, and C. Georgopoulos, *EMBO J.* **15,** 4806 (1996).
[12] M. Zylicz, T. Yamamoto, N. McKittrick, S. Sell, and C. Georgopoulos, *J. Biol. Chem.* **260,** 7591 (1985).
[13] R. W. Hendrix, *J. Mol. Biol.* **129,** 375 (1979).
[14] G. N. Chandrasekhar, K. Tilly, C. Woolford, R. Hendrix, and C. Georgopoulos, *J. Biol. Chem.* **261,** 12414 (1986).
[15] A. Wawrzynow and M. Zylicz, *J. Biol. Chem.* **270,** 19300 (1995).
[16] J. Martin, M. Mayhew, T. Langer, and F. U. Hartl, *Nature (London)* **366,** 228 (1993).
[17] A. Azem, M. Kessel, and P. Goloubinoff, *Science* **265,** 653 (1994).
[18] A. Engel, M. K. Hayer Hartl, K. N. Goldie, G. Pfeifer, R. Hegerl, S. Muller, A. C. da Silva, W. Baumeister, and F. U. Hartl, *Science* **269,** 832 (1995).
[19] Z. Török, L. Vigh, and P. Goloubinoff, *J. Biol. Chem.* **271,** 16180 (1996).





0076-6879/98 $25.00

TABLE I

QUATERNARY STRUCTURES OF CHAPERONES AND CHAPERONE COMPLEXES IN SOLUTION[a]

| Chaperone or complex | Size of monomer (kDa) | Published Structure(s) | Ref. | Analytical methods |
|---|---|---|---|---|
| SecB | 17.2 | 4 to 6-mer | 1 | FP (4-mer), SEC (6-mer) |
| | | 4-mer | 2 | SEC |
| DnaK | 69.0 | 1-mer | 3 | DGS + SEC |
| | | 1-, 2-, 3-, >3-mers | 4 | N-PAGE, SEC |
| | | Equilibrium | 5 | AUC, QLS, SEC |
| BiP | | 1-, >2-mers | 6 | N-PAGE |
| Hsc70 | | 1-, 2-, 3-, >3-mers; equilibrium | 7 | SEC; AUC |
| GrpE | 21.7 | 1-mer | 8 | DGS |
| | | 2-mer | 9 | CL |
| | | 2-mer | 5 | AUC, QLS, SEC |
| | | (3-mer) | 10 | CL |
| | | (6-mer) | 11 | CL, SEC |
| DnaJ | 40.9 | 2-mer | 12 | AUC, DGS |
| | | 2 to 8-mer | 40 | AUC |
| GroEL | 57.2 | 14-mer | 13 | DGS, EM |
| GroES | 10.4 | 7-mer | 14 | DGS, SEC, EM |
| DnaK–GrpE | | $DnaK_1$–$GrpE_1$ | 8 | DGS + SEC |
| | | $DnaK_1$–$GrpE_2$ | 5 | AUC |
| | | ($DnaK_1$–$GrpE_3$) | 10 | SEC |
| DnaK–DnaJ | | $DnaK_1$–$DnaJ_{1\text{-}2}$ | 15 | SEC |
| GroEL–GroES | | $GroEL_{14}$–$GroES_7$ | 16 | SEC |
| | | $GroEL_{14}$–$GroES_7$ + $GroEL_{14}$–$GroES_{14}$ | 17–19 | CL, EM, ED, FS |

[a] FP, Ferguson plot; SEC, size-exclusion chromatography; DGS, density gradient sedimentation; N-PAGE, native polyacrylamide electrophoresis; AUC, analytical ultracentrifugation; QLS, quasi-elastic light scattering; CL, chemical cross-linking; EM, electron microscopy; ED, equilibrium dialysis; FS, fluorescence spectroscopy. Structures listed in parentheses were characterized as "uncertain" in their publications.

publications as "uncertain" and can differ, for a given chaperone, depending on the analytical method.

The best tool for measuring quaternary structures of proteins and their interactions in solution under physiological conditions is the analytical ultracentrifuge. The first analytical ultracentrifuge was constructed in 1923 by Svedberg. In the late 1920s, a theory for determining molecular masses from measurements of sedimentation and diffusion coefficients was formulated and the technique was steadily developed and extensively used in biochemistry and polymer science until around 1980 (for a historical review, see Schachman[20]). With the increasing interest in molecular biology in the early 1980s, interest in the exact measurement of oligomeric structures and equilibrium constants by what was a difficult and laborious technique rapidly declined and many universities even discontinued teaching the theory of analytical ultracentrifugation. The maintenance and operation of the few still operating instruments, mainly the Spinco model E (Beckman, Palo Alto, CA) analytical ultracentrifuge, became more and more time consuming and expensive.

Ironically, it was difficulties associated with the correct folding and function of recombinant proteins produced using molecular biological techniques that rekindled interest in precise methods for measuring oligomeric states and their association–dissociation equilibria, and only the analytical ultracentrifuge provided valid characterization of certain parameters. In 1989 a new instrument, the Optima XL-A (developed by Beckman, Palo Alto, CA), became available. The Optima XL-A was developed from the Optima XL standard preparative ultracentrifuge and is equipped with a new photoelectric scanning absorption optical system enabling exact measurement of concentration profiles at wavelengths of 190–800 nm, markedly increasing the sensitivity compared with older instruments. More recently, the Optima XL-I centrifuge became available, which is equipped with an interference optical system allowing the measurement of concentration gradients of substances without absorbance by fringe displacement. Computer-driven operation and automatic storage of data made analytical ultracentrifugation fast and easy and a renaissance of this technique for biochemical applications was predicted.[21] Nonexperts can operate the new instrument and powerful computer programs for data evaluation are supplied by the manufacturer or are made freely available via the World Wide Web by specialists in the field.

Because of the excellent series of publications,[22–26] it is not the intent

[20] H. K. Schachman, *in* "Analytical Ultracentrifugation in Biochemistry and Polymer Science" (S. E. Harding, A. J. Rowe, and J. C. Horton, eds.), pp. 3–15. The Royal Society of Chemistry, Cambridge, 1992.

[21] H. K. Schachman, *Nature* (*London*) **341,** 259 (1989).

[22] S. E. Harding, A. J. Rowe, and J. C. Horton (eds.), "Analytical Ultracentrifugation in Biochemistry and Polymer Science." The Royal Society of Chemistry, Cambridge, 1992.

[23] T. M. Schuster and T. M. Laue (eds.), "Modern Analytical Ultracentrifugation." Birkhäuser, Boston, 1994.

[24] G. Ralston, "Introduction to Analytical Ultracentrifugation." Beckman Instruments, Fullerton, California, 1993.

[25] D. K. McRorie and P. J. Voelker, "Self-Associating Systems in the Analytical Ultracentrifuge." Beckman Instruments, Fullerton, California, 1993.

[26] J. C. Hansen, J. Lebowitz, and B. Demeler, *Biochemistry* **33,** 13155 (1994).

of this chapter to give an overview of analytical ultracentrifugation or a detailed discussion of specific methods. Rather, after a brief introduction into basic principles of the technique, we use practical examples to illustrate the ability of this methodology to resolve controversial questions in the chaperone field and discuss some results in the context of other methodologies used to determine quaternary structure.

## Basic Principles of Analytical Ultracentrifugation

In an analytical ultracentrifugation experiment an analysis is made of the distribution of molecules according to their masses within a field of gravitation. This analysis can also be done using a preparative centrifuge, where the concentration gradient is measured after the run and drop fractionation of the contents of the tubes; however, in this case only final concentration profiles are obtained from a single centrifugation run and sedimentation must occur in the presence of preformed density gradients (e.g., of glycerol or sucrose). Such gradients influence the distribution of macromolecules within the field of gravity and therefore do not allow the basic mathematical description of the observed concentration profiles as discussed below. Nevertheless, this technique is often used to determine apparent sedimentation coefficients relative to known standards and we refer to this method as "density gradient sedimentation."

An analytical ultracentrifuge has the advantage of allowing continuous observation of the concentration profiles within the cells during the run in the absence of stabilizing components. In the Optima XL-A centrifuge this is achieved by an absorption optic that scans the cells from top to bottom (Fig. 1) and automatically stores the radial distribution curves on computer disks.

Two types of experiments can be performed in the analytical ultracentrifuge that require different mathematical treatments and produce values for different physicochemical parameters. In a sedimentation equilibrium experiment the rotor speed is set to a relatively low value (20,000–5000 rpm for molecular masses of 10–300 kDa). After a given time, depending on the height of the liquid columns within the sample cells (about 16–24 hr for 2- to 3-mm columns), every macromolecule reaches a position within the gravity field where its movement by sedimentation is apparently exactly compensated for by diffusion. From this time on the observed concentration profile does not change (Fig. 1A). The movement of molecules in a field of gravity is described by the partial differential equation of Lamm.[27] A

[27] O. Lamm, *Ark. Mat. Astron. Fys.* **21B,** 1 (1929).

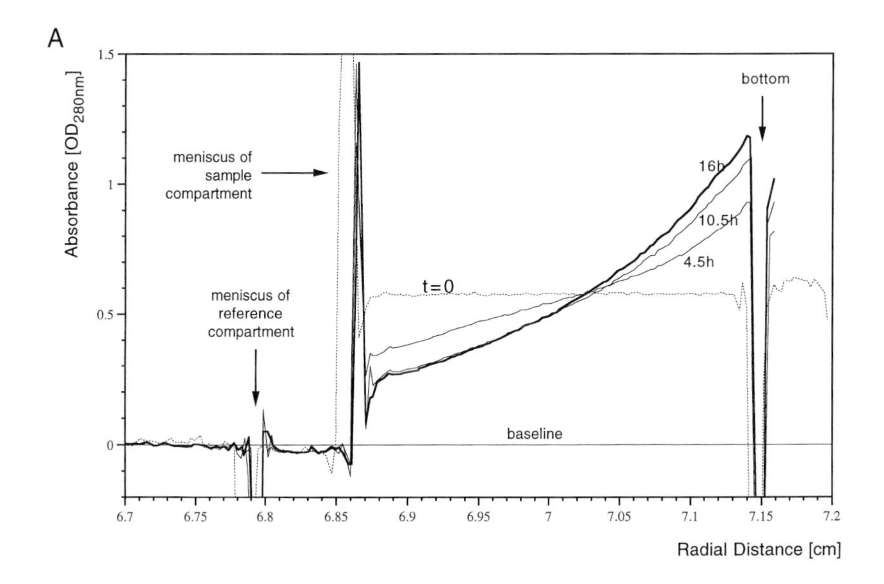


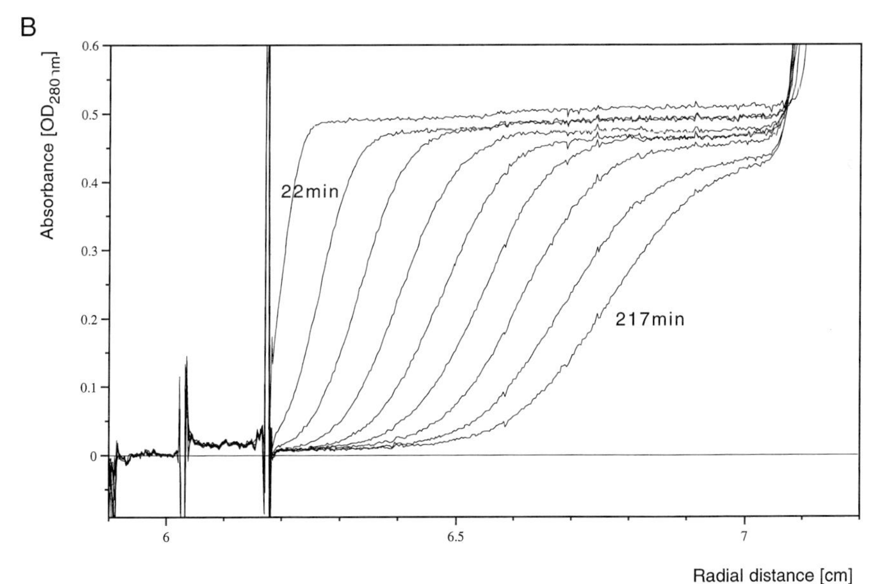


FIG. 1. Radial absorption scans obtained using the Beckman Optima XL-A centrifuge and the chaperone SecB at a concentration of 0.7 mg/ml from a sedimentation equilibrium experiment (A) and a sedimentation velocity experiment (B). (A) Scans were automatically taken immediately at 3000 rpm ($t$ = 0) and after 4.5, 10.5, and 16 hr at 10,000 rpm as indicated. No changes of profiles were observed after 16 hr, indicating an equilibrium between sedimentation and diffusion. Positions of menisci of the reference and sample compartments and of the cell bottom are indicated. (B) After reaching the desired speed of 40,000 rpm, scans were automatically taken about every 24 min. The boundary between the solvent (*left*) and the solute (*right*) moves constantly from the meniscus toward the cell bottom.

solution of the Lamm equation under conditions of sedimentation equilibrium is given by Eq. (1):

$$d(\ln c)/d(r^2) = M(1 - \overline{v}\rho)\omega^2/2RT \tag{1}$$

where $c$ is the concentration that depends on the radial position $r$, $M$ is the molar mass, $\overline{v}$ is the partial specific volume of the macromolecule, $\rho$ is the density of the solution (in most practical cases identical with the density of the solvent), $\omega$ is the angular velocity, $R$ is the gas constant ($R = 8.31 \times 10^7$ g cm$^2$ mol$^{-1}$ K$^{-1}$ sec$^{-2}$), and $T$ is the absolute temperature. When ln $c$ is plotted versus $r^2$ a straight line is obtained for the simplest case of a single molecular species. As $\omega$ and $T$ are kept constant during the centrifugation run and $\rho$ and $\overline{v}$ are known constants for the specific solvent and solute, respectively, the molar mass $M$ can be directly obtained from the slope of this straight line. It is important to emphasize that Eq. (1) contains no parameter describing molecular shape, and that $M$ is exclusively derived from direct measurement of the concentration gradient $c(r)$ and known constants without the need for calibration with reference molecules. Analytical ultracentrifugation is an absolute method that determines accurate molecular mass without employing calibration in solution.

When multiple species of macromolecules are present, Eq. (1) must be formulated for each species and the plot of ln $c$ versus $r^2$ will result in a curved line, as $c$ represents the sum of all partial concentrations $c_i$.

The exponential solution derived from the Lamm equation describes directly the observed concentration profile and is given in Eq. (2):

$$c(r) = c_m \exp[M(1 - \overline{v}\rho)\omega^2(r^2 - r_m^2)/2RT] \tag{2}$$

where $c_m$ is the concentration at the meniscus and $r_m$ is the radial position of the meniscus. For a multiple component system the overall concentration $c$ is described by the sum of exponential functions $c_i(r)$ for all single species $i$. Special treatments of Eq. (2) take into consideration effects of nonideality.

In sedimentation velocity experiments the rotor speed is set to relatively high values (e.g., 40,000 rpm for a protein of 40 kDa). Under these conditions diffusion is surpassed by sedimentation and a constant movement of the boundary formed by solvent and solution from the meniscus toward the cell bottom is observed (Fig. 1B). The velocity $v$ of a molecule is proportional to the centrifugal field $\omega^2 r$ as expressed by the Svedberg equation [Eq. (3)]:

$$v/\omega^2 r = s = M(1 - \overline{v}\rho)/N_{av}f \tag{3}$$

where $s$ is the sedimentation constant (with the unit Svedberg, S; 1S = $10^{-13}$sec), $N_{av}$ is Avogadro's number, and $f$ is the frictional coefficient that relates to the diffusion coefficient $D$ by the Einstein equation [Eq. (4)]:

$$D = RT/N_{av}f \quad (4)$$

Combining Eq. (3) and (4) results in Eq. (5), which allows the determination of $M$ from $s$ and $D$:

$$M = sRT/D(1 - \overline{v}\rho) \quad (5)$$

$D$ can be measured, for example, by quasielastic light scattering. Alternatively, $D$ can also be determined from the spreading of the sedimenting boundary between solvent and solution that is caused by diffusion.[28] However, spreading of the boundary is also caused by sample heterogeneity and asymmetry of the boundary can be caused by concentration-dependent sedimentation and diffusion or by molecular interactions if chemical equilibria are involved. In many cases, molecular masses obtained from sedimentation equilibrium experiments are more accurate than those derived from sedimentation velocity experiments.

When the mass of the macromolecule is known, Eq. (3) can be used to calculate its frictional coefficient, $f$, from a sedimentation velocity result. The ratio $f/f_0$, where $f_0$ is the frictional coefficient of a rigid sphere having an identical volume as the molecule of interest, gives valuable information about how much the shape of the molecule is extended compared with a sphere. $f_0$ can be computed via the Stokes equation [Eq. (6)]:

$$f_0 = 6\pi\eta R_0 \quad (6)$$

where $\eta$ is the viscosity of the solvent and $R_0$ is the radius of a rigid sphere having an identical volume as the molecule of interest according to Eq. (7):

$$R_0 = (3M\overline{v}/4\pi N_{av})^{1/3} \quad (7)$$

## Other Methodologies Frequently Used to Determine Chaperone Quaternary Structures in Solution

The self- and nonself-association of chaperones was extensively investigated by size-exclusion chromatography or by chemical cross-linking with glutaraldehyde followed by sodium dodecyl sulfate (SDS)–polyacrylamide gel analysis (Table I). In some cases apparent molecular masses were determined by Ferguson plots derived from native polyacrylamide electrophoresis and density gradient sedimentation. The stoichiometries of large chaper-

[28] J. Behlke and O. Ristau, *Biophys. J.* **72,** 428 (1997).

onin complexes were determined by equilibrium dialysis and the arrangements of subunits that had been conserved by cross-linking or freeze drying were also investigated by electron microscopy. Some of the contradictory results (Table I) may be related to arbitrary assumptions of one or more parameters that are required by a given method but were not determined experimentally. Table II gives an overview of these well-established methods for determination of protein quaternary structure.

When molecular masses are derived from only one method—size-exclusion chromatography, Ferguson plots, gradient sedimentation, or quasi-

TABLE II
COMPARISON OF METHODS FOR DETERMINATION OF PROTEIN QUATERNARY STRUCTURE[a]

| Method | Chemical modification of analyte | Equilibrium method | Measured parameter(s) | Additional information required to obtain $M$ | Pitfalls and limitations |
|---|---|---|---|---|---|
| Size-exclusion chromatography, Ferguson plot | No | No | $R_s$ | Shape | Calibration; interaction with column support (SEC); disturbance of fast equilibria |
| Chemical cross-linking | Yes | No | Oligomeric state | Calibration by SDS–PAGE | Cross-linking artifacts; disturbance of equilibria |
| Density gradient sedimentation | No | No (when components obtain different densities) | $s$ | $f$ or $D$, resp. shape | Calibration |
| Quasi-elastic light scattering | No | Yes | $D$ | $s$, resp. shape | Calibration |
| Equilibrium dialysis | Depends on detection method | Yes | Stoichiometry | No determination of $M$ | Interactions with membrane, Donnan effect |
| Spectroscopy | Depends on method | Yes | Stoichiometry | No determination of $M$ | Requires suitable spectroscopic tag |
| Osmometry | No | Yes | $\Pi$ | Concentration dependency of $\Pi$ | Interactions with membrane, Donnan effect |
| Electron microscopy | Depends on method | No | Shape of single molecules | Apparent $M$ by scanning transmission EM (STEM) | Disturbance of equilibria |
| AUC/sed. vel. | No | No | $s$ and $D$ | None | Hydrostatic pressure |
| AUC/sed. eq. | No | Yes | $M$ | None | Number of components |

[a] Sed. vel., sedimentation velocity; sed. eq., sedimentation equilibrium; $R_s$, Stokes radius; $\Pi$, osmotic pressure.

elastic light scattering—the necessary calibration is made using a set of reference proteins with known identical shapes and different molecular masses. Clearly, the obtained molecular mass of a protein will be correct only when its shape is similar to that of the reference proteins. In general, reference proteins of globular shape are used. However, if a protein resembles an elongated shape as do the chaperones DnaK[5,29] and GrpE,[5,10] the obtained Stokes radius and sedimentation coefficient are higher and the diffusion coefficient is smaller than expected for its true molecular mass and oligomeric state. Furthermore, for size-exclusion chromatography, hydrophobic interactions of the protein with the column support[30] would lead to elution volumes that are too high, resulting in apparent Stokes radii $R_s$ and proposed molecular masses that are too low.

Chemical cross-linking covalently modifies a protein and as a consequence reversible equilibria can be disturbed. At high protein concentrations, there is an increased probability of cross-linking molecules that do not specifically interact (statistical intermolecular cross-linking), implying oligomeric states that are not of physiological relevance. On the other hand, the absence of cross-linking does not necessarily mean the absence of oligomerization, as appropriate functional groups on both protein subunits may be too distant to undergo chemical reaction with the bifunctional cross-linker. These points are reflected by the general observations that higher cross-link yields are obtained when the length of cross-linker is increased or when its selectivity is decreased.

Electron microscopy visualizes the shape of single molecules of large size and has, therefore, been extensively used to study the interactions of chaperonins. As noted, depending on the electron microscopic method, either cross-linking, staining, or freeze-drying of the complexes is required, with potential unknown effects on equilibria.

Equilibrium dialysis and spectroscopic methods have been used less frequently and osmometry[31,32] has not been used at all in the chaperone field. However, these methods are usually powerful, as they provide results at equilibrium without chemical modification, assuming for equilibrium dialysis that radioactive *in vitro* labeling is not used. In contrast to the other methods the basic information obtained using equilibrium dialysis and spectroscopic methods consists of stoichiometries and association constants of complexes rather than molecular masses.

[29] L. Shi, M. Kataoka, and A. L. Fink, *Biochemistry* **35,** 3297 (1996).

[30] P. L. Dubin, S. L. Edwards, M. S. Mehta, and D. Tomalia, *J. Chromatogr.* **635,** 51 (1993).

[31] M. P. Tombs and A. R. Peacocke, "The Osmotic Pressure of Biological Macromolecules." Clarendon Press, Oxford, 1974.

[32] H. Schönert and B. Stoll, *Eur. J. Biochem.* **176,** 319 (1988).

Analysis of the sedimentation velocity results for a chaperone in the analytical ultracentrifuge provides molecular masses by combining $s$ and $D$ via Eq. (5) without requiring calibration by reference proteins. When there are weak molecular interactions, equilibria may be disturbed as a consequence of high hydrostatic pressures generated by high rotor speeds. Such phenomena are less frequently observed when analytical ultracentrifugation is performed under conditions of sedimentation equilibrium, owing to the much lower gravity forces at lower rotor speeds. Artifactual disturbance of equilibria can be experimentally excluded by showing that identical results are obtained at different rotor speeds.

Each of the methods discussed thus far has been extensively used to provide useful insight into many aspects of biochemistry. However, as noted, it is advantageous to use the simplest methodology that provides accurate results and it is necessary to perform control experiments required to exclude possible artifacts associated with each method. The analysis of sedimentation equilibrium in the analytical ultracentrifuge directly provides molecular mass based on thermodynamic principles without the need for additional assumptions; the main limitations are the accuracy of data acquisition and the number of different molecular species present in the solution under investigation.

## Purification of Chaperones for Quaternary Structure Analysis

Most methodologies that analyze oligomerization, such as analytical ultracentrifugation, require well-characterized, purified proteins. The results should reflect the native state of the protein and, therefore, special care must be taken when selecting the purification strategy. For example, chromatographic methods such as ion-exchange chromatography and affinity chromatography, as far as specific elution of the protein is concerned (e.g., by competition with an excess of ligand present in the elution buffer), normally preserve native structure, whereas reversed-phase chromatography and affinity chromatography using nonspecific elution methods (e.g., extreme values of pH that cause conformational changes) are more susceptible to irreversible denaturation. Concentration of diluted solutions by precipitation using ammonium sulfate is preferable to the use of organic reagents or ultrafiltration that sometimes leads to irreversible aggregation as a consequence of interactions with hydrophobic molecules or membrane surfaces. The purification protocol should minimize the number of concentration steps; this is done by following a step that causes dilution with one that concentrates. Size-exclusion chromatography is a useful final purification step, as it effectively removes traces of nonspecific aggregates and small molecules. Chaperone interactions often depend on the presence of nucleotides. Therefore, knowledge of the ADP and ATP content of the chaperone under investigation can be important and should be analyzed by high-performance liquid chromatography (HPLC) or enzymatic assays.[33] The authenticity of the primary sequence of the purified chaperone should be verified by ion spray mass spectroscopy, particularly when the chaperone was purified after overexpression by using recombinant techniques. When a colorimetric assay is used to measure chaperone concentration the assay must be calibrated with reference to the protein content of a batch of the same chaperone that was determined by amino acid analysis. Finally, quantitative assays that measure chaperone function provide evidence that native structure is preserved.

## Investigation of Quaternary Structure of SecB: A Practical Example

The cytosolic *Escherichia coli* transport chaperone SecB is a multimeric protein composed of identical 17.2-kDa subunits. A molecular mass of 115 kDa was obtained by size-exclusion chromatography using three different gel-filtration supports, suggesting a hexamer, whereas 79 kDa was obtained from native PAGE and Ferguson analyses, suggesting a tetramer.[1] The tetramer was confirmed by size-exclusion chromatography on a Sephacryl S-300 column.[2]

The tetrameric structure appears to be exceptional in the world of chaperones. It was shown that binding of short peptide ligands to SecB caused the exposition of a hydrophobic patch.[34] This change of conformation of SecB during function may be important for its high selectivity in recognizing the unfolded state of a polypeptide chain and its low specificity for the primary sequence. The quaternary structure of SecB has not been determined by analytical ultracentrifugation; we use this chaperone to demonstrate the methodology.

### *Protocol: Determination of Oligomeric State of SecB by Analysis of Sedimentation Equilibrium*

1. Purified SecB is extensively dialyzed against 50 m*M* Tris-HCl (pH 7.7), 100 m*M* NaCl and aliquots of the dialyzed solution are diluted to 0.7, 0.17, and 0.07 mg/ml.

2. Each dilution (110 $\mu$l) is transferred to the sample compartments of one of three double-sector cells. The reference sectors of the cells are filled with 130 $\mu$l of buffer of the last dialysis step. The cells are positioned in a four-hole rotor and centrifugation in an Optima XL-A analytical ultracentrifuge is initiated at a setting of 3000 rpm and 20°.

[33] B. Feifel, E. Sandmeier, H.-J. Schönfeld, and P. Christen, *Eur. J. Biochem.* **237,** 318 (1996).
[34] L. L. Randall, *Science* **257,** 241 (1992).

3. After reaching the desired speed, a wavelength scan of each cell from 200 to 400 nm is taken at a fixed position within the liquid column ($r$ = 7.00 cm). From these spectra optimal wavelengths $\lambda$ are selected for subsequent radial scans (e.g., 275, 280, and 285 nm for the cell with the highest concentration and 230, 235, and 240 nm for the cells with medium and lowest concentrations).

4. The speed is increased to 10,000 rpm (optimal value for an estimated molecular mass of about 70 kDa according to Chervenka[25,35]) and the run is continued overnight to achieve sedimentation equilibria (Fig. 1A).

5. Radial scans of the concentration gradients at various wavelengths are automatically performed.

6. The speed is increased to 40,000 rpm for depletion of the menisci. Radial scans at various wavelengths are performed to enable the experimental determination of baselines, which is extremely important for accurate results.

*Data Evaluation*

7. From the radial scans at 40,000 rpm the solvent baseline absorbancies of solvent within the sample cells are determined at various wavelengths in the depleted regions below the menisci.

8. From the radial scans at 10,000 rpm the positions of menisci are determined and the regions of exponential distribution curves are extracted and stored by the DOS program XLA (supplied with the centrifuge for further data analysis). Data points close to the meniscus and close to the cell bottom are excluded because of scattering effects.

9. Data analysis of the extracted distribution curves is performed using the DOS program EQASSOC (supplied with the centrifuge). This program calculates monomer molecular mass and association constants for different self-association models that best fit an experimental distribution curve. The required partial specific volume for SecB at 20° is taken from Table III as $\overline{v}$ = 0.726 cm$^3$ g$^{-1}$ and the density of the Tris buffer is calculated as $\rho$ = 1.003844 g cm$^{-3}$ according to Laue *et al.*[25,36] By constraining all association constants, $K_i$, to very small numbers ($\log K_i = -20$), the software is forced to calculate the molecular mass of a single component by fitting Eq. (2) to an experimental profile. Baseline absorbancies are constrained to the experimental values obtained from step 7. EQASSOC, as well as various other programs, offers the possibility to treat the baseline as a free parame-

[35] C. H. Chervenka, "A Manual of Methods for the Analytical Ultracentrifuge." Beckman Instruments, Palo Alto, California, 1969.

[36] T. M. Laue, B. D. Shah, T. M. Ridgeway, and S. L. Pelletier, *in* "Analytical Ultracentrifugation in Biochemistry and Polymer Science" (S. E. Harding, A. J. Rowe, and J. C. Horton, eds.), pp. 90–125. Royal Society of Chemistry, Cambridge, 1992.

TABLE III
VALUES OF PARTIAL SPECIFIC VOLUME $\bar{v}$ OF CHAPERONES AT DIFFERENT TEMPERATURES AS CALCULATED FROM THEIR PRIMARY SEQUENCES[a]

| | $\bar{v}$(cm$^3$ g$^{-1}$) | | |
|---|---|---|---|
| Chaperone | 4° | 20° | 37° |
| SecB | 0.720 | 0.726 | 0.734 |
| DnaK | 0.728 | 0.735 | 0.742 |
| DnaJ | 0.718 | 0.725 | 0.732 |
| GrpE | 0.731 | 0.737 | 0.745 |
| GroEL | 0.735 | 0.742 | 0.749 |
| GroES | 0.742 | 0.749 | 0.756 |

[a] Computed according to Laue.[25,36] Values for other temperatures can be obtained by linear interpolation.

ter in the fitting algorithms. The accuracy of the baseline obtained from such an approach depends markedly on the quality of the data. Because an error of 0.01 OD (optical density) in baseline determination would cause an error of about 5% in molecular mass, we strongly recommend always determining baselines experimentally. Table IV shows relative molecular masses $M_r$ that resulted from fits of experimental profiles that we obtained at

TABLE IV
MOLECULAR MASSES OF SecB[a]

| Concentration (mg/ml) | λ (nm) | $M_r$ (kDa) | |
|---|---|---|---|
| 0.7 | 275 | 68.9 | 69.4 ± 0.6 |
| | 280 | 69.2 | |
| | 286 | 70.0 | |
| 0.17 | 276 | 63.5 | 64.4 ± 1.1 |
| | 280 | 64.3 | |
| | 285 | 64.5 | |
| | 230 | 63.0 | |
| | 235 | 66.2 | |
| | 241 | 65.0 | |
| 0.07 | 230 | 61.2 | 62.8 ± 1.7 |
| | 235 | 62.6 | |
| | 241 | 64.5 | |

[a] Obtained using our sedimentation equilibrium protocol and data analysis employing the program EQASSOC.

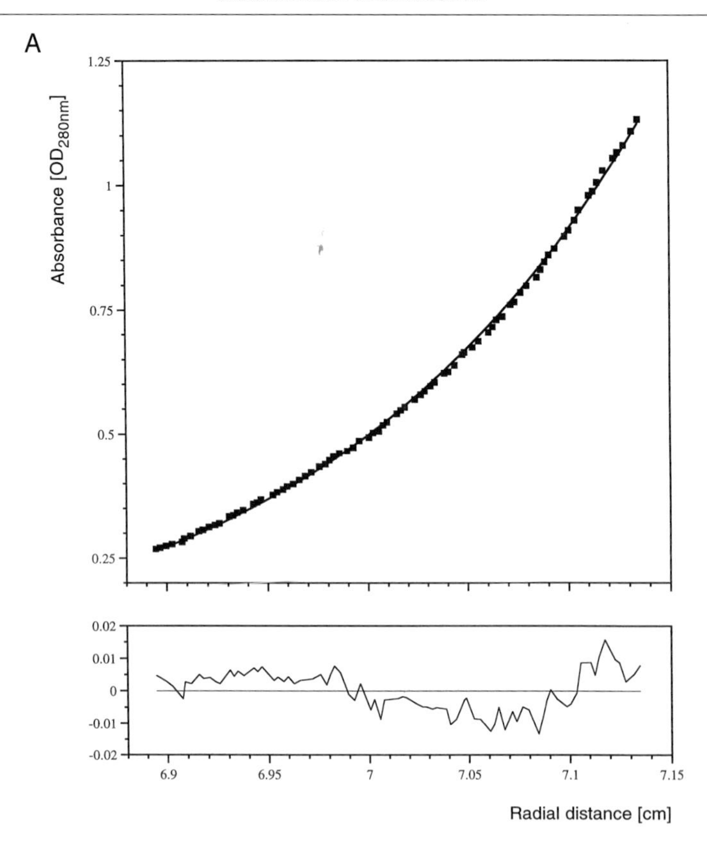


FIG. 2. (A–C) *Top:* Radial absorbance distribution of SecB (squares) in sedimentation equilibrium and best fits with different models (solid lines). *Bottom:* Radial dependent residuals representing deviations of the experimental data from those generated by the model. (A) The sedimentation equilibrium at 0.7 mg/ml was scanned at 280 nm and fitted with the single-component model, returning a molecular mass of 69.2 kDa (see Table IV), close to 68.8 kDa of the SecB tetramer. (B and C) The sedimentation equilibrium at 0.07 mg/ml was scanned at 230 nm and fitted with the tetramer model (B), resulting in relatively large residuals with systematic radial dependence or with the monomer/tetramer equilibrium model (C), resulting in an excellent fit with small, equally distributed residuals.

different wavelengths $\lambda$ by following the above protocol. The experimental profiles obtained at a loading concentration of 0.7 mg/ml as measured at 280 nm and the best fit using the single component model are shown in Fig. 2A.

10. At the highest investigated SecB concentration of 0.7 mg/ml the fitting algorithm almost exactly yielded the molecular mass of a SecB tetramer of 68.8 kDa (Table IV). At lower concentrations the apparent molecu-

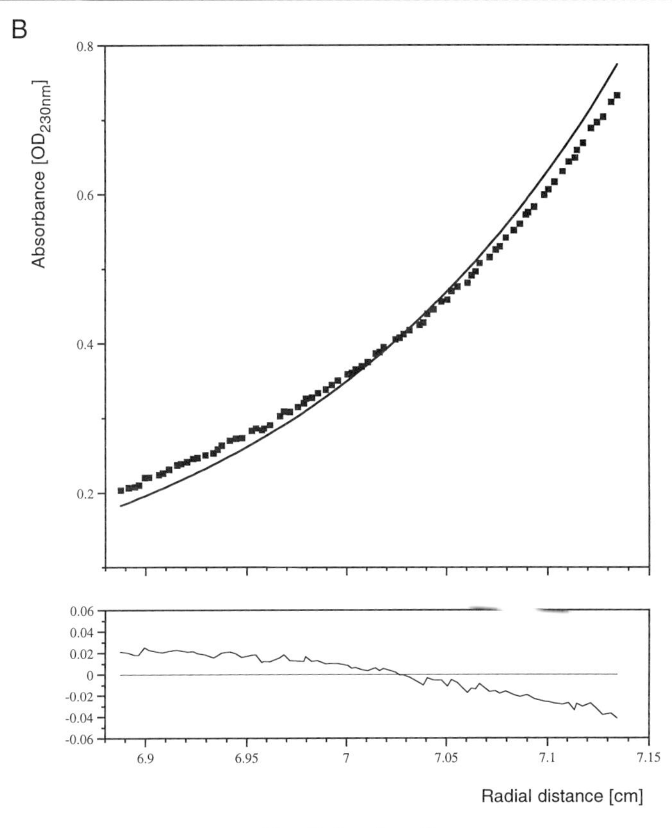


FIG. 2. (*continued*)

lar masses decreased, indicating a dissociation equilibrium. At the SecB concentration of 0.07 mg/ml a fit based on a single species model with 68.8 kDa results in systematic deviations from the experimental data as shown in Fig. 2B. Therefore, profiles obtained at 0.17 and 0.07 mg/ml are reanalyzed by the program EQASSOC, assuming an equilibrium of monomeric and tetrameric SecB.

The fitting algorithm yields the logarithms of association constants that are based on absorption units and can be converted to molarity units using the extinction coefficients of SecB at the various wavelengths. These extinction coefficients can easily be derived from the initial wavelength scans at 3000 rpm. Using this procedure, we obtained association constants of $K_{1,4} = (2.9 \pm 1.5) \times 10^{19}\ M^{-3}$ for the monomer/tetramer equilibrium model, resulting in fits with significantly improved qualities as shown in Fig. 2C for a profile at 0.07 mg/ml.

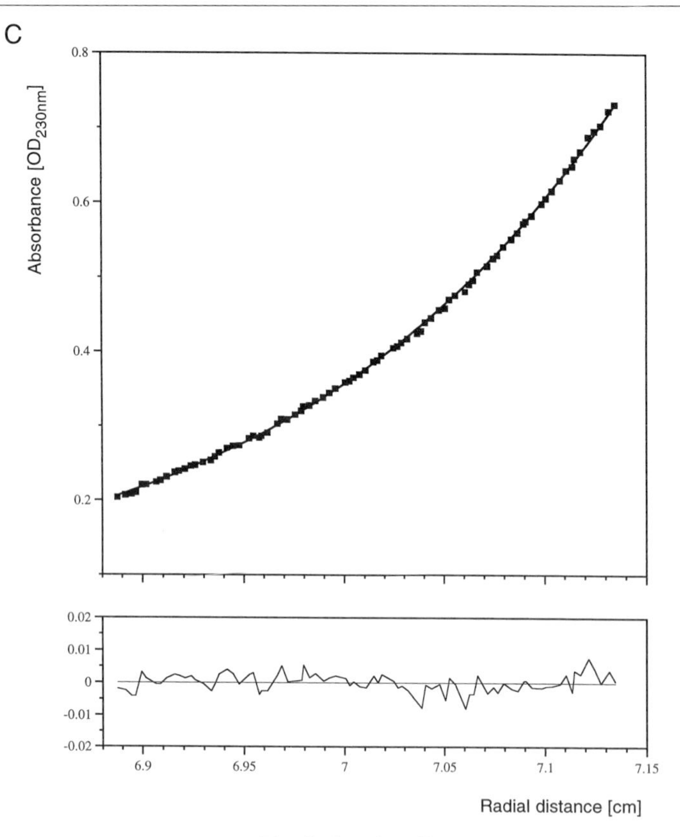


FIG. 2. (*continued*)

*Protocol: Determination of Oligomeric State of SecB by Measuring Diffusion and Sedimentation Velocity*

1. Purified SecB is dialyzed as described for the sedimentation equilibrium experiment above and then diluted to 0.7 mg/ml.

2. Quasi-elastic light scattering of the protein solution is measured at 20° in a DynaPro-801 TC instrument (Protein Solutions Ltd., U.K.) giving a diffusion coefficient of $D = 5.4 \times 10^{-7}$ cm$^2$ sec$^{-1}$ for SecB.

3. Protein solution (400 $\mu$l) is transferred to the sample compartment of a double-sector cell and the reference compartment is filled with about 420 $\mu$l of buffer of the final dialysis step.

4. Centrifugation is initiated at 40,000 rpm.

5. Radial scans at 280 nm are automatically taken every 24 min (Fig. 1B). After about 3.5 hr, when the boundary between solvent and solution is near the cell bottom, centrifugation is stopped.

*Data Evaluation*

6. The position of the meniscus of an early concentration profile and points of inflection of the later profiles are determined using the DOS program XLAVEL (supplied with the centrifuge). From these data the software computes the sedimentation coefficient as $s = 4.28$ S. In combination with $D$ from the light-scattering measurement the molecular mass of SecB is calculated via Eq. (5) as $M_r = 71.2$ kDa, using the same values for $\bar{v}$ and $\rho$ as shown in the sedimentation equilibrium protocol. This mass is close to the theoretical mass of a SecB tetramer of 68.8 kDa.

## Quaternary Structure of GrpE

The cochaperone GrpE undergoes oligomerization, a structural feature that may be important for function. Mutant GrpE proteins that evidence aberrant activity have altered oligomerization properties when analyzed by size-exclusion chromatography or SDS–PAGE after chemical cross-linking.[11]

Published data on the oligomeric states of GrpE are controversial. Initial data were obtained by density gradient centrifugation.[8] The obtained sedimentation coefficient of about 2.5S is consistent with that of a globular protein having the size of a GrpE monomer, and the authors proposed GrpE to be monomeric.

A dimeric structure for GrpE was first suggested on the basis of cross-linking of GrpE at a low concentration of 1.2 $\mu M$ (26 $\mu$g/ml) with glutaraldehyde.[9] We confirmed the dimer at a far higher concentration of 28 $\mu M$ (0.6 mg/ml) by analytical ultracentrifugation techniques.[5] The sedimentation coefficient obtained from a sedimentation velocity experiment was combined with diffusion coefficients from quasi-elastic light-scattering measurements using Eq. (5), yielding molecular masses as expected for the GrpE dimer. This result was confirmed by sedimentation equilibrium experiments. Neither quasi-elastic light-scattering nor analytical ultracentrifugation results indicated bimodality, suggesting that only one species of oligomer was present in solution. Furthermore, GrpE analyzed by size-exclusion chromatography on a Superose 12 column calibrated with globular proteins eluted as a single symmetrical peak at 190 kDa. As the elution volume was too small, the diffusion coefficient too high, and the sedimentation velocity too low for a 43.3-kDa globular protein, we proposed an elongated shape for the GrpE dimer.[5]

The high apparent molecular mass of GrpE in size-exclusion chromatography was confirmed by others using a Bio-SEP 3000 silica column (Phenomenex) and the authors also suggested that the shape of the molecule is highly asymmetric.[10] However, SDS–PAGE analysis of cross-linking experiments with glutaraldehyde in the range of 5–15 $\mu M$ GrpE revealed broad bands at a position corresponding to a mass of 55 $\pm$ 5 kDa; this observation and data gained from the investigation of the interaction of GrpE with DnaK led the authors to conclude that it is "perhaps more likely" that native GrpE is a trimer.

In another publication[11] SDS–PAGE analysis of GrpE cross-linked at various protein concentrations (1–60 $\mu M$; 0.02–1.3 mg/ml) showed most oligomers migrating at the position of a dimer. However, SDS–PAGE analysis of cross-linked GrpE at concentrations above 3.7 $\mu M$ revealed additional bands corresponding to positions of multiples of dimers up to hexamers and higher. To exclude cross-linking artifacts, the protein was separately analyzed by size-exclusion chromatography on a Superdex 200 HR column calibrated with globular proteins. In this analysis GrpE appeared as a single peak at the position of about 135 kDa. The authors concluded that GrpE is "putatively hexameric" in solution; however, it was emphasized that this was a tentative interpretation and that "the confirmation of the oligomeric form as a hexamer must await more rigorous biochemical techniques."

As already discussed, analytical ultracentrifugation under conditions of sedimentation equilibrium is the best method for analyzing the association state of an elongated protein, because this is a direct approach where the result is not influenced by the molecular shape. In a yet unpublished study (H.-J. Schönfeld *et al.*[36a]) we reinvestigated oligomerization of GrpE in a concentration range of 0.04–4 mg/ml following a similar protocol as given for the analysis of the sedimentation equilibrium of SecB.

Figure 3 shows best fits to the experimental data when assuming a dimeric, trimeric, or hexameric structure for GrpE. Clearly, an appropriate fit was obtained only with the dimer model. Furthermore, GrpE dimers did not dissociate on dilution within the investigated concentration range. The GrpE dimer was confirmed by sedimentation analysis following a similar protocol as described for SecB.[36a]

## Quaternary Structure of DnaJ

The *E. coli* chaperone DnaJ represents a member of the highly conserved heat shock protein family HSP40. The molecule contains four domains: a so-called J domain represented by about 70 amino-terminal amino

[36a] H.-J. Schönfeld *et al.*, in preparation (1998).

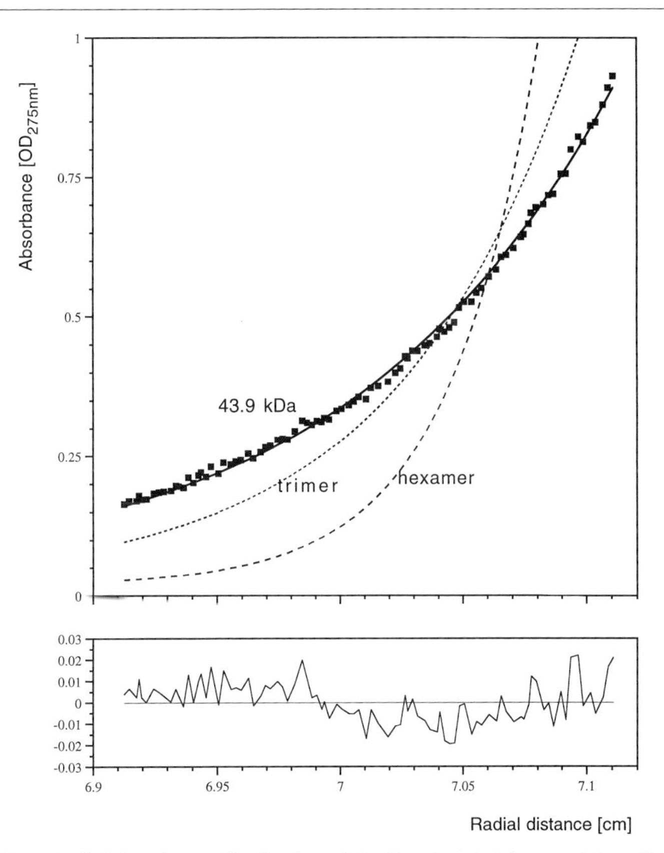


FIG. 3. *Top:* Radial absorbance distribution of GrpE at 4 mg/ml (squares) in sedimentation equilibrium and best fits with the single-component model (solid line), returning a molecular mass of 43.9 kDa (theoretical mass of the GrpE dimer, 43.3 kDa); the trimer model or the hexamer model as indicated. *Bottom:* Radial dependent residuals representing deviations of the experimental data from those generated by the single-component model.

acids, followed by a G/F module (a 35-amino acid-long glycine and phenylalanine reach region), a cysteine-rich region (containing 8 of the 10 cysteines of DnaJ), and a carboxy-terminal region, which is less conserved and probably responsible for substrate binding. The arrangement of cysteines represents a zinc finger motif and the binding of two zinc ions per DnaJ protomer has been reported.[37] DnaJ contains an active dithiol/disulfide group, also giving the molecule the function of a protein-disulfide isomerase.[38] Purifica-

[37] B. Banecki, K. Liberek, D. Wall, A. Wawrzynow, C. Georgopoulos, E. Bertoli, F. Tanfani, and M. Zylicz, *J. Biol. Chem.* **271,** 14840 (1996).

[38] A. de Crouy Chanel, M. Kohiyama, and G. Richarme, *J. Biol. Chem.* **270,** 22669 (1995).

tion of overexpressed DnaJ is more tedious than that of other *E. coli* chaperones because of its hydrophobicity and requirement for detergents.[12,39] The oligomeric state of DnaJ was characterized as a dimer by size-exclusion chromatography and density gradient centrifugation.[12] However, we also observed higher oligomeric forms of DnaJ by analytical ultracentrifugation.[40] These forms may represent nonspecific aggregates as their proportion depended on the preparation and not on the protein concentration. Some of these aggregates may be formed during overexpression and purification by the formation of artifactual intermolecular disulfide bridges; moreover, the release of zinc ions results in destabilization of DnaJ while simultaneously increasing its aggregation state.[37] The elucidation of the native quaternary structure of DnaJ requires further study.

## Quaternary Structures of HSP70s and Interactions of DnaK with GrpE and DnaJ

The *E. coli* chaperone DnaK represents a member of the highly conserved heat shock protein family HSP70. HSP70 proteins consist of at least two functional domains: a more conserved amino-terminal 44-kDa ATPase domain and a more variable carboxy-terminal substrate-binding domain.[41] Although the mechanism of ATP binding, hydrolysis, and release and that of substrate binding and release as well as the interaction of the two domains with each other, with co-chaperones, and with substrate proteins are important for overall chaperone function, these are not yet well understood.

### *Self-Association of DnaK and Homologs*

For native DnaK, a molecular mass of 78.4 kDa was calculated by size-exclusion chromatography and density gradient centrifugation.[3] This value is somewhat higher than the molecular mass of 69.0 kDa as calculated from the amino acid sequence. Size-exclusion chromatography indicated that a minor fraction of DnaK was dimeric. The existence of a slow and reversible equilibrium between monomeric and oligomeric forms of DnaK in the absence of ATP was established by native polyacrylamide electrophoresis,[4] size-exclusion chromatography,[4,5] and analytical ultracentrifugation.[5] Equilibria of oligomeric forms were also described for the mammalian members of the HSP70 family such as bovine BiP[6] and rat Hsc70.[42] Elongated shapes

[39] M. Ohki, F. Tamura, S. Nishimura, and H. Uchida, *J. Biol. Chem.* **261,** 1778 (1986).

[40] H.-J. Schönfeld, D. Schmidt, and M. Zulauf, *Progr. Colloid Polym. Sci.* **99,** 7 (1995).

[41] T. G. Chappell, B. B. Konforti, S. L. Schmid, and J. E. Rothman, *J. Biol. Chem.* **262,** 746 (1987).

[42] N. Benaroudj, B. Fang, F. Triniolles, C. Ghelis, and M. M. Ladjimi, *Eur. J. Biochem.* **221,** 121 (1994).

for the DnaK molecules were first reported on the basis of relatively low peak elution volumes in size-exclusion chromatography, relatively low sedimentation coefficients from analytical ultracentrifugation, and high diffusion coefficients from quasi-elastic light scattering[5] and were more recently confirmed by small-angle X-ray scattering.[29]

A rigorous investigation of self-association of mammalian Hsc70 was performed by analytical ultracentrifugation.[7] In sedimentation velocity experiments sedimentation coefficients corresponding to three different species were distinguished. Their value ratios were as expected for Hsc70 monomers, dimers, and trimers. Furthermore, only the mass fraction of each sedimenting species and not the sedimentation coefficients depended on the protein concentration up to 1.6 mg/ml. Such behavior is expected for a mass-action law equilibrium. The frictional ratio values $f/f_0$ suggested that all oligomeric species were asymmetrical. Analysis of sedimentation equilibrium experiments at different concentrations and rotor speeds confirmed the equilibrium and, interestingly, suggested that self-association of Hsc70 may be unlimited, as it can be well described by an isodesmic model where the association constants of each stage are equal. This model is also consistent with a DnaK peak in the exclusion volume of a Superose 12 column (≥2000 kDa) that appeared only at high concentrations.[5]

Analytical ultracentrifugation of recombinant DnaK fragments indicated that the active site for self-association of DnaK is located on the substrate-binding domain adjacent to the ATPase domain (H.-J. Schönfeld and B. Bukau, unpublished data).

It should be emphasized that oligomerization is a specific feature of HSP70 protein molecules that must be distinguished from nonspecific aggregation. The difference is well illustrated by the observation that DnaK and Hsc70 oligomerization is favored at low temperatures,[5,43] whereas DnaK and Hsc70 aggregation occurs at temperatures above 42°.[42,44]

*Interaction of DnaK with GrpE*

DnaK forms complexes with GrpE in the absence of ATP that are stable in up to 2 *M* KCl and, on the basis of density gradient sedimentation analysis, a 1:1 stoichiometry was initially proposed for this interaction.[3] We reported a 1:2 (DnaK:GrpE) stoichiometry, on the basis of analytical ultracentrifugation of the complex and on titration experiments using size-exclusion chromatography and native polyacrylamide electrophoresis.[5] In

[43] N. Benaroudj, F. Triniolles, and M. M. Ladjimi, *J. Biol. Chem.* **271,** 18471 (1996).

[44] S. Sadis, K. Raghavendra, T. M. Schuster, and L. E. Hightower, *in* "Current Research in Protein Chemistry" (J. J. Villafranca, ed.), p. 339. Academic Press, San Diego, California, 1990.

the same study we observed deoligomerization of DnaK by GrpE, yielding single molecular species in solution after mixing of DnaK with GrpE at a molar ratio of 1:2 and resulting in monomodal distribution in analytical centrifugation and quasi-elastic light scattering experiments.

On the basis of similar titration experiments others have suggested a 1:3 stoichiometry for the DnaK–GrpE complex.[10] However, GrpE was purified under denaturing conditions for this study and the higher amounts of GrpE needed for the saturation of the heterocomplex may be explained by incomplete refolding of GrpE after purification.

### *Interaction of DnaK with DnaJ*

The stimulation of the ATPase activity of DnaK by DnaJ[45] reflects an interaction between the two chaperones. Probably, the substrate-binding unit of DnaK[15] binds to the amino terminus of DnaJ.[46] Stable complexes of DnaK and DnaJ in the presence of ATP have been isolated[15] but the molecular details of these interactions are unknown. In *Thermus thermophilus* the interaction of the respective homologs, T.DnaK and T.DnaJ, appeared to be more stable as these two chaperones copurified.[47] The complex was characterized as a hexameric ring of three DnaK and three DnaJ molecules. The analysis was done by size-exclusion chromatography, Ferguson plots, amino-terminal amino acid sequencing, and electron microscopy and it would be interesting to characterize the hydrodynamic features of this structure further in the analytical ultracentrifuge. The authors have identified three molecules of a 8-kDa peptide, T.DafA, as further components of the heterocomplex that are essential for its assembly.[48] Interestingly, it was reported that YDJ1, the yeast DnaJ homolog, in substoichiometric amounts induces reversible association (in the report termed "polymerization") of yeast or bovine brain Hsp70 in the presence of ATP.[49]

### *Implications of Self- and Heteroassociation for Function of DnaK–DnaJ–GrpE Machine*

ATP induces a major conformational change in DnaK that causes the release of bound substrate.[50,51] Probably, the binding of ATP to the ATPase

[45] K. Liberek, J. Marszalek, D. Ang, C. Georgopoulos, and M. Zylicz, *Proc. Natl. Acad. Sci. U.S.A.* **88,** 2874 (1991).

[46] D. Wall, M. Zylicz, and C. Georgopoulos, *J. Biol. Chem.* **269,** 5446 (1994).

[47] K. Motohashi, H. Taguchi, N. Ishii, and M. Yoshida, *J. Biol. Chem.* **269,** 27074 (1994).

[48] K. Motohashi, M. Yohda, I. Endo, and M. Yoshida, *J. Biol. Chem.* **271,** 17343 (1996).

[49] C. King, E. Eisenberg, and L. Greene, *J. Biol. Chem.* **270,** 22535 (1995).

[50] K. Liberek, D. Skowyra, M. Zylicz, C. Johnson, and C. Georgopoulos, *J. Biol. Chem.* **266,** 14491 (1991).

[51] D. R. Palleros, K. L. Reid, L. Shi, W. J. Welch, and A. L. Fink, *Nature* (*London*) **365,** 664 (1993).

region disturbs a physical interaction between this region[52] and the substrate-binding region of about 150 amino acids adjacent to the ATPase domain.[53]

Deoligomerization of DnaK is also observed on binding of ATP[4] or GrpE[5] to its ATPase region. Furthermore, binding of substrates to eukaryotic DnaK homologs also results in their deoligomerization.[43,54] As noted, at least one oligomerization site on DnaK is located within its substrate-binding region. Taken together, the phenomenon of interdomain communication is similar to DnaK oligomerization. Interdomain communication is the result of intramolecular interaction and oligomerization may be the result of intermolecular interaction of the same two functional domains.

Monomeric DnaK is probably the active molecule in DnaK-mediated protein folding *in vivo* and oligomeric DnaK may constitute a pool of latent chaperone that is activated during heat stress (temperature increase, increasing amounts of unfolded protein, and increase in GrpE expression).[5] DnaJ binds to the carboxy terminus of DnaK and probably stabilizes the DnaK–substrate interaction.[55]

## Quaternary Structures and Interactions of GroEL and GroES

The GroEL–GroES complex from *E. coli* is the most thoroughly investigated chaperonin system. However, the quaternary structures that are stable under given conditions and are actively involved in the different steps of chaperonin-mediated protein folding are controversial and a topic of several publications.[18,19,56–58]

GroEL forms a complex of 14 identical 57.3-kDa subunits arranged in 2 stacked heptameric rings whereas GroES forms a heptameric ring of identical 10.4-kDa subunits. The ring structures were demonstrated by electron microscopy[13,14] and, more recently, high-resolution X-ray structures of both chaperonins became available.[59–61] GroEL obtains weak

[52] A. Buchberger, A. Valencia, R. McMacken, C. Sander, and B. Bukau, *EMBO J.* **13,** 1687 (1994).

[53] A. Buchberger, H. Theyssen, H. Schroder, J. S. McCarty, G. Virgallita, P. Milkereit, J. Reinstein, and B. Bukau, *J. Biol. Chem.* **270,** 16903 (1995).

[54] B. Gao, E. Eisenberg, and L. Greene, *J. Biol. Chem.* **271,** 16792 (1996).

[55] X. Zhu, X. Zhao, W. F. Burkholder, A. Gragerov, C. M. Ogata, M. E. Gottesman, and W. A. Hendrickson, *Science* **272,** 1606 (1996).

[56] M. K. Hayer Hartl, J. Martin, and F. U. Hartl, *Science* **269,** 836 (1995).

[57] A. Azem, S. Diamant, M. Kessel, C. Weiss, and P. Goloubinoff, *Proc. Natl. Acad. Sci. U.S.A.* **92,** 12021 (1995).

[58] O. Llorca, J. L. Carrascosa, and J. M. Valpuesta, *J. Biol. Chem.* **271,** 68 (1996).

[59] K. Braig, Z. Otwinowski, R. Hegde, D. C. Boisvert, A. Joachimiak, A. L. Horwich, and P. B. Sigler, *Nature (London)* **371,** 578 (1994).

[60] W. A. Fenton, Y. Kashi, K. Furtak, and A. L. Horwich, *Nature (London)* **371,** 614 (1994).

TABLE V
PARAMETERS OBTAINED FROM ANALYTICAL ULTRACENTRIFUGATION EXPERIMENTS WITH GROEL AND GROES

| Parameter[a] | GroEL | GroES |
|---|---|---|
| $s_{20,w}$ (S) | 22.13 ± 0.16 | 3.92 ± 0.04 |
| $D_{20,w} \times 10^7$ cm²/sec | 2.59 ± 0.09 | 5.30 ± 0.14 |
| $M_{sD}$(kDa) | 802.6 ± 16 | 72.4 ± 2.1 |
| $M_{eq}$(kDa) | 795.8 ± 12 | 71.9 ± 2.3 |
| $f/f_0$ | 1.30 | 1.44 |

[a] $s_{20,w}$, Sedimentation coefficient corrected for water at 20°; $D_{20,w}$, diffusion coefficient corrected for water at 20°; $M_{sD}$, molecular mass from sedimentation velocity via the Svedberg equation; $M_{eq}$, molecular mass from sedimentation equilibrium.

ATPase activity, whereas GroES modulates the ATPase activity of GroEL[13] but has no affinity for ATP by itself, as chemical cross-linkage that was obtained with azido-ATP[62] turned out to be nonspecific.[63]

The ultracentrifuge has been used relatively infrequently to study the GroE system, in comparison with other biochemical methods. Most of the ultracentrifugation data concern the size and stability of the chaperonin homocomplexes. A sedimentation coefficient for GroES heptamers of 4.5S was estimated by the relatively inaccurate gradient sedimentation technique.[14] More exact sedimentation and diffusion coefficients (Table V) were determined by us, using analytical ultracentrifugation.[64] The flat and asymmetrical structure of the co-chaperonin has a marked effect on the results. The GroES oligomer possesses only limited stability and in diluted solutions below 1 $\mu M$ partial dissociation was observed.[65] This behavior could cause misinterpretation of the GroEL–GroES interaction, especially in extremely dilute solutions.[18] Sedimentation velocity experiments with truncated GroES revealed that oligomerization occurs via the carboxy terminus of the molecule.[66] Sedimentation velocity analyses also showed that GroES is capable of refolding and reassembling from a urea-denatured

[61] J. F. Hunt, A. J. Weaver, S. J. Landry, L. Gierasch, and J. Deisenhofer, *Nature* (*London*) **379,** 37 (1996).

[62] J. Martin, S. Geromanos, P. Tempst, and F. U. Hartl, *Nature* (*London*) **366,** 279 (1993).

[63] M. J. Todd, O. Boudkin, E. Freire, and G. H. Lorimer, *FEBS Lett.* **359,** 123 (1995).

[64] J. Behlke, O. Ristau, and H.-J. Schönfeld, *Biochemistry* **36,** 5149 (1997).

[65] J. Zondlo, K. E. Fisher, Z. Lin, K. R. Ducote, and E. Eisenstein, *Biochemistry* **34,** 10334 (1995).

[66] J. W. Seale and P. M. Horowitz, *J. Biol. Chem.* **270,** 30268 (1995).

state, indicating that chaperonins may not require other chaperones for their assembly.[67] Furthermore, it was demonstrated by the sedimentation velocity method that the stability of the GroEL complex is enhanced by binding of the small protein substrate rhodanese[68] but decreased by binding of ADP.[69]

GroEL and GroES form asymmetrical "bullet-shaped" 1:1 and symmetrical "American football-shaped" 1:2 heterocomplexes.[17] The structures of these complexes under different conditions were studied by chemical cross-linking and electron microscopy[18] as well as by fluorescence methods.[19] The functional significance of the two shapes of chaperonin heterocomplexes is controversial. The electron microscopic study[18] attributed the occurrence of the symmetrical complex mainly to nonphysiological conditions, such as high magnesium concentrations (50 versus 5 m*M*) and a high pH (pH 8.0 vs. pH 7.2).

Using analytical ultracentrifugation in the absence of any nucleotides, we found that GroEL and GroES form a weak complex mostly with a 1:1 stoichiometry.[64] If one assumes equal binding sites for GroES on GroEL then it is expected that mostly asymmetric bullet-like structures would occur in such solutions. By adding increasing amounts of ATP analogs (ATP-$\gamma$-S, adenosine 5'-*O*-(3-triphosphate) or AMP–PNP, adenosine 5'-($\beta$,$\gamma$-iminotriphosphate)) the affinity of GroEL for GroES increases by two to three orders of magnitude (Fig. 4). This is attributed to movement of the apical domain in GroEL.[59] Alteration of the GroES-to-GroEL ratio results in the appearance of symmetrical 2:1 (American football-type) complexes in addition to asymmetrical structures. In the presence of 1 m*M* ATP analog and GroES-to-GroEL ratios of 3–4, on the average up to 1.8 GroES heptamers are observed in complexes with GroEL tetradodecamers (Fig. 5). This number does not depend on the magnesium concentration (50 versus 5 m*M*) or the pH (pH 8.0 vs. pH 7.2) (Fig. 5) in contrast to a previous report.[18] The transition between the two types of complexes was demonstrated by the titration of GroEL with increasing amounts of GroES and/or ATP analogs. In contrast to the ATP analogs, ADP inhibits GroES binding to the second GroEL ring. As noted above, the formation of symmetrical heterocomplexes as intermediates under protein-folding conditions has also been observed using other methodologies. However, using sedimentation equilibrium it was possible to clearly derive the transition between the two different forms, including the association constants of the undisturbed equilibria, results that will permit a better understanding of the protein-folding cycle in solution.

[67] J. W. Seale, B. M. Gorovits, J. Ybarra, and P. M. Horowitz, *Biochemistry* **35,** 4079 (1996).
[68] J. A. Mendoza, B. Demeler, and P. M. Horowitz, *J. Biol. Chem.* **269,** 2447 (1994).
[69] B. M. Gorovits and P. M. Horowitz, *J. Biol. Chem.* **270,** 28551 (1995).

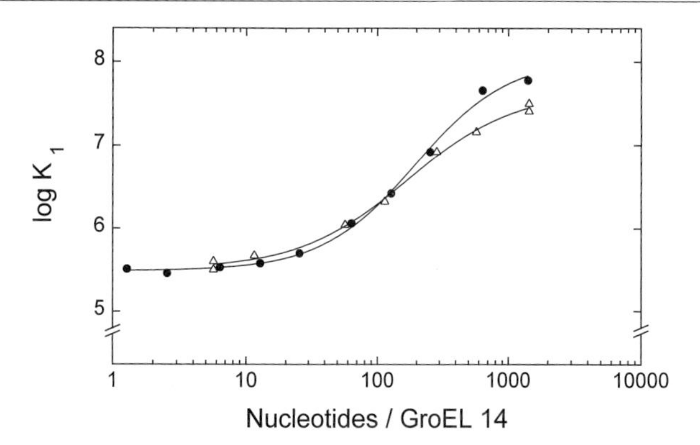


FIG. 4. Analysis of the interaction between GroEL and GroES by analytical ultracentrifugation. GroEL (0.71 $\mu M$) and 1.59 $\mu M$ GroES were mixed in 50 m$M$ Tris-HCl (pH 7.7), 100 m$M$ KCl, and 10 m$M$ $MgCl_2$. The effects of different concentrations of nucleotides ATP$\gamma$S (●) and AMP-PNP (△) on the association constant $K_1$ are shown. Nucleotide concentrations are presented as number of nucleotide molecules per GroEL tetradodecamer molecule. Measurements were made with the analytical ultracentrifuge Optima XL-A, using six-channel cells. Using the program Polymol[64] there was an excellent fit of radial concentration distributions at sedimentation equilibrium [recorded at three different wavelengths (280, 285 and 290 nm) after 30 to 40 hr at 6000 rpm and 10°], assuming two possible binding sites for GroES. From these fits the partial concentrations of free reactants (GroEL, $M_r$ 800 kDa; GroES, $M_r$ 72.7 kDa) and complexes [$GroEL_{14}GroES_7$, $M_r$ 873 kDa; $GroEL_{14}(GroES_7)_2$, $M_r$ 944 kDa] and the association constants were computed. Assuming only one binding site for GroES resulted in bad fits and unusually high association constants ($>10^{20}$ $M^{-1}$).

## Concluding Remarks

As discussed, determination of protein quaternary structure should never be based exclusively on relative methods. The overview in Table II points out that only a few methodologies, such as analytical ultracentrifugation techniques, unambiguously deliver molecular masses of homo- or heterocomplexes in solution without the need for arbitrary assumptions. Our protocols for measuring the sedimentation equilibrium and sedimentation velocity of chaperone complexes represent simple procedures that novices in the field of analytical ultracentrifugation can easily apply to obtain reliable data on oligomeric states and undisturbed equilibria, assuming access to a Beckman Optima XL-A centrifuge.

Application of the protocols that have been presented here revealed that the tetrameric structure assumed by various investigators for SecB is correct and is the first confirmation of the initial data obtained from size-exclusion chromatography (Table I). A more thorough analysis of SecB sedimentation equilibrium results at low protein concentrations provided the first observation of a monomer/tetramer equilibrium.

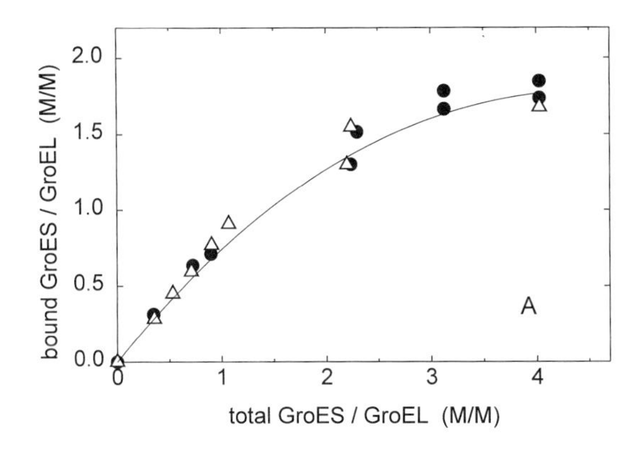


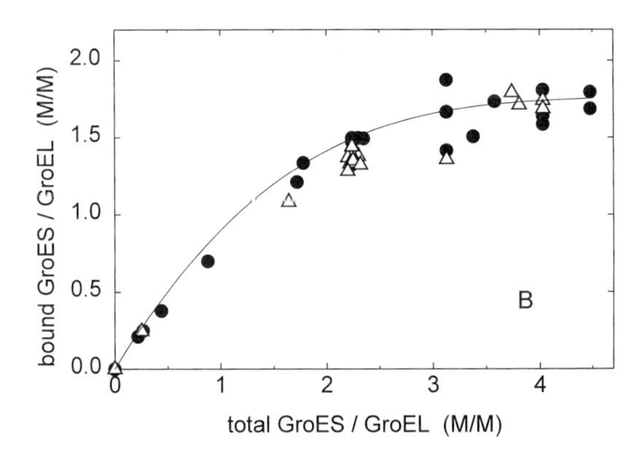


FIG. 5. The interaction of GroEL and GroES in different buffer systems. GroEL oligomers (0.71 $\mu M$) were mixed with different ratios of GroES oligomers as indicated in high magnesium salt buffer [△; 50 m$M$ Tris-HCl (pH 8.0), 50 m$M$ KCl, 50 m$M$ $MgCl_2$] or low magnesium salt buffer [●; 20 m$M$ MOPS–KOH (pH 7.2), 50 m$M$ KCl, 5 m$M$ $MgCl_2$] in the presence of 1 m$M$ AMP–PNP (A) or 1 m$M$ ATP$\gamma$S (B). Measurements were made as described in Fig. 4.

In the case of the chaperone GrpE, for which structures from monomer to hexamer have been proposed (Table I), only dimers were found within a broad concentration range when analyzed by analytical ultracentrifugation.

The equilibria of HSP70 oligomers depend on protein concentration, temperature, and other parameters, results that are well established in the literature; however, the influence of distinct oligomeric forms on the mode of chaperone action still remains obscure.

The lack of knowledge of DnaJ oligomers and DnaK–DnaJ interactions is related to the difficulty in obtaining structurally homogeneous preparations of DnaJ. Investigation of these and of DnaK–substrate and DnaJ–

substrate interactions will certainly be useful subjects for future analytical ultracentrifugation experiments.

We demonstrated the practicality of investigating the comparably large chaperonin heterocomplexes under equilibrium. Of course, this technique cannot decide what species, bullets or American footballs or both, are the active species in chaperonin-mediated protein folding. However, as an equilibrium method it is best suited to determine the conditions under which the different species exist.

Finally, the investigation of chaperonin–substrate interactions will be a productive field for future analytical ultracentrifugation experiments.

## Acknowledgments

The authors thank E. Kusznir and F. Müller for providing analytical ultracentrifugation runs with SecB and GrpE, B. Pöschl for excellent technical support in chaperone purification, and H. Etlinger for reading the text and engaging in many encouraging discussions.

# [24] Probing Conformations of GroEL-Bound Substrate Proteins by Mass Spectrometry

*By* CAROL V. ROBINSON, MICHAEL GROSS, and SHEENA E. RADFORD

## Introduction

Molecular chaperones represent one of the most important new concepts in cell biology. Their genetics and cell biology have been studied in great detail[1] and their roles in heat shock as well as protein folding, assembly, degradation, and trafficking in the cell have been well described.[2] Despite the great interest and the central importance of these molecules in biology, relatively little is understood at the atomic level about the mechanism of their action. One of the central questions in this field, therefore, is the mechanism by which protein folding *in vivo* is facilitated by the chaperonins GroEL and GroES. Significant advances have been made toward addressing this question, using a number of different methods, including cryoelectron microscopy,[3] protein biochemistry,[1] site-directed

[1] F. U. Hartl, *Nature* (*London*) **381,** 571 (1996).
[2] R. J. Ellis, "The Chaperonins." Academic Press, San Diego, California, 1996.
[3] A. M. Roseman, S. Chen, H. White, K. Braig, and H. R. Saibil, *Cell* **87,** 241 (1996).



0076-6879/98 $25.00

mutagenesis,[4] and X-ray crystallography,[5,6] which together have produced a model for the mechanism of GroEL-assisted protein folding.[1] The complexity of the chaperonin system, however, which involves interactions between GroEL and GroES, binding and hydrolysis of ATP, asymmetrical GroEL:GroES "bullets," symmetrical GroEL:GroES "footballs," and protein substrate bound *cis* and *trans* to GroES, poses a formidable challenge to biophysicists who want to study the functional details of the chaperonin reaction cycle on the molecular scale.

To gain a molecular knowledge of the mechanism of chaperonin-assisted protein folding, details of the conformation of the bound protein substrate at different stages of the chaperonin reaction cycle are required. This is notoriously difficult to study by classic structural methods because the sheer size of the assembly (>800,000 Da) rules out most types of high-resolution nuclear magnetic resonance (NMR) experiments, and the relatively disordered bound substrate is not amenable to X-ray crystallography. Developments in electrospray ionization mass spectrometry (ESI-MS), particularly the ability to study large proteins from aqueous solutions, have provided new opportunities to obtain structural information about these complex molecular systems. Here we describe the methodology we have developed to elucidate by ESI-MS the conformational properties of substrate proteins bound to GroEL. To provide an overview of the information obtained with this methodology, selected examples from previous publications[7,8] are used along with unpublished data. We also discuss the general application of ESI-MS in the study of noncovalent protein complexes and the future potential of this rapidly advancing technique.

## Mass Spectrometric Methods for Studying Native and Partially Folded States of Proteins

The essential features of the conventional electrospray process are summarized in Fig. 1. Briefly, the protein solution is introduced into the electrospray ion source through a glass or stainless steel needle [labeled (1) in Fig. 1] held at a high voltage. This results in charged microdroplets of protein solution that, with the aid of nitrogen nebulizer gas [(2) in Fig. 1],

[4] W. A. Fenton, Y. Kashi, K. Furtak, and A. L. Horwich, *Nature (London)* **371,** 614 (1994).

[5] K. Braig, Z. Otwinowski, R. Hegde, D. Boisvert, A. Joachimiak, A. Horwich, and P. Sigler, *Nature (London)* **371,** 578 (1994).

[6] G. H. Lorimer and M. J. Todd, *Nature Struct. Biol.* **3,** 116 (1996).

[7] C. V. Robinson, M. Gross, S. J. Eyles, J. J. Ewbank, M. Mayhew, F. U. Hartl, C. M. Dobson, and S. E. Radford, *Nature (London)* **372,** 646 (1994).

[8] M. Gross, C. V. Robinson, M. Mayhew, F. U. Hartl, and S. E. Radford, *Protein Sci.* **5,** 2506 (1996).

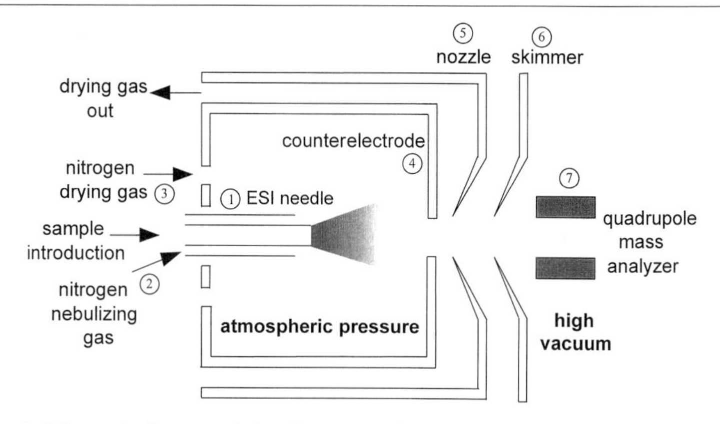


FIG. 1. Schematic diagram of the electrospray instrument, highlighting the experimental parameters used for acquiring mass spectra of proteins under native conditions. Samples were incubated at 4° (on ice) or precooled before their introduction into the mass spectrometer by continuous infusion via the ESI probe. The ESI source is operated without source heating. Nitrogen drying gas (3) was admitted at a flow rate of 250 liters/min. Samples were introduced through a rheodyne injector and were pumped at a flow rate of 10 liters/min into the electrospray interface with a fluid delivery module (Michrom Bioresources). Solvent delivery was by pressurized helium bottle; this was also immersed in an ice–salt bath to cool the delivery solvent and hence to minimize sample heating and hydrogen exchange. The ESI interface was also equilibrated overnight to ensure the interface was at the same pH and isotopic content as the sample being analyzed. Nitrogen nebulizer gas (2) (flow rate, 20 liters/hr) was cooled by inserting a copper coil into the gas line, which was then immersed in an ice–salt bath. The nebulizer gas surrounds the electrospray capillary/needle and thus cooling the nebulizer gas maintains a low temperature in the electrospray capillary and hence a low temperature in the protein sample for analysis. In general, a needle voltage of 3.5 kV was used to produce mass spectra of GroEL and its complexes. The voltage was lowered, typically to around 2.5 kV, in experiments where noncovalent complexes were required to persist in the gas phase. A voltage of 0.7 kV on the counterelectrode (4) was used to maintain a stable electrospray beam. For analysis of chaperone complexes a cone voltage (5) of 50–70 V was applied. A skimmer offset (6) of 10 V was used to maintain noncovalent interactions while an increase in the offset aids dissociation of the chaperone complex (typical values between 20 and 50 V). For our experiments on GroEL complexes by hydrogen exchange, the quadrupole mass analyzer was not adjusted for high resolution because the efficiency of the dissociation process is not sufficient to provide an intense ion beam under high-resolution conditions. All mass spectra of the chaperonin complex represent the average of 10 scans over an $m/z$ range of 1100–1900.

produce a fine electrospray of protein-containing droplets. These droplets are evaporated with nitrogen drying gas [(3) in Fig. 1] and pass though the counterelectrode [(4) in Fig. 1]. As the droplet size diminishes, protein ions enter the gas phase either by desorption from the droplet or by evaporation of solvent. These gas-phase protein ions then pass through the nozzle and skimmer electrode [(5) and (6) in Fig. 1, respectively], and are transferred into a quadrupole analyzer [(7) in Fig. 1] and analyzed according to their mass-to-charge ratio. Traditionally, ESI mass spectra of proteins are acquired using solutions of proteins in organic solvents [50% (v/v) methanol or acetonitrile], under acidic conditions [1% (v/v) formic acid], and with high source temperatures (50°).[9] Under these conditions proteins are usually disassembled and denatured, and although mass spectra with high mass accuracy can be obtained, information about protein conformation is lost. A major breakthrough in the use of mass spectrometry to analyze protein conformations, therefore, came with the ability to ionize proteins from solutions in which they retain their native conformations [mildly acidic conditions (pH 5–6), low source temperature (20° or below), and in the absence of organic solvents], which we refer to hereafter as "native" conditions. In combination with measurements of hydrogen exchange, ESI-MS has now become a powerful technique for the determination of new features of protein structure and macromolecular interactions.[7,8,10,11]

One important attribute of ESI-MS from native conditions is the ability to preserve noncovalent protein–ligand interactions within the mass spectrometer, and many examples now exist of protein–ligand interactions that have survived in the gas phase.[12] Studies suggest that variation of the conditions in the ESI interface (see caption to Fig. 1) and the precise nature of the protein–ligand interaction affect the amount of complex detected in the mass spectra of these complexes.[11] Thus, whereas ionic and van der Waals interactions are enhanced in the gas phase and in the lower pressure conditions of the mass analyzer, it is likely that hydrophobic interactions are generally less favored in the gas phase. An interesting example of the latter is seen in the ESI mass spectrum of the complex between GroEL and human dihydrofolate reductase (DHFR) (Fig. 2c). Thus, although the sample was introduced as a stable GroEL : DHFR complex (see below), the complex dissociates in the gas phase of the mass spectrometer to produce GroEL monomers (mass, 57,197 Da) and the bound DHFR is released (mass, 21,321 Da), consistent with the view that the interaction between the chaperonin and its substrate protein involves hydrophobic interactions (reviewed in Ref. 1). By contrast, under the same solution and mass spectrometry conditions, at least a large proportion of DHFR introduced into the mass spectrometer from its native state (Fig. 2a) retains bound dihydrofolate, consistent with the known importance of ionic interactions in the

[9] J. B. Fenn, M. Mann, C. K. Meng, S. F. Wong, and C. M. Whitehouse, *Science* **246,** 64 (1989).

[10] P. Gervasoni, W. Staudenmann, P. James, P. Gehrig, and A. Plückthun, *Proc. Natl. Acad. Sci. U.S.A.* **93,** 12289 (1996).

[11] C. V. Robinson, E. W. Chung, B. B. Kragelund, J. Knudsen, R. T. Aplin, F. M. Poulsen, and C. M. Dobson, *J. Am. Chem. Soc.* **118,** 8646 (1996).

[12] K. J. Light-Wahl, B. E. Winger, and R. D. Smith, *J. Am. Chem. Soc.* **115,** 5869 (1993).

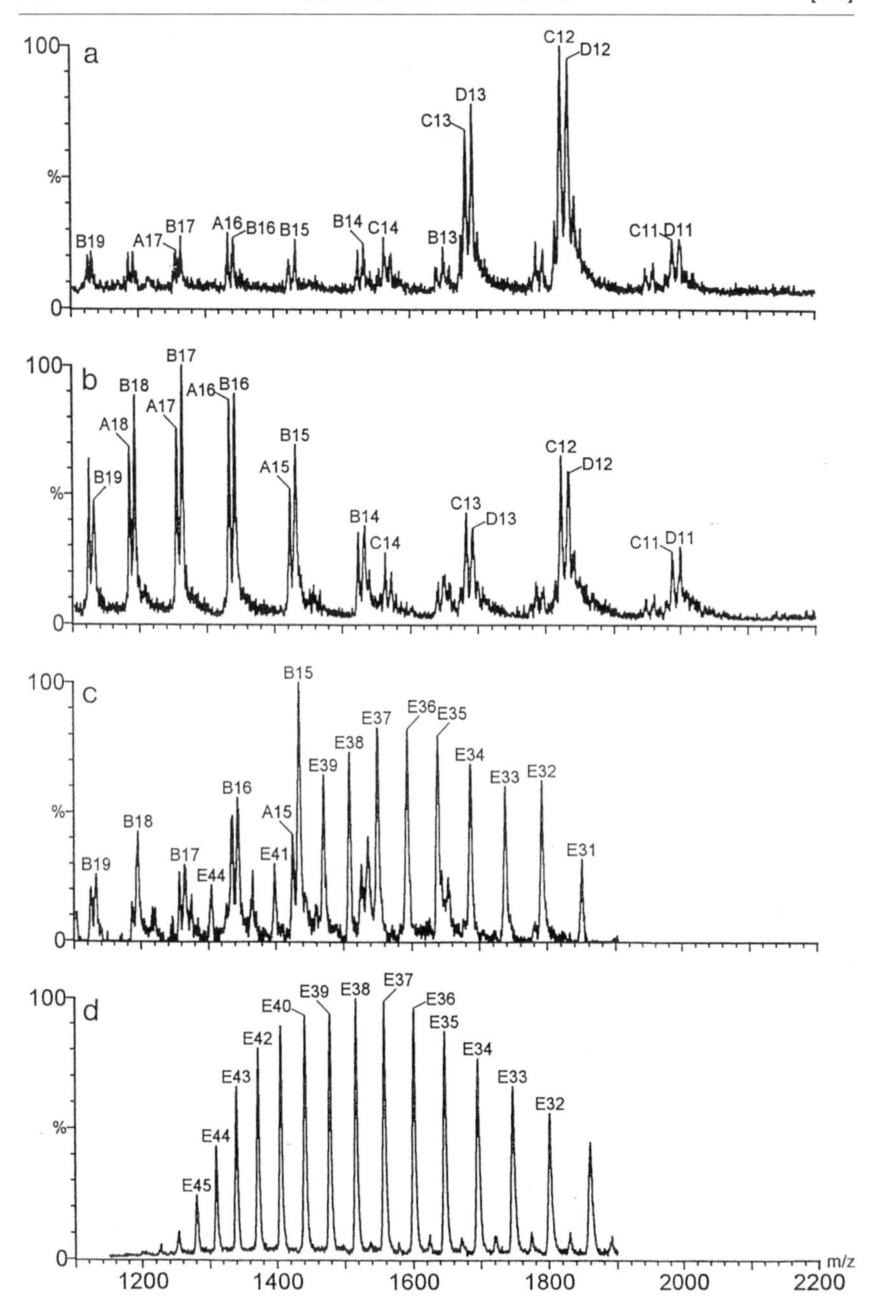
a
b
c
d
100
%
0
C12
D12
D13
C13
B19
A17
B17
A16
B16
B15
B14
C14
B13
C11
D11
B18
A18
A15
E44
E41
E39
E38
E37
E36
E35
E34
E33
E32
E31
E45
E43
E42
E40
1200
1400
1600
1800
2000
2200
m/z

DHFR active site.[8] In the context of hydrogen exchange experiments, the ability to dissociate the chaperonin complex in the gas phase (where no further hydrogen exchange takes place) provides a unique opportunity to monitor the hydrogen exchange properties of the substrate protein, without the need first to dissociate the complex, as required, for example, in NMR measurements of hydrogen exchange.[13,14]

## Measurements of Hydrogen Exchange by Electrospray Ionization Mass Spectrometry

Historically, hydrogen exchange as a tool for the characterization of protein structure and dynamics was developed by Linderstrøm-Lang at the Carlsberg Laboratory in the early 1950s (reviewed in ref. 15) when densitometry was the only method available to follow the isotope content of a sample. The renaissance of the technique came in the late 1980s when two-dimensional $^1H$ NMR of proteins made site-specific observation of hydrogen exchange feasible.[16] Thus, the resilience of a protein to exchange of amide hydrogen atoms (for example, with solvent deuterons) is an exquisitely sensitive probe for determining the structure and dynamics of proteins in solution. Methods were developed for the study of hydrogen exchange both in stable states and in transient intermediates of protein folding, giving insights into the folding pathways of a number of small proteins.[17] These methods exploit the fact that amide hydrogen atoms that are buried in the native structure of a protein, and/or participate in the hydrogen-bonding network of secondary structure elements, exchange with solvent deuterons much more slowly than those in unstructured regions. As the difference in mass between hydrogen and deuterium is 1 Da, ESI-MS can also be used

[13] R. Zahn, C. Spitzfaden, M. Ottiger, K. Wüthrich, and A. Plückhun, *Nature* (*London*) **368,** 261 (1994).

[14] R. Zahn, S. Perrett, G. Stenberg, and A. R. Fersht, *Science* **271,** 642 (1996).

[15] F. M. Richards, *Protein Sci.* **1,** 1721 (1992).

[16] S. W. Englander and L. Mayne, *Annu. Rev. Biophys. Biomol. Struct.* **21,** 243 (1992).

[17] C. K. Woodward, *Curr. Opin. Struct. Biol.* **4,** 112 (1994).

FIG. 2. Electrospray mass spectra of (a) native DHFR (obtained from $H_2O$, pH 5.0); (b) partially unfolded DHFR (at pH 5.0, 50°); (c) DHFR bound in a stable complex with GroEL; and (d) GroEL in the absence of its protein substrate (both at pH 5.0, 20°). All mass spectra were obtained under identical mass spectrometry and solution conditions, with the exception of (b), where the solution and interface temperature were set at 50°. In all the spectra shown, charge state series arising from GroEL are labeled E; those labeled A, B, C, and D arise from DHFR. (Reproduced with permission from M. Gross, C. V. Robinson, M. Mayhew, F. U. Hartl, and S. E. Radford, *Protein Sci.* **5,** 2506 (1996).)

to monitor the rate and extent of hydrogen exchange. Furthermore, the advent of ESI from aqueous solution conditions has now enabled hydrogen exchange measurements of conformational states of proteins, ranging from the fully denatured to the native states, by MS.[7,18,19] By combining ESI-MS with pulsed quench flow methods,[20] information about protein folding reactions *in vitro* is also possible.[18,21]

The application of ESI-MS in studying the conformation of a substrate protein bound to a molecular chaperone offers new experimental challenges. The 800,000 Da complex must be introduced into the mass spectrometer under native conditions, the instrument tuned such that the protein complex dissociates in the gas phase, and hydrogen exchange of the complex during sample introduction must be negligible, even for unstable states. Using the protocol outlined in the caption to Fig. 1, this can be achieved. We have used this approach to investigate the conformational properties of two different substrate proteins bound to GroEL (a three-disulfide derivative of bovine $\alpha$-lactalbumin, [3-SS]BLA,[7] and two different folding intermediates of human DHFR.[8]). The general concept of our experimental approach is represented schematically in Fig. 3, which also includes examples of potential future applications (see caption to Fig. 3 for details).

## Preparing Protein Samples for Hydrogen Exchange Electrospray Ionization Mass Spectrometry

The essential procedures used to prepare complexes of GroEL with [3-SS]BLA or DHFR for analysis by ESI-MS are summarized in Fig. 4. The method involves exchanging all exchange-labile hydrogens for deuterium in the substrate protein. In the case of [3-SS]BLA, this is achieved by incubating bovine $\alpha$-lactalbumin with $D_2O$ containing EGTA (0.5 m*M*) and dithiothreitol (DTT, 0.5 m*M*) for several hours (usually overnight) at 20°, pD 7.0. For DHFR, exchange of all labile sites for deuterium is performed by denaturing the protein in 6 *M* deuterated guanidium chloride (dGdmCl) (prepared by recrystallizing GdmHCl 10 times in $D_2O$) for 10 min at 20°. Complexes of the deuterated substrate proteins with GroEL are then made by diluting the substrate protein into a solution containing a 1.2-fold molar excess of the GroEL oligomer relative to the substrate protein, in an unbuffered $D_2O$ solution, which has been adjusted to pD 7.0

[18] A. Miranker, C. V. Robinson, S. E. Radford, and C. M. Dobson, *FASEB J.* **10,** 93 (1996).

[19] E. W. Chung, E. J. Nettleton, C. J. Morgan, M. Gross, A. Miranker, S. E. Radford, C. M. Dobson, and C. V. Robinson, *Protein Sci.* **6,** 1316 (1997).

[20] R. L. Baldwin, *Curr. Opin. Struct. Biol.* **3,** 84 (1993).

[21] A. Miranker, C. V. Robinson, S. E. Radford, R. T. Aplin, and C. M. Dobson, *Science* **262,** 896 (1993).

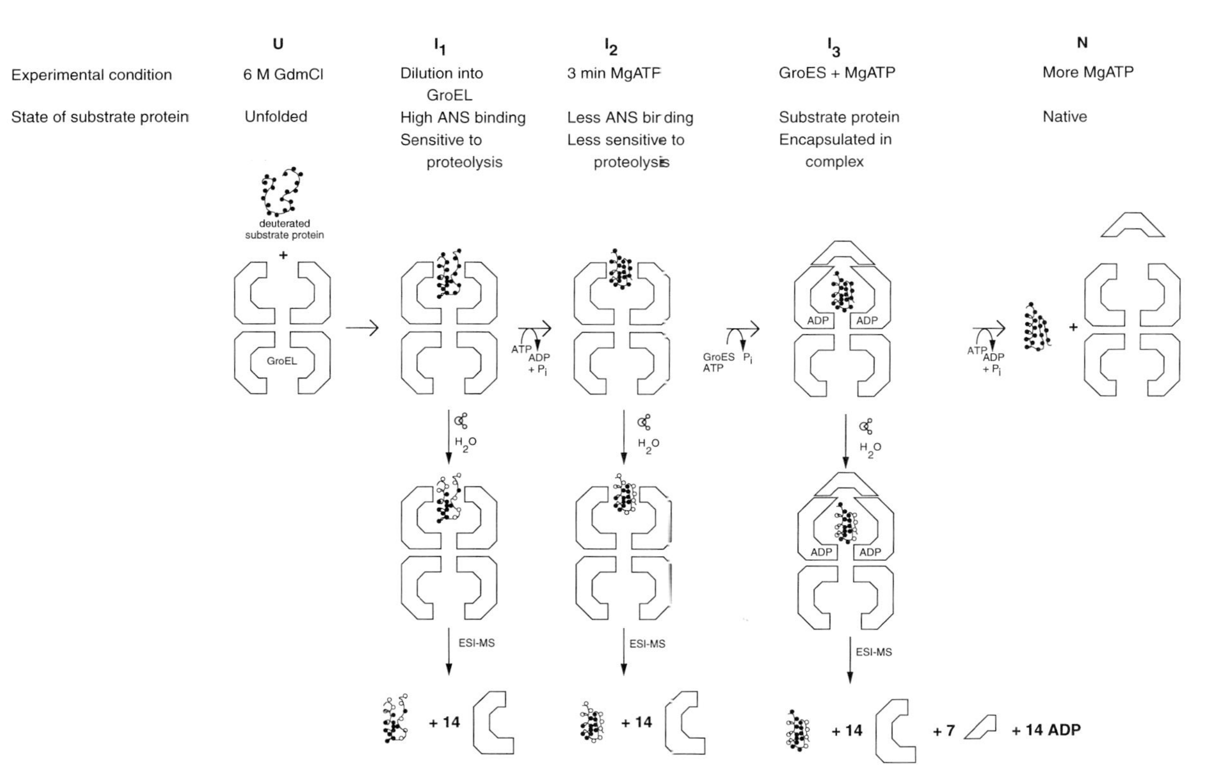


FIG. 3. Schematic diagram of experiments designed to measure hydrogen exchange in GroEL : substrate complexes by ESI-MS at different stages of chaperonin-assisted protein folding. A generalized pathway of chaperonin-assisted folding is represented by the horizontal arrows, whereas the procedures of sampling and analysis by mass spectrometry, which can, in principle, start from any stage of the pathway, are referred to by vertical arrows. The methods used for preparation of protein samples and for acquisition of ESI mass spectra are given in text.

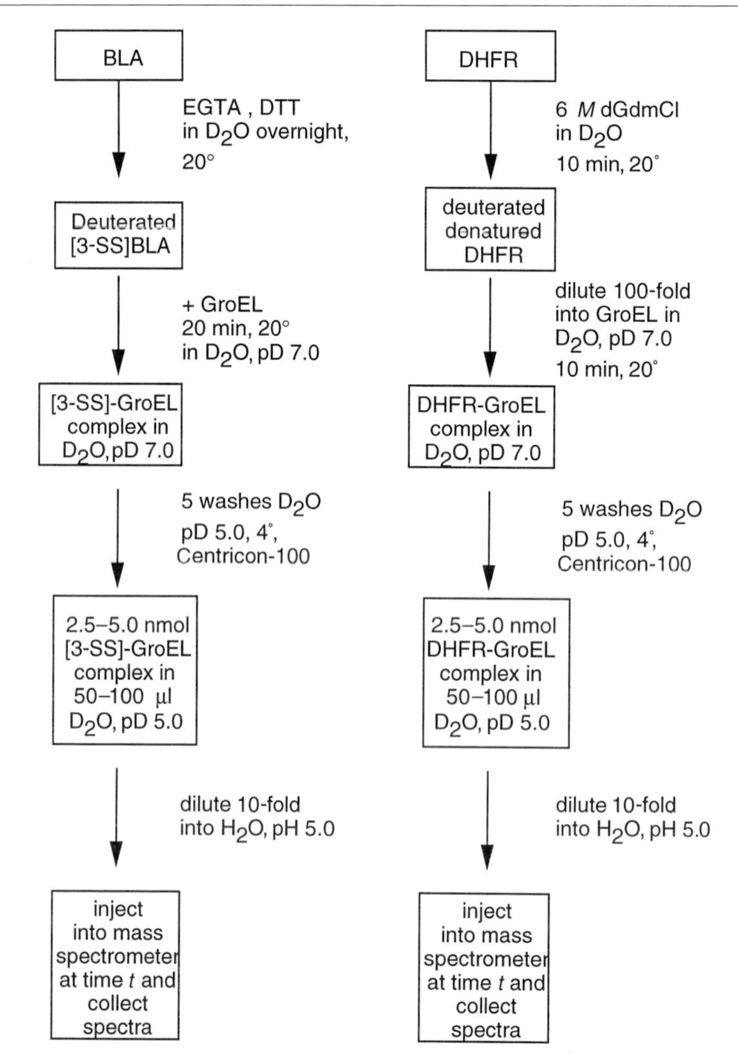


FIG. 4. Schematic representation of the methods used to prepare complexes of GroEL and substrate proteins for analysis of hydrogen exchange by ESI-MS.

by the addition of formic acid. Because ESI-MS is sensitive to the presence of buffer salts, it is essential to avoid their use during preparation of protein samples. The final stage is to reduce the pD of the complex to pD 5.0 (at which pH the complex is stable, the intrinsic rate of hydrogen exchange reduced 100-fold relative to that at pH 7.0,[22] and ESI mass spectra of excellent quality can be obtained). Residual dGdmCl, EGTA, and DTT are then removed from the sample by sequential dilution into $D_2O$, pD 5.0, and reconcentration of the samples, using a Centricon-100 concentrator (Amicon, Danvers, MA). In all cases, the integrity and stability of complexes used for ESI-MS are verified by size-exclusion chromatography, using a miniature gel-filtration column (packed with S-300 resin) equilibrated in water (adjusted to pH 5.0 with formic acid). Column fractions are analyzed by sodium dodecyl sulfate-polyacrylamide gel electrophoresis (SDS–PAGE) following standard procedures.[23] Under these conditions the complex is stable for at least 24 hr at 4°. To initiate hydrogen exchange the samples are diluted 10-fold into $H_2O$ solution at pH 5.0 and ESI mass spectra are recorded at various time points after dilution, as described in the caption to Fig. 1. About 100 $\mu$l of a solution containing 5 nmol of the GroEL complex provides sufficient protein for 10 time points after the 1-in-10 dilution to initiate hydrogen exchange.

Complexes of substrate proteins bound to GroEL at different stages of ATP-dependent protein folding, and ternary GroEL : GroES : protein complexes, can all be prepared following the preceding principles. For example, DHFR rebound to GroEL after several rounds of ATP-driven substrate cycling was prepared by incubation of the DHFR–GroEL complex described above (in $D_2O$, pD 7.0), with 1 m*M* ATP–2 m*M* magnesium acetate for 3 min at 20°. The reaction was then quenched by addition of 1.5-fold molar excess of EDTA (also in $D_2O$). DHFR that had rebound to GroEL was then reisolated by ultrafiltration using a Centricon-100 membrane (Amicon) and desalted as described previously. In this case, hydrogen exchange experiments were carried out immediately after the sample was prepared.

## Information Contained within the Electrospray Mass Spectrum

In the context of the experiments described here, ESI-MS provides three different kinds of information. First, changes in the charge state distributions of proteins can be used to identify conformational changes; second, by monitoring changes in mass with time, information about hydrogen–deuterium exchange may be obtained; and finally, the line widths of individual charge states in a hydrogen exchange experiment contain

[22] Y. Bai, J. S. Milne, L. Mayne, and S. W. Englander, *Proteins Struct. Funct. Genet.* **17,** 75 (1993).

[23] M. Gross, I. J. Kosmowsky, R. Lorenz, H. P. Molitoris, and R. Jaenicke, *Electrophoresis* **15,** 1559 (1994).

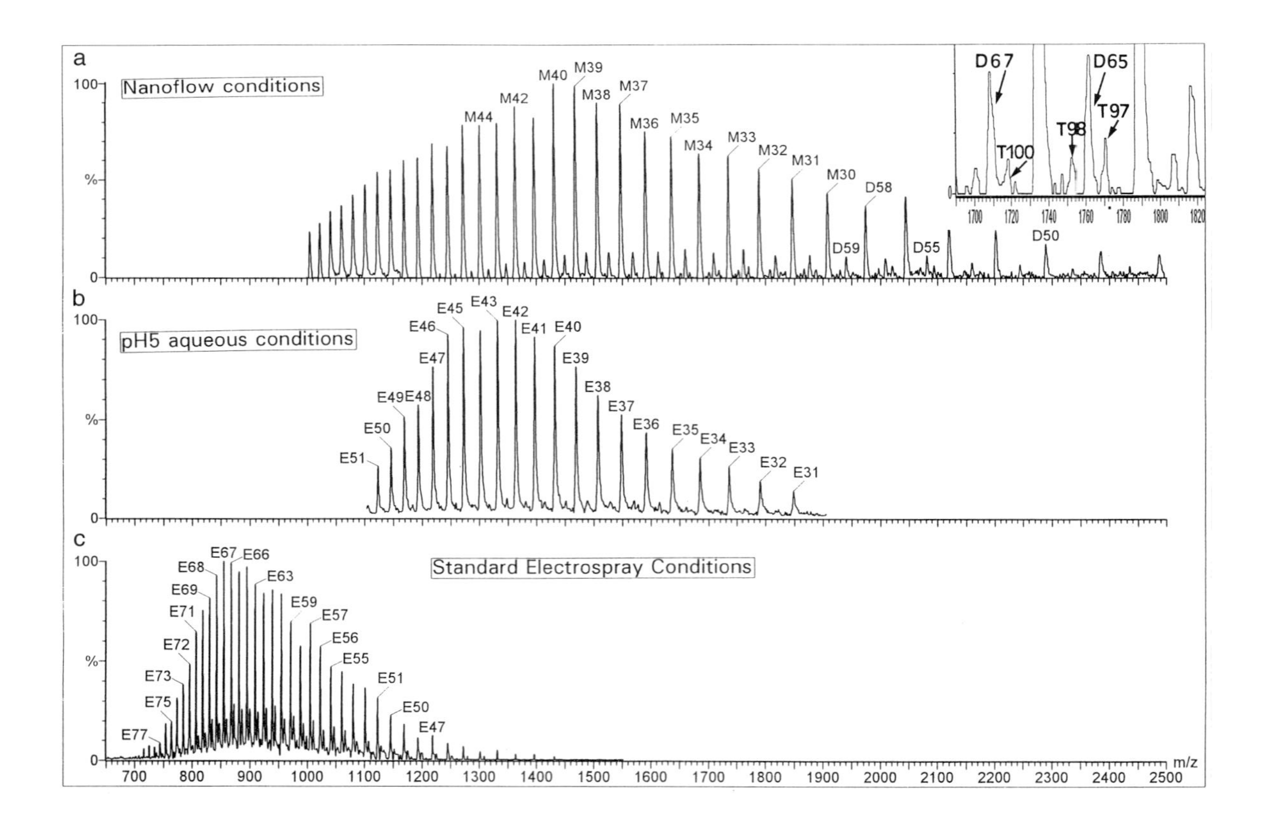
a
Nanoflow conditions
M44
M42
M40
M39
M38
M37
M36
M35
M34
M33
M32
M31
M30
D58
D59
D55
D50
D67
T100
T98
D65
T97
b
pH5 aqueous conditions
E51
E50
E49
E48
E47
E46
E45
E43
E42
E41
E40
E39
E38
E37
E36
E35
E34
E33
E32
E31
c
Standard Electrospray Conditions
E77
E75
E73
E72
E71
E69
E68
E67
E66
E63
E59
E57
E56
E55
E51
E50
E47
100
%
0
m/z
700
800
900
1000
1100
1200
1300
1400
1500
1600
1700
1800
1900
2000
2100
2200
2300
2400
2500

information about the distribution of molecules with different exchange properties. The combination of these three aspects of the mass spectrum makes this approach particularly powerful, and all three parameters have been used to analyze the interaction between GroEL and its substrate proteins.[7,8]

### *Charge State Distribution*

In ESI-MS, proteins are analyzed according to their mass-to-charge ratio. Because proteins have multiple sites that can carry either positive or negative charges, a typical mass spectrum of a single protein contains a series of peaks corresponding to populations of molecules with different charges. As well as determining the exact mass of the protein from an optimized fit to the charge state distribution, additional information can be obtained by analysis of the distribution of charges in the protein molecule,[24] which, in a qualitative manner, reflects the different susceptibility of amino acid side chains to protonation in different conformational states. A clear illustration of this is shown by analysis of the ESI mass spectrum obtained from GroEL under standard (denaturing) and native ESI-MS conditions (Fig. 5b and c). The much higher charge state distribution in the former (which ranges from +47 to +77) reflects the increased solvent accessibility of side chains in the denatured state.

A second example is shown in the ESI mass spectrum of human DHFR (Fig. 2a), which demonstrates how both mass and conformational differences can be resolved within the same ESI mass spectrum. The splitting of the peaks into doublets of comparable intensity (labeled A–B and C–D

[24] S. K. Chowdhury, V. Katta, and B. T. Chait, *J. Am. Chem. Soc.* **112,** 9012 (1990).

FIG. 5. Comparison of mass spectra obtained from the GroEL oligomer under different electrospray conditions. In (a), nanoflow conditions were used to acquire the spectrum. Under these conditions, oligomeric species including dimers and trimers (labeled D and T) are observed. In (b), standard electrospray conditions were used at pH 5 in aqueous solution. In (c), standard electrospray conditions in 50% acetonitrile–50% water at pH 3 were used. These effect complete dissociation to GroEL monomers (the small additional peaks in this spectrum arise from a higher molecular weight protein impurity in this sample). The nanoflow ESI mass spectrum was recorded on a Platform II mass spectrometer (Micromass), using a nanoflow probe equipped with a gold-coated borosilicate capillary at a needle voltage of 1 kV without the counterelectrode. The source was operated without nebulizer gas and with a reduced drying gas flow rate (3 liters/hr). All other spectrometer settings were identical to those listed in the caption to Fig. 1. Note that the charge states in the nanoflow spectrum extend from +62 to +23; this may reflect the difference in instrument conditions required in changing from conventional to nanoflow conditions.

in Fig. 2) is due to a mass difference attributed to the absence of the N-terminal methionine residue in ~50% of the protein molecules. In addition, two separate charge state series can be observed for each population, one at higher *m/z* (labeled C, D in Fig. 2) and the second at lower *m/z* (labeled A, B in Fig. 2). The masses of the A, B charge states confirm the expected mass of the protein with and without an N-terminal methionine. By contrast, the mass of the C, D charge states shows noncovalent binding of dihydrofolate used to elute DHFR from an affinity column during its preparation. The two charge state series are interpreted in terms of ~20% of the DHFR preparation being relatively unfolded under these conditions (giving rise to series A, B), and 80% of the DHFR being folded and able to bind dihydrofolate (C, D). This is substantiated by the extended nature of the charge state distribution of series A, B, a feature commonly observed in ESI mass spectra of unfolded proteins.[25] The spectrum shown in Fig. 2b, recorded at 50° under the same instrument and solution conditions, confirms this interpretation.

The ESI mass spectrum of the GroEL–DHFR complex (Fig. 2c) shows that the charge state series corresponding to DHFR and the GroEL monomers are readily resolvable in the mass spectrum of the complex, and that under the mass spectrometry conditions used (see caption to Fig. 1) the complex has dissociated within the mass spectrometer. The peaks corresponding to DHFR that has been released from the complex appear at low *m/z*, suggesting that the conformation of GroEL-bound DHFR is at least partially denatured. By contrast, the charge state distribution of GroEL in isolation and in the complex are similar, suggesting, in accord with cryoelectron microscopy (cryo-EM) data,[26] that the conformation of the chaperonin does not change substantially when it binds its substrate protein.

Although changes in charge state distributions provide an indication of the conformation of proteins in different environments, their interpretation can be misleading because other factors can also have a profound effect on the appearance of the mass spectrum. For example, even though the acid molten globule state of $\alpha$-lactalbumin is known to be expanded relative to the native $Ca^{2+}$-bound form,[27] the former actually appears at a higher *m/z* (i.e., is less charged) than the latter.[7] This apparent anomaly arises because the high concentration of formate ions used to generate the acid molten globule state (5%, v/v) results in "charge stripping" of the protein.[28]

[25] M. Przybylski and M. O. Glocker, *Angew. Chem., Int. Ed. Engl.* **35,** 806 (1996).

[26] S. Chen, A. Roseman, A. Hunter, S. Wood, S. Burston, N. Ranson, A. Clarke, and H. Saibil, *Nature (London)* **371,** 261 (1994).

[27] K. Kuwajima, *FASEB J.* **10,** 102 (1996).

[28] U. A. Mirza and B. T. Chait, *Anal. Chem.* **66,** 2898 (1994).

This observation highlights the importance of ensuring identical solution and instrument conditions wherever possible when interpreting charge state distributions in structural terms.

*Hydrogen Exchange Experiments and Electrospray Ionization Mass Spectrometry*

The hydrogen exchange kinetics of [3-SS]BLA bound to GroEL are shown in Fig. 6.[7] These results show that deuterons in the GroEL-bound substrate protein are significantly more protected from hydrogen exchange than predicted for a random coil, but substantially less protected than observed for native, $Ca^{2+}$-bound, BLA. Indeed, the rate of hydrogen exchange in GroEL-bound [3-SS]BLA resembles closely that observed for the weakly protected, but well-defined, molten globule state of BLA,[27] suggesting that like the latter, at least some regions of the substrate protein are persistently structured in the GroEL-bound protein. We have demonstrated a similar result for GroEL-bound DHFR[8] (although this substrate protein is significantly more protected than GroEL-bound [3-SS]BLA). A similar approach has been used to detect hydrogen exchange protection in a complex of GroEL and $\beta$-lactamase.[10] The advantage of the ESI-MS approach over traditional NMR methods of measuring hydrogen exchange, lies in the ability to detect small numbers of weakly protected sites (as a

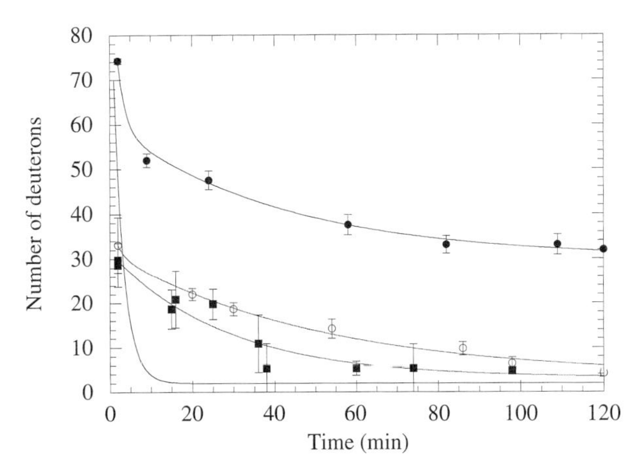


FIG. 6. Kinetic profiles of hydrogen exchange at pH 5.0 and 4° monitored by ESI-MS of the calcium-bound native state of BLA (●), the free molten globule [3-SS]cam-BLA[32] (○), GroEL-bound [3-SS]BLA (■), and a simulation of the hydrogen exchange profile expected for completely unstructured BLA (line), calculated from the intrinsic rate of hydrogen exchange and taking into account near-neighbor inductive effects.[22] (Reproduced with permission from C. V. Robinson, M. Gross, S. J. Eyles, J. J. Ewbank, M. Mayhew, F. U. Hartl, C. M. Dobson, and S. E. Radford, *Nature* (*London*) **372,** 646 (1994).)

consequence of the high mass resolution of the technique) and to measure exchange in real time without the need first to dissociate the complex. The disadvantages of the method are that site-specific information about the location of the protected sites cannot be readily obtained (although possibilities exist by either proteolytic fragmentation before mass analysis[29] or by collisions within the mass spectrometer[18,30]), and that a contribution toward protection by binding to the chaperonin surface cannot be completely ruled out.

*Peak Width Analysis*

The peak width of a molecular ion in the ESI mass spectrum of a protein arises from the contribution of the natural abundance isotopes to the individual charge states, principally from 1% of $^{13}C$, and the resolution capabilities of the mass spectrometer. For small model proteins, under optimal resolution conditions, the peak width of a charge state at half-height compares closely with the value calculated theoretically assuming a binomial distribution of isotopes.[19] The peak shape of the charge states arising from a protein containing deuterium atoms is further broadened and information about the distribution of deuterons in the protein molecules can be obtained from its analysis. The analysis of peak widths for protein–chaperone complexes, although not obtained under optimal resolution conditions, still contains vital information about the population of states with different hydrogen exchange properties that are bound to GroEL. For example, the two peaks shown in Fig. 7 were recorded at an early time point (11 min) during exchange of GroEL-bound [3-SS]BLA and much later (92 min) in the exchange profile of GroEL-bound DHFR, both obtained under identical experimental conditions. The widths of the two peaks clearly demonstrate that the bound protein represents a distinct, weakly protected population of molecules, and that a wide range of species with differing extents of protection are not found within the GroEL central cavity. This is an important and unique facet of the ESI-MS approach, which can be used to distinguish between different mechanisms of hydrogen exchange and to resolve different conformational states of proteins populated simultaneously.[7,8,18]

[29] Z. Zhang and D. L. Smith, *Protein Sci.* **2,** 522 (1993).

[30] A. D. Miranker, G. H. Kruppa, C. V. Robinson, R. T. Aplin, and C. M. Dobson, *J. Am. Chem. Soc.* **118,** 7402 (1996).

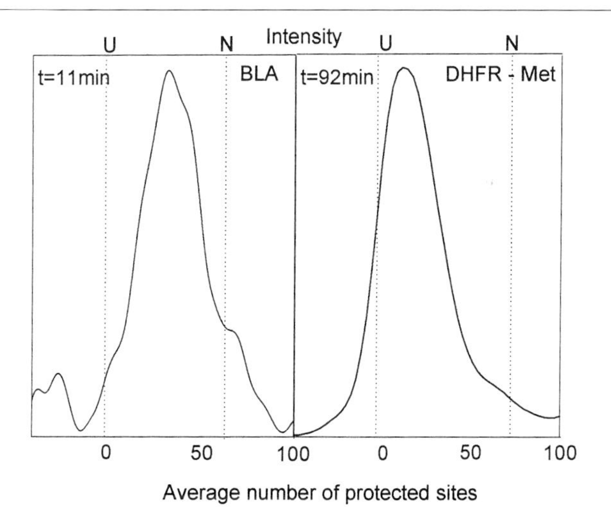


FIG. 7. Comparison of the peak widths of a single charge state of GroEL-bound [3-SS]BLA (+12) and DHFR (+16) after 11 and 92 min of isotope exchange, respectively. Both peak widths are superimposed on a scale of average number of protected sites and the positions of the native (N) and unfolded (U) states of both proteins are shown after the same exchange time and under the same experimental conditions. (Reproduced with permission from C. V. Robinson, M. Gross, S. J. Eyles, J. J. Ewbank, M. Mayhew, F. U. Hartl, C. M. Dobson, and S. E. Radford, *Nature* (*London*) **372,** 646 (1994) and M. Gross, C. V. Robinson, M. Mayhew, F. U. Hartl, and S. E. Radford, *Protein Sci.* **5,** 2506 (1996).)

## Interaction of GroEL with Nucleotides

The experiments presented have focused on the conformation of the substrate protein bound to GroEL. A natural extension of this methodology is to study the conformation of the chaperonin itself and conformational changes in both the chaperonin and substrate protein in the presence of GroES, and/or different nucleotides. Our current studies of GroEL have shown that instrumental conditions affect the extent of dissociation of the GroEL oligomer observed in the mass spectrum. Figure 5, for example, shows ESI mass spectra obtained under standard (denaturing) electrospray conditions, as well as under aqueous conditions using a standard electrospray interface, and under nanoflow conditions.[31] The data from standard electrospray under denaturing conditions (Fig. 5c) show that GroEL has dissociated within the mass spectrometer to its monomeric subunits. In aqueous solution using a standard electrospray conditions (Fig. 5b), small amounts of GroEL dimers are observed. By contrast, under the nanoflow conditions (Fig. 5a), significant amounts of GroEL dimers (labeled

[31] M. Wilm and M. Mann, *Int. J. Mass Spectrom. Ion Proc.* **136,** 167 (1994).

D, 114,394.1 ± 15 Da), and even trimeric species (labeled, T, 171,596.5 ± 28 Da) are clearly observed. An unequivocal assignment of these peaks can be made, because peaks arising from GroEL dimers carrying an odd number of charges lie between monomer charge states. Similarly, trimeric species are assigned as the small peaks between monomer and dimer charge states. These results demonstrate that the nanoflow technique preserves more of the quaternary structure of GroEL and this ability, together with the reduction in sample quantities and tolerance to buffer salts, ensure an important role for nanoflow in future studies of this nature.

The interaction of GroEL with GroES and ADP in the presence of an excess of $Mg^{2+}$ has also been studied by ESI-MS and preliminary spectra have been obtained (Fig. 8). The spectrum of this complex shows the presence of two charge state series, one (labeled A, Fig. 8) arising from GroES (mass, 10,386 Da) and the other from GroEL monomers (labeled B, Fig. 8). Significantly, the mass of the GroEL monomer increases under these conditions; the mass determined for the leading edge of the peak (57,606.8 ± 15 Da) is consistent with binding of one molecule of ADP

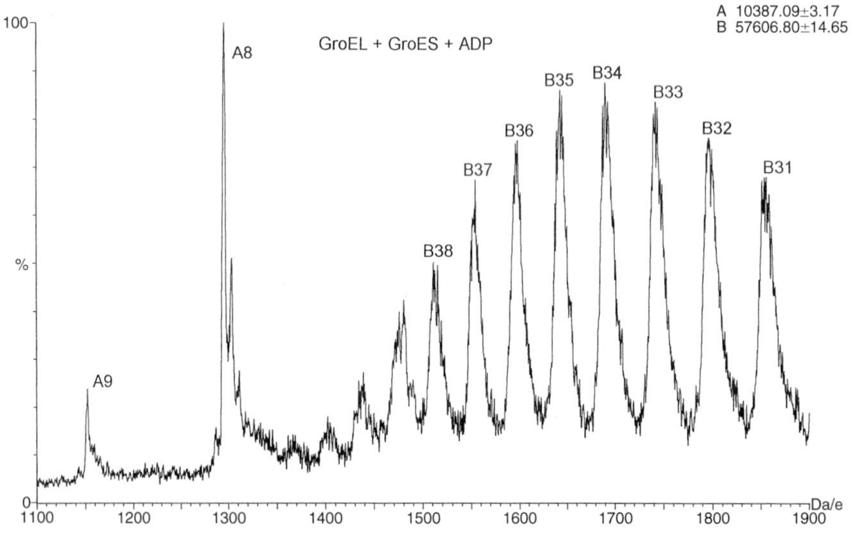


FIG. 8. ESI mass spectrum of the complex formed between GroEL and GroES in the presence of ADP/ATP and an excess of $Mg^{2+}$. Charge states arising from both GroES (labeled A) and GroEL (labeled B) are observed in the same spectrum. The broad nature of the GroEL peaks arises from binding of ADP and ATP as well as from differing amounts of $Mg^{2+}$ and phosphate. The mass spectrometry conditions were identical to those detailed in the caption to Fig. 1 for dissociating noncovalent complexes.

(an additional mass of, 427 Da) to each monomer. The charge states arising from the GroEL monomers are also significantly broader in this spectrum than observed in previous spectra of GroEL; this arises from binding of multiple copies of $Mg^{2+}$ in addition to the nucleotide. No $Mg^{2+}$ or ADP bound to the co-chaperonin under these conditions. This spectrum clearly demonstrates the potential to monitor a wide range of conformational states of GroEL by ESI-MS, even in the presence of magnesium ions and nucleotides, and hence opens the door to the analysis of an entire GroEL-assisted folding pathway by this technique. Moreover, the rapidly advancing capabilities of mass spectrometry, including nanoflow methods of sample introduction and the ability to fragment protein ions within the spectrometer, mean that ESI-MS will undoubtedly play an increasingly important role both in the study of chaperonins and in the analysis of macromolecular interactions in general.

## Acknowledgments

The work described here was carried out in the Oxford Centre for Molecular Sciences, which is funded by the BBSRC, EPSRC, and MRC. S.E.R. is currently at the University of Leeds. M.G. is a David Phillips Research Fellow (BBSRC). C.V.R. and S.E.R. acknowledge the Royal Society for support. We thank Chris Dobson and Ulrich Hartl for fruitful collaborations and discussions. We also acknowledge with thanks the contributions made to this work by Andrew Miranker and Robin Aplin. We thank Yashushi Kuwata for providing the GroEL sample used in Fig. 5a. Finally, we thank Robin Procter, and Brian Green (Micromass) for technical support and useful discussions.

# [25] Fluorescence Anisotropy Method for Investigation of GroEL–GroES Interaction

*By* Boris M. Gorovits and Paul M. Horowitz

## Introduction

The molecular chaperones GroEL and GroES have been shown to assist proteins in their folding and assembly *in vivo* and *in vitro*.[1] GroEL is a homotetradecamer ($GroEL_{14}$) of 57-kDa subunits arranged in two stacked heptameric rings.[2] The activity of GroEL is modulated by the

[1] R. E. J. Ellis and S. M. Van der Vies, *Annu. Rev. Biochem.* **60,** 321 (1991).
[2] K. Braig, Z. Otwinowski, R. Hegde, D. C. Boisvert, A. Joachimiak, A. L. Horwich, and P. B. Sigler, *Nature* (*London*) **371,** 578 (1994).



0076-6879/98 $25.00

heptameric co-chaperonin GroES ($GroES_7$). Mutagenesis studies have indicated that hydrophobic residues located in the apical domains of GroEL play important roles in the binding of both GroES and unfolded polypeptide.[3] This arrangement permits the possibility of 2:1 stoichiometry in the GroES–GroEL interaction. In fact, electron microscopy and cross-linking experiments have provided evidence for the existence of both symmetrical ($GroES_7$–$GroEL_{14}$–$GroES_7$) and asymmetrical ($GroES_7$–$GroEL_{14}$) complexes under various conditions.[4,5] Formation of the symmetrical $GroES_7$–$GroEL_{14}$–$GroES_7$ complexes was suggested to play an important role in the mechanism of the chaperonin-assisted refolding.[6,7] Therefore, the ability to investigate stoichiometry of the $GroEL_{14}$–$GroES_7$ interaction in solution under physiological conditions is important. Here we describe a direct method of detecting GroEL–GroES interaction in solution by using fluorescently labeled co-chaperonin GroES and applying a fluorescence anisotropy assay method.

## Principles of Method

Fluorescence anisotropy measurements have been widely used to detect change in the rotational diffusion properties of many proteins. This method has been applied to quantify protein–ligand and protein–protein association reactions. The investigated protein is initially labeled with a fluorophore with appropriate fluorescence lifetime. The sample is then excited with vertically polarized light. The value of anisotropy ($r$) is calculated by determining intensity of the horizontally and vertically polarized emission light by using Eq. (1):

$$r = \frac{(I_v - B_v) - G(I_h - B_h)}{(I_v - B_v) + 2G(I_h - B_h)} \tag{1}$$

where $I_v$, $I_h$, $B_v$, and $B_h$ are the intensities of the vertically and horizontally polarized emissions of sample ($I$) and blank ($B$) with vertically polarized excitation light. $G$ is a correction factor equal to $I_v/I_h$, where the excitation light is horizontally polarized.[8,9] The relationship between anisotropy and

[3] W. A. Fenton, Y. Kashi, K. Furtak, and A. L. Horwich, *Nature (London)* **371,** 614 (1994).

[4] H. Saibil, *in* "The Chaperonins" (R. J. Ellis, ed.), p. 245. Academic Press, San Diego, California, 1996.

[5] A. Azem, M. Kessel, and P. Goloubinoff, *Science* **265,** 653 (1994).

[6] A. Azem, S. Diamant, M. Kessel, C. Weiss, and P. Goloubinoff, *Proc. Natl. Acad. Sci. U.S.A.* **92,** 12021 (1995).

[7] Z. Torok, L. Vigh, and P. Goloubinoff, *J. Biol. Chem.* **271,** 16180 (1996).

[8] J. R. Lakowicz, "Principles of Fluorescence Spectroscopy." Plenum, New York, 1983.

[9] T. Azumi and S. P. McGlynn, *J. Chem. Phys.* **37,** 2413 (1962).

rotation diffusion of the particle carrying the fluorophore is described by the Perrin–Weber equation:

$$\frac{r_0}{r} = 1 + \frac{R_g T}{\eta V} \tau \qquad (2)$$

where $r$ is the measured anisotropy, $R_g$ is the gas constant, $\tau$ is the lifetime of the excited state, $\eta$ is the viscosity of the solution, and $V$ is the effective volume of the particle carrying the fluorescent probe. This law predicts a linear relationship between $1/r$ and the ratio $T/\eta$. The intercept of $1/r$ vs $T/\eta$ on the ordinate determines $r_o$, the anisotropy in the absence of molecular motion. This limiting anisotropy should, in principle, only be a function of the fluorophore.

Assuming that fluorophore-labeled ligand can exist in either the free (f) or the bound (b) form, the final value of the observed anisotropy can be described as

$$r = f_f r_f + f_b r_b \qquad (3)$$

where $r_f$ and $r_b$ refer to the anisotropies of the free and bound fluorophore-labeled compound, respectively; $r$ is observed anisotropy value; and $f_f$ and $f_b$ refer to the fraction of the total fluorescence that is generated by free and bound fluorophore-labeled compound, respectively ($f_f + f_b = 1$). Equation (3) is correct when the quantum yield of the fluorophore is not affected owing to the protein–protein interaction.

The $r_f$ and $r_b$ values can be easily measured by examining the fluorophore-labeled ligand in the absence of the protein and under conditions of complete binding (high protein concentration), respectively. Equation (3) can be rearranged to yield

$$f_b = \frac{r - r_f}{r_b - r_f} \qquad (4)$$

allowing one to analyze quantitatively the ligand–protein interaction. The concentration of bound $GroES_7$ is then calculated by Eq. (5):

$$[GroES_{7,bound}] = f_{bound}[GroES_7]_0 \qquad (5)$$

where $[GroES_{7,bound}]$ is the concentration of the bound fraction of $GroES_7$, and $[GroES_7]_0$ is the total concentration of $GroES_7$.

*Materials*

Reagents include succimidyl-1-pyrene butyrate (pyrene; purchased from Molecular Probes, Eugene, OR), nucleotides ADP and ATP (purchased from Sigma, St. Louis, MO), triethanolamine hydrochloride (TEA-HCl), Tris, HCl, $MgCl_2$, and KCl (all reagents of analytical grade). GroES and

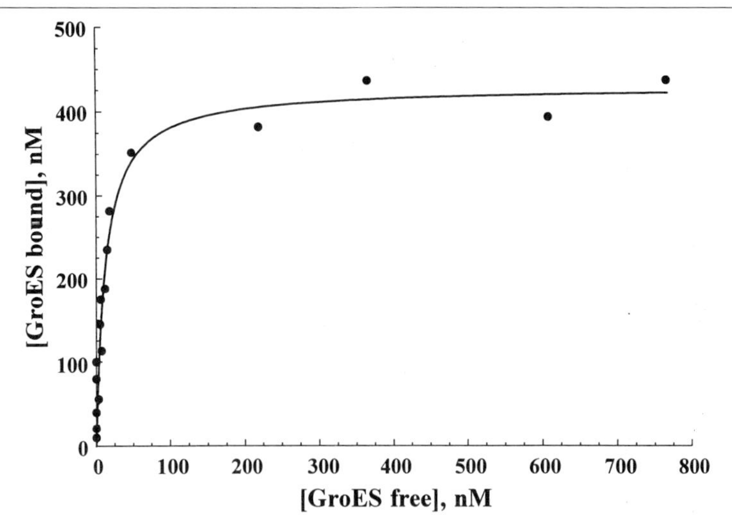


FIG. 1. Direct measurement of the binding of $GroES_7$ to $GroEL_{14}$, using anisotropy of the pyrene-labeled $GroES_7$ fluorescence. The anisotropy of the pyrene-labeled $GroES_7$ was measured as described in Methods. The $GroEL_{14}$ concentration was 200 n*M*. Measurements were made in buffer containing 10 m*M* KCl, 10 m*M* $MgCl_2$, and 1 m*M* ATP.

GroEL are purified as described previously.[10,11] Protein concentrations are determined by the method of Bradford.[12]

## Methods

### *GroES Labeling with Succimidyl-1-pyrene Butyrate*

For labeling of GroES with the pyrene derivative succimidyl-1-pyrene butyrate, GroES (3.2 mg/ml) is dissolved in 50 m*M* TEA-HCl, pH 7.8. To 500 $\mu$l of the protein solution, a 5-$\mu$l solution of succinimidyl-1-pyrene butyrate in dimethylformamide (2.62-mg/ml initial concentration) is added under vigorous mixing. This provides a 1.5 molar excess of succinimidyl-1-pyrene butyrate (68 $\mu M$ final concentration) relative to GroES oligomer concentration (45 $\mu M$). The solution is vortexed gently in a dark tube for 2 hr at room temperature. The reaction is stopped by addition of 55 $\mu$l of 1 *M* Tris-HCl, pH 7.8, providing an excess of Tris (final concentration of

[10] R. A. Staniforth, A. Cortes, S. G. Burston, T. Atkinson, J. J. Holbrook, and A. R. Clarke, *FEBS Lett.* **344,** 129 (1994).

[11] A. C. Clark, E. Hugo, and C. Frieden, *Biochemistry* **35,** 5893 (1996).

[12] M. M. Bradford, *Anal. Biochem.* **72,** 248 (1976).

100 m$M$) relative to the initial concentration of label. The solution is then dialyzed against 50 m$M$ Tris-HCl (pH 7.8), 10 m$M$ $MgCl_2$ overnight at 4° to remove the free label.

To quantify the amount of label that is covalently incorporated into the GroES the concentration of pyrene is measured by determining the absorbance of the sample at 340 nm and assuming the extinction coefficient[13] to be 43,000 $M^{-1}$ $cm^{-1}$. The protein concentration is measured as stated above. The final value is determined by the ratio of the molar concentration of pyrene label to the molar concentration of GroES oligomer.

*GroES Binding to GroEL*

Various concentrations of $GroES_7$ are incubated with 160–200 n$M$ $GroEL_{14}$ in 50 m$M$ Tris-HCl (pH 7.8), 10 m$M$ $MgCl_2$. The binding condition can be changed by including a low (0.5 m$M$) or high concentration (10 m$M$) of KCl, or different nucleotides (ADP, ATP). To prevent significant ATP degradation, the nucleotide should be added to the sample solution just before the measurement. The anisotropy is recorded within 1–3 min after addition of specified nucleotide (ATP, ADP, or AMP–PNP).

The anisotropy of the pyrene-labeled GroES in the presence of GroEL is recorded with excitation at 340 nm and the emission monitored at 395 nm.

To estimate the stoichiometry of the GroEL–GroES interaction under different conditions of binding, pyrene-labeled $GroES_7$ oligomer is titrated to the solution containing a fixed known amount of $GroEL_{14}$. The fraction of the $GroEL_{14}$–bound $GroES_7$ is then calculated as described above by using Eqs. (3)–(5). Anisotropy values for the free ($0.029 \pm 0.001$) and bound GroES ($0.072 \pm 0.001$) are measured in the absence of GroEL and in the presence of a fourfold excess of the chaperonin. Figure 1 shows the result of a titration in which the concentration of the bound GroES [bound $GroES_7$] is plotted versus concentration of the free GroES [free $GroES_7$] in the solution. Initial concentration of $GroEL_{14}$ is 200 n$M$. Using such a plot one can then estimate the stoichiometry of the interaction of the molecular chaperones under particular conditions. In this example $GroEL_{14}$ is able to bind two oligomeric $GroES_7$ in the presence of 1 m$M$ ATP.

[13] R. P. Haugland, "Handbook of Fluorescent Probes and Research Chemicals," 6th Ed. Molecular Probes, Eugene, Oregon, 1996.

# [26] Photoincorporation of Fluorescent Probe into GroEL: Defining Site of Interaction

By JEFFREY W. SEALE, BILL T. BRAZIL, and PAUL M. HOROWITZ

## Background

Evidence has highlighted the importance of hydrophobic interactions in the mechanism of substrate recognition by the chaperonin GroEL.[1-3] A thermodynamic model has been proposed that invokes hydrophobic partitioning of unfolded polypeptide substrates onto GroEL. This mode of partitioning has been confirmed for the substrate $\beta$-lactamase.[4] This idea has evolved from the notion that a common intermediate in the folding pathway of proteins is the "molten globule."[5] The molten globule is characterized by the presence of near-native secondary structure, but the lack of many tertiary interactions that determine the native state. In fact, molten globules are often identified by their ability to bind hydrophobic probes. Our laboratory has made extensive use of the hydrophobic probe 1,1-bis(4-anilino)naphthalene-5,5′-disulfonic acid (bisANS) to study the modulation of hydrophobic surfaces in GroEL, with the goal of identifying sites of interaction and conformational changes associated with the binding of chaperonin substrates.[6-8]

BisANS was first described by Rosen and Weber.[9] In an aqueous environment, the fluorescence of bisANS is quenched and has a maximum at approximately 550 nm. On transfer to a hydrophobic environment, the fluorescence of the probe increases sharply and the maximum is shifted to approximately 500 nm. These properties make the binding of bisANS an attractive method for studying the modulation of hydrophobic surfaces in proteins. It has been shown, however, that bisANS can influence its own

[1] B. T. Brazil, J. L. Cleland, R. S. McDowell, N. J. Skelton, K. Paris, and P. M. Horowitz, *J. Biol. Chem.* **272,** 5105 (1997).

[2] Z. Lin, F. P. Schwartz, and E. Eisenstein, *J. Biol. Chem.* **270,** 1011 (1995).

[3] R. Zahn, S. E. Axmann, K.-F. Rücknagel, E. Jaeger, A. A. Laminet, and A. Plückthun, *J. Mol. Biol.* **242,** 150 (1994).

[4] R. Zahn and A. Plückthun, *J. Mol. Biol.* **242,** 165 (1994).

[5] O. B. Ptitsyn, R. H. Pain, G. V. Semisotnov, E. Zerovnik, and O. I. Razgulyaev, *FEBS Lett.* **262,** 20 (1990).

[6] B. M. Gorovits, C. S. Raman, and P. M. Horowitz, *J. Biol. Chem.* **270,** 2061 (1995).

[7] D. L. Gibbons and P. M. Horowitz, *J. Biol. Chem.* **270,** 7335 (1995).

[8] P. M. Horowitz, S. Hua, and D. L. Gibbons, *J. Biol. Chem.* **270,** 1535 (1995).

[9] C.-G. Rosen and G. Weber, *Biochemistry* **8,** 3915 (1969).



0076-6879/98 $25.00

binding by shifting the equilibrium between the bisANS-free and bisANS-bound forms of the protein.[10] The major disadvantage of using bisANS until now has been the noncovalent nature of the interaction between the probe and the protein surface. We have described a method for the covalent incorporation of bisANS into GroEL with subsequent identification of the site of interaction.[11] A solution of protein is illuminated with ultraviolet (UV) light in the presence of bisANS. The bisANS adduct is still fluorescent and remains responsive to its environment.[11] The protein is subsequently subjected to proteolysis and the site of bisANS binding can be determined by peptide purification and sequencing. Although we developed the method specifically for GroEL, we have also shown that it is applicable to any protein that can bind bisANS.[11] In this chapter we describe general conditions for the photoincorporation of bisANS into proteins and the subsequent identification of the peptide fragments.

## Methods

### *General Strategy*

1. Identify the conditions for bisANS binding.
2. Using the conditions for bisANS binding, illuminate with UV light (254 nm).
3. Establish the optimal time for labeling by performing a time course for labeling.
4. Confirm photoincorporation by sodium dodecyl sulfate-polyacrylamide gel electrophoresis (SDS–PAGE) of the reaction products. Fluorescent proteins can be visualized by illumination of the gel on a UV light box.
5. Treat labeled protein with the protease of choice to generate peptide fragments.
6. Separate the peptide fragments using a suitable method, such as SDS–PAGE, high-performance liquid chromatography (HPLC), and fast protein liquid chromatography (FPLC). Labeled peptides will exhibit fluorescence at 500 nm when excited at a wavelength of 397 nm.
7. Determine the sequence/identity of the labeled peptide(s) by sequencing methods or mass spectrometry.

This general method is expanded in the following sections for the photolabeling of GroEL in the presence of $Mg^{2+}$.

[10] L. Shi, D. R. Palleros, and A. L. Fink, *Biochemistry* **33,** 7536 (1994).

[11] J. W. Seale, J. L. Martinez, and P. M. Horowitz, *Biochemistry* **34,** 7443 (1995).

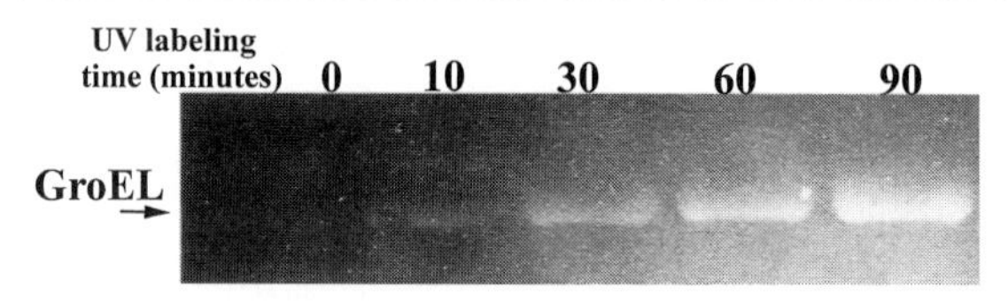


FIG. 1. Time course of photoincorporation of bisANS into GroEL in the presence of $Mg^{2+}$. BisANS was photoincorporated into GroEL as described in text. GroEL is indicated by the arrow. The time of incorporation (in minutes) is noted above each lane. After photolabeling, the protein was subjected to SDS–PAGE and the fluorescent protein visualized using a UV-light box. Free bisANS runs near the dye front (not shown).

*Conditions for bisANS Binding to GroEL.* The first step in locating the bisANS-binding site through photoincorporation involves identifying conditions under which bisANS binds to the protein of interest. In the case of GroEL, it has been shown that changes in the quaternary structure of the protein facilitate the binding of bisANS.[12] Building on this information, we then identified other factors that facilitated the binding of bisANS to GroEL. These conditions include the addition of polyvalent cations such as spermidine,[8] acidic pH,[7] substrate peptide binding,[1] or increased hydrostatic pressure.[6] Here, we describe the conditions for photoincorporation of bisANS into GroEL in the presence of $Mg^{2+}$.

*Photolabeling Conditions.* In a total volume of 500 $\mu$l,

Tris-HCl (pH 7.8), 50 m$M$
BisANS, 100 $\mu M$
GroEL (promoter), 20 $\mu M$
$MgCl_2$, 10 m$M$

Aliquots (50 $\mu$l) of the labeling mixture are added to a round-bottom well of a microtiter plate and placed on ice. A hand-held UV lamp (model UVS-11 Mineralight lamp, 115 V, 60 Hz, 0.16 A, 254 nm; Ultra-violet Products, San Gabriel, CA) is placed approximately 1 cm above the samples and the illumination is carried out for up to 90 min (see the next section). It is important to note that we have been unable to achieve photoincorporation of bisANS into GroEL under acidic conditions, where bisANS has been shown to bind.[7] Because bisANS does bind under these conditions, the acidic pH must interfere with the mechanism of photoincorporation.

*Time Course of Photolabeling.* To ensure that the maximum photoincorporation of bisANS has been attained, a time course of the photolabeling reaction should be performed. Figure 1 shows the results of such a time course for the labeling of GroEL in the presence of $Mg^{2+}$. For this experiment, aliquots of the labeling reaction are removed and analyzed by SDS–

[12] J. A. Mendoza, B. Demeler, and P. M. Horowitz, *J. Biol. Chem.* **269,** 2447 (1994).

PAGE. After boiling in the presence of SDS, all of the noncovalently bound bisANS is removed and is indicated by a bright, fluorescent band running near the dye front on the gel (data not shown). As shown in Fig. 1, maximum bisANS photoincorporation occurs in the presence of $Mg^{2+}$ after approximately 90 min.

*Identification of bisANS-Binding Site.* Several methods can be used to identify the binding site of bisANS. We have routinely used limited proteolytic digestion of GroEL with chymotrypsin to follow conformational changes in GroEL, and thus have a peptide map available. These results have been previously described in detail.[11] GroEL is resistant to proteolysis by chymotrypsin in the native state, and thus the proteolysis is routinely performed in the presence of 2.5 *M* urea. GroEL (1 mg/ml) is incubated at room temperature with 0.1% (w/w) chymotrypsin at various times. Proteolysis is stopped by the addition of phenylmethylsulfonyl fluoride (PMSF) to a final concentration of 2.9 m*M* on ice for 10 min. The digestion products are then analyzed by SDS–PAGE using 12% (w/w) acrylamide gels. Peptide fragments are identified by viewing the gel on a UV light box (Chromatovue transilluminator model TM-40; Ultraviolet Products).

For identification of the fluorescent peptides, electroblotting and peptide sequencing are employed. The peptides from the gels described above are electrophoretically transferred to an Immobilon-P membrane (Millipore Corp., Bedford, MA), using a semidry technique. After transfer, the fluorescent peptides are once again visualized on a UV light box and are marked with a pencil. The membrane is then stained with 0.1% Amido Black-10B in 10% (v/v) methanol and 2% (v/v) acetic acid until the peptides are just visible. Peptides that correspond to fluorescent bands are excised, washed with 20% (v/v) methanol, and air dried. The peptides are then subjected to approximately 12 rounds of sequencing using an Applied Biosystems (Foster City, CA) 477A protein sequence analyzer. The sequences of the peptides are compared with the published sequence of GroEL to identify the bisANS-binding site. Using this method, we have determined that the site of covalent bisANS photoincorporation lies within a peptide comprising residues F204–I249. This peptide lies within the apical domain of GroEL,[13] which has been implicated in the binding of substrate polypeptides as well as GroES.[14]

*Applications of Photolabeling.* The ability to link bisANS covalently to proteins makes the fluorescent probe much more useful for the study of protein structure and dynamics. We have been able to incorporate bisANS

[13] K. Braig, Z. Otwinowski, R. Hedge, D. C. Boisvert, A. Joachimiak, A. L. Horwich, and P. B. Sigler, *Nature* (*London*) **371,** 578 (1994).

[14] W. A. Fenton, Y. Kashi, K. Furtak, and A. L. Horwich, *Nature* (*London*) **371,** 614 (1994).

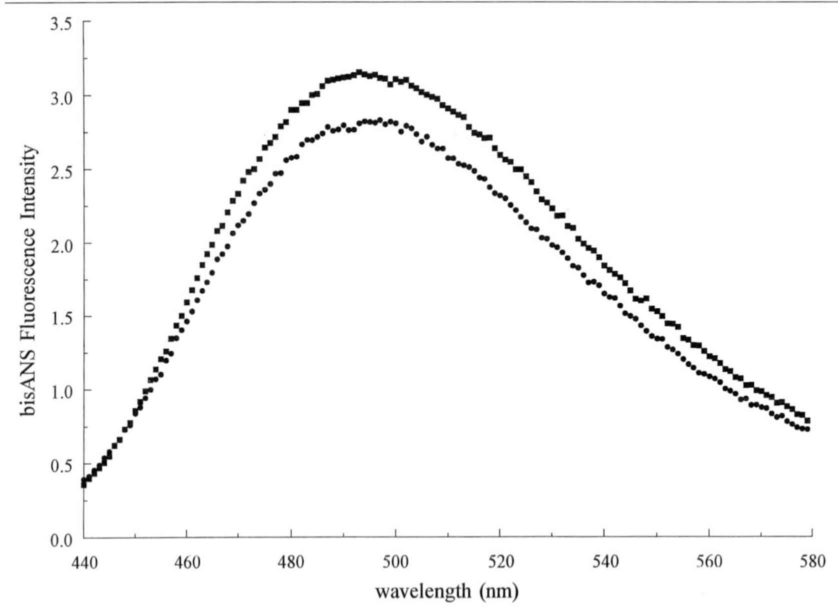


FIG. 2. Response of bisANS–GroEL to GroES binding. GroEL was labeled with bisANS in the presence of $Mg^{2+}$ as described in text. Unincorporated bisANS was removed from bis-GroEL using a PD-10 column. Samples of bis-GroEL were mixed with GroES in the presence of ADP. The fluorescence spectrum was recorded with an excitation wavelength of 394 nm. Solid squares represent bis-GroEL in the absence of any ligand. Solid circles represent bis-GroEL complexed with GroES in the presence of ADP.

into dissimilar proteins such as bovine serum albumin, rhodanese, ribonuclease A, lactate dehydrogenase, and GroES,[11] indicating that the phenomenon is not confined to GroEL. One such benefit from photoincorporation of bisANS, as described here, is the mapping of hydrophobic surfaces in proteins. We have been able to show that bisANS can bind in a region that has been implicated in binding to substrate polypeptides through hydrophobic interactions. This result supports the hypothesis that substrate proteins bind in the apical domain, as well as the idea that hydrophobic interactions are involved in the recognition of substrates.

Photoincorporation of bisANS also allows for the introduction of a fluorescent probe to an identifiable region of the protein. We now know that bisANS can be photoincorporated into GroEL under a number of different conditions, with the label being incorporated at the same site under all conditions.[11] We have made use of this observation to study conformational changes in the apical domain of GroEL using bisANS as a reporter. GroEL in which bisANS has been incorporated under native conditions remains active as judged by its ability to assist the refolding of rhodanese (data not shown). Figure 2 shows the response of photoincorporated bisANS to the binding of GroES to GroEL. GroEL was labeled with bisANS in the presence of $Mg^{2+}$ as described above. All of the unincorporated bisANS was removed by passing the labeling mixture through a PD10 column. GroES was then added to a solution of bis-GroEL in the presence of ADP. It has been shown that the binding of GroES to GroEL is associated with a movement in the apical domain of GroEL.[15] In response to this conformational change, the fluorescence of bisANS is quenched. The physical nature of this change is now being investigated.

## Summary

We have elucidated conditions for the covalent incorporation of a nonspecific hydrophobic probe, bisANS, into various proteins. Using this method, we are able to map hydrophobic surfaces in proteins. In addition, we have shown that for GroEL, we are able to use the fluorescence of the incorporated bisANS to monitor conformational changes in a defined region of the protein in response to various effectors. This method should be useful for studying both protein structure and dynamics.

## Acknowledgments

This research was supported by research Grants GM 25177 and ES 05729 from the NIH and research Grant AQ 723 from the Robert A. Welch Foundation.

[15] A. M. Roseman, S. Chen, H. White, K. Braig, and H. R. Saibil, *Cell* **87,** 241 (1996).

# [27] Analysis of Chaperone Function Using Citrate Synthase as Nonnative Substrate Protein

*By* JOHANNES BUCHNER, HOLGER GRALLERT, and URSULA JAKOB

## Introduction

Molecular chaperones are characterized by their remarkable ability to recognize selectively and bind nonnative proteins, which are often aggregation sensitive.[1] They suppress unspecific side reactions and promote the

[1] R. I. Morimoto, A. Tissières, and C. Georgopoulos, "The Biology of Heat Shock Proteins and Molecular Chaperones." Cold Spring Harbor Press, Cold Spring Harbor, New York, 1994.



0076-6879/98 $25.00

(re)folding of proteins to their native state under nonpermissive conditions. Various chaperones differ not only in their substrate specificity but also in their molecular mechanisms.[2] To study the function of specific chaperones, a number of different substrate proteins have been used that differ in their quaternary structure, their rate of folding, and their tendency to undergo irreversible side reactions during (un)folding.[3] We describe here the use of citrate synthase (CS) as a chaperone substrate. Citrate synthase is well suited, because unfolding and refolding can be easily monitored and both the thermal unfolding as well as the folding pathway of chemically denatured CS are known in some detail. Furthermore, the influence of the four major classes of molecular chaperones [small heat shock proteins (Hsps), GroEL, Hsp70, and Hsp90] on CS folding has been characterized.[2,4-6] The use of CS as a substrate protein for molecular chaperones allows the researcher to address whether chaperones suppress aggregation, undergo stable or transient interactions with the substrate protein, interact with early- or late-unfolding intermediates, act in an ATP-dependent manner, and resemble a member of one of the major classes of molecular chaperones.

## Substrate Protein Citrate Synthase

Citrate synthase (EC 4.1.3.7) is a commercially available, dimeric, mitochondrial protein, composed of two identical subunits (48.969 kDa each). The nuclear-encoded $\alpha$-helical protein is imported into mitochondria posttranslationally. The three-dimensional structure is known at a resolution of 2.7 Å.[7] In *Escherichia coli*, CS was identified as an *in vivo* substrate of GroE[8]; CS catalyzes the first step of the citric acid cycle, the condensation of oxaloacetic acid and acetyl-CoA to citrate and coenzyme A.[9] In the absence of any substrates, CS is readily inactivated on incubation at higher temperatures with a midpoint of transition at 48°.[10-12] The loss of activity

[2] J. Buchner, *FASEB J.* **10,** 10 (1996).

[3] R. Jaenicke, *Curr. Opin. Struct. Biol.* **3,** 104 (1993).

[4] J. Buchner, M. Schmidt, M. Fuchs, R. Jaenicke, R. Rudolph, F. X. Schmid, and T. Kiefhaber, *Biochemistry* **30,** 1586 (1991).

[5] U. Jakob, H. Lilie, I. Meyer, and J. Buchner, *J. Biol. Chem.* **270,** 7288 (1995).

[6] M. Ehrnsperger, S. Gräber, M. Gaestel, and J. Buchner, *EMBO J.* **16,** 221 (1997).

[7] S. Remington, G. Wiegand, and R. Huber, *J. Mol. Biol.* **158,** 111 (1982).

[8] A. L. Horwich, L. Brocks Low, W. A. Fenton, I. N. Hirshfeld, and K. Furtak, *Cell* **74,** 909 (1993).

[9] P. A. Srere, H. Brazil, and L. Gonen, *Acta Chem. Scand.* **17,** 129 (1963).

[10] O. Wieland, L. Weiss, and I. Eger-Neufeldt, *Biochem. Z.* **339,** 501 (1964).

[11] P. A. Srere, *J. Biol. Chem.* **241,** 2157 (1966).

[12] W. Zhi, P. Srere, and C. T. Evans, *Biochemistry* **30,** 9281 (1991).

$$\mathrm{N} \rightleftarrows \mathrm{I_1} \rightleftarrows \mathrm{I_2} \rightleftarrows \mathrm{I_n} \rightarrow \rightarrow \mathrm{A}$$

FIG. 1. Thermal unfolding pathway of CS at 43°. N, Native state; $I_1$ and $I_2$, unfolding intermediates; A, aggregates. Note that the reaction from N to $I_1$ and the reaction from $I_1$ to $I_2$ are reversible, whereas the subsequent formation of aggregates is an irreversible reaction.

is preceded by major structural changes in CS. Citrate synthase can be effectively stabilized by the addition of the substrates oxaloacetic acid and/or acetyl-CoA under thermal unfolding conditions.[10,11] This "induced fit" mechanism of substrate binding to CS has been analyzed in detail using crystal structures of CS in the absence and presence of the substrates.[7,13,14] The substrates seem to lock CS in an active conformation and shift the midpoint of the thermal transition to 66.5°.[12]

## Influence of Molecular Chaperones on the Thermal Unfolding Pathway of Citrate Synthase

The thermal unfolding of CS can be induced *in vitro* at temperatures similar to heat shock temperatures *in vivo*. At 43°, the enzyme is readily inactivated and subsequently aggregates. The analysis of the thermal unfolding pathway of CS has revealed the presence of at least two dimeric unfolding intermediates, $I_1$ and $I_2$, which are in equilibrium with the native state (Fig. 1).[5] The presence of these two unfolding intermediates became apparent when stabilization experiments with the substrate oxaloacetic acid were performed during the inactivation reaction. The substrate not only stabilizes the still native species,[10,11] but also allows the refolding of two intermediates to the native, active state (see Ref. 5 and below). Depending on the individual chaperones and their molecular mechanisms, changes can be observed in the (1) inactivation kinetics, (2) reactivation kinetics, and (3) yield of reactivated species and (4) aggregation kinetics of thermally unfolding CS.

### *Influence of Molecular Chaperones on Thermal Aggregation of Citrate Synthase*

Like many other thermally unfolding proteins, CS shows the tendency to aggregate on incubation at higher temperatures.[5] This is based on unspecific hydrophobic interactions between unfolding intermediates and results in the formation of high molecular weight particles. These particles render the protein solution turbid because they scatter light, a process that can

[13] G. Wiegand, S. Remington, J. Deisenhofer, and R. Huber, *J. Mol. Biol.* **174,** 205 (1984).
[14] M. Karpusas, B. Branchaud, and S. Remington, *Biochemistry* **29,** 2213 (1990).

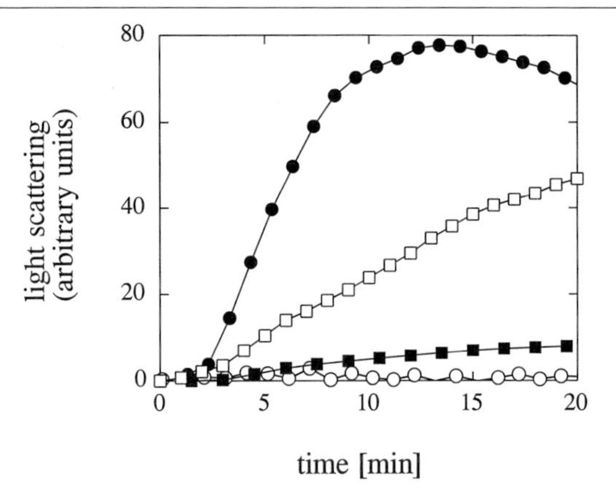


FIG. 2. Influence of different chaperones on the aggregation of CS. Light-scattering measurements of CS (0.15 $\mu M$ monomer) were performed in 40 m$M$ HEPES–KOH, pH 7.5, at 43° in the absence of chaperones (●), and in the presence of Hsp25 (□) (0.03 $\mu M$), Hsp90 (■) (0.6 $\mu M$), or (○) GroEL (0.15 $\mu M$).

be monitored using spectrometry. Molecular chaperones are able to suppress this aggregation process effectively.[4] Chaperones recognize and subsequently bind these aggregation-sensitive unfolding intermediates and therefore keep the local concentration of intermediates free in solution low.[15] This results in a significantly decelerated aggregation process. The function of molecular chaperones in suppressing aggregation is therefore easily detected by the suppression of light scattering of nonnative proteins. This method represents a specific but rather qualitative assay for molecular chaperones.

Within 2 min of incubation of CS at 43°, light scattering can be detected. It increases in an exponential manner to reach a maximum after approximately 12 min (Fig. 2). During the course of the reaction both the number and size of the aggregates increase, therefore the kinetics cannot be analyzed quantitatively. Depending on the concentration of the chaperone present in the incubation reaction, the light-scattering signal will be reduced. Controls with equal volumes of buffer lacking the respective chaperone must be included, because the buffer composition is known to affect aggregation processes.[16] Furthermore, control reactions with nonchaperone proteins [e.g., immunoglobulin G (IgG), bovine serum albumin (BSA), and lysozyme] and known chaperones should be performed.

[15] T. Kiefhaber, R. Rudolph, H. H. Kohler, and J. Buchner, *Bio/Technology* **9,** 825 (1993).

[16] R. Rudolph, G. Böhm, H. Lilie, and R. Jaenicke, *in* "Folding Proteins." Protein Function—A Practical Approach (T. E. Creighton, ed.), 2nd Ed. Oxford University Press, New York, pp. 57–100 (1997).

Depending on the functional mechanism of the individual chaperone, aggregation can be affected differently.

1. The presence of stoichiometric amounts of GroEL leads to a complete suppression of aggregation because GroEL forms apparently stable complexes with the unfolding intermediates (Fig. 2). Also, the addition of GroES and ATP does not lead to an increase in the light-scattering signal (H. Grallert and J. Buchner, unpublished data, 1997).
2. In the presence of Hsp90, light scattering is largely reduced (Fig. 2). A slow and linear increase in the light-scattering signal is observed in Fig. 2, based on the more transient interaction of Hsp90 with CS compared with GroEL.[5] After the release of the substrate proteins, only part of the folding intermediates is able to rebind, while the remaining part is susceptible to aggregation. At any given time, however, the concentration of folding intermediates free in solution is low and aggregation is therefore significantly reduced.
3. Hsp70 is able to suppress the aggregation of thermally unfolding CS. In contrast to Hsp90, where no influence of ATP can be detected, the action of Hsp70 is strongly influenced by ATP. Addition of ATP after formation of the complex results in a sudden release of the aggregation-prone folding intermediates, which is reflected by the instant increase in light-scattering signal (data not shown).
4. In the presence of substoichiometric amounts of small Hsps, the light-scattering signal of CS is also largely reduced (Fig. 2). In the presence of stoichiometric amounts of small Hsps, no increase in the light-scattering signal can be observed in the time frame of the measurements, suggesting the formation of a stable complex between small Hsps and unfolding intermediates of CS.

*Methods*

1. *Instrument setting:* A fluorescence spectrophotometer equipped with a stirrable, thermostatted cell holder should be used to monitor light scattering. Excitation and emission wavelengths are both set to 500 nm (to detect smaller aggregates, 360 nm may be used) and both emission and excitation slits are set to 2 nm. Recording data points every 0.2 to 0.5 sec is recommended.

2. *Preparation of citrate synthase:* An ammonium sulfate suspension of CS from porcine heart mitochondria (Boehringer Mannheim, Indianapolis, IN) is dialyzed against TE buffer (50 m*M* Tris, 2 m*M* EDTA, pH 8.0) overnight at 4° and concentrated to about 17 mg/ml using Centricon-30 microconcentrators (Amicon, Danvers, MA). To remove precipitated proteins the solution is then centrifuged at 14,000 rpm in a microcentrifuge

for 30 min at 4° and the exact protein concentration of the supernatant is determined, using the published extinction coefficient of 1.78 for a 1-mg/ml solution in a 1-cm quartz cuvette.[17] Aliquots are prepared and stored at −20°. Thawed aliquots of CS must not be refrozen. For each set of experiments, an aliquot of CS is supplemented with TE buffer to yield a final concentration of 30 $\mu M$ CS (monomer). This sample is kept on ice.

3. *Preparation of molecular chaperones:* The concentration of the individual chaperones should be on the order of at least 2 mg/ml to prevent significant changes in the incubation temperature after addition of the protein. Ideally, the molecular chaperones should be dialyzed against 40 m$M$ HEPES–KOH, pH 7.5, before the measurements. If this is not possible due to stability problems of the respective proteins, buffer controls are needed. The same amount of buffer without protein is added to the incubation reaction to account for influences of various buffer components on the aggregation of CS (e.g., glycerol, ammonium sulfate).

4. *Incubation reaction:* The incubation buffer (40 m$M$ HEPES–KOH, pH 7.5) is preincubated at 43° and pipetted into the thermostatted quartz cuvette. Molecular chaperones and/or other components are then added. Potential changes in light scattering due to the chaperone protein are monitored for 5 min before CS is diluted 1:200 into the incubation reaction (0.15 $\mu M$ monomer) under stirring and light scattering is monitored for at least 20 min.

*Comments*

1. Filtration of the incubation buffer before the measurements reduces the background signal of the buffer. Buffer substances other than *N*-2-hydroxyethylpiperazine-*N'*-2-ethanesulfonic acid (HEPES) can be used. In this case the effect of the buffer on the stability of CS and the respective chaperone must be thoroughly tested.

2. Continuous temperature control in the cuvette is required to monitor the sample temperature. Even small changes in the incubation temperature may result in major kinetic changes.

3. The light-scattering signal should not change significantly as a result of the addition of the chaperone proteins, and it should remain constant for at least 5 min before adding CS. Spinning for 30 min at 48,000 *g* at 4° is recommended to remove precipitated proteins. A continuous increase in the light-scattering signal in the presence of the chaperone alone indicates that the protein unfolds and is therefore not suited for measurements at elevated temperatures.

[17] S. M. West, S. M. Kelly, and N. C. Price, *Biochim. Biophys. Acta* **1037,** 332 (1990).

4. The presence of trace amounts of proteases in chaperone preparations can result in an apparent chaperone-like effect on aggregation. To test for proteases, aliquots of 20 $\mu$l are withdrawn at different time points during the reaction, supplemented with 5 $\mu$l of reducing 5× Laemmli buffer, and boiled for 5 min at 95°. The samples are analyzed using 15% (w/v) sodium dodecyl sulfate-polyacrylamide gel electrophoresis (SDS–PAGE), and CS is detected using silver staining. In the absence of proteases, the same amount of CS should be present at all time points tested.

5. The labeling for changes in the light-scattering signal should be given in arbitrary units and not as percent aggregation, because there is no linear correlation.

### *Influence of Molecular Chaperones on Thermal Inactivation of Citrate Synthase*

Citrate synthase readily inactivates on incubation at 43°. This inactivation process can be monitored by measuring the remaining activity of the enzyme during the incubation at elevated temperatures. Within seconds after start of the incubation at 43°, the activity decreases and after 15 min no active CS can be detected (Fig. 3). The kinetics of inactivation follow an exponential curve, which can be fitted with a first-order reaction ($k_{app}$ = ~9 × $10^{-3}$ $sec^{-10}$).[5]

The following influences of molecular chaperones on CS inactivation can be observed.

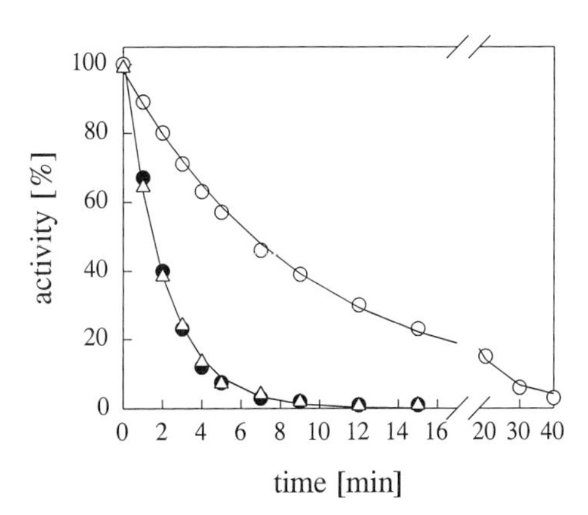


Fig. 3. Inactivation kinetics of CS at 43° in the presence of Hsp90 and Hsp25. Inactivation of CS (0.15 $\mu M$ monomer) was measured in 40 m$M$ HEPES–KOH, pH 7.5, in the absence (●) of chaperones, and in the presence of 0.3 $\mu M$ Hsp25 (△), or 0.6 $\mu M$ Hsp90 (○).

1. Stoichiometric concentrations of GroEL do not influence the inactivation process. At higher concentrations, however, it increases the rate constant of inactivation, owing to the high association constant of GroEL (H. Grallert and J. Bucher, unpublished data, 1997). Inactive folding intermediates are bound rapidly by GroEL and are no longer in equilibrium with the native state. It therefore seems that GroEL inactivates CS, a phenomenon that is also observed in the case of various other substrate proteins with which GroEL is stably associating.[18,19]
2. Hsp90 apparently "stabilizes" CS by slowing the rate constant of thermal inactivation by up to a factor of 10 (Fig. 3). This "thermoprotective" effect is due to the transient interaction between Hsp90 and CS conformers, which are native-like or able to refold during the activity assay.[5] This is similar to the observed interaction of GroEL with thermally unfolding rhodanese.[20]
3. Hsp70 does not influence the inactivation process of CS (U. Jakob and J. Buchner, unpublished data, 1994), independent of the Hsp70 concentration used and also independent of the presence of ATP in the reaction. This observation, together with the unchanged reactivation rate of CS in the presence of Hsp70, suggests that Hsp70 does not interact stably with early unfolding intermediates of CS, which are still highly structured and dimeric and therefore not suited to serve as substrates for Hsp70.
4. Small heat shock proteins, like Hsp25, show no influence on the inactivation kinetics of CS (Fig. 3). However, in contrast to Hsp70, they increase the yield of reactivatable CS intermediates and decelerate the reactivation rate.[6] Thus they seem to form apparent stable chaperone–substrate complexes with early CS unfolding intermediates during inactivation.

*Methods*

1. *Preparation of CS and chaperones:* CS and chaperones are prepared as described above, except that stock solutions of 15 $\mu M$ CS (monomer) are used for the experiments.

2. *Activity assay:* CS activity is determined according to Srere *et al.*[9] The activity assay is based on the first step of the citric acid cycle, in which

[18] A. A. Laminet, T. Ziegelhoffer, C. Georgopoulos, and A. Plückthun, *EMBO J.* **9,** 2315 (1990).
[19] P. V. Viitanen, G. K. Donaldson, G. H. Lorimer, T. H. Lubben, and A. A. Gatenby, *Biochemistry* **30,** 9716 (1991).
[20] J. A. Mendoza, G. H. Lorimer, and P. M. Horowitz, *J. Biol. Chem.* **267,** 17631 (1992).

CS catalyzes the condensation of oxaloacetic acid (OAA) and acetyl-CoA to citrate and coenzyme A. The side product coenzyme A reduces stoichiometrically the Ellman's reagent dithio-1,4-nitrobenzoic acid (DTNB), a reaction that goes along with an increase in absorption at 412 nm.[9]

For each activity assay, the reaction mixture consists of 930 $\mu$l of TE buffer (50 m*M* Tris, 2 m*M* EDTA, pH 8.0), 10 $\mu$l of 10 m*M* oxaloacetic acid (in 50 m*M* Tris, pH not adjusted), 10 $\mu$l of 10 m*M* DTNB (in TE buffer), and 30 $\mu$l of 5 m*M* acetyl-CoA (in TE buffer). This reaction mixture, which can be prepared in advance for at least 20 test assays (when used within 1 hr of preparation), is preincubated at 25°. Before the activity test, 980 $\mu$l of the reaction mixture is pipetted into 1.5-ml plastic cuvettes and the reaction is started by the addition of 20 $\mu$l of 0.15 $\mu M$ CS (monomer). The activity assay is performed at 25° and the change in absorbance is recorded online. The linear slope of the initial increase in absorption is used to calculate the specific activity of CS and represents the amount of active CS present in the incubation reaction. The specific activity can be calculated as follows:

$$\text{Specific activity (U/mg)} = \Delta E/\text{min} \times V/(\varepsilon d v c)$$

where $V$ is the test volume (ml), $\varepsilon$ is the molar extinction coefficient of DTNB (13,600 $M^{-1}$ cm$^{-1}$), $d$ is the pathlength of the cell (cm), $v$ is the sample volume (ml), and $c$ is the enzyme concentration (mg/ml). The specific activity of native CS from pig heart is 150 U/mg.

3. *Thermal inactivation:* CS (15 $\mu M$) is diluted 100-fold with stirring into 40 m*M* HEPES–KOH, pH 7.5 (final volume, 300 $\mu$l), which is preincubated at 25° in a 2-ml Eppendorf tube equipped with a small stirring bar. The activity of CS is determined and set to 100%. The inactivation reaction is started by placing the test tube in a 43° water bath. Within 30 sec, the temperature is adjusted and an aliquot is withdrawn to determine the remaining CS activity. At least eight more aliquots are taken during the time course of inactivation and tested. Chaperones are added before CS at 25°.

### *Influence of Molecular Chaperones on Reactivation of Thermally Inactivated Citrate Synthase*

Addition of oxaloacetic acid (OAA) at any given time during the thermal inactivation of CS leads to the reactivation of a certain amount of folding intermediates.[5] The number of molecules reactivated depends on the time of addition of OAA and peaks at about 6 min after start of the

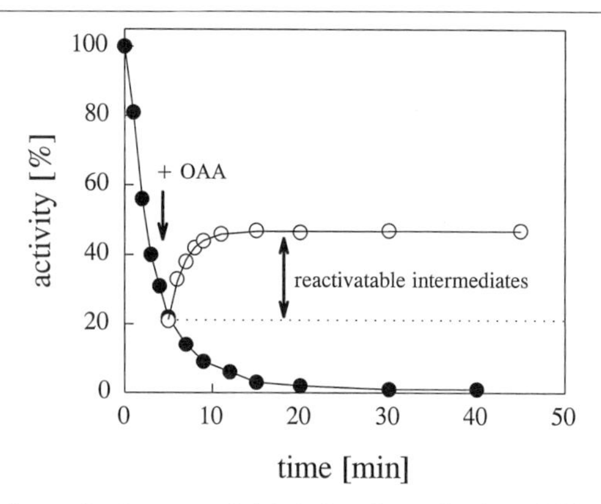


FIG. 4. Inactivation and subsequent OAA-induced reactivation of CS at 43°. The inactivation of CS (0.15 $\mu M$ monomer) was performed in 40 m$M$ HEPES–KOH, pH 7.5. After 5 min of incubation at 43° (as indicated by the arrow), either inactivation was continued (●) or reactivation was initiated by adding 1 m$M$ OAA to the inactivation reaction (○). The double-headed arrow indicates the amount of intermediates present at the time point of OAA addition. The dotted line represents the amount of native CS stabilized by OAA.

inactivation with about 20% reactivated species (Figs. 4 and 5A). Later in the time course of inactivation, the number of folding intermediates, which are in equilibrium with the native state, declines and no reversible folding intermediates are detectable about 15 min after the start of the inactivation. This is due to aggregation processes. The reactivation reaction of CS at 43° can be described by two first-order reactions with a fast phase ($k_1 = 50 \times 10^{-3}$ sec$^{-1}$) and a slow phase ($k_2 = 4.2 \times 10^{-3}$ sec$^{-1}$) corresponding to intermediate $I_1$, which is in fast equilibrium with the native state, and $I_2$, which is in a slow equilibrium with intermediate $I_1$ (Fig. 1). The intermediates still seem to be dimeric because the kinetics of reactivation are independent of the CS concentration used. Reactivation can also be induced by a temperature shift to 25°. Depending on the chaperone present during the reactivation, the following effects have been observed.

1. No reactivation of CS is observed in the presence of GroEL. This is due to the stable interaction of GroEL with unfolding intermediates. GroES and ATP are needed for efficient reactivation. Under reducing conditions, GroEL is able to increase the yield of reactivated CS to about 80%. (H. Grallert and J. Buchner, unpublished observations, 1997).
2. Reactivation experiments in the presence of Hsp90 lead to significantly changed rates of refolding. Both rates are dramatically slowed

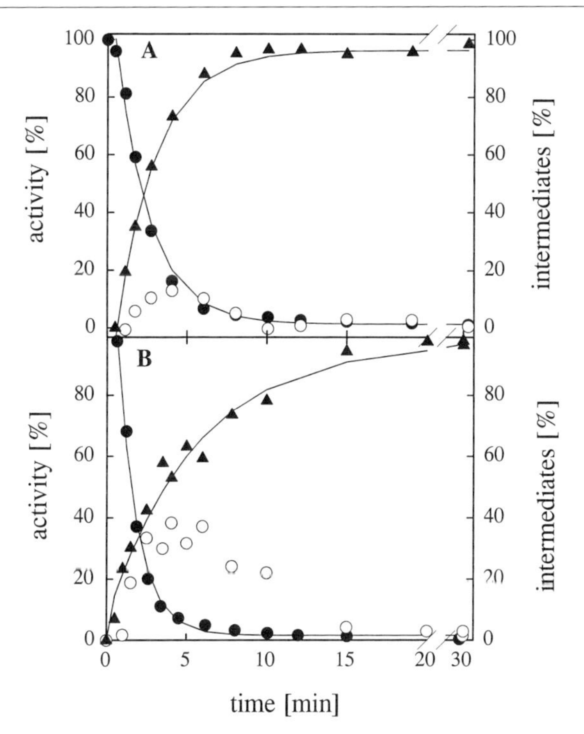


FIG. 5. Dissection of the unfolding pathway of CS. CS (0.15 $\mu M$ monomer) was incubated at 43°. Inactivation (●) was monitored by activity measurements at the time points indicated. In a subsequent experiment, reactivation was started with OAA (1 m$M$) at the same time points. The amount of intermediate (○) present at a given time point was calculated by subtracting the activity at the beginning of reactivation from the activity 60 min after addition of OAA (compare with Fig. 4). The amount of irreversibly denatured CS (▲) represents the difference between the native protein at the beginning of the reaction and the sum of native CS and intermediates at the given time point. Formation of unfolding intermediates in the absence of chaperone (A) and in the presence of 0.3 $\mu M$ Hsp25 (B). The maximum deviation of the data points was <5%.

in the presence of Hsp90, suggesting that Hsp90 is able to interact transiently with both $I_1$ and $I_2$. The yield of reactivated species is only slightly increased, but folding intermediates are detectable for longer periods of time in the presence of Hsp90 compared with spontaneous inactivation.[5]

3. Hsp25 traps up to 40% of inactivated CS as refoldable intermediates (Fig. 5). For an effective release of the bound intermediates from Hsp25 in the absence of OAA a temperature shift to 25° and the addition of the molecular chaperone Hsp70 and ATP are required.[6]

*Methods*

1. *Preparation of CS, chaperones, and activity assay:* See above.

2. *Reactivation reaction:* After the start of the inactivation reaction at 43°, the remaining activity is monitored up to the time point at which refolding of the folding intermediates should be induced. Then, OAA (0.1 *M* in 50 m*M* Tris, pH 8.0) is diluted 1 : 100 into the incubation reaction and incubation is continued at 43°. At various time points after start of the refolding reactions, aliquots (20 $\mu$l) are removed and assayed for CS activity (compare with Fig. 4). Owing to the presence of both folding intermediates and the distinct rate constants of the individual folding reactions, a parallel reactivation mechanism can be applied to distinguish between intermediates $I_1$ and $I_2$. Measurements in the presence of molecular chaperones are performed as described above. Alternatively, reactivation can be induced by a temperature shift to 25°.

*Comments*

1. All reagents for the activity assay should be freshly prepared and kept on ice.

2. Oxaloacetic acid is very acidic and the stock solution should be titrated with NaOH to pH 8.0. It is also possible to titrate the pH of OAA with reagents other than NaOH to pH 8.0.

3. The amount of intermediates present at a given time can be calculated by subtracting the activity at the beginning of reactivation from the activity 60 min after addition of OAA. The amount of irreversibly denatured CS is the difference between the native protein at the beginning of the reaction and the sum of native CS and intermediates at the given time point.

## Influence of Molecular Chaperones on Renaturation of Chemically Denatured Citrate Synthase

The term *chemically denatured* is used here to set unfolding in the presence of denaturants apart from the effects exerted by elevated temperatures. Citrate synthase monomerizes and unfolds readily on incubation in 6 *M* guanidinium chloride. Early refolding experiments revealed that only a small fraction of molecules reaches the native state due to immediate aggregation of refolding CS molecules.[4] Optimizing the refolding conditions (low CS concentration, low refolding temperatures), however, leads to substantially higher refolding yields, a necessary prerequisite for studying the folding pathway of CS (U. Jakob, H. Grallert, and J. Buchner, unpublished data, 1995).

$$2U \longrightarrow 2M \longrightarrow D$$

FIG. 6. Folding pathway of chemically denatured CS. U, The unfolded state; M, the folded, association-competent monomer; D, the active dimer. Folding on the dimer level cannot be excluded.

The folding pathway of CS has been analyzed using activity assays, chemical cross-linking, and size-exclusion chromatography (Fig. 6). The refolding of CS follows a Uni–Bi molecular folding reaction, with a rate-determining folding step on the monomer level followed by a concentration-dependent second-order association reaction. Simulation of the reactivation kinetics gives a rate constant for the folding reaction ($k_1$) of $1.3 \times 10^{-3}$ $sec^{-1}$ and a rate constant for the association reaction of $5.0 \times 10^3 M^{-1} sec^{-1}$ (U. Jakob and J. Buchner, unpublished data, 1994). Folding reactions on the dimer level cannot be excluded.

### *Influence of Molecular Chaperones on Aggregation of Chemically Denatured Citrate Synthase*

The effect of chaperones on the folding of CS has been studied by monitoring the influence of the chaperones on the aggregation of renaturating CS. Most of the CS folding intermediates aggregate within the first 6 min after initiation of the renaturation process at 25°. This reaction can be monitored directly by light-scattering measurements.

Similar to the thermal aggregation of CS, different molecular chaperones show different influences on this aggregation process (Fig. 7).

1. Stoichiometric amounts of GroEL lead to a complete suppression of the light-scattering signal. Thus GroEL seems to bind early folding intermediates of CS and forms stable substrate–chaperone complexes.[4] Addition of GroES and ATP results in a slight increase in light scattering over time due to aggregation of part of the released CS molecules from GroEL.
2. Hsp90 suppresses significantly the aggregation of renaturating CS. However, excess amounts of Hsp90 must be present to observe this effect. In contrast to the situation in the presence of GroEL, where the light-scattering signal does not increase over the time course of the measurement, aggregation occurs in the presence of Hsp90. This is probably due to the transient interaction of Hsp90 with refolding CS molecules, which on release from Hsp90 can either rebind to Hsp90 or aggregate.[21]

[21] H. Wiech, J. Buchner, R. Zimmermann, and U. Jakob, *Nature (London)* **358,** 169 (1992).

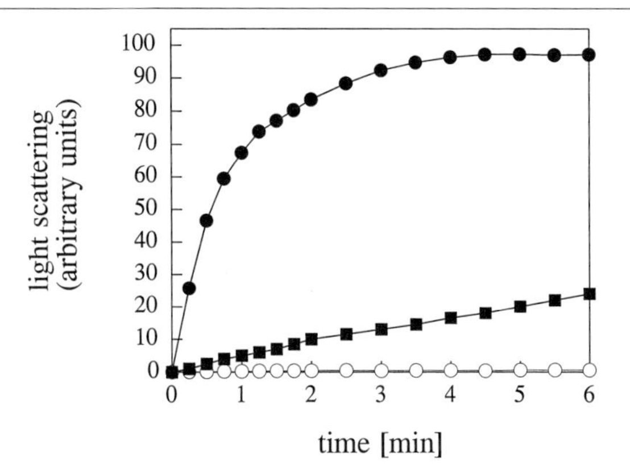


FIG. 7. Influence of GroEL and Hsp90 on the aggregation of chemically denatured CS at 25°. Denatured CS was diluted to a final concentration of 0.15 $\mu M$ (monomer) into 50 m$M$ Tris-HCl, pH 8.0. Light scattering was monitored in the presence of 0.15 $\mu M$ GroEL (○), in the presence of 0.90 $\mu M$ Hsp90 (■), and in the absence of chaperones (●).

3. Small heat shock proteins, like Hsp25, show no effect on the aggregation of chemically denatured CS even with a high excess of oligomeric small Hsp (M. Ehrnsperger and J. Buchner, unpublished data, 1997). Compared with the strong influence of small Hsps on the thermal aggregation of CS, this implies that small Hsps may form stable complexes with early unfolding intermediates only.

*Influence of Molecular Chaperones on Reactivation of Chemically Denatured Citrate Synthase*

Reactivation of guanidinium-denatured CS is strongly influenced by the concentration of the nonnative protein[4] and the folding conditions (e.g., temperature and additives[12,22]). Accordingly, reactivation yields for the spontaneous folding reaction vary from 0 to 60%. As mentioned above, reactivation of chemically denatured CS involves a folding step preceding the concentration-dependent association reaction. However, intermediates in this folding pathway have not been characterized yet.

Differences in the folding kinetics and the yield of reactivated CS can be observed depending on the chaperone used.

[22] M. Schmidt, J. Buchner, M. Todd, G. H. Lorimer, and P. Viitanen, *J. Biol. Chem.* **269,** 10304 (1994).

1. GroEL forms apparently stable complexes with CS folding intermediates. Therefore, no reactivation is observed in the absence of either ATP or the co-chaperone GroES. Under permissive folding conditions for the spontaneous folding of CS, the presence of ATP alone is sufficient for reactivation of CS from complexes with GroEL. In this case, the folding kinetics are slower than in the absence of the chaperone owing to rebinding to GroEL. In addition, a slightly increased yield of reactivated molecules compared with the folding in the absence of chaperones can be observed.[22]

   Under more nonpermissive conditions, the presence of both GroES and ATP is required to allow efficient folding. Reactivation occurs in the time range of 120 min and with a final yield of about 80% at 25° (Fig. 8).
2. Hsp90 decelerates the folding kinetics of CS with increasing chaperone concentrations. This again can be explained by the transient binding of Hsp90 to CS. With increasing concentrations of Hsp90, rebinding of folding intermediates to Hsp90 is faster than the refolding of CS, which is reflected by the decreased overall rate constant of reactivation. The yield of reactivated CS molecules is increased slightly to about 40%.[21]
3. Small heat shock proteins such as Hsp25 and Hsp26 do not show any influence on the folding kinetics or the yield of reactivation of chemically denatured CS (M. Ehrnsperger, S. Walke, and J. Buchner, unpublished data, 1997).

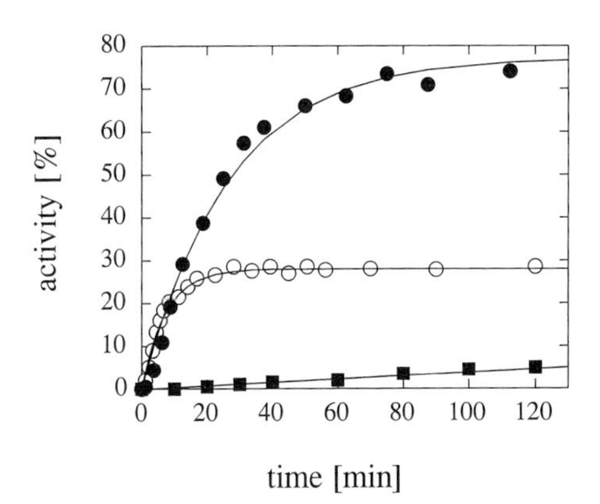


FIG. 8. Reactivation of chemically denatured CS in the presence or absence of GroE at 25°. Denatured CS was diluted to a final concentration of 0.15 $\mu M$ (monomer) into 50 m$M$ Tris-HCl, pH 8.0, 5 m$M$ $NgCl_2$, 10 m$M$ KCl. Activity assays were performed in the presence of 0.15 $\mu M$ GroEL and 2 m$M$ ATP (■), in the presence of 0.15 $\mu M$ GroEL, 0.3 $\mu M$ GroES, and 2 m$M$ ATP (●), or in the presence of 2 m$M$ ATP (○).

*Methods*

1. *Preparation of chaperones and activity assay:* See above.

2. *Preparation of chemically denatured CS:* For each set of experiments, an aliquot of CS is denatured in 6 *M* guanidinium chloride (in 50 m*M* Tris-HCl, pH 8.0) and 20 m*M* dithioerythritol (DTE). The final concentration of denatured CS (monomer) should be 30 $\mu M$ for light-scattering measurements and 15 $\mu M$ for reactivation reactions. The samples are incubated in the presence of denaturants for at least 2 hr at room temperature to allow complete unfolding. Denatured samples are kept on ice.

3. *Renaturation of CS:* For light-scattering measurements chemically denatured CS is diluted 1:200 into refolding buffer (50 m*M* Tris-HCl, pH 8.0, or 40 m*M* HEPES, pH 7.5) at 25° under vigorous stirring. The parameter settings correspond to the ones used for monitoring the light scattering of thermally aggregating CS (see previous section on light scattering).

Reactivation of denatured CS is initiated by diluting CS 100-fold into renaturation buffer (50 m*M* Tris-HCl, pH 8.0, or 40 m*M* HEPES, pH 7.5) under vigorous stirring. The renaturation sample is kept at 25°. At the time points indicated, aliquots (20 $\mu$l) are with drawn and the activity is determined using the test assay described previously in this chapter. The activity of 0.15 $\mu M$ native CS (monomer) is set to 100%.

*Comments*

1. At lower temperatures visible aggregation is significantly reduced. Below 15° light scattering can no longer be observed under the conditions used.

2. The residual guanidinium concentration in the folding reaction should be kept as low as possible to avoid interference with folding and the interaction between nonnative protein and chaperone. Routinely, 60 m*M* guanidinium is present after the dilution step. Using the appropriate protein concentrations in the denaturation step, the guanidinium concentration in the refolding reaction can be further reduced to 10–20 m*M*.

3. Mutants of CS that differ in their binding affinity to GroEL in the nonnative state can be employed to study the interaction of different chaperones.[23]

4. For the reactivation experiments rapid mixing of CS on dilution in the refolding buffer is required. Otherwise high local concentrations of aggregation-sensitive folding intermediates will lead to rapid and quantitative aggregation.

[23] R. Zahn, P. Lindner, S. E. Axmann, and A. Plückthun, *FEBS Lett.* **380,** 152 (1996).

# [28] Purification and Characterization of Small Heat Shock Proteins

*By* JOHANNES BUCHNER, MONIKA EHRNSPERGER, MATTHIAS GAESTEL, and STEFAN WALKE

## Introduction

Small heat shock proteins (sHsps) form an abundant and ubiquitous family of stress proteins that have been found in all organisms studied so far. Characteristically, sHsps range in monomer size from 15 to 30 kDa but form oligomeric complexes of 200–800 kDa,[1-4] comprising 9 to about 50 subunits. The number of sHsps members varies in different organisms. While plants express several classes in the cytosol as well as in organelles,[5] mammals possess only one cytosolic member in addition to the lens protein $\alpha$-crystallin, which has been reported to show functional and sequence homology to sHsps.[6-8] The structurally and functionally most thoroughly investigated mammalian sHsps are human Hsp27 and its murine homolog Hsp25. *Saccharomyces cerevisiae* expresses two members of the sHsp family, Hsp42 and Hsp26.[9,10] Although Hsp42 and Hsp26 are strongly induced on heat shock, double-deletion mutants or deletion of either protein does not lead to an altered phenotype in stress response.[9-11] In *Escherichia coli* proteins related to the sHsp family (IbpA and IbpB) have been found in association with insolubilized recombinant protein,[12] indicating that the Ibps might be involved in protein folding and prevention of protein aggregation.

In contrast to other Hsp families, sHsps show little over all primary sequence homology, with conserved sequences restricted mostly to the

[1] R. Chiesa, M. J. McDermott, and A. Spector. *FEBS Lett.* **268,** 222 (1990).
[2] J. Behlke, G. Lutsch, M. Gaestel, and H. Bielka, *FEBS Lett.* **288,** 119 (1991).
[3] N. J. Bentley, I. T. Fitch, and M. F. Tuite, *Yeast* **8,** 95 (1992).
[4] P. J. Groenen, K. B. Merck, W. W. de Jong, and H. Bloemendal, *Eur. J. Biochem.* **225,** 1 (1994).
[5] R. Waters, G. J. Lee, and E. Vierling, *J. Exp. Bot.* **47,** 325 (1996).
[6] T. D. Ingolia and E. A. Craig, *Proc. Natl. Acad. Sci. U.S.A.* **79,** 2360 (1982).
[7] J. Horwitz, *Proc. Natl. Acad. Sci. U.S.A.* **89,** 10449 (1992).
[8] U. Jakob, M. Gaestel, K. Engel, and J. Buchner, *J. Biol. Chem.* **268,** 1517 (1993).
[9] D. Wotton, K. Freeman, and D. Shore, *J. Biol. Chem.* **271,** 2717 (1996).
[10] R. E. Susek and S. Lindquist, *Mol. Cell. Biol.* **9,** 5625 (1989).
[11] L. Petko and S. Lindquist, *Cell* **45,** 885 (1986).
[12] S. P. Allen, J. O. Polazzi, J. K. Gierse, and A. M. Easton, *J. Bacteriol.* **174,** 6938 (1992).



0076-6879/98 $25.00

C-terminal half of the protein.[13–16] Apart from these homologies, sHsps are grouped together on the basis of their small monomeric molecular weight, their ability to form large oligomers, and their heat inducibility.

Small Hsps have been found in varying concentrations in all organisms and tissues investigated so far, even under physiological conditions, where they can amount to about 0.1% of total cellular protein.[17–19] In response to elevated temperatures, the expression of sHsps is strongly increased.[20–22] In addition, artificial overexpression of sHsps on stress has repeatedly been correlated with the acquisition of thermotolerance,[23–26] indicating a general thermoprotective role of the proteins. These protective properties are probably mediated by the ability of sHsps to support proper folding and refolding of nonnative cellular protein. Experiments *in vitro* have demonstrated that sHsps can function as molecular chaperones by preventing irreversible aggregation of other proteins and increasing the yield of renaturation after heat or chemical denaturation.[7,8,27,28] In contrast to other classes of chaperones this protective function is not associated with ATP binding or hydrolysis.[29] Furthermore, phosphorylation, which is a common feature of mammalian sHsps, seems to lead to changes in the oligomeric structure of the proteins but does not interfere with their *in vitro* chaperone activity.[30] The

[13] G. Wistow, *FEBS Lett.* **181,** 1 (1985).

[14] N. Plesofsky-Vig, J. Vig, and R. Brambl, *J. Mol. Evol.* **35,** 537 (1992).

[15] W. W. de Jong, J. A. Leeunissen, and C. E. Voortner, *Mol. Biol. Evol.* **10,** 103 (1993).

[16] U. Jakob and J. Buchner, *Trends Biochem. Sci.* **19,** 205 (1994).

[17] P. Bhat and C. N. Nagineni, *Biochem. Biophys. Res. Commun.* **158,** 319 (1989).

[18] K. Kato, H. Shinohara, N. Kurobe, Y. Inaguma, K. Shimizu, and K. Ohshima, *Biochim. Biophys. Acta* **1074,** 173 (1991).

[19] R. Ciocca, S. Oesterreich, G. C. Chamness, W. L. McGuire, and S. A. Fuqua, *J. Natl. Cancer Inst.* **85,** 1558 (1993).

[20] S. Oesterreich, R. Benndorf, and H. Bielka, *Biomed. Biochim. Acta* **49,** 219 (1990).

[21] Y. Inaguma, H. Shinohara, S. Goto, and K. Kato, *Biochem. Biophys. Res. Commun.* **182,** 844 (1992).

[22] A.-P. Arrigo and J. Landry, *in* "The Biology of Heat Shock Proteins and Molecular Chaperones" (R. I. Morimoto, ed.), p. 335. Cold Spring Harbor Laboratory Press, Plainview, New York, 1994.

[23] J. Landry, P. Chrétien, H. Lambert, E. Hickey, and L. A. Weber, *J. Cell Biol.* **109,** 7 (1989).

[24] E. Rollet, J. N. Lavoie, J. Landry, and R. M. Tanguay, *Biochem. Biophys. Res. Commun.* **185,** 116 (1992).

[25] J. N. Lavoie, G. Gingras-Breton, R. M. Tanguay, and J. Landry, *J. Biol. Chem.* **268,** 3420 (1993).

[26] P. R. L. A. van den Jissel, P. Overcamp, U. Knauf, M. Gaestel, and W. W. de Jong, *FEBS Lett.* **355,** 54 (1994).

[27] M. Ehrnsperger, S., Gräber, M. Gaestel, and J. Buchner, *EMBO J.* **16,** 221 (1997).

[28] G. J. Lee, A. M. Roseman, H. R. Saibil, and E. Vierling, *EMBO J.* **16,** 659 (1997).

[29] R. Jaenicke and T. E. Creighton, *Curr. Biol.* **3,** 234 (1993).

[30] U. Knauf, U. Jakob, K. Engel, J. Buchner, and M. Gaestel, *EMBO J.* **13,** 54 (1994).

interaction of sHsps with several nonnative proteins has been described as stable and long-lived.[28,31–33] It was even suggested that sHsp-bound substrate might be destined for degradation.[32] However, complexes formed between sHsps and unfolded substrate have been reported to be able to dissociate in the presence of at least one cytosolic factor, which was suggested to be Hsp70.[27,28] This discovery has led to the notion that under stress conditions sHsps might create a reservoir of nonnative protein bound to the chaperones, which allows refolding in cooperation with ATP-dependent chaperones such as Hsp70 once physiological conditions are restored.

Bearing in mind that sHsps have been associated with several diseases and a variety of cellular functions such as signal transduction,[34] it remains to be seen how the chaperone functions of sHsps are used in these contexts.

## Purification of Inclusion Body Protein B from *Escherichia coli*

### *Methods*

The method described here allows the purification of approximately 2.5–5 mg of inclusion body protein B (IbpB) from 1 liter of *Escherichia coli* culture, with a purity of >95%. An average of 3 g of cells (wet weight) is obtained from 1 liter of bacterial culture. Estimation of the final purity is obtained from densitometric scanning of Coomassie-stained sodium dodecyl sulfate (SDS) gels. After each purification step fractions are pooled according to the appearance of the respective band on SDS-polyacrylamide gels[35]; if the band pattern is ambiguous Western blot analysis[36] is performed, using specific antibodies against the sHsp.

*Protein Expression.* For expression of the inclusion body protein B the plasmid pbf3-IbpB (a generous gift from R. Beck and H. Burtscher, Boehringer Mannheim, Penzberg, Germany) is used. This plasmid contains the *IbpB* gene cloned into the *Eco*RI and *Hin*dIII restriction sites of the multiple cloning site of the pbf3 plasmid. The gene is under the control of a *tacP* promoter sequence. Protein overexpression can thus be induced by addition of isopropyl-$\beta$-D-thiogalactopyranoside (IPTG). The plasmid also encodes chloramphenicol acetyltransferase, which mediates the chloram-

[31] P. V. Rao, J. Horwitz, and J. S. Zigler, Jr., *J. Biol. Chem.* **269,** 13266 (1994).

[32] G. J. Lee, N. Pokala, and E. Vierling, *J. Biol. Chem.* **270,** 10432 (1995).

[33] K. P. Das, J. M. Petrash, and W. K. Surewicz, *J. Biol. Chem.* **271,** 10449 (1996).

[34] M. Ehrnsperger, M. Gaestel, and J. Buchner, *in* "Structure and Function of Molecular Chaperones: The Role of Chaperones in the Life Cycle of Proteins" (A. Fink, ed.), p. 533 Marcel Decker, New York, 1997.

[35] U. K. Laemmli, *Nature (London)* **227,** 680 (1970).

[36] H. Towbin, T. Staehelin, and J. Gordon, *Proc. Natl. Acad. Sci. U.S.A.* **76,** 4350 (1979).

phenicol resistance used for selection. *Escherichia coli* strain JM83 has been transformed with the plasmid pbf3-IbpB by standard methods.[37]

For large-scale IbpB expression, Luria broth (LB) medium containing chloramphenicol (30 μg/ml) is inoculated with a 1 : 500 dilution of a stationary overnight culture. Cells are grown at 37° with vigorous shaking at 250 rpm. At an optical density of 0.8 at 595 nm, expression of IbpB is induced by addition of IPTG to a final concentration of 1 m*M*. Cells are harvested 4 hr after induction by centrifugation (2500 *g*, 5 min) and washed once with buffer A [40 m*M* *N*-2-hydroxyethylpiperazine-*N*′-2-ethanesulfonic acid (HEPES)–KOH (pH 7.5), 1 m*M* ethylenediaminetetraacetic acid (EDTA), 1 m*M* dithioerythritol (DTE)] plus 150 m*M* KCl.

*Purification.* Purification of recombinant IbpB is achieved in a three-step procedure, using DE52, a gel-filtration step, and an additional anion-exchange column step. All purification steps are performed at 4°.

1. The cell pellet of a 4-liter culture is resuspended in 50 ml of buffer A plus 150 m*M* KCl, including a protease inhibitor mix containing phenylmethylsulfonyl fluoride (PMSF, 1 m*M*), leupeptin (1 μ*M*), *p*-aminobenzoic acid (PABA, 2.5 m*M*), pepstatin (1 μ*M*) and 4-(2-aminoethyl)-benzenesulfonyl fluoride hydrochloride (Pefabloc, 1 m*M*).
2. Cells are lysed twice, using a French press operated at 18,000 psi. The crude extract is clarified by centrifugation (20,200 *g*, 45 min, 4°).
3. The supernatant is applied to a 50-ml DE52 column equilibrated with buffer A plus 150 m*M* KCl. After washing with buffer A plus 150 m*M* KCl, IbpB is developed with a 500-ml linear KCl gradient ranging from 150 to 500 m*M* in buffer A. IbpB elutes at about 200 to 300 m*M* KCl. IbpB-containing fractions are pooled and concentrated using an Amicon (Danvers, MA) cell with a YM30 membrane (30-kDa cutoff range).
4. The concentrated solution is applied to an S300 HR gel-filtration column (Pharmacia, Uppsala, Sweden) equilibrated in buffer A plus 200 m*M* KCl. IbpB-containing fractions elute around 60-ml elution volume. These fractions are pooled and dialyzed against buffer A plus 150 m*M* KCl.
5. The protein is further purified using a 6-ml Resource-Q anion-exchange column (Pharmacia) equilibrated with buffer A plus 150 m*M* KCl. The dialyzed protein solution is applied and the column washed with buffer A plus 150 m*M* KCl. IbpB elutes between 270 and 330 m*M* KCl in a linear KCl gradient from 150 to 500 m*M* in buffer A. The flow rate is 3 ml/min. The fractions are checked by SDS-polyacrylamide gel electrophoresis (SDS–PAGE) and pooled according to their purity.

[37] J. Sambrook, E. F. Fritsch, and T. Maniatis, *in* "Molecular Cloning" (C. Nolan, ed.), 2nd Ed. Cold Spring Harbor Laboratory Press, Cold Spring Harbor, New York, 1989.

6. The pooled fractions are dialyzed against buffer A plus 150 m*M* KCl, concentrated by ultrafiltration with an Amicon cell as described above to a final concentration of 0.5–0.7 mg/ml, and centrifuged for 45 min at 20,200 *g* at 4°. The supernatant is aliquoted, frozen in liquid nitrogen, and stored at −80°.

*Comments*

1. The most significant purification step is the gel-filtration column step. Because IbpB forms large oligomeric complexes typical for all members of the small heat shock family, most contaminating proteins can be removed in this purification step.

2. A serious problem when working with IbpB is its tendency to aggregate at relatively low protein concentrations. Thus the final concentration does not exceed 0.7 mg/ml to minimize loss of protein as a result of aggregation. The presence of at least 150 m*M* KCl seems to stabilize IbpB in solution and reduces the amount of aggregation during storage (P. Goloubinoff, personal communication, 1997). This is in contrast to the highly homologous protein IbpA, which aggregates under all buffer conditions tested. Furthermore, IbpB elutes from the chromatography columns in distinct peaks and does not smear over all the fractions as observed for IbpA.

3. Depending on the fractions pooled, the purified protein might contain nucleotide impurities. This can be circumvented by adding DNase I (final concentration, 10 $\mu$g/ml), $MgCl_2$ (10 m*M*), and $MnCl_2$ (1 m*M*) and incubating the solution for >30 min at room temperature before loading the crude extract onto the DE52 column.

4. The protein concentration is determined either by the Bradford protein determination assay[38] or by using a theoretical extinction coefficient of 0.86 at 280 nm for a 1-mg/ml protein solution at an optical path length of 1 cm. The protein concentration determined according to the absorbance at 280 nm is corrected for light scattering.[39]

## Purification of Hsp26 from *Saccharomyces cerevisiae*

### *Methods*

The purification method described here allows the overexpression and purification of the small heat shock protein Hsp26 from *S. cerevisiae*. A typical yield to be achieved is about 3–5 mg of purified protein from 1 liter of cell culture. One liter of cell culture amounts to an average of 4 g of

[38] M. M. Bradford, *Anal. Biochem.* **72,** 248 (1976).

[39] A. F. Winder and W. L. Gent, *Biopolymers* **10,** 1243 (1971).

cells (wet weight). The purity of the protein is estimated to be >95% by densitometric scanning. After each purification step fractions are pooled according to the appearance of the respective band on SDS–PAGE.[35] If the band pattern is ambiguous, Western blot analysis[36] using specific antibodies against the sHsp is performed.

*Protein Expression.* High-level expression of Hsp26 is achieved using the shuttle vector pJV517, which is a derivative of the $2\mu$ plasmid pRS464. The high copy number ensures good expression levels. The plasmid contains the *Ura3* gene for selection of yeast cells and the $\beta$-lactamase gene, which allows selection in *E. coli*. For protein expression, the Ura-yeast strain JT(DIP) GPD26(A) is used. The yeast strain and the plasmid are a generous gift from S. Lindquist (University of Chicago, Chicago, IL). Details on the strain and the plasmid have been described previously.[40] The Hsp26 gene is cloned into the *Bgl*II and *Pst*I restriction sites of the multiple cloning site. Four liters of yeast minimal medium without uracil[41] are inoculated with a 1:100 dilution of a stationary liquid culture of the strain JT(DIP) GPD26(A) transformed with the plasmid pJV517. Cells are grown at 30° to late logarithmic phase, to an optical density of about 0.8 at 595 nm, and harvested by centrifugation (2500 *g*, 5 min, 4°).

*Purification.* Purification of Hsp26 is accomplished in a three-step purification protocol, using two different anion exchangers and a gel-filtration column. During the purification all buffers and reaction vessels are precooled on ice and the purification steps are always carried out at 4°.

1. The cell pellet is washed with ice-cold buffer A [40 m*M* HEPES–KOH (pH 7.5), 1 m*M* DTE, 1 m*M* EDTA] plus 50 m*M* NaCl. The buffer contains a protease-inhibitor mix [1 $\mu M$ leupeptin, 2.5 m*M* *p*-aminobenzoic acid (PABA), 1 m*M* Pefabloc (Boehringer Mannheim, Mannheim, Germany), 1 $\mu M$ pepstatin] to avoid protease degradation during the purification procedure.
2. The cell pellet is resuspended 1:2 in buffer A plus 50 m*M* NaCl plus protease-inhibitor mix and cells are broken open with glass beads (0.5-mm diameter) in a MSK300 rotator (Braun Melsungen AG, Melsungen, Germany) for 3 min. For 15 g of cells (wet weight), 75 g of glass beads are used in a 100-ml glass vessel. The vessel is cooled every 10 sec with a short pulse of liquid $CO_2$. The crude extract is separated from the glass beads by filtration through a filter paper under slight vacuum conditions. Extensive foaming of the protein solution should be avoided. Glass beads are washed once with 10–20 ml of buffer A plus 50 m*M* NaCl.

[40] R. E. Susek and S. Lindquist, *Mol. Cell. Biol.* **10,** 6362 (1990).

[41] F. M. Ausubel (ed.), "Short Protocols in Molecular Biology," 3rd Ed., p. 13-3. John Wiley & Sons, New York, 1995.

3. The cell lysate is centrifuged (20,200 *g*, 45 min, 4°), fresh protease inhibitors are added, and the soluble extract is applied to a 50-ml DE52 ion-exchange chromatography column equilibrated in buffer A plus 50 m*M* NaCl. The column is washed with buffer A plus 50 m*M* NaCl to remove unspecifically bound protein. Hsp26 is developed with a 500-ml NaCl gradient in buffer A, ranging from 50 to 500 m*M* NaCl. Hsp26-containing fractions, which elute at 200–300 m*M* NaCl, are pooled and dialyzed against buffer A plus 50 m*M* NaCl.

4. The protein solution is loaded onto a 6-ml Resource-Q ion-exchange column (Pharmacia), equilibrated in buffer A plus 50 m*M* NaCl. The column is washed with buffer A plus 50 m*M* NaCl and a linear NaCl gradient (50–500 m*M*) is used for elution. The flow rate is 3 ml/min. Hsp26 elutes under these conditions in the range of 200–250 m*M* NaCl as determined by SDS–PAGE.

5. The Hsp26-containing fractions are pooled and further purified on a 90-ml Superdex S200 pg column (Pharmacia) equilibrated in buffer B [40 m*M* HEPES–KOH (pH 7.5), 1 m*M* DTE, 1 m*M* EDTA] plus 200 m*M* NaCl. The column is operated at a flow rate of 0.5 ml/min. Hsp26 elutes at about 60-ml buffer volume, indicating a large oligomeric species. Fractions containing pure Hsp26 are pooled, dialyzed against buffer A plus 50 m*M* NaCl, and concentrated to approximately 3 mg/ml by ultrafiltration using an Amicon (Danvers, MA) cell with a YM30 membrane and Centricon microconcentrators with 30-kDa cutoff (Amicon). The protein solution is centrifuged at 20,200 *g* at 4°, aliquoted, frozen in liquid nitrogen, and stored at −80°.

*Comments*

1. The major difficulty concerning the purification of Hsp26 is partial proteolysis as verified by Western blot analysis using a polyclonal antibody against Hsp26. Therefore, growth into stationary phase is avoided to reduce the expression of additional yeast proteases.[42] The purification protocol should be carried out as quickly as possible and fresh protease inhibitors should be added at each step if significant proteolytic degradation occurs. The partly degraded Hsp26 is not separated from the oligomeric Hsp26 complexes during gel-filtration chromatography, indicating that the truncated subunits are still incorporated in the Hsp26 complexes.

2. The Hsp26 concentration is determined using either the Bradford protein determination assay[38] or a theoretical extinction coefficient of 0.86 at 280 nm for a 1-mg/ml Hsp26 solution at 1-cm path length.

[42] E. W. Jones, *Methods Enzymol.* **194,** 428 (1991).

3. The protein can be concentrated to at least 10 mg/ml without significant aggregation.

4. Using a Q-Sepharose Fast Flow column (Pharmacia) in the first purification step instead of the DE52 column results in slightly better purification yields at this stage, but does not have a significant influence on the final purity of the protein preparation.

5. Figure 1 shows a Coomassie-stained SDS–polyacrylamide gel of the described Hsp26 purification steps. The Western blot shows no degraded Hsp26. The polyclonal Hsp26 antibody cross-reacts with a contaminant of approximately 60 kDa, which is easily removed during the purification procedure.

## Purification of Recombinant Murine Hsp25 and Human Hsp27

### *Methods*

The method described here allows the purification of about 10 mg of mammalian recombinant sHsp (mouse Hsp25, human Hsp27) per liter of

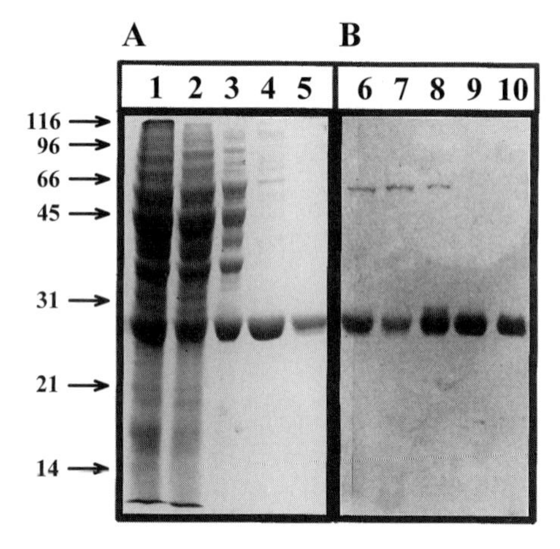


FIG. 1. Purification of Hsp26 from *Saccharomyces cerevisiae*. (A) Coomassie blue-stained SDS–polyacrylamide gel (12.5%, w/v) fractionation of proteins from each step of the purification of Hsp26. Lane 1, crude extract after cell lysis with glass beads; lane 2, supernatant of crude extract; lane 3, pooled fractions after DE52-cellulose column treatment; lane 4, pooled fractions after Resource-Q column treatment; lane 5, pooled fractions after Superdex S200 gel-filtration column treatment. (B) Western blot analysis of the proteins from each step of the Hsp26 purification. A polyclonal antibody from rabbit (a kind gift from S. Lindquist, Chicago) was used to identify Hsp26 during the purification and the blot was developed using an ECL (enhanced chemiluminescence) detection kit (Amersham); lanes 6–10 correspond to lanes 1–5 on the SDS–polyacrylamide gel. The molecular weight ($\times 10^{-3}$ kDa) is indicated on the left-hand side.

cell culture. One liter of cell culture amounts to a wet cell pellet of about 3 g. The purification leads to a purity of ≥90%. After each purification step fractions are pooled according to the appearance of the respective band on SDS–PAGE[35]; if the band pattern is ambiguous Western blot analysis[36] is performed, using specific antibodies against the sHsp.

*Protein Expression.* For recombinant expression of Hsp25 and Hsp27 we use the strain *E. coli* BL21(DE3)[43] and the plasmids pAK3038Hsp25 and pAK3038Hsp27,[8,44] respectively.

A 50-ml preculture in LB medium containing ampicillin (200 $\mu$g/ml) is inoculated with single colonies of *E. coli* BL21(DE3) harboring the plasmid pAK3038Hsp25/27 grown overnight on an LB agar plate containing ampicillin (100 $\mu$g/ml). The preculture is grown at 37° for 135 min.

To inoculate each of six 1-liter cultures [LB medium containing 0.4% (w/v) glucose and ampicillin (200 $\mu$g/ml)], 5 ml of the preculture is used. After about 4 hr of shaking at 37° and 300 rpm the expression of Hsp25/27 is induced at an optical density of 1 (1 cm, 601 nm) by addition of 400 $\mu$l/liter of 1 *M* isopropylthiogalactoside (IPTG; final concentration, 0.4 m*M*). The cells are incubated in the shaker for another 2 hr at 37° and subsequently harvested and pooled by centrifugation for 10 min at 2600 *g* and 4°.

The wet cell pellet can be stored at −80° if necessary.

*Purification.* The purification of recombinant mammalian sHsp is achieved by ammonium sulfate precipitation and anion-exchange chromatography.

1. The cell pellet is resuspended in 50 ml of lysis buffer [50 m*M* Tris-HCl (pH 8.0), 100 m*M* NaCl, 1 m*M* EDTA] and centrifuged for 10 min at 4° and 4000 *g*.

2. For lysis, 60 ml of lysis buffer, 160 $\mu$l of 50 m*M* PMSF (freshly dissolved in methanol; final concentration, 0.13 m*M*), and 2 ml of lysozyme (25 mg/ml in lysis buffer) are added to the pellet and redissolved cells are incubated on ice for 20 min. Subsequently, 80 mg of solid sodium deoxycholate is added and cells are transferred to 37°. After a 15-min incubation under constant stirring with a glass rod, about 1–2 mg of lyophilized DNase I (Boehringer GmbH, Mannheim, Germany) is added and this step is repeated every 5 min until the viscosity of the solution decreases significantly (about 15 min after the first addition of DNase I). The solution is then centrifuged for 15 min at 0° at 20,400 *g*.

[43] F. W. Studier, A. H. Rosenberg, J. J. Dunn, and J. W. Dubendorff, *Methods Enzymol.* **185,** 60 (1990).

[44] M. Gaestel, B. Gross, R. Benndorf, M. Strauss, W. H. Schunk, R. Kraft, A. Otto, H. Böhm, J. Stahl, H. Drabsch and H. Bielka, *Eur. J. Biochem.* **179,** 209 (1989).

3. The supernatant is transferred to room temperature and a saturated solution of ammonium sulfate (pH 7.0) is added dropwise under constant stirring until a final saturation of 40% is reached. The stirring is continued for a further 30 min and the sample is then centrifuged for 10 min at 20,400 *g* and 20°. The supernatant is discarded and the Hsp25/27-containing pellet is redissolved in 20 ml of buffer 1 [20 m*M* Tris-HCl (pH 7.6), 10 m*M* $MgCl_2$, 30 m*M* ammonium chloride, 0.5 m*M* dithiothreitol, 0.05 m*M* $NaN_3$, 2 $\mu M$ PMSF]. The redissolved pellet is dialyzed three times for at least 6 hr altogether at 4° against 600 ml of buffer 1 and then again centrifuged as described.

4. Ion-exchange chromatography is carried out on a 135-ml DEAE-Sepharose CL-6B column (3 × 17 cm; (Pharmacia), using a flow rate of 3 ml/min. After equilibration of the column with 300 ml of buffer 1, the dialyzed sample is applied, and the column is washed with 100 ml of buffer 1 and developed with a 500-ml gradient from 0 to 200 m*M* NaCl in buffer 1. Hsp25 elutes as a single peak at about 100 m*M* NaCl whereas Hsp27 elutes at about 120 m*M* NaCl.

5. Pooled peak fractions containing Hsp25/27 are precipitated overnight at 4° after adding solid ammonium sulfate to a final saturation of 50%. The precipitated protein is pelleted by centrifugation for 10 min at 20,400 *g* and 20°.

6. The pellet is redissolved in 2 ml of buffer 1 and dialyzed three times against 600 ml of buffer 1 at 4°. The dialyzed sample is again centrifuged as described. The supernatant contains Hsp25/27, which can be stored in aliquots at −80°.

*Comments*

1. An important step that influences the yield of the method is the successive and complete lysis of the cells expressing Hsp25/27.

2. If a purity greater that 90% is required, a final purification step using size-exclusion chromatography on Superose 6 (in which the Hsp25/27 oligomers elute shortly after the void volume of the column) should be introduced.

3. The concentration of the protein obtained is usually 10–30 mg/ml and can be determined using the method of Bradford[38] or according to the calculated extinction coefficients of 1.87 and 1.99 for Hsp25 and Hsp27, respectively (0.1% solution at 280 nm and 1-cm path length).[45]

*Chaperone Assay*

The aggregation of the thermolabile enzyme citrate synthase (CS) at elevated temperatures is an excellent tool with which to determine the

[45] D. B. Wetlaufer, *Adv. Protein Chem.* **17,** 303 (1962).

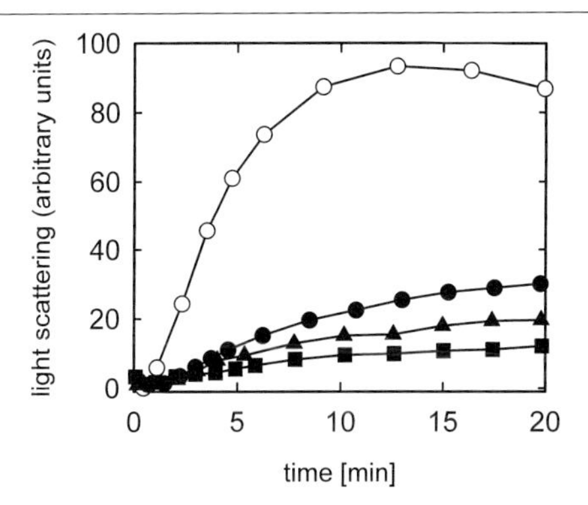


FIG. 2. Suppression of the thermal aggregation of porcine heart citrate synthase (CS) in the presence of different small heat shock proteins. The kinetics of aggregation were determined by light scattering measurements. CS was diluted 1:200 to a final concentration of 100 n*M* into buffer (40 m*M* HEPES–KOH, pH 7.5) at 43°, in the absence of additional protein (○), and in the presence of 100 n*M* Hsp26 (●), 50 n*M* human recombinant Hsp27 (▲), and 50 n*M* murine recombinant Hsp25 (■). Light scattering was measured in a Perkin-Elmer (Norwalk, CT) MPF-44A luminescence spectrophotometer in stirred and thermostatted quartz cells. Both the emission and excitation wavelengths were set to 500 nm with spectral bandwidths of 2 nm. The concentrations of CS and Hsp26 are calculated for a dimer and a 24-mer, respectively. Both Hsp27 and Hsp25 concentrations refer to hexadecamers.

ability of chaperone proteins to interact with unfolding proteins and suppress their aggregation. The assay has been used repeatedly to characterize the chaperone activity of different sHsps.[8,27,28,46] The experimental procedure is described by Buchner *et al.* (see [27] in this volume[47]).

Figure 2 shows the suppression of the thermal aggregation of CS by the small heat shock proteins Hsp26, Hsp25, and Hsp27. Spontaneous aggregation of CS starts 2 min after incubation at 43° and reaches a plateau at about 10 min. In the presence of the different sHsps the aggregation of CS is significantly reduced. Hsp25 and Hsp27 suppress the CS aggregation almost completely even at substoichiometric ratios. In the case of Hsp26, equimolar amounts of sHsp are necessary to achieve comparable effects.

As there are no data concerning the oligomeric state of IbpB available yet, quantification of chaperone effects of this sHsp is difficult and not included here.

[46] Z. Chang, T. P. Primm, J. Jakana, I. H. Lee, I. Serysheva, W. Chiu, H. F. Gilbert, and F. A. Quiocho, *J. Biol. Chem.* **271,** 7218 (1996).

[47] J. Buchner, H. Grallert, and U. Jakob, *Methods Enzymol.* **290,** Chap. 27 (1998) (this volume).

# [29] Expression, Purification, and Molecular Chaperone Activity of Plant Recombinant Small Heat Shock Proteins

*By* GARRETT J. LEE and ELIZABETH VIERLING

## Introduction

Plant small heat shock proteins (sHSPs) have been shown to have chaperone activity *in vitro*.[1,2] The sHSPs in plants are categorized into five gene families based on amino acid sequence similarity, immunological cross-reactivity, and intracellular localization.[3] There are two classes of cytosolic proteins (class I and II) and three families of organelle-localized proteins that are targeted either to the endoplasmic reticulum, chloroplast, or mitochondrion. All of these proteins contain the conserved carboxyl-terminal domain identified in other eukaryotic sHSPs and in the $\alpha$-crystallin proteins[4–6]; however, they lack the consensus phosphorylation sites of the mammalian sHSPs.[7] Chaperone activity of plant sHSPs has been studied almost exclusively with the cytosolic class I proteins, but presumably the approaches discussed here could be used to study any plant sHSP.

The sHSPs range in mass from ~15 to 42 kDa, but in their native state they exist as large oligomers.[8–10] As is true for any functional study, an important consideration in preparation of sHSPs for chaperone studies is the preservation of the native protein complex. For several plant sHSPs it has been shown that, when expressed in *Escherichia coli*, the individual monomers will assemble into oligomers with an apparent molecular weight similar to that of sHSP complexes detected in plant extracts[9] (E. Basha, G. J. Lee, and E. Vierling, unpublished). Therefore, we have worked exclusively with recombinant sHSPs that are expressed from cDNAs in *E. coli* and then purified by conventional methods in high yield. However, the

[1] G. J. Lee, N. Pokala, and E. Vierling, *J. Biol. Chem.* **270,** 10432 (1995).
[2] G. J. Lee, A. M. Roseman, H. R. Saibil, and E. Vierling, *EMBO J.* **16,** 659 (1997).
[3] E. R. Waters, G. J. Lee, and E. Vierling, *J. Exp. Bot.* **296,** 325 (1996).
[4] W. W. deJong, J. A. Leunissen, and C. E. Vooter, *Mol. Biol. Evol.* **10,** 103 (1993).
[5] N. Plesofsky-Vig, J. Vig, and R. Brambl, *J. Mol. Evol.* **35,** 537 (1992).
[6] E. Waters, *Genetics,* **141,** 785 (1995).
[7] M. Gaestel, W. Schroder, R. Benndorf, C. Lippman, and J. Buchner, *J. Biol. Chem.* **266,** 14721 (1991).
[8] K. W. Helm, P. R. LaFayette, R. T. Nagao, J. L. Key, and E. Vierling, *Mol. Cell. Biol.* **13,** 238 (1993).
[9] K. W. Helm, G. J. Lee, and E. Vierling, *Plant Physiol.* **114,** 1477 (1997).
[10] T.-L. Jinn, Y.-M. Chen, and C.-Y. Lin, *Plant Physiol.* **108,** 693 (1995).


METHODS IN ENZYMOLOGY, VOL. 290
0076-6879/98 $25.00

chaperone assays described here could be performed equally well using sHSPs purified directly from plant material.

### Construction of Expression Vectors

Small HSP cDNAs are best expressed from a high-copy plasmid such as pJC20,[1,11] which directs T7 polymerase-dependent transcription in *E. coli* BL21(DE3) cells. As it is unknown to what extent the N-terminus contributes to sHSP activity, to enable synthesis of the sHSP from its own start methionine codon, site-directed mutagenesis by PCR (polymerase chain reaction) or other standard techniques is used to engineer the start codon to an *Nde*I(CATATG) restriction site. Subsequent ligation of the modified cDNA to the *Nde*I site in the pJC20 multiple cloning site results in synthesis of the sHSP with the authentic start methionine. It may also be possible to express the sHSP with an N-terminal affinity tag, provided that subsequently the tag can be cleanly excised from the recombinant sHSP and the purified monomer reassembled into the native oligomer. This approach has not, however, yet been used with sHSPs.

### Expression and Initial Analysis of Recombinant Proteins

We have found that sHSP expression constructs can be unstable in BL21(DE3) cells. To minimize plasmid loss, sHSP expression should be performed with cells directly propagated from a recent transformant, preferably less than 1 week old. Initial experiments with the recombinant sHSP should be aimed at determining the level of target protein expression, solubility, and monomeric and oligomeric size. This is best accomplished with late log-phase bacterial cell cultures of 2 to 50 ml. Small HSP induction should first be assessed by analyzing total cell proteins by sodium dodecyl sulfate-polyacrylamide gel electrophoresis (SDS–PAGE) on 12–14% (w/v) polyacrylamide gels, which allow resolution of low molecular mass polypeptides. About 100 μl of isopropyl-β-D-thiogalactopyranoside (IPTG)-induced cells is removed from the culture, and after brief centrifugation in a microcentrifuge, the cell pellet is resuspended in 50 μl of 1× SDS–PAGE sample buffer, boiled for 3 min, then electrophoresed and stained with Coomassie blue. For comparison, it is best to run samples of cells removed before and at time points up to 6 hr after the addition of 1 m*M* IPTG. Small HSPs can typically comprise 10–20% of the total *E. coli* protein, and may even be highly expressed in the absence of IPTG. Parallel samples from cells transformed with vector only should also be analyzed

[11] J. Clos and S. Brandau, *Protein Expression Purif.* **5,** 133 (1994).

as a negative control. By SDS–PAGE, the apparent subunit molecular weight of the recombinant protein should be similar to that deduced from the cDNA sequence. However, if antibodies are available, it is even more desirable to show comigration between the authentic protein from a tissue sample and its recombinant counterpart by Western blotting. A lower than expected molecular weight, or complete absence of the recombinant protein, could indicate proteolysis, which may arise from improper folding or assembly within *E. coli* cells.

Once optimal expression conditions have been determined, sHSP solubility can be investigated by preparing crude, soluble cell extracts. To do this, late log-phase cells from a 50-ml culture expressing the sHSP are pelleted by centrifugation, resuspended in approximately 3 ml of cold 25 m*M* Tris-HCl–1 m*M* EDTA (TE), supplemented with 5 m*M* ε-aminocaproic acid, and 1 m*M* benzamidine. With the sample in an ice jacket, the cells are sonicated for 30 sec using a microprobe, then supplemented with 1 m*M* phenylmethylsulfonyl fluoride. After centrifuging in a microcentrifuge at top speed, the supernatant is removed and the resulting pellet is solubilized by resuspending and boiling in a volume of 1× SDS sample buffer equal to that of the supernatant. The distribution of soluble and insoluble sHSP can then be determined by running equal volumes of supernatant and pellet samples on an SDS gel. In general, approximately 30 μg of total protein within the supernatant fraction is suitable for visualization on minigels stained with Coomassie blue.

## Oligomeric Size Determination by Nondenaturing Pore-Exclusion Polyacrylamide Gel Electrophoresis

Most plant sHSPs assemble into oligomeric complexes of molecular mass between 200 and 300 kDa.[1,8–10] If the recombinant sHSPs are recovered in the soluble fraction of the *E. coli* cell lysate, then the oligomeric state of the protein in crude extracts can be investigated by nondenaturing PAGE. Note that we have found that some recombinant plant sHSPs expressed in *E. coli* are found in insoluble inclusion bodies (our unpublished results). Remarkably, these sHSPs can be resolubilized by urea denaturation and then reassembled into a soluble, oligomeric form by subsequent dialysis. Typically, insoluble cell pellets containing the sHSP are resuspended with fresh, TE-buffered 7*M* urea, then exhaustively dialyzed against TE buffer. After dialysis, the sample is centrifuged to remove insoluble material, and the supernatant is analyzed for the sHSP by SDS–PAGE and nondenaturing PAGE. Because most *E. coli* proteins are soluble in crude extracts, resurrecting the sHSP from the insoluble fraction often results in severalfold enrichment of the sHSP, which can be purified to

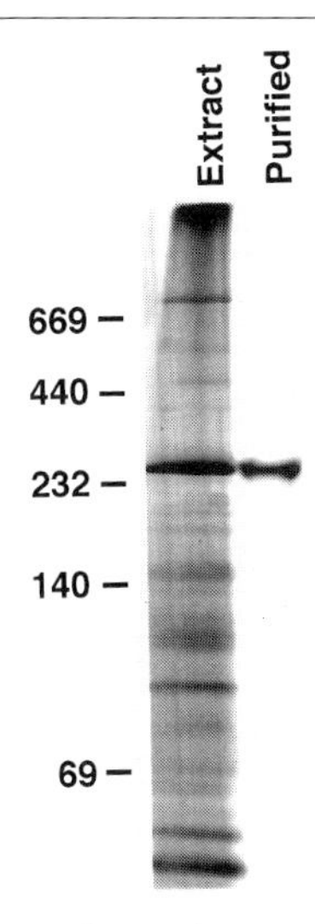


FIG. 1. Analysis of HSP18.1 by nondenaturing pore-exclusion PAGE. Protein (70 $\mu$g) from an extract of *E. coli* cells expressing recombinant HSP18.1 (left lane) or 7 $\mu$g of purified HSP18.1 (right lane) were separated as described in text, then stained with Coomassie blue. The positions of molecular weight markers ($\times$ 10−3) are shown on the left-hand side.

homogeneity in combination with the techniques described below for the purification of recombinant HSP18.1.

Nondenaturing pore-exclusion PAGE is performed using a modification of Anderson *et al.*[12] (Fig. 1). With this method the native molecular weight of the protein can be estimated by analysis on a single gel using standard electrophoresis materials and equipment. Proteins are separated on a polyacrylamide gradient gel and electrophoresis is carried out for an extended period, such that proteins migrate until they become trapped by the matrix. Thus, size and shape only, not charge, contribute to relative protein migration in this system. With globular standards of known molecular weight, a logarithmic relationship can be derived to estimate sHSP native molecular weights. Clearly, this technique is useful only for proteins with a net negative charge (as is the case for most plant sHSPs) under these conditions. When used in combination with Western blotting, pore-exclusion PAGE is extremely useful for estimating and comparing the apparent sizes of the authentic and recombinant sHSP oligomers. The following gel system is suitable for resolving proteins between 69 and 669 kDa on a 16 cm $\times$ 12 cm $\times$ 1.5 mm gel.

[12] L. O. Anderson, H. Borg, and M. Mikaelsson, *FEBS Lett.* **20,** 199 (1972).

*Solutions*

Gel buffer (8×)
Tris base, 363 g
HCl (12 *N*), 42 ml
Bring to 1 liter with $H_2O$, adjust to pH 8.8

Gel recipe

| | Acrylamide (22.5%, w/v) | Acrylamide (4.0%, w/v) |
|---|---|---|
| Acrylamide (30%, w/v), bisacrylamide (0.8%, w/v) | 12 ml | 2.4 ml |
| Gel buffer (8×) | 2.0 ml | 2.0 ml |
| Glycerol | 2.0 ml | 0 |
| $H_2O$ | 0 | 11.6 ml |
| *N,N,N′,N′*-tetramethylethylenediamine (TEMED) | 6 μl | 8.4 μl |
| Ammonium persulfate (10%, w/v) | 36 μl | 36 μl |

Tank buffer (5×)

Tris base, 75.6 g
Glycine, 360 g
Bring to 2.5 liters with $H_2O$, adjust to pH 8.3

Nondenaturing sample buffer (4×)
Tris-HCl (pH 8.0), 240 m*M*
Sucrose (60%, w/v)
ε-Aminocaproic acid, 20 m*M*
Benzamidine, 4 m*M*
Bromphenol blue (0.04%, w/v)

Standards: High molecular weight electrophoresis calibration kit (Pharmacia, Piscataway, NJ); reconstitute lyophilized cake with 500 μl of 1× sample buffer and load 30 μl/lane

*Procedure*

Connect an 18-gauge needle to the end of the tubing coming from the gradient maker and peristaltic pump. Place the needle at the top edge of the assembled gel plates with the end of the needle between the plates. Initiate gel polymerization by mixing in TEMED and ammonium persulfate and add the gel mix to the chambers of the gradient maker. Start pumping the gradient with the 20% solution entering the gel apparatus first, allowing the progressively less dense acrylamide to layer over the top. Overlay the finished gel with water and allow to polymerize. When polymerized, remove the water overlay, pour a 4% spacing gel with the bottom of the comb inserted 1 cm from the start of the running gel, and allow to polymerize.

Set up the gel apparatus with tank buffer and load the samples and standards in nondenaturing sample buffer. Run the gel at 100 V for 40–48 hr in a cold room, then stain. Approximately 50 μg of total protein from a crude extract per lane is sufficient for Coomassie blue staining. Small HSPs such as HSP18.1, which assemble into discretely sized oligomers and approach 20% of the total soluble protein (as judged by SDS–PAGE), appear as prominent bands against the *E. coli* protein background (Fig. 1). Those sHSPs that express poorly or do not assemble into a single species can be identified by comparison with a lane containing a vector-only negative control, or by Western blotting.

## Purification of Recombinant HSP18.1

Although various plant sHSPs possess different physical characteristics, including monomeric and oligomeric size and isoelectric point,[13] we have found that such differences are relatively subtle such that the following purification procedure for recombinant HSP18.1 from *Pisum sativum* (garden pea) can serve as a good prototype for purifying other sHSPs. HSP18.1 purification can be readily accomplished by a combination of ammonium sulfate precipitation, sucrose gradient centrifugation, and anion-exchange chromatography.[1] Ammonium sulfate fractionation takes advantage of the fact that many plant sHSPs precipitate at high ammonium sulfate concentrations relative to most *E. coli* proteins. Sucrose gradient fractionation in a vertical rotor is used next, as a rapid technique to separate sHSPs efficiently on the basis of their large oligomeric size. As a final step, DEAE anion-exchange chromatography is used to isolate the purified protein. Except where noted, cold solutions and conditions should be used. In general, small samples should be retained after each step to monitor the presence and purity of the sHSP by SDS–PAGE and its oligomeric state by nondenaturing pore-exclusion PAGE. This procedure yields approximately 15 mg of pure protein, but can vary depending on the level of sHSP expression.

### *Reagents and Materials Required for HSP18.1 Purification*

All reagents are of the highest grade available.

- LB medium (1500 ml) supplemented with ampicillin (100 μg/ml) in a 4-liter flask
- IPTG

[13] E. Vierling, *Annu. Rev. Plant Physiol. Plant Mol. Biol.* **42,** 579 (1991).

Tris base
Ethylenediaminetetraacetic acid (EDTA)
Aminocaproic acid,
Benzamidine
Phenylmethylsulfonyl fluoride
Ammonium sulfate
Sucrose
Urea
NaCl
DEAE-Sepharose Fast Flow resin (Sigma, St. Louis, MO)
Hydroxyapatite (Bio-Rad, Hercules, CA) (optional)
Sonicator
Centrifuge with GSA and SS34 rotors (Sorvall, Norwalk, CT) or equivalent
Ultracentrifuge with VC53 or VTi50 rotor (Beckman, Fullerton, CA)
Polyallomer Quick Seal tubes (Beckman), 40 ml
Stirred-cell concentrator with YM10 membrane (Amicon, Danvers, MA), 50 ml

*Cell Growth and Small Heat Shock Protein Induction*

Pick one colony from a plate of BL21(DE3) cells freshly transformed with the expression vector and inoculate 1500 ml of LB medium supplemented with ampicillin. Grow the cells overnight at 37° for 12–14 hr with shaking. Add 1 m*M* IPTG and continue to grow for four more hours. After 4 hr of induction, withdraw 100 $\mu$l of cell culture and verify the presence of the sHSP by SDS–PAGE.

*Cell Harvest and Lysis*

At the end of the 4-hr induction period, harvest the cells by centrifuging in a GSA rotor for 8 min at 5000 rpm. Pour off the LB and resuspend the cells in a total volume of 50 ml of cold 25 m*M* Tris-HCl–1 m*M* EDTA, pH 7.5 (TE), then transfer the cells to a 50-ml polypropylene beaker. Add 5 m*M* $\varepsilon$-aminocaproic acid and 1 m*M* benzamidine to the cell suspension. With the beaker in an ice jacket, sonicate the sample with four 30-sec sonication treatments, removing the sonicator tip and waiting 1 min between each treatment to prevent overheating. After the first and second sonication treatments, mix in 1 m*M* phenylmethylsulfonyl fluoride from a 100 m*M* stock in 2-propanol. At the end of the four sonication cycles, the sample should appear darker and more translucent than the initial unbroken cell suspension. Centrifuge the cell lysate for 30 min at 12,000 rpm in an SS34 rotor, transfer the supernatant to a glass beaker, then bring the supernatant to 60 ml with TE.

*Ammonium Sulfate Fractionation*

An ammonium sulfate fraction enriched in the sHSP of interest can be determined empirically. In general, we have found that plant sHSPs are often precipitated with ammonium sulfate concentrations greater than 60%. For HSP18.1, the crude extract is brought to 60% ammonium sulfate by slowly adding solid ammonium sulfate to the ice-jacketed sample on a stir plate. Following centrifugation for 10 min at 12,000 rpm in an SS34 rotor, proteins in the supernatant are next precipitated with 95% ammonium sulfate, then centrifuged. The resulting pellet containing the sHSP is gently resuspended with 8 ml of TE, then dialyzed against 1 liter of TE for 2 hr.

*Sucrose Gradient Fractionation*

While the sHSP fraction is dialyzing, four 0.2–0.8 *M* linear sucrose gradients made in TE are poured in 40-ml Beckman Quick Seal tubes. The total volume of each gradient is 37 ml, leaving the upper 3 ml of each tube for the sHSP sample. The dialyzed sample is brought to approximately 13 ml (total volume) with TE, then 3 ml is applied over each sucrose gradient. The tubes are sealed, then centrifuged for either 3 hr in a VC53 rotor or for 2 hr, 45 min in a VTi50 rotor at 50,000 rpm with the brake and slow acceleration turned on and the temperature set at 4°. After centrifugation, tubes are punctured, and the bottom 20 ml (for VC53) or 10 ml (for Vti50) is collected and pooled. Alternatively, smaller fractions can be collected and analyzed for sHSP content by SDS–PAGE. For convenience, sucrose pools can be stored for months at −80°.

*DEAE Chromatography*

Because a small number of *E. coli* proteins remain strongly bound to HSP18.1 during purification (presumably because of its molecular chaperone activity), urea is added to the sucrose gradient pool to dissociate the bound *E. coli* proteins before chromatography. Urea denaturation also improves the separation of the sHSP from other proteins by decreasing sHSP retention on the DEAE column, making HSP18.1 the first protein to elute. We have found that 3 *M* urea or greater dissociates dodecameric HSP18.1 into monomers (our unpublished data) that are fully competent for reassembly on removal of urea by dialysis. However, to prevent unwanted chemical modification to the sHSP, care should be taken to use freshly prepared urea solutions and to minimize sHSP exposure to the denaturant.

Denaturation is accomplished by dissolving solid urea into the sucrose pool to a concentration of 3 *M* and allowing the solution to sit for 15 min on ice. The sample is then loaded at 0.5 ml/min onto a 20-ml bed of DEAE-Sepharose Fast Flow resin in a 2.5-cm-diameter column equilibrated with TE. After loading, the column is washed at 1 ml/min with TE supplemented with 3 *M* urea while collecting 5-ml fractions. When the 280-nm ultraviolet (UV) absorbance of the column effluent reaches baseline, proteins are eluted at 0.5 ml/min with 120 ml (total volume) of a 0–120 m*M* NaCl linear gradient in TE plus 3 *M* urea, while collecting 5-ml fractions. The presence and purity of the sHSP in the fractions are checked by running 25 $\mu$l of UV-absorbing fractions on SDS–PAGE, which is compatible with the presence of urea. HSP18.1 is found in both the wash and the first peak to elute off the column. When analyzed by SDS–PAGE plant sHSPs from samples containing urea or high concentrations of buffer salts often appear smeared. This appearance normally disappears with dialysis, or can be minimized by placing an additional 0.5-cm zone of 10% (w/v) stacking gel between the 5% (w/v) stacking gel and the running gel. Pure fractions are pooled and dialyzed overnight against a minimum of 500 vol of TE, then concentrated to approximately 1–2 mg/ml using a 50-ml stirred-cell concentrator fitted with a YM10 membrane (Amicon). Finally, the protein is divided into approximately 100-$\mu$l aliquots and stored frozen at $-80°$ for up to several months.

*Further Purification by Hydroxyapatite Chromatography*

For sHSP preparations other than HSP18.1 in which contaminating proteins remain, hydroxyapatite chromatography under native conditions often brings the sHSP to homogeneity. Following DEAE chromatography, the sample is dialyzed against TE to remove urea and NaCl, then loaded onto a 10- to 20-ml bed of hydroxyapatite (Bio-Rad) equilibrated with TE. After washing with TE until the UV absorbance returns to baseline, the sHSP is eluted with five column volumes of a 0–0.5 *M* sodium phosphate gradient, pH 7.5, in TE. Because the 0.5 *M* phosphate solution becomes insoluble at 4°, the elution step is carried out at room temperature. Ultraviolet-absorbing fractions are checked for the presence of sHSP by SDS–PAGE, then the pure fractions are pooled, dialyzed against TE, concentrated if necessary, and frozen in aliquots.

## Assaying Molecular Chaperone Activity

One of the emerging characteristics of sHSPs is their ability to interact with a variety of nonnative proteins *in vitro.* At temperatures in the range of 40–45°, plant sHSPs such as HSP18.1 prevent the aggregation of heat-denatured model substrates in an ATP-independent manner.[1,2] For HSP18.1, the basis for this aggregation protection is the stable binding of nonnative substrate to the sHSP.[2] In addition to the conformational changes that occur within substrates on heat denaturation, HSP18.1 also appears to undergo a temperature-dependent conformational change that exposes regions important for substrate binding.[2] In contrast to what is observed at temperatures above 40°, below this temperature most model substrates interact transiently with HSP18.1, so that substrates denatured with mild heat or guanidine hydrochloride can refold spontaneously with the passive assistance of the sHSP.[1] Detailed methods to assay thermal aggregation and the refolding of mild heat- or guanindine hydrochloride-denatured substrates in the presence of sHSPs are described elsewhere.[14] Here, we present methods to monitor stable substrate–sHSP complexes and subsequent substrate refolding in conjunction with cell-free extracts.

## Formation of Substrate–Small Heat Shock Protein Complexes

In general, the strength and quantity of substrate binding to HSP18.1 are dependent on three factors: temperature, time, and the heat lability of the particular substrate. In principle, any heat-labile protein can be utilized as a model substrate. We have used porcine mitochondrial malate dehydrogenase (homodimer of 35-kDa subunits) (Boehringer Mannheim, Indianapolis, IN), porcine citrate synthase (CS) (homodimer of 50-kDa subunits) (Sigma), porcine glyceraldehyde-3-phosphate dehydrogenase (homotetramer of 35-kDa subunits) (Sigma), and firefly luciferase (61-kDa monomer) (Promega, Madison, WI).[2] All are readily available and have convenient enzymatic activity assays. Buffers suitable for experiments at elevated temperatures include phosphate and *N*-2-hydroxyethylpiperazine-*N*′-2-ethanesulfonic acid (HEPES), which exhibit only small changes in a p$K_a$ with increasing temperature.

For HSP18.1, the standard substrate-binding reaction (200-$\mu$l total volume) is set up by combining 1 $\mu M$ HSP18.1 dodecamer with a given concentration of substrate in 50 m$M$ sodium phosphate, pH 7.5, in a microcentrifuge tube at room temperature. In the case of firefly luciferase, to prevent adherence of the luciferase to the microcentrifuge tube walls, the tube is first blocked with bovine serum albumin (1 mg/ml) for at least 15 min, then rinsed well with water. The sample is then incubated in a water bath at 22° or at an elevated temperature. In general, for any given substrate, two of three parameters—substrate concentration,

[14] G. J. Lee, *Methods Cell. Biol.* **50,** 325 (1996).

incubation temperature, and incubation time—should be held constant while the third is varied. For most substrates, a concentration of 1–3 $\mu M$ protomer and temperatures ranging from 40 to 45° for 15–90 min are suitable.[2] Following incubation, the sample is chilled on ice for 1 min, centrifuged for 15 min at top speed in a refrigerated microcentrifuge, then examined carefully for the presence of a pellet indicative of substrate aggregation. By observing the presence or absence of pelleted protein, this procedure can provide qualitative data concerning the ability of the sHSP to protect against substrate aggregation. However, in the case of HSP18.1 and high concentrations of GAPDH, we have found that very large, insoluble GAPDH–HSP18.1 complexes can result from the association of soluble GAPDH–HSP18.1 precursors.[2] Nonetheless, the composition of pellets can be analyzed by resuspending them with SDS sample buffer and visualizing the solubilized material by SDS–PAGE. The supernatant of the binding reaction can then be analyzed by size-exclusion high-performance liquid chromatography (HPLC) or nondenaturing PAGE as described in the following sections.

## Monitoring Substrate–Small Heat Shock Protein Complex Formation by Size-Exclusion High-Performance Liquid Chromatography

Size-exclusion HPLC has proved to be an extremely valuable method for assaying the formation of substrate–sHSP complexes. Not only is the technique rapid, but it is particularly well suited for monitoring the size distribution and abundance of substrate–sHSP complexes[2] (Fig. 2). Because the size of substrate–sHSP complexes can be extremely large, we have found sizing columns with very large void volumes to work the best. Although substrate–sHSP complexes can be prone to nonspecific adsorption to size-exclusion matrices, the problem can be overcome with the use of silica-based columns in conjunction with phosphate mobile phases. One column, the TSK G4000SWXL by TosoHaas (Supelco, Bellefonte, PA), fulfills the preceding criteria, and has excellent resolution for proteins in the range of 100 kDa up to 1 MDa or more.

The HPLC system should be fitted with a 200-$\mu$l sample loop, followed in series by an inline filter (Upchurch, Oak Harbor, WA) fitted with a 0.25-$\mu$m replaceable frit, the column, and finally the UV detector set at 220 nm. The system should be equilibrated with a mobile phase consisting of 0.2 $M$ sodium phosphate–0.2 $M$ NaCl, pH 7.3, running at 1 ml/min. At this flow rate, runs can be completed in less than 15 min. We have found the mobile phase composition to be critical, as other buffers such as Tris, HEPES, or lower concentrations of phosphate can often result in the nonspecific adsorption of substrate–sHSP complexes.

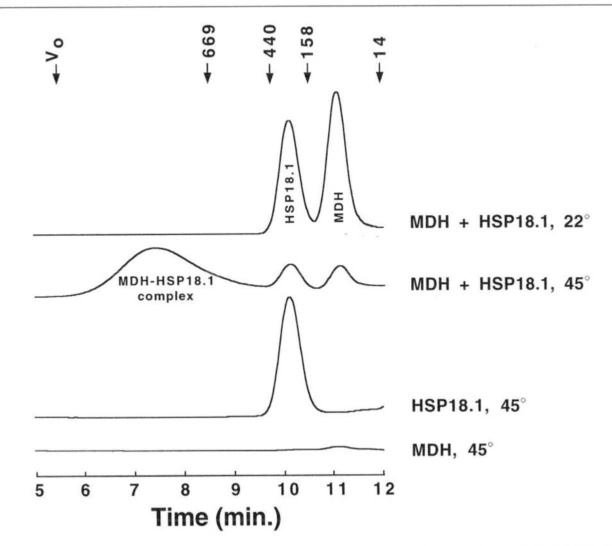


FIG. 2. Separation of HSP18.1, malate dehydrogenase (MDH), and MDH–HSP18.1 complexes by size-exclusion HPLC. MDH (2 $\mu M$), 1 $\mu M$ HSP18.1, or a mixture of the two proteins, was incubated for 60 min at the indicated temperatures. After centrifugation, the supernatants were subjected to size-exclusion HPLC as described in text. After centrifugation, insoluble protein was recovered only from the sample in which MDH was incubated alone. In this sample almost all of the MDH formed pelletable aggregates such that no absorbance is detected in this sample on HPLC analysis (bottom trace). The elution times of molecular weight standards are shown. The column void volume ($V_0$) was determined by the elution of Dextran blue, which has a molecular weight of $2 \times 10^6$.

After substrate–sHSP complex formation, the sample is centrifuged for 15 min at top speed in a refrigerated microcentrifuge and the supernatant is immediately injected into the HPLC system. The removal of debris by centrifugation before injection is superior to sample filtration through membranes, which usually results in a significant loss of substrate–sHSP complexes. Any remaining particulates injected into the HPLC system are efficiently trapped by the inline filter. If desired, the composition of substrate–sHSP complexes can be determined by collecting appropriate fractions from the HPLC run and examining them by SDS–PAGE.[2]

## Analysis of Substrate–Small Heat Shock Protein Complexes by Nondenaturing Polyacrylamide Gel Electrophoresis

In the absence of an HPLC system, stable substrate–sHSP complexes can be detected by nondenaturing PAGE on 4% (w/v) polyacrylamide gels. In general, we have found that such complexes, because of their large size, do not readily enter the more concentrated acrylamide matrix in the pore-exclusion gels described above. Whereas the apparent sizes of sHSPs can be estimated by pore-exclusion PAGE, and substrate–HSP complex size can be estimated by size-exclusion HPLC, electrophoresis on gels of constant acrylamide concentration can only qualitatively demonstrate the formation of substrate–sHSP complexes. On standard nondenaturing electrophoresis complex formation is inferred on the basis of the appearance of a new species with a lower relative mobility after sHSP and substrate are heated. In the case of a CS–HSP18.1 complex, such a species appears on a 4% (w/v) gel as a broad band that is retarded relative to free CS or HSP18.1 (Fig. 3). Analysis of malate dehydrogenase–sHSP interactions by this technique cannot be performed because malate dehydrogenase retains a net positive charge under these electrophoretic conditions. The behavior of glyceraldehyde 3-phosphate–HSP18.1 or luciferase–HSP18.1 complexes by this technique has not been evaluated.

Polyacrylamide (4%, w/v) gels are poured in standard fashion in a 16 cm × 12 cm × 1.5 mm format, using twice the volume of the 4% (w/v) gel recipe shown above for the nondenaturing pore-exclusion gels, but with the following exceptions: (1) the final gel buffer concentration should be

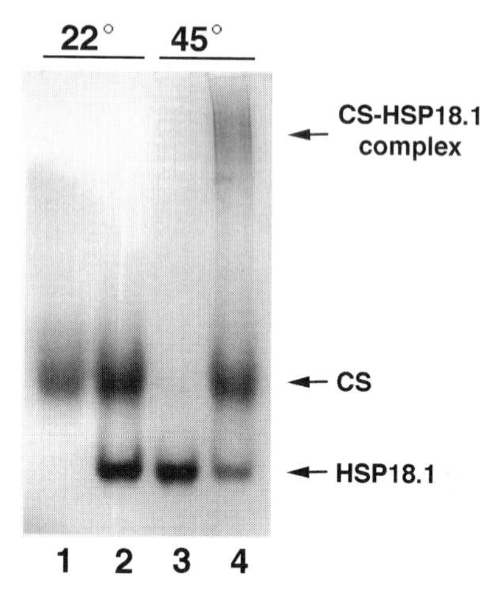


FIG. 3. Separation of a citrate synthase–HSP18.1 complexes by nondenaturing PAGE. Citrate synthase (CS, 0.2 mg/ml), HSP18.1 (0.1 mg/ml), or a mixture of the two proteins was incubated for 30 min at 22 or 45° as indicated, subjected to nondenaturing PAGE on a 4% (w/v) polyacrylamide gel as described in text, and then stained with Coomassie blue. Lane 1, CS only; lane 3, HSP18.1 only; lanes 2 and 4, mixtures of CS and HSP18.1. The region labeled as CS–HSP18.1 complex can be excised from the gel and shown to contain both CS and HSP18.1 by reelectrophoresis on an SDS–polyacrylamide gel (not shown).

one-half that in the pore-exclusion gel, and should be adjusted to pH 8.0; (2) no spacer gel is necessary—the comb can be inserted when the running gel is poured; and (3) TEMED and 10% (w/v) ammonium sulfate should be increased to 15 and 300 $\mu$l, respectively, for 32 ml of gel mix.

Citrate synthase–sHSP complexes should be formed similarly to those prepared for size-exclusion HPLC analysis, then supplemented with native sample buffer. Electrophoresis should be carried out in a cold room for approximately 6 hr at 100 V. Once electrophoresis is complete, care should be taken when handling the delicate gel. It is often convenient to place the gel while still bound to one of the electrophoresis plates directly into Coomassie stain, and allow the gel to free itself under gentle shaking.

The position of the CS–sHSP band varies with the size of the complex. In the case of very large complexes, which barely enter the gel, stained complexes may be seen only near the wells of the gel. To increase migration into the gels, one or more of the following variables should be considered: (1) decreasing complex size by reducing the time, temperature, or substrate concentration during complex formation, (2) decreasing the concentration of acrylamide to 3.5% (w/v), and (3) increasing the electrophoresis time.

After nondenaturing electrophoresis, the citrate synthase and sHSP content in the complexes can be analyzed by excising from an unfixed, duplicate gel the region corresponding to the stained CS–HSP complex. The excised band can then be placed in the well of an SDS gel, electrophoresed, and stained with Coomassie blue or silver. When citrate synthase–HSP complex bands are very broad, the excised band can be cut into approximately 1-cm segments, or simply placed horizontally onto the SDS gel.

## Monitoring Substrate-Refolding Competence

Data suggest that a heat-denatured substrate stably bound to HSP18.1 can be refolded with the assistance of ATP-dependent molecular chaperones present in cell extracts.[2] When preformed luciferase–HSP18.1 complexes (Fig. 4A) are added to either rabbit reticulocyte lysate or wheat germ extracts, luciferase refolding can be monitored by the regain of luciferase enzymatic activity (Fig. 4B). Essentially no refolding is observed in the absence of the extracts or ATP, or if luciferase that has been aggregated in the absence of HSP18.1 is added to the folding reaction. These data suggest that only luciferase bound to the sHSP is folding competent. Data have demonstrated that HSP70 is a critical component within the extracts (G. J. Lee and E. Vierling, unpublished), and similar conclusions have

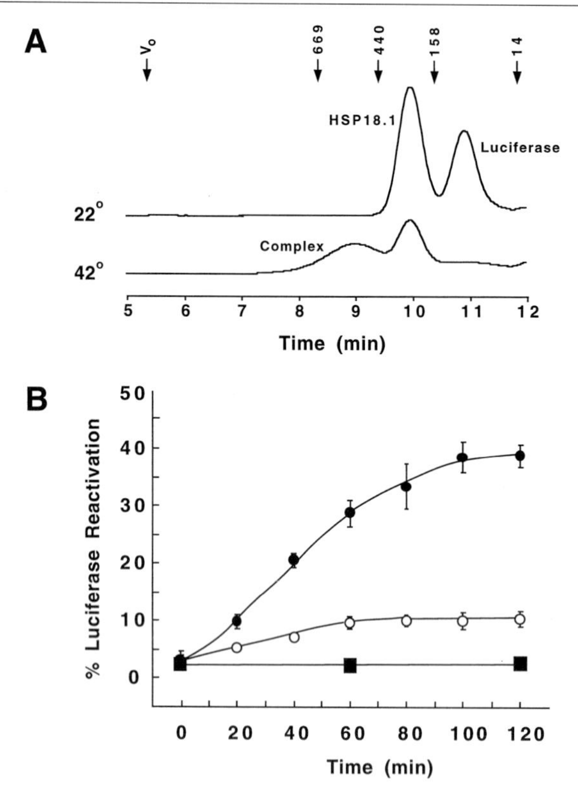


FIG. 4. Heat-denatured firefly luciferase bound to HSP18.1 can refold in the presence of cell extracts. In (A), a mixture of 1 $\mu M$ luciferase and 1 $\mu M$ HSP18.1 was incubated for 15 min at 22 or 42° to form luciferase–HSP18.1 complexes, then analyzed by size-exclusion HPLC. In (B), luciferase–HSP18.1 complexes from an identical sample heated to 42° were added to rabbit reticulocyte lysate (●) or wheat germ extract (○) supplemented with ATP, then assayed for luciferase activity. (■) Samples in which the luciferase–HSP18.1 complexes were added to ATP-supplemented buffer that lacked extracts, or to extracts that lacked ATP.

been reached by Buchner and colleagues[15] studying mammalian sHSPs and refolding of citrate synthase.

*Assay Procedure*

1. Form luciferase–HSP18.1 complexes with 1 $\mu M$ firefly luciferase and 1 $\mu M$ HSP18.1 dodecamer as described (see Formation of Substrate-Small Heat Shock Protein Complexes), with the exception that the 200-$\mu$l mixture is divided into two 100-$\mu$l aliquots in bovine serum albumin-blocked 0.65-

[15] M. Ehrnsperger, S., Gräber, M. Gaestel, and J. Buchner, *EMBO J.* **16,** 221 (1997).

ml tubes. Incubate one tube for 15 min at 42° to form the complexes and the other at 22° to serve as the native luciferase control. Store on ice.

2. In 0.65-ml tubes, supplement 30 μl of untreated rabbit reticulocyte lysate (Green Hectares, Oregon, WI; or Promega) or wheat germ extract (Promega) with 12.5 μl of 4× refolding buffer (100 m*M* HEPES–KOH, 8 m*M* ATP, 20 m*M* $MgCl_2$, 40 m*M* KCl, 4 m*M* dithiothreitol, pH 7.5), and adjust to 48.75 μl with $H_2O$. For ATP-deficient reactions, omit ATP, and supplement the mixtures with 1 μl of potato apyrase (0.5 unit/μl) in 25 m*M* HEPES–KOH, pH 7.5 (Sigma).

3. Preincubate the samples in a 30° water bath for 5 min.

4. Initiate the reactions by adding 1.25 μl of the complexed or native luciferase–sHSP samples (25 n*M* luciferase final) and return to the 30° bath. After initiation, immediately monitor luciferase activity by mixing 2.5 μl of the refolding reaction with 50 μl of the substrate mixture supplied in the luciferase assay system (Promega) and measure in a luminometer. Alternatively, luciferase activity can be similarly measured using a liquid scintillation counter according to the luciferase assay system instructions. Because of the high sensitivity of scintillation counters, first rapidly dilute a small aliquot of the refolding reaction 1:500 in 25 m*M* HEPES–KOH, pH 7.5, supplemented with bovine serum albumin (1 mg/ml) in a 1.7-ml microcentrifuge tube. Immediately mix 2.5 μl of the diluted sample with 50 μl of the substrate mixture and measure in the scintillation counter.

5. Over time, measure and calculate luciferase reactivation at 30° relative to the activity of the native, unheated sample. Initially the luciferase activity from the heated samples should be less than 5% of the native luciferase control. If not, the length of the luciferase–sHSP binding reaction should be increased accordingly.

# [30] Lens α-Crystallin: Chaperone-Like Properties

*By* JOSEPH HORWITZ, QING-LING HUANG, LINLIN DING, and MICHAEL P. BOVA

## Introduction

α-Crystallin is one of the abundant structural proteins of the vertebrate eye lens, where it can account for about 40% of the total soluble mass.[1,2]

[1] H. Bloemendal, "Molecular and Cell Biology of the Eye Lens." Wiley, New York, 1981.
[2] J. J. Harding and M. J. C. Crabbe, *in* "The Eye" (H. Davson, ed.), Vol. IB, p. 207. Academic Press, New York, 1984.



0076-6879/98 $25.00

Being a member of the small heat shock protein family, it possesses some chaperone-like properties.[3–7] The $\alpha$-crystallin family consists of two genes, $\alpha$A and $\alpha$B. The $\alpha$A gene encodes a 173-amino acid residue polypeptide, and the $\alpha$B gene encodes a 175-amino acid residue polypeptide. There is a 57% sequence similarity between the two polypeptides. Under physiological conditions, $\alpha$-crystallin is always isolated as a heterogeneous high molecular weight aggregate of $\alpha$A and $\alpha$B, with molecular weights ranging from 300,000 to more than 5 million. The average molecular weight of most $\alpha$-crystallin preparations is ~800,000. Until 1985, it was generally believed that $\alpha$-crystallin was a lens-specific protein. However, $\alpha$B-crystallin has now been found in numerous tissues of the body, such as heart, skeletal muscle, brain, lung, skin, and kidney.[8] $\alpha$A-Crystallin is much less abundant outside the eye lens. An important finding is the fact that $\alpha$B-crystallin is overexpressed in many degenerative diseases.[9] The biochemical, biophysical, gene regulation, expression, and evolutionary properties of $\alpha$-crystallin have been studied extensively.[1,2,9–13]

In this chapter, we present a protocol for preparing calf or cow lens $\alpha$-crystallin as well as the preparation and purification of recombinant human $\alpha$B-crystallin. Several assays for the chaperone-like properties of $\alpha$-crystallin are presented.

## Purification of Lens $\alpha$-Crystallin

The most common source for lens $\alpha$-crystallin has been cow or calf lens. Numerous methodologies for fractionation and purification of the various crystallin species have been described.[1,2] Lens $\alpha$-crystallin is most commonly prepared using gel-exclusion chromatography. The $\alpha$-crystallin obtained under physiological conditions is always a tight complex of $\alpha$A and $\alpha$B. To obtain separated $\alpha$A or $\alpha$B, a high concentration of urea or other denaturing

[3] T. D. Ingolia and E. A. Craig, *Proc. Natl. Acad. Sci. U.S.A.* **79,** 2360 (1982).

[4] R. Klemenz, E. Frohle, R. H. Steiger, R. Schafer, and A. Aoyama, *Proc. Natl. Acad. Sci. U.S.A.* **88,** 3652 (1991).

[5] J. Horwitz, *Proc. Natl. Acad. Sci. U.S.A.* **89,** 10449 (1992).

[6] U. Jakob, M. Gaestel, K. Engel, and J. Buchner, *J. Biol. Chem.* **268,** 7414 (1993).

[7] K. B. Merck, P. J. T. A. Groenen, C. E. M. Voorter, W. A. De Haard-Hoekman, J. Horwitz, H. Bloemendal, and W. W. de Jong, *J. Biol. Chem.* **268,** 1046 (1993).

[8] S. P. Bhat and C. N. Nagineni, *Biochem. Biophys. Res. Commun.* **158,** 319 (1989).

[9] C. M. Sax and J. Piatigorsky, *Adv. Enzymol. Relat. Areas Mol. Biol.* **69,** 155 (1994).

[10] J. Piatigorsky and G. J. Wistow, *Cell* **57,** 197 (1989).

[11] W. W. de Jong, J. A. M. Leunissen, and C. E. M. Voorter, *Mol. Biol. Evol.* **10,** 103 (1993).

[12] P. J. T. A. Groenen, K. B. Merck, W. W. de Jong, and H. Bloemendal, *Eur. J. Biochem.* **225,** 1 (1994).

[13] J. I. Haynes, M. K. Duncan, and J. Piatigorsky, *Dev. Dynamics* **207,** 75 (1996).

conditions are used in conjunction with various chromatographic methods.[14-19] A typical methodology for obtaining native, total α-crystallin is given in the following sections.

*Procedure*

*Material.* Fetal calf eyes obtained from freshly slaughtered animals (Pel Freez Biologicals, Rogers, AK) are kept at −80° until use. On thawing, the eye is cut open and the lens is carefully removed. All of the following procedures are performed over ice or in a cold room at 4–6°. Alternatively, freshly removed calf or cow eyes can be obtained in local slaughter houses.

*Homogenization.* Two fetal lenses weighing approximately 1.6 g are homogenized in a Dounce (Wheaton, Millville, NJ) homogenizer with 4.8 ml of buffer containing 20 m*M* Tris-HCl, 0.1 *M* NaCl, and 1 m*M* $NaN_3$, pH 7.9. The homogenization is performed gently to minimize foaming. The homogenate is then spun at 18,000 *g* for 15 min and the supernatant is collected. The amount of protein in the supernatant can be estimated by measuring the absorbance of the supernatant at 280 nm (dilute 1:100). One optical density (OD) unit is approximately 0.5 mg/ml of total lens soluble proteins. Typical yields are 450 mg of lens soluble proteins.

*Gel-Permeation Chromatography.* One milliliter (~90–100 mg) is applied to a column (2.5 × 90 cm of Sepharose CL-6B; Pharmacia, Piscataway, NJ) equilibrated with the homogenization buffer. The sample is eluted with the same buffer at a flow rate of 0.4 ml/min at 4 to 6°. Approximately 3.5 ml/tube is collected. Figure 1 shows the elution profile obtained from such a preparation. The first peak to elute contains the high molecular weight α-crystallin (HMWα). The second peak contains the low molecular weight α-crystallin (average molecular weight ~800,000), followed by high molecular weight β-crystallin (βH), low molecular weight β-crystallin (βL), and γ-crystallin. A highly purified sample of α-crystallin can be obtained by pooling the fractions under the low molecular weight α peak (see arrows in Fig. 1). These fractions contain a total of ~18 mg of α-crystallin. The purity of that α-crystallin preparation was assessed by sodium dodecyl sulfate–polyacrylamide gel elecrophoresis (SDS–PAGE), as shown in Fig. 2. Shown are the αB fraction, αA fraction, and a truncated form of αA

[14] A. Spector, L. K. Li, R. C. Augusteyn, A. Schneider, and T. Freund, *Biochem. J.* **124,** 337 (1971).

[15] W. W. de Jong, F. S. M. van Kleef, and H. Bloemendal, *Eur. J. Biochem.* **48,** 271 (1974).

[16] H. Bloemendal and G. Groenewoud, *Anal. Biochem.* **117,** 327 (1981).

[17] P. J. M. Van Den Oetelaar, R. B. Bezemer, and H. J. Hoenders, *J. Chromatogr.* **398,** 323 (1987).

[18] A. Stevens and R. C. Augusteyn, *Curr. Eye Res.* **6,** 739 (1987).

[19] M. S. Swamy and E. C. Abraham, *Curr. Eye Res.* **10,** 213 (1991).

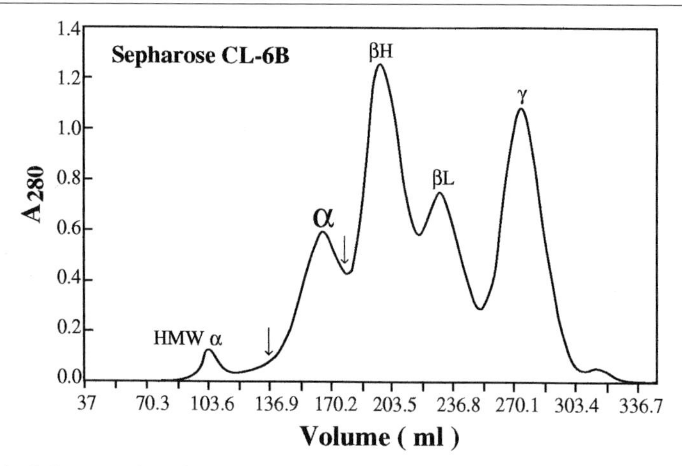


Fig. 1. Gel-permeation chromatography of soluble proteins from fetal calf lenses. The arrows point to the total $\alpha$-crystallin fractions that were collected (see text for details).

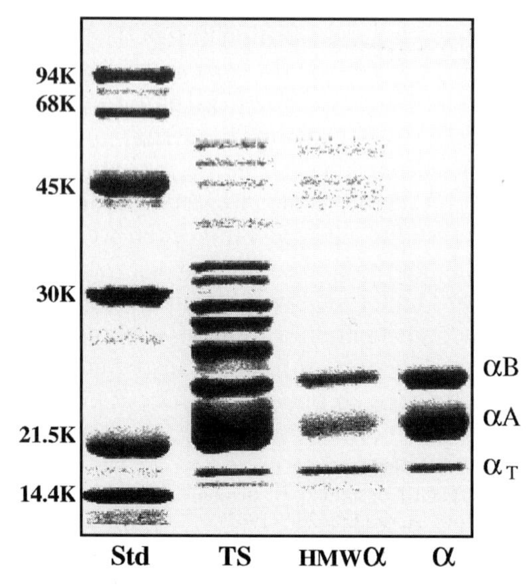


Fig. 2. SDS–polyacrylamide gel electrophoresis (12.5%, w/v) of fetal calf lens proteins. Std, Standard marker proteins; TS, total soluble protein fraction; HMW$\alpha$, high molecular weight $\alpha$-crystallin eluting at ~103 ml; $\alpha$, total $\alpha$-crystallin peak consisting of $\alpha$A, $\alpha$B, and a truncated $\alpha$A designated $\alpha_T$.

designated $\alpha_T$. Densitometric analysis of these bands reveals that the ratio of αA to αB is about 3:1, which is commonly found in many preparations. The truncated form of αA, $\alpha_T$, for this preparation comprises about 6% of the total α-crystallin. The densitometric analysis revealed that this preparation is at least 98% pure.

*Concentration of Samples and Storage.* To achieve high protein concentration (50 mg/ml), pooled fractions from the column chromatography can be concentrated by using Amicon concentrators (Centriplus 30; Amicon, Beverly, MA) in a preparative centrifuge at ~5000 rpm and 4°. Samples can be stored at −80° for relatively long periods. Repeated freezing and thawing of samples will cause degradation and insolubilization of the protein.

*Effects of Lens Age on α-Crystallin Preparation.* α-Crystallin undergoes many posttranslational modifications during the normal growth and aging of the lens.[1,2,12] Many of the modifications involve truncation. As shown in Fig. 2, even relatively young fetal lenses exhibit a significant amount of truncated αA. Therefore, for many applications whole fetal calf lenses may not be an optimum source of material. For most applications investigators use calf lenses or cow lenses that can be obtained readily at local slaughterhouses. In general, when older lenses are used, before homogenization, the central part of the lens (~50%) is removed with an appropriate cork borer, and only the cortical material is used. In cortical preparations, the amount of truncated α-crystallin is minimal, whereas in nuclear preparations of old lenses it can reach values of 26% of the total α-crystallin.[20,21]

## Expression of Recombinant α-Crystallin

It was first shown by Merck *et al.*[22] that α-crystallin can be expressed in its native conformation and in relatively high yields. Recombinant α-crystallin has properties similar to those of native α-crystallin isolated from eye lens. Recombinant α-crystallin is amenable to manipulations such as truncation, insertion, or site-directed mutagenesis.[7,23–25] These manipula-

[20] W. W. de Jong, F. S. M. van Kleef, and H. Bloemendal, *Eur. J. Biochem.* **48,** 271 (1974).

[21] F. S. M. van Kleef, M. J. C. M. Nijzink-Maas, and H. J. Hoenders, *Eur. J. Biochem.* **48,** 563 (1974).

[22] K. B. Merck, W. A. De Haard-Hoekman, B. B. Oude Essink, H. Bloemendal, and W. W. de Jong, *Biochim. Biophys. Acta* **1130,** 267 (1992).

[23] R. H. P. H. Smulders, K. B. Merck, J. Aendekerk, J. Horwitz, L. Takemoto, C. Slingsby, H. Bloemendal, and W. W. de Jong, *Eur. J. Biochem.* **232,** 834 (1995).

[24] R. H. P. H. Smulders, I. G. von Geel, W. L. H. Gerards, H. Bloemendal, and W. W. de Jong, *J. Biol. Chem.* **270,** 13916 (1995).

[25] R. H. P. H. Smulders, J. Carver, R. A. Lindner, M. A. M. van Boekel, H. Bloemendal, and W. W. de Jong, *J. Biol. Chem.* **271,** 29060 (1996).

tions provide a clear advantage for studying the function and structure of α-crystallin.

*Cloning of Human αB-Crystallin*

cDNA of αB-crystallin is amplified by polymerase chain reaction (PCR) from a human lens epithelial λ gt11 cDNA library using the following primer set:

Sense primer 5′ ATGGACATCGCCATCCACCACCCCTGGAT 3′
Antisense primer 5′ CATCTATTTCTTGGGGGCTGCGGTGACAGC 3′

The PCR reaction is carried out in a 50-μl volume containing 10 m*M* Tris-HCl (pH 8.3), 50 m*M* KCl, 1.5 m*M* $MgCl_2$, 50 m*M* dNTP, 1.0 μl of each primer, 2 μl of cDNA library, and 5 units of *Taq* polymerase. Amplification is performed for 35 cycles with conditions for denaturation at 94° for 1 min, annealing at 54° for 2 min, and extension at 72° for 3 min. The 531-base PCR product is gel purified and ligated into the pT7 blue vector (Noragen, Madison, WI). This construct introduces a single point mutation, changing His-7 to threonine. This is reverted to wild-type human αB-crystallin by using the following primer set:

Sense primer
5′ GGAATCTCATATGGACATCGCCATCCACCACCCCTGGAT 3′
Antisense primer
5′ GTATCGGATCCCATCTATTTCTTGGGGGCTGCGGTGACAGC 3′

Amplification is for 25 cycles using the preceding conditions and the resulting construct is digested with *Nde*I and *Xho*I, gel purified, and ligated into the pET20b(+) expression vector (Novagen, Madison, WI). The resulting 549-bp PCR product is sequenced with an Applied Biosystems (Foster City, CA) 3600 sequencer and verified to be human wild-type αB-crystallin. The flow chart for the expression of αB in pET-20b(+) is shown in Fig. 3. The purification of recombinant αB is presented in Fig. 4.

Figure 5 (lane 1) shows the protein distribution of the total cell lysate. The protein profile of the αB-rich fraction that is eluted with 0.1 *M* NaCl from the Mono Q column is shown in Fig. 5 (lane 2). This fraction is then concentrated, filtered, and chromatographed on a Pharmacia FPLC (fast protein liquid chromatography) system using a Superose 6 HR 10/30 column. The elution profile obtained from this fraction is shown in Fig. 6. The peak appearing between 10 and 15 ml contains the purified αB-crystallin. The purity of this fraction is assessed by SDS–PAGE using Coomassie blue staining as well as silver staining (see Fig. 5, lanes 3 and 4; and Fig. 6). Densitometric analysis of the purified αB-crystallin shows it to be at least 99% pure.

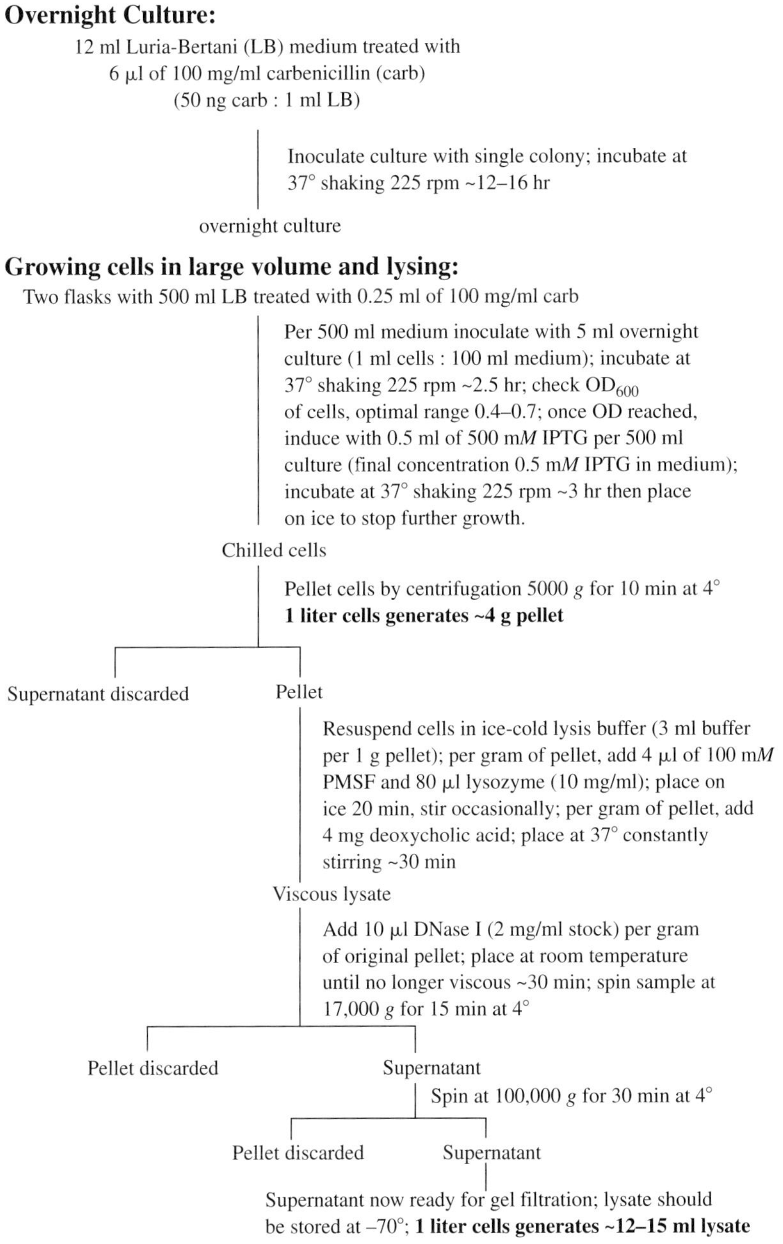

**Overnight Culture:**

12 ml Luria-Bertani (LB) medium treated with
6 μl of 100 mg/ml carbenicillin (carb)
(50 ng carb : 1 ml LB)

Inoculate culture with single colony; incubate at 37° shaking 225 rpm ~12–16 hr

overnight culture

**Growing cells in large volume and lysing:**

Two flasks with 500 ml LB treated with 0.25 ml of 100 mg/ml carb

Per 500 ml medium inoculate with 5 ml overnight culture (1 ml cells : 100 ml medium); incubate at 37° shaking 225 rpm ~2.5 hr; check $OD_{600}$ of cells, optimal range 0.4–0.7; once OD reached, induce with 0.5 ml of 500 m*M* IPTG per 500 ml culture (final concentration 0.5 m*M* IPTG in medium); incubate at 37° shaking 225 rpm ~3 hr then place on ice to stop further growth.

Chilled cells

Pellet cells by centrifugation 5000 *g* for 10 min at 4°
**1 liter cells generates ~4 g pellet**

Supernatant discarded | Pellet

Resuspend cells in ice-cold lysis buffer (3 ml buffer per 1 g pellet); per gram of pellet, add 4 μl of 100 m*M* PMSF and 80 μl lysozyme (10 mg/ml); place on ice 20 min, stir occasionally; per gram of pellet, add 4 mg deoxycholic acid; place at 37° constantly stirring ~30 min

Viscous lysate

Add 10 μl DNase I (2 mg/ml stock) per gram of original pellet; place at room temperature until no longer viscous ~30 min; spin sample at 17,000 *g* for 15 min at 4°

Pellet discarded | Supernatant

Spin at 100,000 *g* for 30 min at 4°

Pellet discarded | Supernatant

Supernatant now ready for gel filtration; lysate should be stored at –70°; **1 liter cells generates ~12–15 ml lysate**

FIG. 3. Flow diagram for the expression of αB in a pET20b(+) vector. IPTG, Isopropylthiogalactoside; PMSF, phenylmethylsulfonyl fluoride.

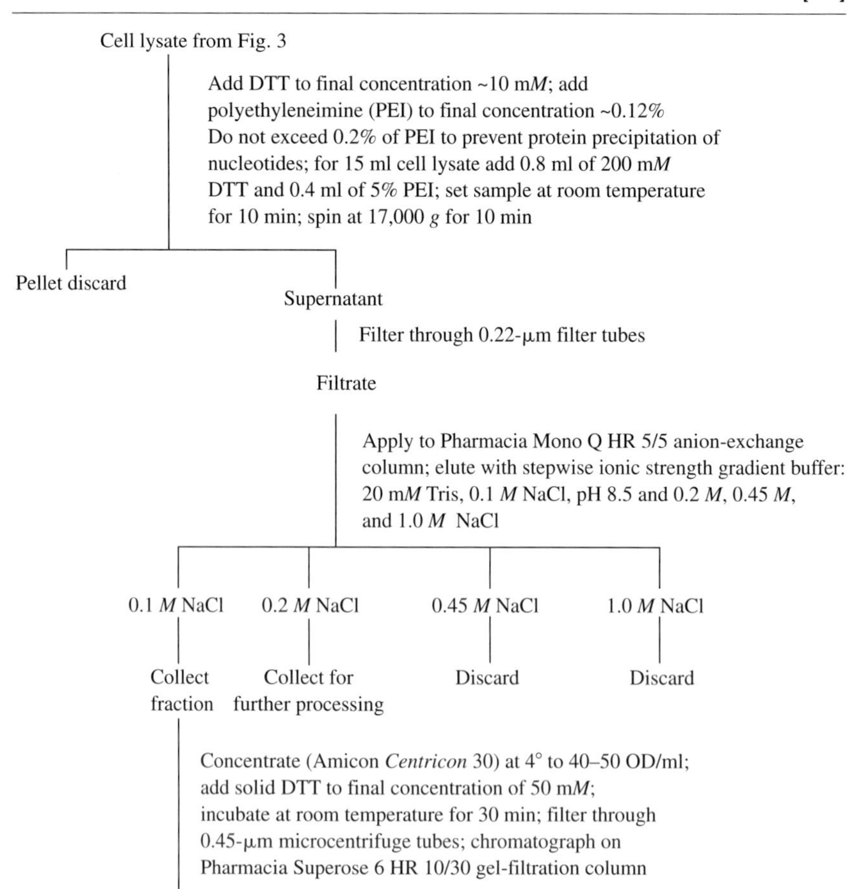


FIG. 4. Flow diagram for the purification of recombinant αB-crystallin.

## *Comments on Purification Procedures*

The amount of αB-crystallin obtained at 0.1 *M* NaCl is about 45–50% of the total recombinant αB. An additional 30–40% can be eluted at the second step of the gradient (0.2 *M* NaCl). However, in this fraction there is relatively more contamination from bacterial proteins. Therefore, for obtaining highly purified preparations, the use of the first gradient fraction is preferred.

Nucleotides that bind the recombinant α-crystallin preparations could be a major source of contamination. This contamination will not show on the SDS gels. Therefore, absorption spectra of the preparation should be

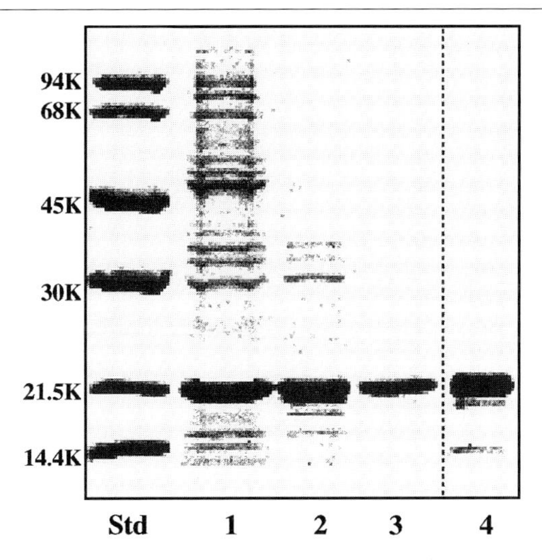


FIG. 5. SDS–polyacrylamide gel electrophoresis (12.5%, w/v) of the recombinant proteins expressed in the pET20b(+) vector. Std, standard marker proteins; lane 1, total cell lysate; lane 2, protein fraction eluted with 0.1 *M* NaCl from the Mono Q HR 5/5 column; lane 3, purified αB after chromatography on the Superose 6 HR 10/30 column (1.5 μg of protein was loaded in lane 3; lanes 1, 2, and 3 were stained with Coomassie blue); lane 4, silver staining of 1 μg of purified recombinant αB-crystallin. (See text for details.)

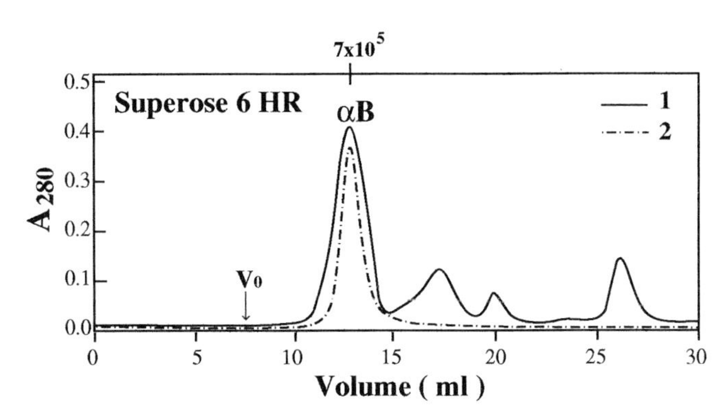


FIG. 6. Gel-permeation chromatography on a Superpose 6 HR 10/30 column of the proteins eluted with 0.1 *M* NaCl from the Mono Q HR 5/5 column. Purified αB-crystallin is collected between 10 and 15 ml. Curve 1 (solid line), 0.1 *M* NaCl fraction from Mono Q; curve 2 (dashed line), rechromatography of the αB fraction taken from the 10- to 15-ml fraction of curve 1.

taken and the ratio of $A_{280}$ to $A_{260}$ should be determined. A clean preparation of recombinant $\alpha$B should have a $A_{280}/A_{260}$ ratio of $\geq$1.5.

### *Selected Properties of Recombinant $\alpha$B and $\alpha$A*

Figure 7 represents the molecular weight distribution of recombinant $\alpha$B, recombinant $\alpha$A, and calf lens total $\alpha$-crystallin. The average molecular weight of native lens total $\alpha$-crystallin is always higher than that of the recombinant $\alpha$B and $\alpha$A. In comparing the width of the elution profiles (at one-half their peak heights), we note that recombinant $\alpha$B is ~35% less polydispersed than native or recombinant $\alpha$A-crystallin. The far- and near-ultraviolet circular dichroism (CD) spectra, reflecting the secondary and tertiary properties of the various $\alpha$-crystallin fractions, are given in Fig. 8. The secondary structure of recombinant $\alpha$A and $\alpha$B are similar. The near-ultraviolet circular dichroism spectra reflecting contributions from tryptophan, tyrosine, and phenylalanine side chains are significantly different (see Fig. 8B). It is interesting to note that the near-ultraviolet circular dichroism spectrum of calf eye lens total $\alpha$-crystallin (where the ratio of $\alpha$A to $\alpha$B is ~3 : 1) is similar to that of pure recombinant $\alpha$B. Table I summarizes some of the biochemical and biophysical properties of these protein fractions.

## Assays for Chaperone-Like Activity of $\alpha$-Crystallin

One of the functions of chaperones is to suppress aggregation of nonnative forms of proteins during the refolding or unfolding processes. The appearance of high molecular weight nonspecific aggregates during un-

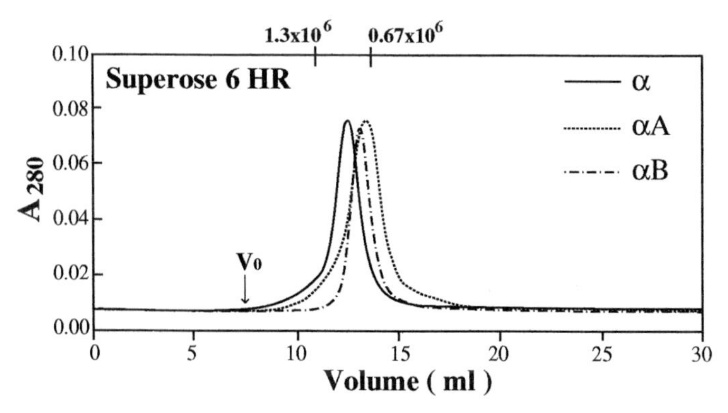


FIG. 7. Gel-permeation elution profiles of native calf lens total $\alpha$-crystallin, recombinant $\alpha$A, and recombinant $\alpha$B. Note that recombinant $\alpha$B is less polydispersed.

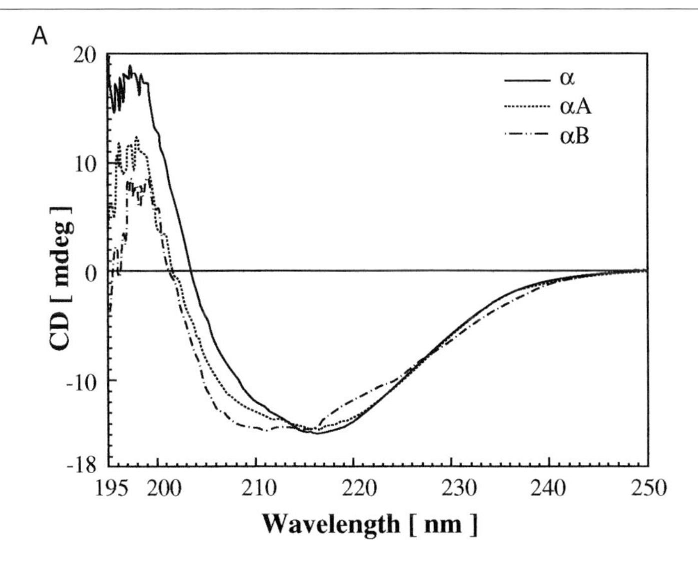


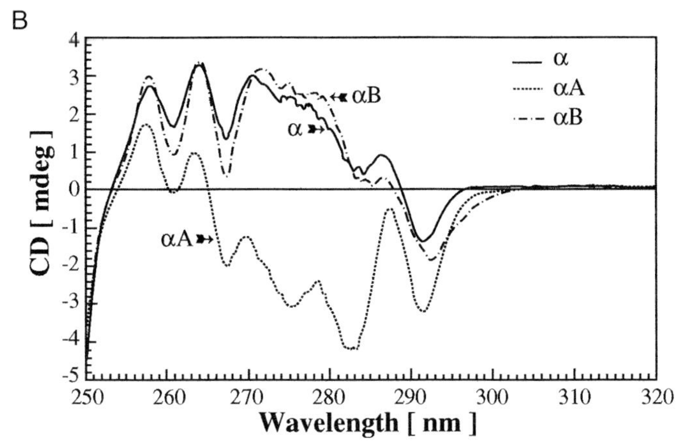


FIG. 8. (A) Far-ultraviolet CD spectra of calf total $\alpha$-crystallin and recombinant $\alpha$A and $\alpha$B. Each spectrum represents the average of 16 scans. The protein concentration of each sample was ~1.8 mg/ml. The path length was 0.2 mm. (B) Near-ultraviolet CD spectra of calf total $\alpha$-crystallin and recombinant $\alpha$A and $\alpha$B. Each spectrum represents the average of 32 scans. The protein concentration of each sample was ~1.8 mg/ml. The path length was 10 mm.

TABLE I
PROPERTIES OF RECOMBINANT HUMAN αB-CRYSTALLIN AND RECOMBINANT BOVINE αA-CRYSTALLIN

| Average apparent native molecular mass[a] (subunit molecular mass) | Apparent p$I$[b] | Molar absorptivity[c] $\varepsilon_{280}$ ($M^{-1}$ cm$^{-1}$) | $A_{280}/A_{260}$ | Secondary structure from CD measurements[d] | | | |
|---|---|---|---|---|---|---|---|
| | | | | α Helix | β Sheet | β Turns | Other |
| αA[e] | 5.3 | 16,500 | 1.6 | I. 0.14 | 0.40 | 0.16 | 0.30 |
| 0.65 MDa | | | | II. 0.17 | 0.37 | 0.24 | 0.23 |
| (19,790 Da) | | | | III. 0.19 | 0.51 | 0.18 | 0.12 |
| αB | 7.5 | 19,000 | 1.5 | I. 0.11 | 0.46 | 0.13 | 0.29 |
| 0.70 MDa | | | | II. 0.13 | 0.41 | 0.24 | 0.22 |
| (20,159 Da) | | | | III. 0.15 | 0.60 | 0.14 | 0.10 |

[a] Native molecular mass was estimated by gel-permeation chromatography on Pharmacia Superose 6 HR 10/30 column using the following standards: thyroglobulin dimer, 1300 kDa; thyroglobulin, 669 kDa; ferritin, 440 kDa.

[b] At 20°.

[c] Molar absorptivities were calculated acccording to the method of Gill and von Hippel [S. C. Gill and P. H. von Hippel, *Anal. Biochem.* **182,** 319 (1989)].

[d] CD measurements were performed on a Jasco model 600 spectropolarimeter using a Softsec (Softwood Co., Brookfield, CT) file conversion program. Estimation of the secondary structure parameters was done by using the self-consistent method of analysis. [N. Sreerama and R. W. Woody, *Anal. Biochem.* **209,** 32 (1993). The submethods were as follows: I, Hennessey and Johnson; II, Kabsch and Sander; III, Levvit and Greer.

[e] Recombination αA-crystallin was purified according to the scheme outlined in Fig. 4, except that the expression vector of Merck *et al.* was used [K. B. Merck, W. A. de Haard-Hoekman, B. B. Oude Essink, H. Bloemendal, and W. W. de Jong, *Biochem. Biophys. Acta* **1130,** 267 (1992)].

folding or denaturation can be readily monitored by measuring their increase of light scattering.[26,27] Therefore, a scattering apparatus, a fluorometer, or a spectrophotometer can be used to measure the scattering properties of the system in question. In general, a spectrophotometer can be used at wavelengths at which there is little or no absorption of the proteins. Because the scattering intensity is proportional to $1/\lambda^4$, the measuring system will be more sensitive at shorter wavelengths. For proteins with no prosthetic groups a wavelength of ~360 nm is commonly used.

### *Unfolding Proteins by Heat*

A common procedure for unfolding a target protein is by using elevated temperature. α-Crystallin can suppress nonspecific aggregation of unfolded

[26] J. Buchner, M. Schmidt, M. Fuchs, R. Jaenicke, R. Rudolph, F. X. Schmid, and T. Kiefhaber, *Biochemistry* **30,** 1586 (1991).

[27] J. Buchner, *FASEB J.* **10,** 10 (1996).

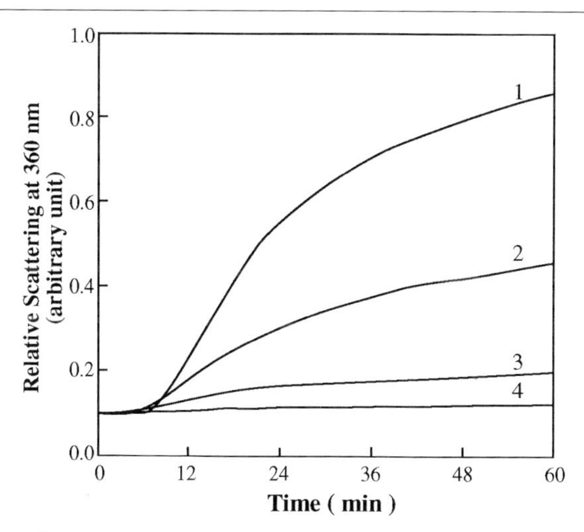


FIG. 9. Aggregation of unfolded yeast alcohol dehydrogenase at 37°. Yeast alcohol dehydrogenase was dissolved in a buffer containing 50 m*M* sodium phosphate, 0.1 *M* NaCl (pH 7), and 1 m*M* 1,10-phenanthroline. The total volume in each experiment was 0.4 ml. Curve 1, 0.2 mg of alcohol dehydrogenase; curve 2, 0.2 mg of alcohol dehydrogenase plus 0.02 mg of $\alpha$B; curve 3, 0.2 mg of alcohol dehydrogenase plus 0.04 mg of $\alpha$B; curve 4, 0.2 mg of alcohol dehydrogenase plus 0.2 mg of $\alpha$B. The scattering was recorded in a 1.0-cm cell in a spectrophotometer at 360 nm and 37°.

proteins at temperatures as high as 66°. In general, the higher the temperature the more effective $\alpha$-crystallin seems to be.[28] However, there is evidence that suggests that at temperatures higher than 45° $\alpha$-crystallin undergoes conformational changes that are not reversible.[29–31] This may complicate the interpretation of the experiments because, at high temperatures, both $\alpha$-crystallin and the target protein undergo conformational changes.

### *Choosing Buffers for Heat Denaturation Experiments*

The pH of many buffers varies significantly with temperature. This may add additional complications in interpreting the experiment. Tris(hydroxymethyl)aminomethane buffer has a $\Delta pH/dt$ of $-0.028$ (units per degree Celsius) and this may not be a suitable buffer for experiments in which the temperature is changed over a wide range. Phosphate buffer, on the other

[28] B. Raman, T. Ramakrishna, and Ch. Mohan Rao, *FEBS Lett.* **365,** 133 (1995).
[29] A. Trdieu, D. Laporte, P. Licinio, B. Krop, and M. Delaye, *J. Mol. Biol.* **192,** 711 (1986).
[30] W. K. Surewicz and P. R. Olsen, *Biochemistry* **34,** 9655 (1995).
[31] K. P. Das and W. K. Surewicz, *FEBS Lett.* **369,** 321 (1995).

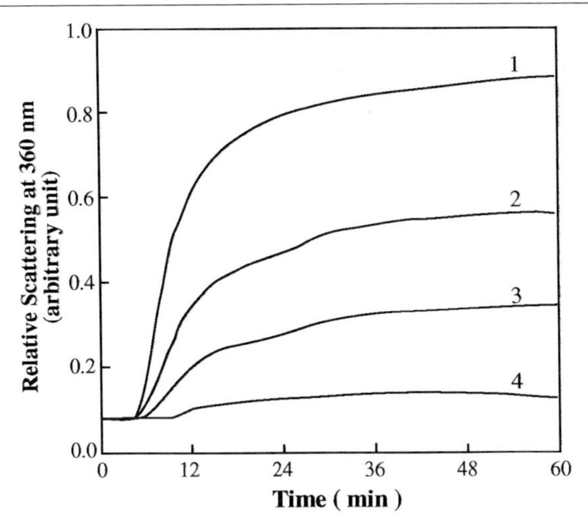


FIG. 10. Aggregation of insulin B chain at 37°. Insulin samples were prepared as discussed in text. At time 0, 20 m*M* DTT is added. Curve 1, 0.1 mg of insulin; curve 2, 0.1 mg of insulin plus 0.025 mg of $\alpha$B; curve 3, 0.1 mg of insulin plus 0.033 mg of $\alpha$B; curve 4, 0.1 mg of insulin plus 0.05 mg of $\alpha$B. The scattering was recorded in a 1.0-cm cell in a spectrophotometer at 360 nm and 37°.

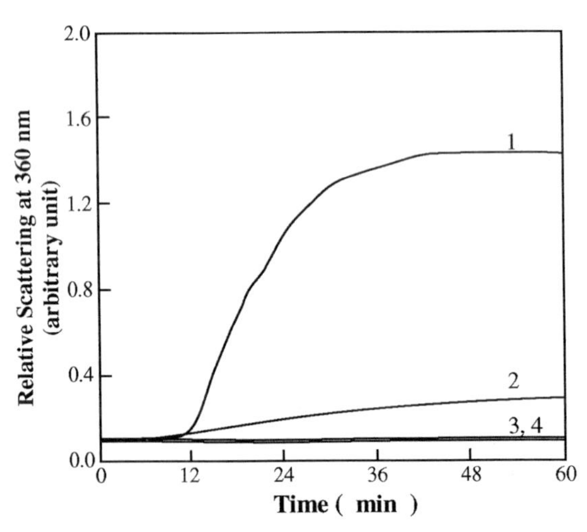


FIG. 11. Aggregation of reduced $\alpha$-lactalbumin at 37°. Lactalbumin samples are prepared as described in text. At time 0, DTT is added (10 to 50 m*M*). Curve 1, 0.5 mg of lactalbumin; curve 2, 0.5 mg of lactalbumin plus 1.0 mg of $\alpha$A; curve 3, 0.5 mg of lactalbumin plus 1.0 mg of total calf lens $\alpha$-crystallin; curve 4, 0.5 mg of lactalbumin plus 0.5 mg of recombinant $\alpha$B. The total volume in each experiment was 0.4 ml. The scattering was recorded in a 1.0-cm cell in a spectrophotometer at 360 nm and 37°.

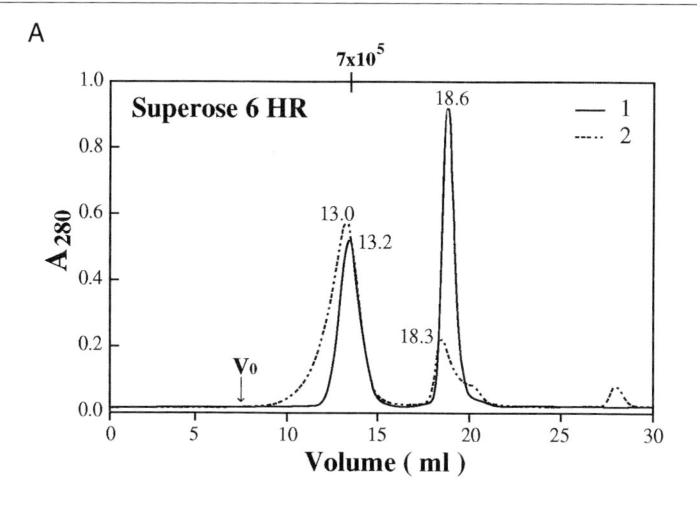


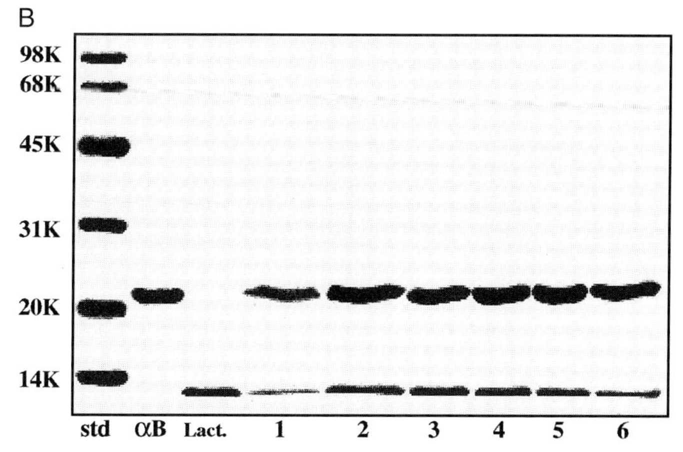


FIG. 12. (A) Gel-permeation chromatography on a Superose 6 HR 10/30 column of a mixture of recombinant $\alpha$B-crystallin and $\alpha$-lactalbumin (2:1, w/w) with and without and addition of DTT. Curve 1 (solid line), without DTT; curve 2 (dashed line), with 10 m$M$ DTT and after incubation at 37° for 2 hr. See text for details. (B) SDS–polyacrylamide gel electrophoresis (12.5%, w/v) of fractions collected from curve 2 in (A). Lancs 1 through 6 represent the fractions collected between 7.9 and 14 ml. See text and Table II for details.

hand, has a $\Delta$pH/$dt$ of $-0.0028$ and is better suited for heat denaturation experiments.

*Choice of Proteins*

The choice of a particular protein for studying its interaction with $\alpha$-crystallin will depend on the specific question asked. The interactions of

TABLE II
DENSITOMETRIC ANALYSIS OF COMPLEXES FORMED BETWEEN $\alpha$B AND $\alpha$-LACTALBUMIN[a]

| Lane | Volume (ml) | Molar ratio ($\alpha$B : lactalbumin) |
|---|---|---|
| 1 | 7.9–8.9 | 1 : 1 |
| 2 | 9.9 | 1 : 1 |
| 3 | 10.8 | 1 : 1 |
| 4 | 11.8 | 1.6 : 1 |
| 5 | 12.8 | 2.3 : 1 |
| 6 | 13.8 | 2.3 : 1 |

[a] As shown in Fig. 12A and B.

many enzymes and structural proteins with $\alpha$-crystallin were studied under a variety of conditions and over a wide temperature range.[5,6,32–35]

*Interaction of Unfolded Yeast Alcohol Dehydrogenase with $\alpha$B*

A typical experiment showing the unfolding of a protein at 37° is given in Fig. 9. Yeast alcohl dehydogenase (ALDH) can be rendered labile at this temperature by adding 1 m$M$ 1,10-phenanthroline to the buffer. Figure 9 shows that under these conditions the yeast alcohol dehydrogenase will start to aggregate within 10 min. Titration of the enzyme with $\alpha$B-crystallin for example, will arrest the nonspecific aggregation. The stable complex that is formed between $\alpha$-crystallin and alcohol dehydrogenase can be analyzed by using other biochemical and biophysical methodologies.

*Unfolding of Proteins by Breaking Disulfide Bonds*

Several proteins can be unfolded by reducing the interchain disulfide bond. Because $\alpha$-crystallin does not possess any disulfide bonds, these proteins are suitable target proteins for experimentation at physiological or lower temperatures. Two readily available proteins for such studies are insulin[36] and $\alpha$-lactalbumin. On reduction of the disulfide bonds of insulin with dithiothreitol, for example, the insulin B chain will aggregate and precipitate while the A chain remains in solution. With $\alpha$-lactalbumin, the

[32] P. V. Rao, J. Horwitz, and S. J. Zigler, *Biochem. Biophys. Res. Commun.* **190,** 786 (1993).
[33] L. Takemoto, T. Emmons, and J. Horwitz, *Biochem. J.* **294,** 435 (1993).
[34] K. Wang and A. Spector, *J. Biol. Chem.* **269,** 13601 (1994).
[35] P. V. Rao, J. Horwitz, and S. J. Zigler, *J. Biol. Chem.* **269,** 132668 (1994).
[36] Z. T. Farabakhsh, Q. L. Huang, L. L. Ding, C. Altenbach, J. H. Steinhoff, J. Horwitz, and W. L. Hubbell, *Biochemistry* **34,** 509 (1995).

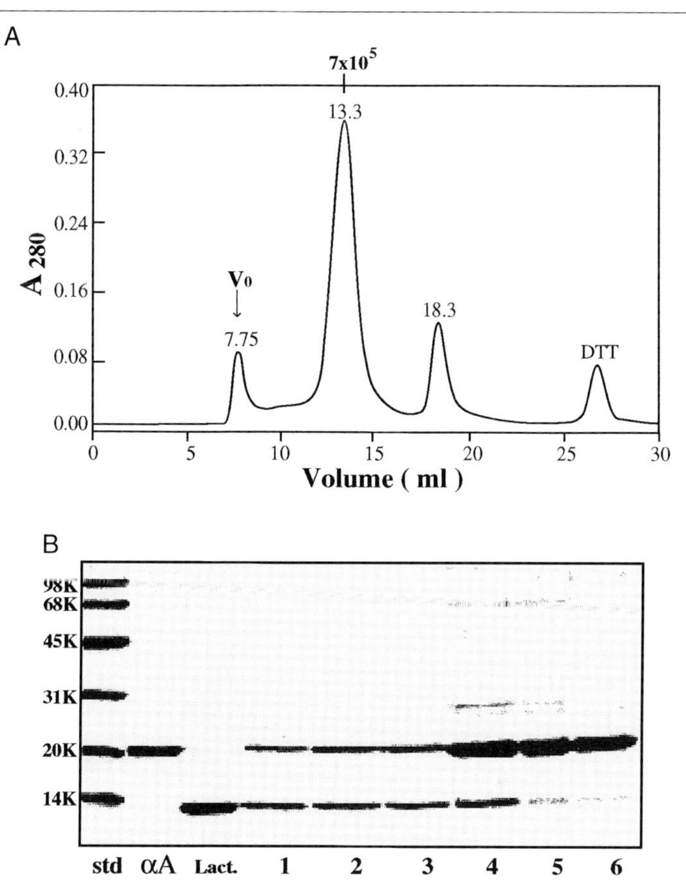


FIG. 13. (A) Gel-permeation chromatography on a Superose 6 HR 10/30 column of a mixture of recombinant $\alpha$A-crystallin and $\alpha$-lactalbumin (4:1, w/w) after the addition of 50 m*M* DTT and incubation at 37° for 2 hr. (B) SDS–polyacrylamide gel electrophoresis (12.5%, w/v) of fractions collected from (A). Lanes 1 through 6 represent the fractions collected between 7.5 and 14 ml. See text and Table III for details.

reduction of the disulfide will lead to nonspecific aggregation and precipitation. This unfolding process can be monitored by measuring the apparent absorbance due to the increase in scattering.

## Preparation of Stock Solutions of Insulin and $\alpha$-Lactalbumin

### *Insulin*

Dissolve 10 mg of insulin (Sigma, St. Louis, MO) in 1–2 ml of 0.1 *N* NaOH; rapidly bring up the pH of the solution to pH 7.2 with 0.5 *M* sodium phosphate buffer, pH 6.8. For insulin, 1 OD is 1 mg/ml. This stock solution can be kept over ice. Fresh solutions of insulin should be made daily. For the reduction of the insulin interchain disulfide, freshly prepared dithiothreitol (DTT) solution is used. The incubation mixture contains 50 m*M* sodium phosphate, 100 m*M* NaCl, pH 7.0 in a final volume of 0.4 ml. Insulin, α-crystallin, and DTT are added as needed.

*Bovine α-Lactalbumin*

Bovine α-lactalbumin (type III; Sigma) is dissolved in reaction buffer containing 50 m*M* sodium phosphate, 0.1 *M* NaCl (pH 7.0), 2 m*M* ethylenediaminetetraacetic acid (EDTA) and kept over ice. α-Crystallin and freshly made DTT are added as needed. Fresh solutions of lactalbumin should be prepared daily. For α-lactalbumin, 1 OD is ~0.5 mg/ml.

*Interaction of Reduced Insulin B Chain with αB-Crystallin*

Figure 10 shows the scattering due to the aggregation of insulin B chain at 37°, and the suppression of the scattering by titrating the system with recombinant αB. The insulin B chain forms a stable complex with αB-crystallin.[36]

*Interaction of α-Lactalbumin with α-Crystallin*

The aggregation of α-lactalbumin at 37° on the addition of DTT is shown in Fig. 11. Lens α-crystallin as well as recombinant αA and αB were capable of suppressing the scattering. The complex formed by such an experiment can be examined using gel-permeation chromatography and SDS–polyacrylamide gel electrophoresis. Figure 12A shows the elution

TABLE III
DENSITOMETRIC ANALYSIS OF COMPLEXES FORMED BETWEEN αA AND α-LACTALBUMIN[a]

| Lane | Volume (ml) | Molar ratio (αA : lactalbumin) |
|---|---|---|
| 1 | 7.5 | 1:2 |
| 2 | 10 | 1:1 |
| 3 | 11 | 1:1 |
| 4 | 12 | 4:1 |
| 5 | 13 | 16:1 |
| 6 | 14 | — |

[a] As shown in Fig. 13A and B.

profile of a mixture of α-lactalbumin and recombinant αB-crystallin under native conditions, and after incubating with 10 m*M* DTT for 2 hr at 37°. There is no interaction between the two proteins without the DTT reduction. Following the reduction of lactalbumin, stable higher molecular weight complexes are formed between αB-crystallin and the unfolded lactalbumin. From measuring the area under the curves we can estimate about 55% of the lactalbumin is bound to αB-crystallin following the treatment with DTT. The fractions eluting between 10 and 12 ml have an estimated molecular weight of ~1 to 2 × $10^6$. SDS–polyacrylamide gel electrophoresis of the fractions containing the complex formed between αB-crystallin and α-lactalbumin is shown in Fig. 12B. The molar ratios between αB-crystallin monomers and α-lactalbumin can be calculated from densitometric analysis of the stained gel. These ratios are given in Table II.

### *Interaction of α-Lactalbumin with Recombinant αA*

Recombinant αA-crystallin was less effective than recombinant αB or lens total α-crystallin in suppressing the aggregation of the unfolded α-lactalbumin (see Fig. 11). In general, higher amounts of recombinant αA-crystallin were needed to achieve complete suppression of the scattering.

Figure 13A shows the complex formed between αA and α-lactalbumin following reduction of the disulfide bridges. A significant amount of the complex eluted at the void volume (molecular weight ≥5 × $10^6$). SDS–polyacrylamide gel electrophoresis of the fraction containing the complex is shown in Fig. 13B. The molar ratio of the complex between α-lactalbumin and αA-crystallin monomer is given in Table III.

On the other hand, total lens α-crystallin (with an αA-to-αB ratio of ~3 : 1) behaved in a manner identical to that observed for recombinant αB (Fig. 11). Therefore, even though αA and αB are related proteins with 57% sequence homology they do interact differently with this target protein.

## Acknowledgments

The authors thank Professor W. W. de Jong for the expression clone pET8c αA and Arlene Horwitz and Stacey Lowman for help with the manuscript. Supported by National Eye Institute Merit Award R37-EY03897 to J.H. and by a Shojin Research Grant to Q.L.H.

# [31] Purification and Properties of BiP

*By* MATHIEU CHEVALIER, LASHAUNDA KING, and SYLVIE BLOND

## Introduction

BiP is a member of the heat shock protein 70 (HSP70) family localized in the endoplasmic reticulum of eukaryotic cells. BiP is an abundant and essential protein involved in polypeptide translocation across the endoplasmic reticulum (ER) membrane, and in folding and assembly of secreted or membrane proteins.[1] BiP is a peptide-binding protein with an associated ATPase activity. It consists of three domains: (1) A 44-kDa catalytic N-terminal domain encompassing the ATPase activity, whose primary sequence is highly conserved among the members of the HSP70s and whose three-dimensional structure is similar to those of actin and hexokinase,[2,3] (2) an 18-kDa peptide-binding domain that has an all-$\beta$ sheet three-dimensional structure,[4–6] and (3) a least conserved C-terminal tail that contains a regulatory motif[7] and carries some information for the oligomerization of the protein (our data, unpublished, 1995). The minimum length of peptidic substrates required to fully stimulate the ATPase activity has been determined to be seven residues rich in hydrophobic amino acids, especially tryptophan, phenylalanine, and leucine residues.[8–10] It is still not clear if the preference of BiP for hydrophobic stretches of amino acids specifies an active role of the chaperone in translocation and folding of newly synthesized polypeptide chains. Two models are currently available: (1) One model assumes that subtle differences in substrate specificity among the

[1] M.-J. Gething, S. Blond-Elguindi, K. Mori, and J. F. Sambrook, *in* "The Biology of Heat Shock Proteins and Molecular Chaperones" (R. I. Morimoto, A. Tissieres, C. Georgopoulos, eds.), p. 111. Cold Spring Harbor Laboratory Press, Cold Spring Harbor, New York, 1994.

[2] K. Flaherty, C. DeLuca-Flaherty, and D. McKay, *Nature* (*London*) **346,** 623 (1990).

[3] K. M. Flaherty, D. B. McKay, W. Kabash, and K. C. Holmes, *Proc. Natl. Acad. Sci. U.S.A.* **88,** 5041 (1991).

[4] T. Wang, J. Chang, and C. Wang, *J. Biol. Chem.* **268,** 26049 (1993).

[5] R. C. Morshauser, H. Wang, G. C. Flynn, and E. R. P. Zuiderweg, *Biochemistry* **34,** 6261 (1995).

[6] X. Zhu, X. Zhao, W. F. Burkholder, A. Gragerov, C. M. Ogata, M. E. Gottesman, and W. A. Hendrickson, *Science* **272,** 1606 (1996).

[7] B. Freeman, M. Myers, R. Schumacher, and R. Morimoto, *EMBO J.* **14,** 2281 (1995).

[8] G. C. Flynn, T. G. Chappell, and J. E. Rothman, *Science* **245,** 385 (1989).

[9] G. C. Flynn, T. G. Chappell, and J. E. Rothman, *Nature* (*London*) **353,** 726 (1991).

[10] S. Blond-Elguindi, S. E. Cwirla, W. J. Dower, R. J. Lipshutz, S. R. Sprang, J. F. Sambrook, and M.-J. Gething, *Cell* **75,** 717 (1993).



0076-6879/98 $25.00

HSP70s are the support for their distinct function. Indeed, although all HSP70s bind to hydrophobic peptides, differences in the spectrum and affinity of recognized substrates have been observed among members of the family.[11–14] Furthermore, different HSP70s are not functionally interchangeable in yeast[15]; (2) a second model proposes that the function of HSP70s would be regulated via interactions with other cellular components, such as DnaJ-related proteins,[16] and that the peptide-binding specificity would play little or no role in HSP70 functions.[17] Both models are supported by a distinct set of data and more experimental evidence is needed to determine if peptide specificity plays a role in HSP70 biological function and how the ATPase and substrate recognition activities of individual members of the family are being regulated in each of the cellular compartments and organisms.

This chapter focuses on different approaches that can be used to biochemically characterize mammalian BiP. Several of the methods can be adapted to any particular protein [polymerase chain reaction (PCR) amplification and gene subcloning into prokaryotic expression vectors, protein purification from inclusion bodies], multidomain proteins (isolation and characterization of BiP domains by direct expression or engineering of an enterokinase site in between the two domains), and secreted proteins (preparation of *Escherichia coli* periplasmic space and mammalian microsomes), to the applications and limitations of histidine and epitope tagging in the purification of recombinant proteins; and of course to the characterization of HSP70-related proteins. We have also included a detailed protocol for the characterization of peptide specificity by using affinity panning of phage display libraries that includes biotinylation of BiP and the use of the avidin–biotin system to immobilize protein on enzyme-linked immunosorbent assay (ELISA) microtitration plates. This approach has been used successfully[10,18,19] and has proved to be a fast and relatively economical

[11] A. M. Fourie, J. F. Sambrook, and M.-J. Gething, *J. Biol. Chem.* **269,** 30470 (1994).

[12] M.-J. Gething, S. Blond-Elguindi, J. Buchner, A. Fourie, G. Knarr, S. Modrow, L. Nanu, M. Segal, and J. Sambrook, *in* "Protein Kinesis: The Dynamics of Protein Trafficking and Stability," Vol. LX of Cold Spring Harbor Symposia on Quantitative Biology, p. 417. Cold Spring Harbor Laboratory Press, Cold Spring Harbor, New York, 1995.

[13] A. Gragerov and M. E. Gottesman, *J. Mol. Biol.* **241,** 133 (1994).

[14] M.-J. Gething, *Curr. Biol.* **6,** 1573 (1996).

[15] J. L. Brodsky, S. Hamamoto, D. Feldheim, and R. Schekman, *J. Cell Biol.* **120,** 95 (1993).

[16] A. J. Caplan, D. M. Cyr, and M. G. Douglas, *Mol. Biol. Cell* **4,** 555 (1993).

[17] P. James, C. Pflund, and E. A. Craig, *Science* **275,** 387 (1997).

[18] I. M. Takenaka, S.-M. Leung, S. J. McAndrew, J. P. Brown, and L. E. Hightower, *J. Biol. Chem.* **270,** 19839 (1959).

[19] A. Gragerov, L. Zeng, X. Zhao, W. Burkholder, and M. E. Gottesman, *J. Mol. Biol.* **235,** 848 (1994).

approach for the identification of high-affinity substrates for HSP70s. We have also used this technique to identify epitopes recognized by monoclonal antibodies and used the information to tag recombinant proteins (S. Blond, unpublished, 1994). Such peptide libraries are now commercially available (New England Biolabs, Beverly, MA) or can be prepared and customized to the needs of the experimentor by using one of the kits currently available (Novagen, Madison, WI), or by following the original method described by Smith and Scott.[20] Finally, peptides can be prepared directly from the expressing phages, at a much lower cost than via chemical synthesis, by inserting a protease cleavage site after the sequence of interest. The method given here can be adapted to any protein that exhibits a peptidic substrate recognition activity (molecular chaperones, antibodies, and other peptide acceptors, peptidic hormone receptors, enzyme inhibitors) and to other applications of phage display libraries.[21]

## Purification of BiP from Animal Tissues

### *Buffers and Solutions*

Buffer A: 50 m*M* triethanolamine (TEA, pH 7.5 with acetic acid), 50 m*M* potassium acetate, 6 m*M* magnesium acetate, 1 m*M* ethylenediaminetetraacetic acid (EDTA), 250 m*M* sucrose, 1 m*M* dithiothreitol (DTT), 0.5 m*M* phenylmethylsulfonyl fluoride (PMSF)

Buffer B: 250 m*M* sucrose, 50 m*M* TEA (pH 7.5), 1 m*M* DTT, 0.5 m*M* PMSF

Buffer X: Buffer B, 100 m*M* KCl, 5 m*M* EDTA

Sucrose cushion: 1.3 *M* in 50 m*M* TEA (pH 7.5), 1 m*M* DTT, 0.5 m*M* PMSF (index of refraction ~960 at 20°)

Sucrose cushion: 0.5 *M* in 50 m*M* TEA (pH 7.5), 1 m*M* DTT, 0.5 m*M* PMSF (index of refraction ~574 at 20°)

Buffer C: 25 m*M* Tris-HCl (pH 7.5), 50 m*M* KCl, 5 m*M* $MgCl_2$, 250 m*M* sucrose, 5 $\mu M$ leupeptin (Sigma, St. Louis, MO), 100 mg of PMSF per liter

Buffer D: 25 m*M* Tris-HCl (pH 7.5), 500 m*M* KCl, 2.5 m*M* EDTA, 5 m*M* $MgCl_2$, 250 m*M* sucrose, 100 mg of PMSF per liter

Purification buffer (2×): 50 m*M* 4-(2-hydroxyethyl)-1-piperazine-ethanesulfonic acid (HEPES, pH 7.0), 100 m*M* KCl, 10 m*M* $MgCl_2$, 1 m*M* DTT, 0.5 m*M* PMSF. Add 1 *M* NaCl in 2× purification buffer if bovine microsomes are the source used for BiP purification

[20] G. P. Smith and J. K. Scott, *Methods Enzymol.* **217,** 228 (1993).

[21] K. T. O'Neil and R. H. Hoess, *Curr. Opin. Struct. Biol.* **5,** 443 (1995).

ATP–Agarose (Sigma)
DEAE-Sephacel (Pharmacia, Piscataway, NJ): Mono Q or Resource Q (Pharmacia) can substitute
Protein assay reagent (Bio-Rad, Hercules, CA): BCA reagent (Pierce, Rockford, IL) can substitute

*Microsome Preparations from Dog Pancreas*

We use a method adapted from Walter and Blobel.[22] The purity of BiP obtained from stripped dog pancreas is high, and the yields obtained usually range from 1 to 2 mg of BiP per pancreas or from 0.1 to 0.4% of the total microsomal proteins (calculations made after microsome stripping). We recommend working on pancreas taken from healthy, unstressed animals as we observed that stress, such as lengthy exercise, does not allow proper sedimentation of the microsomes, probably because most of the ribosomes dissociate from the outer surface of the endoplasmic reticulum (S. Blond, unpublished, 1992). Dogs as a source for pancreas should be used only if the animals are being sacrificed for other studies. Medical or veterinarian schools may sometimes have a limited number of animals available for their students to use in practicing cardiovascular surgery.

Excise the pancreas immediately after sacrifice of the animal, rinse three times with 50 ml of ice-cold buffer A, and immerse in 50 ml of buffer A on ice. Proceed rapidly and work in a cold room for all subsequent steps. Place the pancreas on a metal plate on ice, clean away connective tissues and blood vessels, mince well with a razor blade, and homogenize the pancreas (resuspended 5 g at a time in 4 ml of buffer A) in a glass–Teflon Potter–Elvehjem. Five slow strokes up and down are sufficient to obtain a satisfactory homogenate. Centrifuge for 10 min at 5000 *g* at 4°. Remove the lipids floating at the surface by aspiration and recentrifuge for 10 min at 10,000 *g* at 4°. Remove the cloudy upper phase and gently apply 20 ml of the lower limpid phase onto 10 ml of a 1.3 *M* sucrose cushion in a 30-ml ultracentrifuge tube. Centrifuge for 150 min at 140,000 *g* at 4°. Carefully remove most of the supernatant without disrupting the pellet. Wash the pellet with ice-cold buffer B. To prevent contamination by cytosolic HSP70s, it is necessary to strip the microsomes with 5 m*M* EDTA and a high concentration of KCl salt (100 m*M*) or as follows. Resuspend the rough microsomes in 20 ml of ice-cold buffer X and load on 10 ml of a 0.5 *M* sucrose cushion. Centrifuge for 1 hr at 140,000 *g* at 4°. Resuspend the pellets in buffer B at about 20 mg of total protein per milliliter, freeze in liquid nitrogen, and store at −70° (solutions remain stable for several years).

[22] P. Walter and B. Blobel, *Methods Enzymol* **104,** 84 (1983).

*Microsome Preparations from Beef Liver*

We use a protocol set up in the laboratory of M.-J. Gething by L. Nanu (University of Melbourne, Australia), derived from a method sent to us by A. Puig and H. F. Gilbert (Baylor College of Medicine, Houston, TX). Several milligrams of BiP can be purified from 500 g of bovine liver. Purified BiP can make up 1 to 2% of the total microsomal proteins (after stripping). Work in a cold room, with all buffers and instruments prechilled. Weigh 500 g of bovine liver (obtained immediately after sacrifice of the animal and place on ice). Cut the liver into small pieces, and remove the connective tissue; mince well. Wash the minced liver in 0.9% (w/v) NaCl solution for 30 min at 4°. Homogenize the tissue in 500 ml of buffer C in a blender, avoiding heat formation (five pulses for 5 sec at medium or low speed). Filtrate the homogenate through cheese cloth. Centrifuge at 15,000 *g* for 20 min at 4°. Filter the supernatant through a precooled filter paper. Centrifuge at 100,000 *g* for 90 min at 4°. Resuspend each pellet in 15–20 ml of buffer D, and incubate for 30 min at 4°. Carefully place 20 ml of stripped microsomes onto 10 ml of a 15% (w/v) sucrose cushion [in 5 m*M* Tris-HCl (pH 7.5), 100 m*M* KCl], directly into 30-ml ultracentrifuge tubes. Centrifuge at 30,000 rpm for 75 min, discard the supernatant, and pool the interface and pellet that contains the stripped microsomes. Dilute with buffer C to achieve a final total protein concentration of about 20 mg/ml. Freeze in liquid nitrogen and store at −70°.

*BiP Purification from Stripped Dog or Bovine Microsomes*

Thaw the stripped microsomes and dilute twofold in 2× purification buffer. Add Triton X-100 to 1% (w/v) or Nikkol to 0.1% (w/v) to the suspension, which should become translucent. Centrifuge at 30,000 rpm for 75 min at 4°. If dog microsomes are being used, the addition of 0.5 *M* NaCl is not necessary. Because bovine microsomes tend to be dirtier and highly contaminated by hemoglobin, add NaCl (0.5 *M* final concentration) to the microsomal extract before proceeding with the affinity chromatography on ATP–agarose. Pack a 5- or 10-ml ATP–agarose column at 4° and equilibrate with 1× purification buffer supplemented with 0.5 *M* NaCl. Apply microsomal extract on ATP–agarose at a flow rate of 10 ml/hr. Wash with 50 ml of purification buffer supplemented with 0.5 *M* NaCl or until all hemoglobin is eluted, then with 50 ml of purification buffer supplemented with 1 m*M* NaCl, and finally with the same buffer supplemented with 3 m*M* GTP to remove non-BiP nucleotide-binding proteins. This last, costly wash is not necessary when ion-exchange chromatography on DEAE-Sephacel or Resource Q (Pharmacia) is done as a second purification step. BiP is eluted from the ATP–agarose column with purification buffer freshly supplemented with 3 m$M$ ATP (adjust the pH after addition of ATP and protect the buffer from light). Collect 2-ml fractions and monitor the elution by the Bradford protein assay method,[23] as ATP strongly absorbs light at 280 nm. If ion-exchange chromatography is being performed to further purify BiP, connect the ATP–agarose, after the washes and before the ATP elution, directly to the DEAE-Sephacel column equilibrated with purification buffer. Elute BiP from ATP–agarose as described previously, disconnect the two columns, and wash the ion-exchange column with 5 vol of 25 m$M$ Tris-HCl (pH 7.5), 20 m$M$ KCl, 5 m$M$ $MgCl_2$ or until the baseline comes back to 0 absorbance at 280 nm. Elute from the DEAE-Sephacel column with a 20–500 m$M$ KCl gradient ($V_{\text{total gradient}}$ = 40 ml for a 10-ml column). Mammalian BiP usually elutes at approximately 200 m$M$ KCl. Analyze the fractions by 10% (w/v) sodium dodecyl sulfate-polyacrylamide gel electrophoresis (SDS–PAGE)[24,25] and pool the BiP-containing fractions. If BiP is going to be used immediately for ATPase or peptide-binding assays, dialyze against assay buffer (described below). However, we recommend precipitating the BiP pool at 85% saturation of ammonium sulfate in order to concentrate the protein and remove the bound ATP from the BiP molecules. For best results, slowly add 5.59 g of finely ground ultrapure ammonium sulfate to 10 ml of the BiP solution. Incubate for 30 min with gentle stirring in a salted ice–water bath. Centrifuge for 30 min at 10,000 rpm at 4°, discard the supernatant, and resuspend the pellet so that the BiP concentration will be about 10–15 mg/ml. Dialyze exhaustively against 20 m$M$ HEPES (pH 7.0), 75 m$M$ KCl, 5 m$M$ $MgCl_2$ until no ammonium ions are detected by Nessler's reagent (Aldrich, Milwaukee, WI). Centrifuge the dialysate at 14,000 rpm for 30 min at 4° in a microcentrifuge, determine the BiP concentration,[23] aliquot small volumes in cryotubes (20 to 100 $\mu$l), freeze in liquid nitrogen, and store at −70°.

## Polymerase Chain Reaction Amplification and Subcloning into Expression Vectors: Generalities

The gene of interest can be subcloned into an expression vector by using single-site restriction digest and conventional molecular biology techniques.[26] However, in many cases, new restriction sites need to be engineered in order to insert an ATG starting codon (use *Nco*I or *Nde*I

[23] M. M. Bradford, *Anal. Biochem.* **72,** 248 (1976).

[24] U. K. Laemmli, *Nature* (*London*) **227,** 680 (1970).

[25] "Hoefer Protein Electrophoresis Applications Guide." Hoefer Scientific Instruments, San Francisco, 1994.

[26] J. Sambrook, E. F. Fritsch, and T. Maniatis, *in* "Molecular Cloning: A Laboratory Manual," 2nd Ed. Cold Spring Harbor Laboratory Press, Cold Spring Harbor, New York, 1989.

restriction sites) or to obtain in-frame fusion with a signal sequence, an epitope or histidine tag, a protease cleavage site, or a fusion partner that has a specific activity that will be used for purification or detection purposes. We use PCR amplification to efficiently subclone all of the inserts. Although we have never found any mutations resulting from the lack of fidelity of the enzyme used (*Pfu* DNA polymerase; Stratagene, La Jolla, CA), we recommend sequencing all inserts entirely by using any of the kits commercially available or core facilities when accessible. We refer the experimentor to the instruction manual delivered with the enzyme for all of the experimental designs related to PCR-based cDNA amplification. The two oligonucleotides should have a similar $T_m$ (between 55 and 80°). Any nonannealing restriction site can be added at the 5′ and 3′ ends as well as a few additional flanking bases to increase the efficiency of restriction digest when cleavage is close to the 5′ or 3′ end (refer to New England Biolabs catalog for information). We recommend purifying the obtained fragments before setting up the digestions overnight at 37°. The vector should be dephosphorylated after digestion and purified by one of the numerous available methods[26] or commercially available kits. We favor the technique that uses low-melting agarose,[26] as yield and DNA quality are far superior to any of the other methods we have tried. Estimate the concentrations of the cleaved, dephosphorylated, purified vector and the cleaved, purified insert and set up the ligations overnight at 16° (10 to 20 fmol of vector for a 1, 3, or 5 molar excess of insert in 10 $\mu$l of ligation buffer). Transform competent *E. coli* cells and analyze transformants as described.[26]

## Purification of Murine BiP Expressed in *Escherichia coli*

To express BiP or its constitutive domains, we recommend the pET expression system, which is based on the transcription of the gene of interest through the T7 promoter.[27] The gene encoding the T7 RNA polymerase is under the control of the *lacUV5* promoter, inducible by isopropyl-$\beta$-D-thiogalactopyranoside (IPTG) in BL21(DE3) cells. Many versions of pET vectors are commercially available (Novagen, Stratagene) and several of them have been successfully used in our laboratory and others. The only problem we have encountered, specific to BiP[28] and other HSP70s,[29] is the lethal effect observed when full-length BiP or its peptide-binding domain is expressed in *E. coli.* To maximize the level of expression,

[27] F. W. Studier, A. H. Rosenberg, J. J. Dunn, and J. W. Dubendorff, *Methods Enzymol.* **185,** 60 (1986).

[28] S. Blond-Elguindi, A. M. Fourie, J. F. Sambrook, and M.-J. H. Gething, *J. Biol. Chem.* **268,** 12730 (1993).

[29] W. F. Burkholder, X. Zhao, X. Zhu, W. A. Hendrickson, A. Gragerov, and M. E. Gottesman, *Proc. Natl. Acad. Sci. U.S.A.* **93,** 10632 (1996).

we recommend targeting the protein into the periplasmic space by an N-terminal fusion with a bacterial signal sequence, such as the OmpT sequence (MRAKLLGIVLTTPIAISSFAS), using only freshly transformed BL21(DE3) cells, and growing all cultures in M9 minimal medium that does not contain Bacto-tryptone. We also found that double transformation with a pLysS or pLysE plasmid[27] is an efficient way to inhibit toxic basal expression before the addition of IPTG.

*Purification of Recombinant BiP from Periplasmic Extracts*

We have subcloned the cDNA encoding mature murine BiP in pET12 vector (Novagene), as a fusion with the signal sequence of OmpT. The resulting plasmid, called SecB115, was used in our initial studies on the characterization of recombinant BiP.[28] We found that mouse BiP is less toxic and much more efficiently expressed when secreted in the periplasmic space of BL21(DE3) cells. Transformed BL21(DE3)/SecB115 cells are not stable and need to be freshly prepared each time. Competent BL21(DE3) cells remain efficient for about 6 months stored at −70° when prepared using the RbCl method.[26] Periplasmic extracts are prepared by cold osmotic shock.[30] The purification yields are relatively low (1 to 5 mg of BiP for 20 liters of culture) but the purity of the material obtained can reach 95–98% and is free from contamination by DnaK.

*Materials and Solutions*

M9 minimal medium[26]: To 1 liter of water, add 10 g of M9 base (GIBCO-BRL, Gaithersburg, MD), 0.5 g of NaCl, and 10 g of Bacto-tryptone (optional). Autoclave for 20 min in a 2-liter Erlenmeyer or a 2.5-liter Fernbach flask. When the M9 base has cooled, add 1 ml of 1 *M* $MgSO_4$ (autoclaved), 10 ml of 20% (w/v) glucose (sterilized by filtration through a 0.22-$\mu$m pore size filter; store at 4°), 1 ml of 1% (w/v) thiamine (sterilized by filtration through a 0.22-$\mu$m pore size filter; store at 4°, protect from light), 100 $\mu$l of 1 *M* $CaCl_2$ (sterilized by filtration through a 0.22-$\mu$m pore size filter; store at room temperature), 1 ml of ampicillin or carbenicillin (50 mg/ml; sterilized by filtration through 0.22-$\mu$m pore size filters; store at −20° in 1-ml aliquots)

IPTG (0.4 *M*), sterilized by filtration through a 0.22-$\mu$m pore size filter

Buffer A: 10 m*M* Tris-HCl, pH 8.0

Buffer B: 20% (w/v) sucrose in 30 m*M* Tris-HCl, pH 8.0

Buffer C: 20 m*M* Tris-HCl (pH 7.5), 75 m*M* KCl, 3 m*M* $MgCl_2$, 0.5 m*M* DTT, 0.25 m*M* PMSF 0.5 *M* EDTA (pH 8.0).

[30] H. C. Neu and L. A. Heppel, *J. Biol. Chem.* **240,** 3685 (1965).

Ice-cold 5 m*M* $MgCl_2$

Purification buffer (10×): 200 m*M* HEPES (pH 7.0), 250 m*M* KCl, 50 m*M* $MgCl_2$, 10 m*M* $(NH_4)_2SO_4$, 5 m*M* DTT, 2.5 m*M* PMSF

*Methods.* Inoculate a 1-liter culture with an exponential preculture of transformed BL21(DE3). Grow the cells at 37° under vigorous shaking up to $OD_{600}$ 0.6–0.8. Add IPTG to a 0.4 m*M* final concentration. Harvest the cells after 2 hrs of induction. Wash the cells with 200 ml of buffer A and resuspend in 50 ml of buffer B (or with 80 ml for each gram of cellular wet weight if a fermentor has been used). Add EDTA to a final concentration of 1 m*M*. Stir the cells for 15 min at room temperature. Centrifuge for 30 min, 10,000 rpm at room temperature. Shock the cells by resuspending them in 50 ml of ice-cold 5 m*M* $MgSO_4$. Shake or stir for 10 min in an ice bath, then centrifuge. Recover the supernatant, measure its volume, and add a 1/10 vol of 10× purification buffer. Add the equivalent of 390 g of ammonium sulfate to 1 liter of periplasmic extract (60% saturation of ammonium sulfate, final concentration) slowly, while maintaining pH 7.0. Incubate for 15 min with gentle stirring, and for 15 min without stirring, in an ice-cold bath. Centrifuge for 20 min at 10,000 rpm at 4°. To the supernatant, add slowly the equivalent of 143 g of ammonium sulfate per liter (80% saturation of ammonium sulfate, final concentration), carefully maintaining pH 7.0. Incubate for 15 min with gentle stirring, and for 15 min without stirring, in an ice-cold water bath. Centrifuge at 10,000 rpm for 20 min at 4°. Remove the supernatant and resuspend the pellet in the smallest volume of 1× purification buffer that allows its solubilization. Do exhaustive dialysis in 1× purification buffer at 4°, followed by dialysis overnight against buffer C. Load the dialysate on DEAE-Sephacel equilibrated with buffer C at a 50-ml/hr flow rate. Wash the column with buffer C until the $A_{280}$ measured on flow through equals 0. Elute from the DEAE-Sephacel with a 75–275 m*M* KCl gradient (total volume of gradient is about five column volumes). A good starting flow rate is 10 cm/hr. For a 65-ml, 13-cm-long column, we use a flow rate of 50 ml/hr and collect a fraction every 4 min (about 3.2 ml/fraction). Pool BiP-containing fractions and dialyze overnight against 1× purification buffer. Special attention should be given to contamination by the *E. coli* HSP70 homolog, i.e., DnaK. Even the cleanest periplasmic extract will contain some DnaK that will elute very close to BiP on ion exchange.[28] Load the DEAE-Sephacel pool on ATP–agarose at 4° at a 10-ml/hr flow rate. Wash with 5 vol of buffer C. Elute from the ATP–agarose column with 3 m*M* ATP solution freshly prepared in 20 m*M* Tris-HCl (pH 7.5), 275 m*M* NaCl, 3 m*M* $MgCl_2$, 0.25 m*M* PMSF. Concentrate by ammonium sulfate precipitation (85% saturation), dialyze against assay buffer, aliquot in cryotubes, freeze in liquid nitrogen, and store at −70°.

*Purification of Histidine-Tagged BiP*

The major problem encountered during the purification of recombinant HSP70s is the presence of the *E. coli* homolog DnaK, which has physicochemical properties similar to those of BiP. It is therefore useful to tag the protein of interest with histidine residues to allow rapid, economical, and efficient purification by affinity chromatography. The pET19 vector (Novagen) allows the insertion of a sequence encoding 10 histidine residues at the 5′ extremity of a gene. The purification of histidine-tagged proteins can then be achieved with only one step of affinity chromatography using a nickel column.[31] Applied to BiP and its C-terminal peptide-binding domain, this simple step of purification allows high recovery (5–10 mg of tagged protein for 500 ml of culture) of a protein that is 85–90% pure.

*Materials and Solutions*

His.Bind resin (Novagen)
Binding buffer: 20 m*M* Tris-HCl (pH 7.9), 5 m*M* imidazole, 0.5 *M* NaCl, pepstatin A (1 μg/ml), leupeptin (5 μg/ml)
Charge buffer: 50 m*M* $NiSO_4$
Washing buffer: 20 m*M* Tris-HCl (pH 7.9), 60 m*M* imidazole, 0.5 *M* NaCl
Elution buffer: 20 m*M* Tris-HCl (pH 7.9), 1 *M* imidazole, 0.5 *M* NaCl
Stripping buffer: 20 m*M* Tris-HCl (pH 7.9), 100 m*M* EDTA, 0.5 *M* NaCl

Do not use mercaptoethanol, DTT, or EDTA when using His.Bind resin.

*Method.* Pack 1 ml of His.Bind resin into a 10 × 50 mm column. Wash with 3 ml of sterile water. Charge the column with $Ni^{2+}$ ions, using 5 ml of charge buffer. Equilibrate with 3 ml of binding buffer. Resuspend the cell pellet from the 500-ml culture of histidine-tagged BiP expressing Bl21(DE3) in 20 ml of binding buffer. Break the cells in a French pressure cell prechilled at 4°, or by sonication. Centrifuge the lysate at 4° for 30 min at 20,000 *g*. The histidine-tagged BiP is expressed in the soluble fraction. Filtrate the supernatant with a 0.45-μm pore size filter and load it on the His.Bind resin column. Allow the sample to run through the column at a flow rate of about 10 ml/hr. Wash the column with 15 ml of washing buffer. Elute the histidine-tagged protein with 6 ml of elution buffer. Dialyze the sample twice at 4° against 20 m*M* HEPES (pH 7.0), 75 m*M* KCl, 5 m*M* $MgCl_2$. Wash the column with 5 ml of stripping buffer and store it in 20% (v/v) ethanol at 4°.

[31] M. W. VanDyke, M. Sirito, and M. Sawagado, *Gene* **111,** 99 (1992).

## Purification of BiP Domains

### *Engineering of Enterokinase Site between the Two Domains*

Three highly specific proteolytic enzymes can be used to cleave a protein at a specific site: (1) enterokinase, which cleaves DDDDK↓D,[32] (2) thrombin, which cleaves LVPR↓GS,[33] and (3) factor Xa, which cleaves ID/EGR↓.[34] We use the enterokinase site because it is also recognized by a commercially available murine monoclonal antibody (M2 $IgG_1$; International Biotechnologies (IBI), Inc., New Haven, CT) that can therefore be used both for purification by affinity chromatography and for characterization by immunoblotting of the recombinant protein. We use a vector named pFlag.BiP.Ent (Chevalier *et al.*, submitted), in which an enterokinase site is inserted by M13 site-directed mutagenesis[26] between the residues Gly-389 and Thr-393 of the mature murine BiP, just between the N- and C-terminal domains. Furthermore, a sequence encoding the M2 flag epitope (DYKDDDDK) has been inserted between the signal sequence OmpT and the beginning of the mature BiP sequence. The use of enterokinase to cleave between the two domains of BiP allows, after purification, the recovery of active isolated fragments. This procedure is useful to study the segregation of activities between different domains of a protein, provided that the enzyme site is accessible in the full-length protein, and that the cleavage of the protein allows the recovery of a reasonably large amount of fragments that can be used for further studies. We describe here how to purify Flag-BiP-Ent, how to cleave the fusion protein, and how to isolate the N- and C-terminal fragments (N44 and C30).

#### *Materials and Solutions*

Anion-exchange gel Resource Q (Pharmacia)
Buffer A: 20 m*M* Tris-HCl (pH 8.0), 175 m*M* KCl, 5 m*M* $MgCl_2$, 0.5 m*M* 2-mercaptoethanol
Buffer B: 20 m*M* Tris-HCl (pH 8.0), 325 m*M* KCl, 5 m*M* $MgCl_2$, 0.5 m*M* 2-mercaptoethanol
Anti-Flag M2 affinity gel (IBI)
Phosphate-buffered saline (PBS): 120 m*M* NaCl, 2.7 m*M* KCl, 10 m*M* $Na_2HPO_4$, 1.4 m*M* $K_2HPO_4$ (pH 7.4)
Elution buffer: 0.1 *M* Glycine-HCl (pH 3.0)
Assay buffer: 20 m*M* HEPES (pH 7.0), 50 m*M* KCl, 5 m*M* $MgCl_2$

[32] J. Liepnicks and A. Light, *J. Biol. Chem.* **254,** 1677 (1979).
[33] R. Andreatta, R. Liem, and H. Scheraga, *Proc. Natl. Acad. Sci. U.S.A.* **68,** 253 (1971).
[34] J. Nagai and H. Thorgesen, *Nature* **309,** 810 (1984).

*Method.* A crude extract corresponding to 1 liter of Flag-BiP-Ent-expressing BL21(DE3) cells is clarified by ammonium sulfate precipitation (30% saturation, 0.176 g added to 1 ml). Dialyze the supernatant against buffer A and load the sample at a flow rate of 1 ml/min on a Resource Q column (Pharmacia) that has been equilibrated in buffer A. Wash with buffer A and elute with a KCl linear gradient as described in detail for BiP secreted into the periplasmic space (pSecB115). Flag-BiP-Ent elutes between 230 and 270 m*M* KCl. Pool the fractions that contain BiP and dialyze at 4° against PBS. Load Flag-BiP-Ent onto 5 ml of anti-Flag affinity gel (IBI) equilibrated in PBS, at a flow rate of 10–15 ml/hr. After washing the column with PBS, elute with 30 ml of elution buffer (2 ml/fraction) and immediately neutralize each fraction with 50 $\mu$l of 1 *M* Tris-HCl (pH 8.0). This denaturation step does not affect BiP enzymatic activity when compared with BiP purified by affinity chromatography on ATP–agarose (Chevalier *et al.*, submitted). The fractions are analyzed by 10% (w/v) SDS–PAGE,[24,25] pooled, and dialyzed against assay buffer. After centrifugation, the dialysate can be concentrated with Centricon 50 (Amicon, Danvers, MA) or Biomax 50 (Millipore, Bedford, MA). The price and the relatively quick inactivation of the anti Flag affinity gel (50% efficiency lost in 10 runs) make it a costly method for routine large-scale purification. However, the yield of Flag-BiP-Ent purification by M2 affinity chromatography can reach 5 mg per liter of culture when a new resin is being used. Flag-BiP-Ent aggregates when stored for more than 1 week at 4°, probably owing to the presence of the OmpT signal sequence. This can be prevented by preparing Flag-BiP-Ent from periplasmic extract as described previously in this chapter. The yields are much lower (0.5 mg of Flag-BiP-Ent per liter of culture), but the protein is more stable and can be stored for a longer period of time at −70°. The spheroplasts obtained after cold osmotic shock[30] contain an appreciable amount of nonsecreted soluble OmpT-Flag-BiP-Ent and can be used for purification on an M2 column. To prevent loss of the protein by aggregation, we recommend proceeding with the enterokinase cleavage immediately after purification of Flag-BiP-Ent.

### *Enterokinase Cleavage and Isolation of N- and C-Terminal Fragments (N44 and C30) after Proteolytic Cleavage*

Flag-BiP-Ent is digested with enterokinase (Biozyme, San Diego, CA or Novagen, Madison, WI), 2000 units of enterokinase per milligram of BiP, overnight at 37° in the presence of 10 m*M* $CaCl_2$ and 0.1% (w/v) Triton X-100. The yield of cleavage does not exceed 50%, even at higher enzyme concentrations (5000 units/mg), as the enterokinase site is not totally accessible to the enzyme (Chevalier *et al.*, submitted). The efficiency of enterokinase cleavage may vary according to the individual protein. The enterokinase EK3 (Biozyme) is active within a pH range of pH 5.0 to 8.0. The stability of the enzyme is increased in the presence of a minimum of 2 m*M* $CaCl_2$, even when urea is added up to 4 *M* (Biozyme). This property might be used to increase the accessibility of the enterokinase site through partial denaturation of the sample protein. The flag epitope (DYKDDDDK) of Flag-BiP-Ent is also an enterokinase site and is cleaved with 30–50% efficiency during the digestion step (Chevalier *et al.*, submitted). The purification of the domains is done under native conditions by gel filtration on a Superose 12 HR 10/30 column (Pharmacia). This method allows the separation of the monomeric N-terminal domain from the full-length molecule. The cleaved ATPase N-terminal fragment is fully active after purification (M. Chevalier *et al.*, submitted). The C-terminal domain oligomerizes, and trimers and dimers of C30 copurify with monomers and dimers of the full-length BiP (M. Chevalier, *et al.*, submitted). Further isolation of the C-terminal fragment is achieved by thermic treatment. The C-terminal fragment is thermostable and remains soluble after 15 min at 60° and can be isolated from full-length BiP and the N-terminal fragment, which are thermosensitive (Chevalier *et al.*, submitted). The C-terminal domain retains its native-like structure and peptide-binding properties after heat treatment (Chevalier *et al.*, submitted).

*Direct Expression of N-Terminal ATPase Domain (N44) and Its Purification from Inclusion Bodies*

Enterokinase is a costly enzyme and insertion of a cleavage site in between two structural domains does not guarantee its accessibility to the protease. In some instances, it might be better to express directly the constitutive domains of a protein. Often, recombinant proteins do not fold properly in *E. coli* and form aggregates that accumulate in inclusion bodies. Here we describe such a case, as well as how an active N-terminal ATPase fragment can be recovered from these inclusion bodies, with a method based on the general considerations described by F. A. Marston and D. L. Hartley.[35] A plasmid $pSB_2MT_7$ (S. Blond, unpublished, 1991), which contains the sequence of the mature murine BiP protein under the dependence of the T7 RNA polymerase promoter, was used to amplify the N44 sequence (residues 1 to 390). The PCR fragment was subcloned into the pET11d expression vector. The resulting plasmid, pN44, leads to the high level of expression of the N-terminal domain in inclusion bodies (M. Chevalier, unpublished, 1996).

[35] F. A. Marston and D. L. Hartley, *Methods Enzymol.* **182,** 264 (1990).

*Materials and Solutions*

Buffer A: 20 m$M$ HEPES (pH 7.0), 50 m$M$ KCl, 5 m$M$ $MgCl_2$; add 10 m$M$ DTT and 1 m$M$ ATP just before use

Urea (ultrapure; ICN, Costa Mesa, CA) in buffer A. Guanidium hydrochloride (Sigma), 6 $M$ in buffer A, can also be used as an alternative

Triton X-100 (Sigma)

Nonidet P-40 (Boehringer Mannheim, Indianapolis, IN)

*Method.* pN44-induced cells are stored at $-70°$ if not used immediately. It is preferable to aliquot the cells (per 500 ml of culture) after harvest as N44 is expressed at high levels and much protein can be obtained after resolubilization of inclusion bodies (20–30 mg/500-ml culture). Resuspension and lysis of the cells are performed as described above. Centrifuge the lysate for 30 min at 25,000 $g$ at 4° in 30-ml polypropylene tubes (Nalgene, Rochester, NY). Wash the pellet successively with (1) 20–30 ml of buffer A, (2) 20–30 ml of buffer A plus 1% (w/v) Triton X-100, (3) 20–30 ml of buffer A, (4) 20–30 ml of buffer A plus 0.1% (w/v) Nonidet P-40, and (5) 20–30 ml of buffer A. Resolubilize N44 at 1–5 mg/ml in 8 $M$ urea in buffer A supplemented with 0.1 $M$ 2-mercaptoethanol. The solution should not be too cloudy; if necessary centrifuge and redilute the pellet in 8 $M$ urea to maximize the yield of solubilization. Incubate overnight at 4° with gentle stirring. Centrifuge the sample at 45,000 $g$ for 30 min at 4°. Estimate the protein concentration by the method of Bradford[23] (8 $M$ urea does not interfere with the assay if the volume used in the assay is kept below 5 $\mu$l for 1 ml of total reaction). Renature N44 in buffer A by dilution. Refolding of N44 requires ATP, thus renaturation by dialysis is not recommended. The refolding is initiated by a rapid 50- to 300-fold dilution in ice-cold buffer A supplemented with 10 m$M$ DTT and 1 m$M$ ATP; the final concentration of N44 should be below 1 mg/ml. The yield of renaturation of the N44 fragment is estimated after centrifugation by Bradford protein assay[23] and the ATPase activity is measured as described below. Refolded N44 exhibits an ATPase-specific activity similar to that obtained for full-length BiP (0.30 to 0.35 mol of ATP/min · mol N44 or BiP). Renatured N44 can be concentrated with a Centricon cell (Amicon) and stored at 4°. Ammonium sulfate precipitation (80%) can be performed to strip N44 from bound nucleotides in order to perform ATPase activity assays.

*Characterization of ATPase Activity*

We have used two methods to test the enzymatic activity of BiP and of the N-terminal domain. The first and more widely used assay for the determination of ATPase activity of HSP70s is a thin-layer chromatographic method derived from the method of Shlomai and Kornberg.[36] This method allows the separation of radiolabeled ADP (when [$\alpha$-$^{32}$P]ATP is used as substrate) or radiolabeled $P_i$ (when [$\gamma$-$^{32}$P]ATP is used as substrate) from nonhydrolyzed ATP. The method requires thin-layer chromatography (TLC) plates, which are quite expensive, and is therefore not suitable for experiments with a large number of samples. However, it is a semiquantitative method that is fast and especially convenient when studying the peptide-dependent stimulation of the ATPase activity in the presence of synthetic peptides.[8,28] The second assay, which is a modification of the method developed by J. R. Seals *et al.*,[37] is a direct quantitation of released $P_i$ through the formation of a phosphomolybdate complex, which can be easily extracted in an organic phase and separated from the [$\gamma$-$^{32}$P]ATP (aqueous phase). Both techniques give similar results, the second method being faster and allowing more sensitive results.[38] Kinetic and concentration-dependent assays should be performed with enzyme concentrations at least 100 times inferior to the concentration of the substrate, and the amount of product released after the experiment should not exceed 10% of the initial amount of ATP. These conditions assess that the assay is performed under steady-state conditions.

*Material and Solutions*

Methods 1 and 2

[$\gamma$-$^{32}$P]ATP (3000 Ci/mmol; Amersham, Arlington Heights, IL)
ATP stock (100×): 100 m*M* ATP, pH 7.0 ($\varepsilon_{ATP}$ 15400 $M^{-1}$ cm$^{-1}$ at 259 nm). Store in small aliquots at −70°
ATPase assay buffer: 20 m*M* HEPES (pH 7.0), 50 m*M* KCl, 5 m*M* $MgCl_2$, supplemented with 100 $\mu M$ [$\gamma$-$^{32}$P]ATP (0.5 $\mu$Ci/nmol ATP)

Method 1

TLC plates (cellulose polyethyleneimine fluorescent PEIF; Baker, Phillipsburg, NJ)
Developing chamber for TLC (Fisher, Pittsburgh, PA)
Formic acid (Mallinckrodt, St. Louis, MO)
Lithium chloride (Fisher)
Unlabeled nucleotide mixture of ATP, ADP, and AMP (Pharmacia), 10 m*M* each (pH 6.8–7.0); store in aliquots at −20°

[36] J. Shlomai and A. Kornberg, *J. Biol. Chem.* **255,** 6789 (1980).
[37] J. R. Seals, J. M. McDonald, D. Bruns, and L. Jarett, *Anal. Biochem.* **90,** 785 (1978).
[38] A. S. Loeb, W.-C. Suh, M. A. Lonetto, and C. A. Gross, *J. Biol. Chem.* **270,** 30051 (1995).

Method 2

$\beta$-Radiation scintillation counter
Liquid scintillation vials (20-ml polyethylene; Fisher)
Biodegradable scintillation liquid (Fisher)
*n*-Butyl acetate (Fisher)
Trichloroacetic acid [TCA, 8% (w/v) in water; Fisher]; add $K_2HPO_4$ (pH 7.2) to 1 m*M* final concentration before use
Solution A: 3.75% (w/v) ammonium molybdate, 0.02 *M* silicotungstic acid (optional but recommended) in 3 *N* $H_2SO_4$

*Method 1: Thin-Layer Chromatography Method.* Preheat 20 $\mu$l of ATPase assay buffer at 37° containing 0 to 2 m*M* peptide solution (see below and refer to Table I for peptide sequence). Start the reaction by adding 5 $\mu$l of BiP solution (20 ng to 2 $\mu$g of BIP, total) to the reaction mixture. After incubation at 37° for 15 min, spot 1 $\mu$l of the reaction mixture on the TLC plate, which has been prespotted with 1 $\mu$l of the 10 m*M* nucleotide mixture. Dry immediately and transfer to the developing chamber that has been saturated with 0.5 *M* formic acid–0.5 *M* LiCl (v/v). Allow migration for 15 min (about 1–1.5 in.), dry immediately, and autoradiograph the plate for 15 to 30 min. Eventually, quantify by densitometry.

*Method 2: Molybdate Protocol.* After purification, the BiP protein is dialyzed against the ATPase assay buffer. Centrifuge for 30 min at 12000 *g* at 4° in a microcentrifuge and determine the protein concentration by the method of Bradford.[23] The experiment is carried out at 37° in 20 $\mu$l of ATPase assay buffer containing 0.02 to 2 $\mu$g of BiP. The reaction is linear for 60 min at 10 $\mu$g of BiP/ml under these conditions. After a variable time of incubation (usually a 15-min assay for BiP), quench 10 $\mu$l of the reaction by addition of 200 $\mu$l of ice-cold TCA solution. Mix briefly and put the tubes back on ice for no more than a few minutes. As the ATP is sensitive to acidic degradation (0.05% of hydrolysis per minute after TCA quenching),[37] it is advisable to work with a small number of samples at a time. Add 50 $\mu$l of solution A and 300 $\mu$l of *n*-butyl acetate as described[39] and vortex three times for 8–10 sec each time. The samples are then centrifuged briefly to separate the phases: The radiolabeled phosphomolybdate complex in the upper organic phase is separated from nonhydrolyzable ATP in the lower aqueous phase. Remove quickly a 200-$\mu$l aliquot from the upper organic phase and pour into a labeled liquid scintillation vial in which 6 ml of scintillation liquid has been added. Count the samples in a $\beta$-radiation counter to determine the amount of $^{32}$P released.

*Calculations.* Under the conditions described earlier (i.e., less than 10%

[39] R. L. Post and A. K. Sen, *Methods Enzymol.* **X,** 762 (1967).

of the substrate hydrolyzed) the relationship between the initial velocity and the enzyme concentration is linear. The rate of product formation is therefore derived from the plot of the amount of product formed versus time or versus the quantity of enzyme. The slope of the curve is derived graphically by the best-fit straight line of the least-squares analysis. This allows the calculation of the specific activity of the enzyme according to Eq. (1):

$$v = \frac{[(\text{cpm})(\text{SRA})]/(2.22 \times 10^6)}{tq} \quad (1)$$

where $v$ is the initial velocity in moles of ATP hydrolyzed per minute · milligram of enzyme, SRA is the specific radioactivity of $[\gamma\text{-}^{32}\text{P}]$ATP in the assay buffer in microcuries per mole of ATP, $2.22 \times 10^6$ is the factor of conversion of counts per minute (cpm) into microcuries, $t$ is the reaction time in minutes, and $q$ is the quantity of protein in milligrams.

## Characterization of BiP Substrate Specificity by Using Affinity Panning of Phage-Displayed Peptide Libraries

Initially developed by S. F. Parmley and G. P. Smith,[40] phage-display libraries are based on the expression of foreign random peptides at the surface of bacteriophages. A selection procedure called *affinity biopanning* allows the isolation of phage particles that display peptides specifically recognized by an immobilized receptor. Further sequencing of the phage DNA gives direct access to the peptidic sequences and that information can be used to generate synthetic peptides and determine their affinity for the receptor in *in vitro* binding assays. Affinity panning of phage-display libraries applied to BiP has enabled us to identify a BiP-binding motif shared by the majority of the isolated phages, to analyze our data statistically, and to develop an algorithm for the detection of putative BiP-binding sites in any protein sequence.[10,12,41] We use the biotin–avidin system to immobilize BiP on microtitration plates and to identify phages that display peptides recognized by the molecular chaperone.

### *Materials and Solutions*

Streptavidin (BRL)
Serum albumin [radioimmunoassay (RIA) grade; U.S. Biochemical, Cleveland, OH]
Sulfosuccimidyl-6-(biotinamido)hexanoate (NHS LC-biotin; Pierce)
Pronase solution: 1% (w/v) pronase in 0.1 *M* potassium phosphate buffer (pH 7.0)

[40] S. F. Parmley and G. P. Smith, *Gene* **73,** 305 (1988).
[41] G. Knarr, M.-J. Gething, J. Modrow, and J. Buchner, *J. Biol. Chem.* **270,** 27589 (1995).

2-(4′-Hydroxyazobenzene)benzoic acid (immunopure HABA; Pierce)
Streptavidin solution: 0.15 mg/ml in 50 m*M* $KH_2PO_4$ (pH 6.0), 0.9% (w/v) NaCl (pH 6.0)
Biotin solution: 0.1 m*M* biotin in 50 m*M* $KH_2PO_4$ (pH 6.0), 0.9% (w/v) NaCl
Immunopure avidin conjugated to alkaline phosphatase (Pierce)
PBS: 120 m*M* NaCl, 2.7 m*M* KCl, 10 m*M* $Na_2HPO_4$, 1.4 m*M* $K_2HPO_4$ (pH 7.4)
PBST: PBS supplemented with 0.05% (v/v) Tween 20
Binding buffer: PBS, 3 m*M* $MgCl_2$, 20 m*M* KCl, 0.1% (w/v) bovine serum albumin (BSA)
Acidic elution buffer: 0.1 *N* HCl-glycine (pH 2.2), 0.1% (w/v) BSA
ATP elution buffer: 20 m*M* HEPES (pH 7.0), 20 m*M* KCl, 10 m*M* $(NH_4)_2SO_4$, 2 m*M* $MgCl_2$, 3 m*M* ATP
Neutralization solution: 2 *M* Tris base
PEG solution: 20% (w/v) polyethylene glycol (PEG) 8000 in 2.5 *M* NaCl
Substrate buffer for alkaline phosphatase: 0.1 *M* Tris-HCl (pH 9.5), 0.1 *M* NaCl, 50 m*M* $MgCl_2$
Substrate solution for alkaline phosphatase: Prepare just before use: 44 μl of nitroblue tetrazolium chloride (NBT; Pierce) in 10 ml of substrate buffer. Gently mix by inverting the tube, then add 33 μl of 5-bromo-4-chloro-3′-indolyl phosphate, *p*-toluidine salt (BCIPT; Pierce). Gently mix
Lysis buffer for single-stranded DNA purification: 10 m*M* Tris-HCl (pH 7.6), 0.1 m*M* EDTA, 0.5% (v/v) Triton X-100
LB/tetracycline (20 μg/ml) plates: Three 150 × 15 mm and six 100 × 15 mm culture plates, per library screened
K91 cells: Grow 50 ml of K91 cells[40] in Luria–Bertani (LB) liquid medium.[26] Put on ice when the $A_{600}$ reaches 0.5. Harvest the cells by centrifugation, resuspend in 5 ml of LB medium, and keep on ice until needed

*Biotinylation of BiP*

Dialyze BiP (0.25–1 mg/ml) for 2 hr at room temperature against 0.1 *M* $NaHCO_3$ (pH 8.5). Keep an aliquot for ATPase activity assay. Dissolve 2 mg of NHS LC-biotin in dimethyl sulfoxide. Transfer the dialyzed BiP to an Eppendorf tube and add biotin dropwise (molar ratio of BiP to biotin, 1 : 7). Incubate for 2 hr at room temperature in the dark under gentle stirring. Dialyze overnight against PBS supplemented with 0.1% (w/v) sodium azide. Determine the concentration of dialyzed protein solution by Bradford protein assay.[23] Perform an ATPase activity assay (TLC method) on BiP before and after biotinylation. To quantify spectrophotometrically the number of biotin molecules per molecule of BiP, prepare a calibration curve using HABA as a reagent and a standard solution of biotin as follows: To 0.5 ml of streptavidin solution add 25 $\mu$l of 10 m$M$ HABA in 10 m$M$ NaOH. Add biotin solution in 1-$\mu$l increments. Measure the absorbance at 500 nm (blank, streptavidin only). To quantify BiP biotinylation, add BiP treated with pronase overnight at room temperature (10 to 20 $\mu$g of BiP plus a 1/10 vol of Pronase solution) in 5-$\mu$l increments (30-$\mu$l maximum volume) to 0.5 ml of streptavidin solution; measure the $A_{500}$ and determine the ratio of biotin to BiP by using the HABA calibration curve.

*Affinity Panning*

Coat wells with 50 $\mu$l of streptavidin (10 $\mu$g/ml) in PBS (10 $\mu$g/ml) (six wells for each library, and two wells for background). Incubate for 1 hr at 37°. Wash the wells three times with PBS. Block the wells with 1% (w/v) BSA in PBS for 1 hr at 37°. Wash the wells three times with PBS. Add 50 $\mu$l of biotinylated BiP (20 $\mu$g/ml in PBS) per well. Put 50 $\mu$l of PBS in the background wells. For the first panning, precipitate 1000 library equivalents with a 1/10 vol of 1 $M$ acetic acid (library size transfection units (Tu)/$\mu$l $\times$ 1000 = 1000 equivalents). Incubate for 10 min on ice. Microcentrifuge for 15 min at room temperature. Resuspend the phages in 800 $\mu$l of binding buffer. Keep a 5-$\mu$l aliquot for phage titration. For subsequent pannings, use half of the amplified eluted phages and add binding buffer to 800 $\mu$l (final volume). Add 100 $\mu$l of phages per well, including the background wells. Incubate for 2 hr at 4°. Wash six times with ice-cold PBS. Elute the bound phages with 100 $\mu$l of elution buffer per well. When the acidic elution buffer is being used, the eluates from all wells are pooled after a 10-min incubation at room temperature and immediately neutralized with 2 $M$ Tris base (34 $\mu$l in 600 $\mu$l, or 11 $\mu$l in 200 $\mu$l of background). To elute with the ATP elution buffer, a 2-hr incubation is recommended; no neutralization is required but the phages should be amplified immediately, as we observed that their titer decreases rapidly in the presence of ATP (S. Blond, unpublished, 1992).

*Amplification of Eluted Phages*

Add 1 vol of K91 cells to 1 vol of phages (typically 200 $\mu$l = 1 vol). Incubate at 37° for 20 min (in Eppendorf tubes, no shaking). Plate on large LB/tetracycline plates, three plates per round of panning. Incubate overnight at 37°. Harvest phages from plates by pouring 10 ml of LB medium over each plate. Let the plates stand for 10 min on a shaker. Gently scrape to remove the lawn of bacterial cells. Transfer to a 50-ml disposable sterile tube. If necessary, wash the three plates with more LB (10 ml, total volume). Centrifuge at 12,000 *g* for 15 min at 4°. Transfer the cleared supernatant to a new tube and precipitate the phages by adding 0.2 vol of PEG solution. Mix well and let stand on ice for 1 hr. Pellet the phages by centrifugation at 12,000 *g* for 15 min at 4°. Following centrifugation, remove as much supernatant as possible and resuspend the phages in 800 $\mu$l of binding buffer. To titrate the phages, dilute the library or amplified phages used for the panning $10^7$-, $10^8$-, and $10^9$-fold in binding buffer, and dilute the eluted phages $10^3$-, $10^4$-, and $10^5$-fold. Add 10 $\mu$l from each phage dilution to 50 $\mu$l of exponential K91 cells harvested as described above. Incubate for 20 min at 37°. Add 100 $\mu$l of LB medium and plate on small LB/tetracycline plates. Incubate overnight at 37°. Count the colonies and deduce the phage recovery after panning in the presence or absence of BiP. We observed an enrichment of two orders of magnitude relative to background after two rounds of panning in the presence of BiP.[10] Keep the plates at 4°.

*Colorimetric Assay to Pick up Phages Expressing Peptides Recognized by BiP*

Use one titration plate with 50–200 colonies, one of phages eluted from background wells, and one of phages affinity purified on biotinylated BiP-coated wells. Transfer the colonies to a nitrocellulose filter by simple lifting. Mark the orientation of the filters. Maintain the plates at 37° overnight to allow the transfected cells to multiply again. Wash the filters with PBST supplemented with 1% (w/v) BSA (kept in a squeeze bottle) until no cells are left on the filter. Block the filters for 30 min in PBST, 1% (w/v) BSA at room temperature, on a shaker. Wash the filters three times with PBST. Add 5 ml of biotinylated BiP at 1 $\mu$g/ml in PBST. Incubate for 1 hr at room temperature on a shaker plate. Wash three times with PBST. Add avidin conjugated to alkaline phosphatase (1:2000 in PBST). Incubate for 1 hr at room temperature. Wash with PBST three times and once in substrate buffer. Add 5 ml of substrate solution per filter (use small petri dishes as containers). Incubate at room temperature in the dark. The blue color should develop in 5–15 min. Block the reaction by rinsing in distilled water. Keep the filters in the dark.

*Single-Stranded Phage DNA Isolation*

To isolate single-stranded phage DNA,[26] pick a positive clone and transfer to 1 ml of LB medium/tetracycline (20 $\mu$g/ml). Incubate overnight at 37°. Microcentrifuge the overnight culture for 10 min at 4° to remove bacterial cells. Transfer 1 ml of the supernatant to a tube containing

200 μl of PEG. Incubate for 30 min on ice. Microcentrifuge for 30 min at 4°. Carefully aspirate the supernatant. Return the tubes and let them stand for 10–15 min. Wipe away excess PEG as well as possible. Resuspend the phage pellet in 40 μl of lysis buffer. Incubate for 15 min at 80°. Spin down for 15 min in a microcentrifuge at room temperature. Take the supernatant (about 40 μl) and transfer to a clean Eppendorf tube. Add 50 μl of phenol–chloroform prepared as described[26] and centrifuge for 5 min. Take the supernatant and transfer to a clean tube. Add 100 μl of 95% (v/v) ethanol, 0.12 *M* sodium acetate (final concentration). Incubate for 60 min at −20° to precipitate the DNA. Microcentrifuge for 30 min and remove the supernatant. Wash the DNA pellet with 70% (v/v) ethanol, microcentrifuge for 30 min, and remove the supernatant. Resuspend the pellet in 7 μl of $H_2O$. The amount of single-stranded DNA isolated by this method should be sufficient for one sequencing reaction. Several kits are available for DNA sequencing; all can be used with confidence. Peptide sequences can be analyzed statistically and eventually computed into an algorithm program as described.[10]

## Synthetic Peptide-Binding Assays

### *Peptide Choice, Purification, and Quality Control*

We have used a wide variety of synthetic peptides (some of them reported in Table I) to characterize the specificity of BiP, its constitutive fragments, and single-site and deletion mutants localized in the peptide-binding domain. Because the more soluble and more stable carboxyl-terminal domain (C30) possesses many features of the native, full-length BiP (Chevalier *et al.*, submitted), we have used C30 and its mutants extensively in the laboratory. C30 is efficiently expressed as a soluble and active fragment in *E. coli*. The addition of a histidine tag at the N-terminal extremity does not affect its properties and can thus be used for its purification as described earlier in this chapter for histidine-tagged BiP. We have described two convenient and rapid assays that give accurate estimates for the apparent affinity constants of BiP or its C-terminal domain for peptidic substrates. Other methods such as the peptide-dependent ATPase stimulation (described earlier in this chapter), the binding of biotinylated peptides,[28] the binding of radiolabeled tritiated peptides,[10] or equilibrium dialysis with radiolabeled peptides[29] can also be used. Most of the peptides described in Table I contain the motif, shared by the majority of BiP-binding peptides, of three or four hydrophobic residues in alternating positions.[10] The peptides can be prepared by continuous-flow solid-phase synthesis and analyzed by high-performance liquid chromatography, mass spectrometry, and amino acid analysis as described.[28]

TABLE I
SYNTHETIC PEPTIDES USED TO CHARACTERIZE SPECIFICITY OF BiP AND ITS CARBOXYL-TERMINAL DOMAIN

| Peptide name | Sequence[a] | Molecular mass[b] | Solubility[c] (m$M$) | $\varepsilon^{280}_{M,\text{Gu·Hcl}}$ | BiP score[d,e] | BiP affinity[f] (m$M$) | C30 affinity[f] (m$M$) |
|---|---|---|---|---|---|---|---|
| **Positive class** | | | | | | | |
| pp28 | HWDFAWPW | 1,143 | 0.36 | 15,720[d] | +32 | 0.02[d] | 0.02[g] |
| pp38 | FWGLWPWE | 1,120 | 0.20 | 17,820[d] | +34 | 0.05[d] | ND |
| pp52 | YVDRFIGW | 1,054 | 1.5 | 6,127[d] | +5 | 0.03[d] | ND |
| pp48 | DGVGSFIG | 751 | ND | — | +4 | 0.35[d] | ND |
| pp32 | WTWWEWLA | 1,177.3 | 7.8 | 20,980[d] | +7 | 0.10[g] | ND |
| pep-1 | LSRAPFD | 804.91 | 4.0 | — | −1 | ND | 1.0[g] |
| pep-2 | LSVKFLT | 806.97 | 6.5 | — | −3 | ND | 0.2[g] |
| pep-6 | FASQKFI | 840.01 | 3.0 | — | +3 | ND | 0.3[g] |
| pep-7 | FYRYGVI | 917.08 | 2.2 | 2,560[h] | +5 | ND | 0.05[g] |
| **Negative class** | | | | | | | |
| pep-5 | AWAGSQS | 647.7 | 8.0 | — | −17 | >1.0[g] | >4.0[g] |
| np0.15 | DSNDDWRM | 1,037 | 4.0 | 5,690[h] | −19 | >1.0*d* | >2.0[g] |
| np53 | AGEYYAAL | 983.02 | 6.0 | 2,560[h] | 0 | >1.0[d] | ND |

[a] .One-letter codes denoting the 20 naturally occurring amino acids.
[b] Molecular mass (in Da).
[c] Solubility limit in 20 m$M$ HEPES (pH 7.0), 75 m$M$ KCl, 5 m$M$ $MgCl_2$.
[d] S. Blond-Elguindi, S. E. Cwirla, W. J. Dower, R. J. Lipshutz, S. R. Sprang, J. F. Sambrook, and M.-J. Gething, *Cell* **75,** 717 (1993).
[e] Scores computed using the BiP score algorithm.[d]
[f] Apparent affinity (in m$M$).
[g] Chevalier *et al.*, submitted.
[h] S. C. Gill and P. H. von Hippel, *Anal. Biochem.* **182,** 319 (1989).

## *Peptide-Induced Dissociation of BiP or C30 Oligomers*

Several studies have shown that the degree of oligomerization of BiP *in vivo* varies with the physiological state of the cells.[42–45] In quiescent unstressed cells, BiP is present in various oligomeric states and free from bound polypeptidic substrates. When the level of unfolded or unassembled proteins that accumulate in the ER increases, BiP is found mostly in a monomeric state associated with the unfolded substrates. We have been

[42] L. Carlsson and E. Lazarides, *Proc. Natl. Acad. Sci. U.S.A.* **80,** 4664 (1983).
[43] P. Freiden, J. Gaut, and L. Hendershot, *EMBO J.* **11,** 63 (1992).
[44] B. E. Ledford and G. H. Leno, *Mol. Cell. Biochem.* **138,** 141 (1995).
[45] S. Nakai, S. Kawatani, H. Ohi, T. Kawasaki, Y. Yoshimori, Y. Tashiro, I. Miyata, M. Yahara, M. Satoh, and K. Nagata, *Cell Struct. Funct.* **20,** 33 (1995).

able to show *in vitro* that dissociation of BiP oligomers is induced by binding of specific peptides; this converts BiP inactive oligomers into active monomers able to hydrolyze ATP more efficiently.[28] The effects of the synthetic peptides on BiP oligomerization allowed us to set up an assay that uses native electrophoresis to separate the different species and to follow the peptide-induced dissociation of BiP oligomers at equilibrium. As seen from Table I, the two peptides with the highest BiP scores, pp28 and pp38, do in practice have the highest apparent affinities for BiP and its carboxyl terminal. A less expected finding is that there are some peptides with relatively low BiP scores (e.g., pep-2) that have reasonable affinities for BiP as well as for C30. Our data and those of others[10,12,41] are compatible with the following: Peptides with scores higher than +5, containing at least two of the preferred residues (tryptophan, phenylalanine, and leucine) at alternating positions and no negatively charged residues at any of these positions, have potential as good substrates for BiP. Peptides with very negative values (below −5) have a poor probability of being recognized by the chaperone. Peptides with intermediate scores (between −5 and +5) have about a 50% chance of being a BiP substrate.

*Materials and Solutions*

Binding buffer: 20 m*M* HEPES (pH 7.0), 75 m*M* KCl, 5 m*M* $MgCl_2$, filtrated with a 0.22-μm pore size filter. Keep 5 ml at room temperature and store the remainder at 4°

Native sample buffer (5×): 0.63 *M* Tris, 20% (v/v) glycerol, bromophenol blue (sodium salt), 10% (v/v) 2-mercaptoethanol should be added just before use

Coomassie Brilliant Blue (R-250) staining: Solution 1: 125 ml of 2-propanol, 50 ml of acetic acid, 1.5 ml of 1% (w/v) R-250 solution in deionized water and 325 ml of $H_2O$, at least 45 min. Solution 2: 50 ml of 2-propanol, 50 ml of acetic acid, 1.5 ml of 1% R-250 solution, and 400 ml $H_2O$, at least 30 min. Solution 3: 50 ml of acetic acid, 1.5 ml of 1% R-250, and 450 ml of $H_2O$, at least 2 hr. Solution 4 (destaining): 100 ml of acetic acid and 400 ml of $H_2O$. Gels should be placed on a rotary shaker in just enough solution so that the gels float freely in the tray. Gels may be stored in any of the solutions at room temperature in covered plastic trays. All solutions, except the destaining solution, may be reused up to three times

Peptide solutions: Precisely weigh the desired amount of peptide powder into a 1.5-ml Eppendorf tube and add the appropriate amount of the binding buffer at room temperature to obtain the desired concentration of peptide solution (as determined using the molecular weight of the dry peptide powder and the solubility limit; see Table

I). Place the tubes in a 37° water bath to dissolve the powder, centrifuge briefly (7200 *g*, less than 30 sec), and transfer the supernatant to a clean Eppendorf tube. The pH of the stock solutions is adjusted with 1 *M* KOH. Peptide stock solutions are kept at room temperature for use and are diluted to varying concentrations with binding buffer. Actual concentrations of the peptide stock solutions are determined in one of two ways. For those peptides that contain tyrosine ($\varepsilon_{280\text{ nm}}$ 1280 $M^{-1}$ cm$^{-1}$), tryptophan ($\varepsilon_{280\text{ nm}}$ 5690 $M^{-1}$ cm$^{-1}$), or cysteine ($\varepsilon_{280\text{ nm}}$ 120 $M^{-1}$ cm$^{-1}$), concentrations are determined spectrophotometrically using the specific absorption coefficient described by Gill and von Hippel.[46] All other peptides can be hydrolyzed with 6 *N* HCl, then derivatized with phenyl isothiocyanate using the Millipore-Water Pico·Tag workstation, and analyzed by high-performance liquid chromatography (HPLC) using the Applied Biosystems (ABI, Foster City, CA) 130A HPLC separation sytem with an ABI $C_{18}$ Brownlee column (2.1 × 220 mm) at 37°. Two buffers are used for the separation: a 1.7% (w/v) sodium acetate buffer, pH 5.35 (buffer A), and a 70% (v/v) acetonitrile buffer (buffer B) at a flow rate equal to 300 μl/min, with a 7 to 100% buffer B linear gradient. The amino acid standards used for the concentration calculations can be purchased from Pierce

Protein solutions: C30 protein stock solutions are purified as described above. Rapid affinity purification of histidine-tagged C30 is done with the pET His·Tag system (pET19b; Novagen). Protein solutions should always be dialyzed against binding buffer overnight and stored at 4° before incubation, with binding buffer used for all dilutions. Accurate concentrations of all protein stock solutions are calculated using Bradford protein assays[23]

*Method.* In 0.5-ml Eppendorf tubes, combine 5 μl of peptide solution (concentration range relative to apparent affinity, usually from 10 units below to 10 units above the $K_D$; see Table I) and 5 μl of protein solution (0.5–1.5 mg/ml, depending on the protein used); for some C30 mutants more protein is needed for good visualization with Coomassie staining. Vortex immediately, and microcentrifuge (7200 *g*) for no more than 30 sec to ensure that the entire solution is in the tube bottom. Immediately after centrifugation, place the tubes into a 37° water bath for overnight incubation (12–15 hr). The following morning, spin all of the tubes briefly (7200 *g*, less than 30 sec) and place them back in the water bath for one additional hour. When incubation is complete, microcentrifuge 5 min at 7200 *g*, then add 2 μl of the native sample buffer, spin again, and immediately load the

[46] S. C. Gill and P. H. von Hippel, *Anal. Biochem.* **182,** 319 (1989).

samples on freshly prepared 10% (w/v) native polyacrylamide gels using 10 × 8 cm rectangular glass plates[25] [Hoefer minigels (San Francisco, CA)]. Proceed with native electrophoresis[24,25] at 11–15 mA/gel (constant current) for approximately 2 hr, although run times may differ considerably with various power supplies and percentage of acrylamide chosen. Generally, for proteins used in our laboratory (~19–60 kDa), electrophoresis is complete when the native sample buffer reaches the bottom of the gel. Once the gels are done, protein transfers to cellulose membrane may be done for Western blotting if antibodies are available, or direct staining with Coomassie Brilliant Blue. Either technique may be followed by densitometry analysis. C30–pep-2 complexes (C30 and binding buffer as the control) have shown an immediate and complete dissociation of oligomers to monomeric form with concentrations of peptide as low as 200 $\mu M$ (Chevalier *et al.*, submitted).

### *Binding of Reduced Carboxymethylated Lactalbumin*

The reduced carboxymethylated lactalbumin (RCMLA) assay is used to compare the ability of synthetic peptides to compete with an unfolded protein for binding BiP. This competition binding assay was developed on the basis of observations that hsc70 and DnaK bind to an unfolded form of lactalbumin, the reduced, carboxymethylated form,[16,47,48] and also on the fact that specific synthetic peptides are able to compete with BiP, Hsc70, or DnaK–RCMLA complexes.[11]

#### *Materials and Solutions*

RCMLA stock solution: 30 mg/ml in 50 m$M$ Tris-HCl (pH 7.5), 75 m$M$ KCl. Store in 50-$\mu$l aliquots at −20° (stable for about 1 month)
BiP stock solution: BiP or isolated C30 (1 mg/ml) dialyzed in binding buffer
Peptide solution and binding buffer: See above
Running buffer: 25 m$M$ Tris, 192 m$M$ glycine
Premix: 1 $\mu$l of BiP (1 mg/ml) with 2 $\mu$l of RCMLA (3 mg/ml) in 2 $\mu$l of binding buffer
Native sample buffer: 0.9 g of sucrose, 25 $\mu$l 0.5 $M$ Tris-HCl (pH 6.8), 400 $\mu$l of bromphenol blue (2 mg/ml), 575 $\mu$l of $H_2O$

*Method.* On ice, place 4 $\mu$l of premix, and 4 $\mu$l of peptide solution (binding buffer for control), in 0.5-ml Eppendorf tubes, then vortex and spin immediately (for a maximum of 10 sec at 7200 $g$). Place the tubes in

[47] D. R. Palleros, W. J. Welch, and A. L. Fink, *Proc. Natl. Acad. Sci. U.S.A.* **88,** 5719 (1991).
[48] T. Langer, C. Lu, H. Echols, J. Flanagan, M. K. Hayer, and F. U. Hartl, *Nature (London)* **356,** 683 (1992).

a 37° water bath for 1 or 2 hr, then cool on ice. Add 2 μl of native sample buffer, centrifuge (7200 *g*, 10 to 20 sec), and load on a 6% (w/v) native polyacrylamide gel.[24,25] Run at 25 mA for 1 hr at 4°. Check that the voltage does not exceed 180 V during the run. Stain the gels with Coomassie Brilliant Blue. In the presence of a large excess of RCMLA, BiP or C30 oligomers are no longer present. What should be seen is a shift in the migration when RCMLA (14 kDa) is displaced from BiP or C30 complexes (90 and 44 kDa, respectively) by small peptides (1000 Da). In this case, the BiP or C30–peptide complex should run faster and therefore migrate further than a BiP or C30–RCMLA complex. However, native electrophoresis also depends on the shape and charge of the migrating complex, so some peptides may have an opposite effect on migration if they are positively charged. For examples, see Ref. 11.

# [32] Purification and Characterization of Prokaryotic and Eukaryotic Hsp90

*By* JOHANNES BUCHNER, SUCHIRA BOSE, CHRISTIAN MAYR, and URSULA JAKOB

## Introduction

Heat shock protein 90 (Hsp90) is one of the most abundant proteins in the eukaryotic cell under normal conditions.[1] Under stress conditions its expression level is increased severalfold, suggesting a general protective function. Whereas yeast Hsp90 is essential for survival,[2] *Escherichia coli* Hsp90 can be deleted with only minor effects on growth at elevated temperatures.[3] Generally, Hsp90 seems to be involved in the conformational regulation of key proteins of signaling pathways such as steroid receptors and kinases under physiological conditions and in protein refolding.[4–6] Prokaryotic and eukaryotic Hsp90 protect nonnative substrate proteins from inacti-

[1] S. C. Lindquist and E. A. Craig, *Annu. Rev. Genet.* **22,** 631 (1988).
[2] K. A. Borkovich, F. W. Farrelly, D. B. Finkelstein, J. Taulien, and S. Lindquist, *Mol. Cell. Biol.* **9,** 3919 (1989).
[3] J. C. A. Bardwell and E. A. Craig, *J. Bacteriol.* **170,** 2977 (1988).
[4] D. F. Smith, *Mol. Endocrinol.* **7,** 1418 (1993).
[5] W. B. Pratt, *J. Biol. Chem.* **268,** 21455 (1993).
[6] J. Buchner, *FASEB J.* **10,** 10 (1996).



0076-6879/98 $25.00

vation and subsequent aggregation.[7–9] Stable interaction with folding intermediates might be conferred by the partner proteins with which eukaryotic Hsp90 is known to cooperate. *In vivo,* eukaryotic Hsp90 seems to perform at least part of its function in complex with Hsp70, large immunophilins, and p23 (see Ref. 10 and [33] in this volume[11]).

The structure of an N-terminal domain of Hsp90 has been solved by X-ray crystallography and a binding pocket for ATP and the anti-tumor drug geldanamycin has been identified.[12,13]

## *Escherichia coli* Hsp90 (HtpG)

The *E. coli* protein that is homologous to eukaryotic Hsp90 was discovered by Bardwell and Craig.[14] It is also called HtpG or C62.5. This protein has a molecular mass of 71 kDa and is thus significantly smaller than its eukaryotic relatives, owing to two truncated charged regions.[14] At normal temperatures, *E. coli* Hsp90 constitutes about 0.5% of the total *E. coli* protein.[15] Under heat shock conditions, its expression is increased severalfold.[14] Deletion of *htpG* results in slightly reduced growth at elevated temperatures.[3]

Surprisingly little is known about the function of *E. coli* Hsp90 *in vivo* and extensive searches in the databases reveal that most of the known partner proteins of Hsp90 in higher eukaryotes do not have homologs in *E. coli* (U. Jakob, unpublished, 1997). *In vitro, E. coli* Hsp90 functions as a molecular chaperone, with an activity similar to Hsp90 from yeast or higher eukaryotes.[7]

### *Method*

This method describes the expression and purification of wild-type recombinant Hsp90 from 20 g of *E. coli* cell paste (wet weight). Hsp90 is constitutively expressed and can constitute up to 30% of the total cellular

[7] U. Jakob, H. Lilie, I. Meyer, and J. Buchner, *J. Biol. Chem.* **270,** 7288 (1995).
[8] H. Wiech, J. Buchner, R. Zimmermann, and U. Jakob, *Nature (London)* **358,** 169 (1992).
[9] B. C. Freeman and R. I. Morimoto, *EMBO J.* **15,** 2969 (1996).
[10] D. F Smith and D. O. Toft, *Mol. Endocrinol.* **7,** 4 (1993).
[11] J. Buchner, T. Weikl, H. Bügl, F. Pirkl, and S. Bose, *Methods Enzymol.* **290,** Chap. 33, 1998 (this volume).
[12] C. Prodromou, S. M. Roe, R. O'Brian, J. E. Ladbury, P. W. Piper, and L. H. Pearl, *Cell* **90,** 65 (1997).
[13] C. E. Stebbins, A. A. Russo, C. Schneider, N. Rosen, F. U. Hartl, and N. P. Pavletich, *Cell* **89,** 239 (1997).
[14] J. C. A. Bardwell and E. A. Craig, *Proc. Natl. Acad. Sci. U.S.A.* **84,** 5177 (1987).
[15] S. L. Herendeen, R. A. Van Bogelen, and F. C. Neidhardt, *J. Bacteriol.* **139,**185 (1979).

protein, which results in a yield of approximately 40 mg of highly purified protein.

*Expression. Escherichia coli* Hsp90 [Swiss Protein Sequence Database (SwissProt) Accession Number P10413] is expressed from the plasmid pBJ935 (kind gift of J. C. A. Bardwell, University of Michigan, Ann Arbor, MI), using its own heat shock promoter.[14] The plasmid is transformed into *E. coli* MG1655, resulting in the strain JCB 867. Cells are grown in Luria broth (LB) medium in the presence of ampicillin (100 $\mu$g/ml) at 30°. *Escherichia coli* Hsp90 is constitutively expressed. Growth at 37° or higher temperatures results in cell death, because of the massive overexpression of *E. coli* Hsp90.

After inoculation, the culture is grown in LB–ampicillin (LB–Amp) for 20 hr at 30° until an $OD_{600}$ of 5 is reached. The cells are harvested by centrifugation at 8000 *g* for 10 min at 4°. The bacterial cell pellet is resuspended in 20 ml of ice-cold lysis buffer [50 m*M* potassium phosphate, 20 m*M* KCl, 5% (v/v) glycerol, 1 m*M* ethylenediaminetetraacetic acid (EDTA), pH 7.5] and kept on ice. All further purification steps are performed in a cold room at 4° or on ice. The purification steps of the protein are performed by following the clearly visible bands of overexpressed Hsp90 on a sodium dodecyl sulfate (SDS)–polyacrylamide gel[16] and collecting the appropriate fractions. The identity of the protein is further confirmed by Western blot analysis,[17] using a specific antibody against *E. coli* Hsp90.

*Purification*

1. *Cell lysis and ammonium sulfate precipitation*: The bacterial cell pellet is lysed in an ice-cold French press at 18,000 psi for two cycles. Before each cycle protease inhibitor Cocktail Tablets Complete (one tablet per 50-ml solution, following the manufacturer instructions; Boehringer Mannheim GmbH, Mannheim, Germany) are added to the protein suspension. The crude extract is centrifuged to remove the insoluble material (48,000 *g*, 30 min, 4°) and the supernatant is then supplemented with solid ammonium sulfate to a final concentration of 60% (w/v). After a 30-min incubation on ice followed by a 60-min centrifugation at 48,000 *g* at 4°, the precipitate is collected and dialyzed at 4° against buffer A [50 m*M* *N*-2-hydroxyethylpiperazine-*N*′-2-ethanesulfonic acid (HEPES)–KOH, 100 m*M* KCl, 5% (v/v) glycerol, 1 m*M* EDTA, pH 7.5] overnight.

2. *Anion-exchange chromatography:* After centrifugation (48,000 *g*, 30 min, 4°), the supernatant is loaded onto a 20-ml DE52 column, previously equilibrated in buffer A. Unbound proteins are eluted by washing the

[16] U. K. Laemmli, *Nature* (*London*) **227,** 680 (1970).

[17] H. Towbin, T. Staehelin, and J. Gordon, *Proc. Natl. Acad. Sci. U.S.A.* **76,** 4350 (1979).

column with five column volumes of buffer A. The bound proteins are eluted with a 250-ml gradient from 100 to 500 m*M* KCl in buffer A. The fractions are analyzed for the presence of *E. coli* Hsp90 on a 10% (w/v) SDS–polyacrylamide gel. *Escherichia coli* Hsp90 elutes between 200 and 300 m*M* KCl.

3. *Hydroxyapatite:* The *E. coli* Hsp90-containing fractions are pooled and concentrated to a final volume of about 10 ml (YM30 membrane; Amicon, Danvers, MA). The protein solution is dialyzed overnight against buffer B [20 m*M* potassium phosphate, 5% (v/v) glycerol, pH 6.8] and is subsequently centrifuged to remove precipitated proteins (48,000 *g*, 30 min, 4°). The supernatant is loaded onto a 20-ml hydroxyapatite column, equilibrated in buffer B. After washing the column with 60 ml of buffer B, proteins are eluted with a 20 to 400 m*M* potassium phosphate gradient over 10 column volumes. *Escherichia coli* Hsp90 elutes from the column between 50 and 150 m*M* potassium phosphate. Hsp90-containing fractions are detected by SDS–polyacrylamide gel electrophoresis (SDS–PAGE) and are pooled.

4. *Anion-exchange chromatography at pH 5.5:* The pooled fractions are concentrated to about 5 ml (YM30; Amicon) and the pH of the protein solution adjusted to pH 5.5 by dialysis against buffer C [10 m*M* morpholine-ethanesulfonic acid (MES)–HCl, 20 m*M* KCl, 5% (v/v) glycerol, pH 5.5]. The precipitated proteins are removed by centrifugation (30 min at 48,000 *g* and 4°) and the supernatant is loaded onto a 10-ml DE52 column, previously equilibrated in buffer C. The column is washed with five column volumes of the same buffer and Hsp90 is eluted in a gradient from 20 to 500 m*M* KCl in buffer C. The fractions containing highly purified Hsp90 (180–300 m*M* KCl) are pooled and supplemented with ammonium sulfate (80% final concentration). The ammonium sulfate precipitate can be stored at 4° for several months.

*Comments.* Protein concentrations are determined using the calculated absorbance ($A_{280}$) of 1.25 for a 1-mg/ml solution.[18]

## Yeast Hsp90

Yeast Hsp90 (yHsp90) is an abundant protein that constitutes up to 2% of the total cell protein under normal temperature conditions[1] due to the constitutive expression of the *hsc82* gene. Under heat shock conditions, the amount of Hsp90 increases even more, owing to the additional expression of a second heat shock-regulated Hsp90 gene, *hsp82*. The requirement for the

[18] D. B. Wetlaufer, *Adv. Protein Chem.* **17,** 303 (1962).

high expression rate of yHsp90 under physiological temperature conditions remains elusive. The fact that a yHsp90 mutant expressing only 1/18 of the normal amount of yHsp90 does not show any phenotype is puzzling.[19] However, deletions of both Hsp90 genes together are lethal.[2]

To elucidate the function of yHsp90 in the cell, coexpression experiments of yHsp90 with tyrosine kinases and steroid receptors have been performed in yeast.[20] These studies reveal that yHsp90 is functionally similar to Hsp90 from higher eukaryotes because Hsp90 from higher eukaryotes can substitute in the maturation of tyrosine kinases and the stabilization of steroid receptors. However, because these substrates are nonphysiological, the natural substrates of Hsp90 in yeast remain to be determined. Interaction of Hsp90 with partner proteins is likely to be involved, because homologous partner proteins have been identified.[21] *In vitro* yHsp90 acts as a molecular chaperone in binding aggregation-sensitive folding intermediates, suppressing aggregation, increasing the yield of refolding proteins, and stabilizing thermolabile proteins at elevated temperatures.[7]

### *Method*

This method describes the expression and purification of yHsp90 from 50 g (wet weight cell paste) with a final yield of approximately 50 mg of highly purified heat shock protein.

*Expression.* For expression of yHsp90, an *hsp82* and *hsc82* deletion, protease-deficient yeast strain is used that carries *hsp82* (SwissProt P02829) on a 2$\mu$ plasmid with a GDP promotor (kind gift of S. Lindquist, University of Chicago, Chicago, IL). The level of constitutive expression is about 5% of the total cellular protein. An overnight culture with a volume equal to 1/15 of the final culture is inoculated with a single colony and grown at 25° in yeast–peptone–dextrose (YPD) medium to an $D_{600}$ of 3–4. After addition of the overnight culture to YPD medium, the cells are grown at 25° until late logarithmic phase ($A_{600}$ 2–2.5, ~15 hr). Growth into stationary phase should be avoided. The cells are harvested by centrifugation at 8000 *g* for 20 min at 4°. The cell pellet is washed twice with buffer A [40 m*M* HEPES–KOH, 5 m*M* EDTA, 1 m*M* 1,4-dithioerythritol, pH 7.5, and protease inhibitor cocktail tablets Complete (used as before; Boehringer Mannheim GmbH)]. After centrifugation at 8000 *g* for 20 min at 4° the cell pellet is resuspended in twice the volume of the cell wet weight in ice-cold buffer

[19] D. A. Parsell and S. Lindquist, *in* "The Biology of Heat Shock Proteins and Molecular Chaperones" (R. I. Morimoto, A. Tissieres, and C. Georgopolous, eds.), pp. 457–494. Cold Spring Harbor Laboratory Press, Cold Spring Harbor, New York, 1994.

[20] Y. Xu and S. C. Lindquist, *Proc. Natl. Acad. Sci. U.S.A.* **90,** 7074 (1993).

[21] H.-C. Chang and S. Lindquist, *J. Biol. Chem.* **269,** 24983 (1994).

A and kept on ice. Instead of protease inhibitor Cocktail Tablets Complete from Boehringer Mannheim GmbH, a protease inhibitor mix [pepstatin (40 $\mu$g/ml), leupeptin (1 $\mu$g/ml), and Pefabloc SC (1 $\mu$g/ml); Boehringer Mannheim GmbH] can be used. Protease inhibitors should be added at each purification step because yHsp90 is sensitive to proteolytic digestion. The purification steps of the protein are performed by following the clearly visible bands of overexpressed Hsp90 on a SDS–polyacrylamide gel[16] and by pooling the appropriate fractions. The identity of the protein is further confirmed by Western blot analysis[17] using a specific antibody against yHsp90.

*Purification*

1. *Cell lysis:* The suspension is filled into precooled lysis tubes containing glass beads (0.5-mm diameter). The yeast cells are lysed with glass beads using a Braun MSK 300 bead beater (according to manufacturer instructions) with two 1-min bursts at maximum speed. Lysis should be short (not more than 2 min). During this lysis process the cell suspension is cooled by a stream of compressed $CO_2$. The glass beads are separated from the cracked cells by filtration, using a Büchner funnel. Finally, one Complete tablet is added to the lysed cells and the lysate is supplemented with NaCl to a final concentration of 0.1 *M*. The soluble proteins are separated from the insoluble material by a 60-min centrifugation at 48,000 *g* at 4°.

2. *Anion-exchange chromatography:* The supernatant, supplemented with protease inhibitors, is loaded onto a 40-ml DE52 column equilibrated in buffer A (supplemented with 0.1 *M* NaCl) and the column is washed with 200 ml of the same buffer. The bound proteins are eluted by a linear 400-ml gradient from 0.1 to 0.5 *M* NaCl in buffer A. Fractions are collected and tested for the presence of Hsp90 on a 10% SDS–polyacrylamide gel. Yeast Hsp90 elutes between 0.2 and 0.3 *M* NaCl.

3. *Cation-exchange chromatography at pH 5.5:* The Hsp90-containing fractions are pooled and supplemented with one Complete tablet. The protein solution is concentrated to a volume of approximately 30 ml (YM30 membrane; Amicon) and is then dialyzed against buffer B (20 m*M* sodium phosphate buffer, pH 5.5) at 4°. Insoluble material is removed by centrifugation at 48,000 *g* for 60 min at 4°. The supernatant is then passed through a 20-ml S-Sepharose column, previously equilibrated in the same buffer. Yeast Hsp90 does not bind to the column. One more Complete tablet is added to the flowthrough, which represents the Hsp90 pool. The pH is adjusted to pH 7.5 with 1 *M* HEPES, pH 8.0

4. *Hydroxyapatite:* After ultrafiltration of the yHsp90 pool to a final volume of about 10 ml (YM30; Amicon) and addition of one Complete tablet, the protein is dialyzed overnight against buffer C (0.1 *M* potassium

phosphate bufer, pH 6.8) at 4°. The dialyzed protein solution is then centrifuged (48,000 *g*, 4°, 60 min) and the supernatant is loaded onto a 20-ml hydroxyapatite column, equilibrated in buffer C. Unbound proteins are removed by washing the column with 100 ml of buffer C. The bound proteins are eluted with a 0.1–0.3 *M* potassium phosphate buffer, pH 6.8 gradient over 10 column volumes. Yeast Hsp90 elutes between 0.15 and 0.2 *M* potassium phosphate.

5. *Gel filtration:* The yHsp90-containing fractions are pooled, concentrated to 2 mg/ml, and the salt concentration adjusted to 0.4 *M* KCl and 10 m*M* $MgCl_2$. After a 30-min spin at 48,000 *g* at 4°, 2-ml portions are applied onto a 120-ml Superdex 200 preparatory-grade column (Pharmacia, Piscataway, NJ) equilibrated with buffer D (40 m*M* HEPES–KOH, 0.4 *M* KCl, 10 m*M* $MgCl_2$, pH 7.5). Hsp90-containing fractions (approximate elution volume, 50–70 ml) are pooled and concentrated to the desired protein concentration (YM30, Centricon-30; Amicon). Hsp90 is then dialyzed overnight at 4° against buffer E [40 m*M* HEPES–KOH, 20 m*M* KCl, 5% (v/v) glycerol, pH 7.5]. To remove aggregates, yHsp90 is centrifuged at 48,000 *g* for 60 min at 4° and the supernatant is divided into small aliquots and stored at −80°.

*Comments.* Protein concentrations are determined with the Bradford assay[22] or spectrophotometrically using the calculated absorbance ($A_{280}$) of 0.74 for a 1-mg/ml solution.[18]

## Human Hsp90 from Baculovirus

Mammalian cells express two highly related isoforms of Hsp90, Hsp90$\alpha$ and Hsp90$\beta$, which are identical at 630 of 724 possible amino acid residues.[23] Like other eukaryotic Hsp90s, human Hsp90 (hHsp90) exhibits highly charged regions that are partly missing in *E. coli* Hsp90. Human Hsp90 is found in the cytoplasm as a dimeric protein.[24] Phosphorylation seems to occur at two serine residues.[25] Studies *in vitro* have shown that hHsp90, like other members of the Hsp90 family, exhibits chaperone properties such as suppression of protein aggregation under heat shock conditions.[7] *In vivo,* steroid receptors and certain kinases have been identified as specific substrates of hHsp90 (see Refs. 4 and 5). In this context, Hsp90 seems to perform its function in a complex with a number of partner proteins including Hsp70, large prolyl isomerases, p23, and p60/Hop. Interaction of the Hsp90 complex with ligand-free steroid receptors is a key regulatory step

[22] M. Bradford, *Anal. Biochem.* **72,** 248 (1976).
[23] N. F. Rebbe, J. Ware, R. M. Bertina, P. Modrich, and D. Stafford, *Gene* **53,** 235 (1987).
[24] W. J. Welch and J. R. Feramisco, *J. Biol. Chem.* **257,** 14949 (1982).
[25] S. P. Lees-Miller and C. W. Anderson, *J. Biol. Chem.* **264,** 2431 (1989).

in the signal transduction process. However, because the levels of Hsp90 exceed that of partner proteins, especially under heat shock conditions when Hsp90 expression increases up to 10-fold, there is good reason to assume that some of the important functions of Hsp90 are performed without the help of partner proteins.

*Method*

This method describes the expression and purification of human Hsp90 (hHsp90) from baculovirus-infected Sf9 cells with a final yield of approximately 40 mg from $6 \times 10^8$ cells (corresponding to about 4 g of wet cell pellet).

*Expression.* Hsp90 is prepared by the overexpression of human Hsp90$\beta$ (SwissProt P08238) in Sf9 cells, using the system of Alnemri and co-workers.[26,27] Two days postinfection, cells are harvested by centrifugation at 2000 *g* for 10 min at 4°. The supernatant is discarded and the cell pellets may be stored at −80° until further use (cell culture, infection, and harvest may be performed at the Colorado Cancer Center Tissue Culture Core Facilities). Virus stock containing overexpressing hHsp90 was a generous gift from G. Litwack (Thomas Jefferson University, Philadelphia, PA). The purification steps of the protein are performed by following the clearly visible bands of overexpressed Hsp90 on an SDS–polyacrylamide gel[16] and collecting the appropriate fractions. The identity of the protein is further confirmed by Western blot analysis[17] using a specific antibody against yHsp90.

*Purification*

1. *Cell lysis:* Cell pellets are resuspended in 7.5 ml of buffer A (20 m*M* Tris, 0.1 m*M* EDTA, 100 m*M* NaCl, pH 6.9) per gram of cells. Several protease inhibitors [final concentrations: 40 mg/ml (pepstatin), 1 mg/ml (leupeptin), and 1 mg/ml (Pefabloc SC); Boehringer Mannheim GmbH] are added to the cell suspension. Protease inhibitors should be added preferentially at each purification step because hHsp90 is sensitive to proteolytically digestion. The resuspended cells are left on ice for 30 min, to allow the cells to swell. The cells are lysed by gentle hand homogenization on ice, using a precooled Dounce homogenizer. For cell lysis sonication should be avoided because it may result in significant precipitation of protein. The homogenate is centrifuged at 18,000 *g* for 30 min at 4°. A cocktail of protease inhibitors (as described previously), is added to the supernatant.

[26] E. S. Alnemri and G. Litwack, *Biochemistry* **32,** 5387 (1993).

[27] E. S. Alnemri, A. B. Maksymowych, N. M. Robertson, and G. Litwack, *J. Biol. Chem.* **266,** 3925 (1991).

2. *Anion-exchange chromatography:* The supernatant is loaded onto a 50-ml Q-Sepharose column (Pharmacia) that is preequilibrated with buffer A. Bound proteins are eluted with a gradient ranging from 100 m*M* to 1 *M* NaCl buffer A over five column volumes. Fractions are tested for the presence of hHsp90 on a 10% (w/v) SDS–polyacrylamide gel. Human Hsp90 elutes between 0.4 and 0.5 *M* NaCl.

3. *Hydroxyapatite:* Fractions containing hHsp90 are pooled and dialyzed against buffer B (20 m*M* potassium phosphate, pH 6.8). After dialysis, the sample is applied onto a 2-ml hydroxyapatite fast protein liquid chromatography (FPLC) column (Bio-Rad, Hercules, CA). The protein-binding capacity of this column is about 20 mg; therefore several runs may be necessary. The column is equilibrated in buffer B before sample loading. After washing with buffer B, bound proteins are eluted with a 20–400 m*M* potassium phosphate (pH 6.8) gradient over 10 column volumes. Fractions containing hHsp90 are pooled and concentrated by ultrafiltration (YM30 membranes; Amicon).

4. *Gel filtration:* Samples (2 ml) are applied to a Superdex 200 (Pharmacia) gel-filtration column (120 ml) equilibrated with buffer C [40 m*M* HEPES, 400 m*M* KCl, 5% (v/v) glycerol, pH 7.5]. Fractions containing pure hHsp90 are pooled and concentrated by ultrafiltration (see above). The concentrated sample is dialyzed overnight against buffer D [40 m*M* HEPES (pH 7.5), 5% (v/v) glycerol] at 4°. The dialyzed sample is frozen in liquid nitrogen and stored at −80°.

*Comments.* Protein concentrations are determined with the Bradford assay[22] or spectrophotometrically using the calculated absorbance ($A_{280}$) of 0.7 for a 1-mg/ml solution.[18]

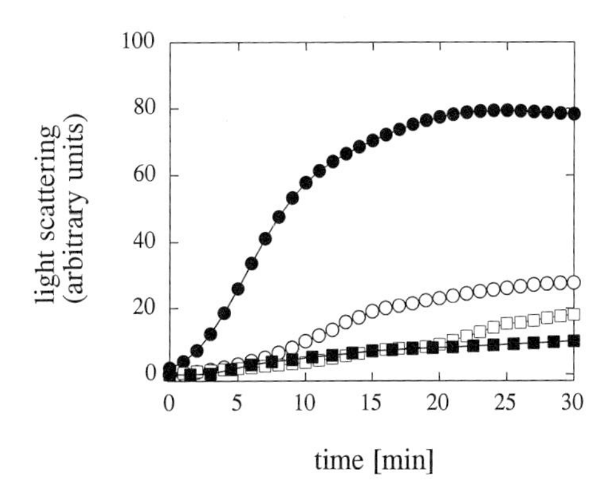


FIG. 1. Influence of Hsp90 from different species on the thermal aggregation of CS at 43°. Conditions are as described in Buchner *et al.*[24] (●) Spontaneous aggregation of CS (0.15 $\mu M$ monomer); (○) HtpG (0.3 $\mu M$); (□) yeast Hsp90 (0.15 $\mu M$); (■) human Hsp90 (0.15 $\mu$M).

## Analysis of Chaperone Activity of Hsp90 *in Vitro*

Prokaryotic and eukaryotic Hsp90s function as molecular chaperones in suppressing unspecific side reactions such as aggregation and supporting the refolding of proteins to the native state *in vitro*.[6,8] A convenient and simple assay for chaperone activity is the thermal inactivation and aggregation of citrate synthase (CS) at 43° (see [27] in this volume[28]). The ability of Hsp90 to recognize and bind unfolding intermediates of CS is reflected by the apparent stabilization of thermally denaturing CS and a significant reduction in aggregation (Fig. 1 and Ref. 7). Both processes can be easily monitored using CS activity assays and light-scattering measurements, respectively. Qualitatively similar results are obtained on studying the influence of Hsp90 on the refolding of chemically denatured CS. Hsp90 suppresses the aggregation of refolding molecules and therefore increases the final yield of reactivation.[8] A detailed description of the chaperone activity assay and the results obtained in the presence of Hsp90 are given in [27] in this volume.[28]

## Acknowledgment

We thank D. Edwards and K. Christensen at the University of Colorado Cancer Center Tissue Culture Care Facilities for Sf9 cell growth, treatment and harvesting.

[28] J. Buchner, H. Grallert, and U. Jakob, *Methods Enzymol.* **290,** Chap. 27, 1998 (this volume).

# [33] Purification of Hsp90 Partner Proteins Hop/p60, p23, and FKBP52

*By* Johannes Buchner, Tina Weikl, Hans Bügl, Franziska Pirkl, and Suchira Bose

## Introduction

Estrogen receptors, in their hormone-free state, exist as large, ~9S complexes in cytosolic cell extracts.[1] Subsequent analysis has revealed that mammalian steroid hormone receptors, in general, exist as heterooligomeric complexes consisting of the receptor associated with Hsp90 and its partner proteins.[2,3] To date, seven Hsp90 partner proteins have been identified and

[1] D. O. Toft and J. Gorski, *Proc. Natl. Acad. Sci. U.S.A.* **55,** 1574 (1966).
[2] D. F. Smith, L. E. Faber, and D. O. Toft, *J. Biol. Chem.* **265,** 3996 (1990).
[3] D. F. Smith, *Sci. Med.* **2,** 38 (1995).



0076-6879/98 $25.00

these include Hsp70,[4] Hip/p48,[5,6] Hop/p60,[7] p23,[8] and one of three large immunophilins: FKBP52 (FKBP59, p59, Hsp56, HBI), FKBP51 (FKBP54), and cyclophilin 40 (Cyp40).[3,7,9,10]

Steroid hormone receptors are soluble intracellular proteins that act as hormone-inducible transcription factors, shuttling between the cytosol and nucleus. The involvement of Hsp90 complexes in this process has been analyzed in most detail for the progesterone aporeceptor (PR).[3,7,8,10,11] During receptor maturation, early and intermediate complexes consisting of the receptor molecule, Hsp90, Hsp70, Hip, and Hop are displaced by a late complex composed of the receptor, Hsp90, p23, and one of three large immunophilins.[12,13] These complexes spontaneously dissociate and their components reenter the assembly pathway, thus allowing a dynamic turnover of receptor complexes.[12] They are required to keep aporeceptors in an inactive yet activatable state. Hormone binding leads to a conformational change in the receptor resulting in the dissociation of the Hsp90 complex and simultaneously allowing receptor dimerization and transformation of the receptor to the active (DNA-binding) state.[3]

In addition, Hsp90 was found in stable complexes with certain kinases.[14] This interaction is thought to stabilize the kinase molecule in a soluble, inactive state until it becomes myristoylated and attached to the plasma membrane.[15] Although these studies suggest specific roles for the Hsp90 complex, a more general function is indicated by the fact that the components of the complex are conserved throughout eukaryotes.[14]

The functional properties of three of the Hsp90 partner proteins, Hop/p60, p23, and one of the large immunophilins, FKBP52 or Cyp40, have been

[4] W. B. Pratt, *Prog. Clin. Biol. Res.* **322,** 119 (1990).

[5] J. Höhfeld, Y. Minami, and F. U. Hartl, *Cell* **83,** 589 (1995).

[6] V. Prapapanich, S. Chen, S. C. Nair, R. A. Rimerman, and D. F. Smith, *Mol. Endocrinol.* **10,** 420 (1996).

[7] D. F. Smith, W. P. Sullivan, T. N. Marion, K. Zaitsu, B. Madden, D. J. McCormick, and D. O. Toft, *Mol. Cell. Biol.* **13,** 869 (1993).

[8] J. L. Johnson and D. O. Toft, *J. Biol. Chem.* **269,** 24989 (1994).

[9] M. Milad, W. Sullivan, E. Diehl, M. Altmann, S. Nordeen, D. P. Edwards, and D. O. Toft, *Mol. Endocrinol.* **9,** 838 (1995).

[10] W. B. Pratt and M. J. Welsh, *Semin. Cell Biol.* **5,** 83 (1994).

[11] D. F. Smith, W. P. Sullivan, J. Johnson, and D. O. Toft, *in* "Steroid Hormone Receptors" (V. K. Moudgil, ed.), p. 247. Birkhäuser, Boston, 1993.

[12] D. F. Smith, L. Whitesell, S. C. Nair, S. Chen, V. Prapapanich, and R. A. Rimerman, *Mol. Cell. Biol.* **15,** 6804 (1995).

[13] S. Chen, V. Prapapanich, R. A. Rimerman, B. Honoré, and D. F. Smith, *Mol. Endocrinol.* **10,** 682 (1996).

[14] U. Jakob and J. Buchner, *Trends Biochem. Sci.* **19,** 205 (1994).

[15] Y. Xu and S. Lindquist, *Proc. Natl. Acad. Sci. U.S.A.* **90,** 7074 (1993).

characterized *in vitro*.[16,17] Here, we describe the purification of recombinant Hop/p60, p23, and FKBP52 and include a description of their chaperone function.

### *FKBP52*

FKBP52 was originally identified when a monoclonal antibody prepared against partially purified untransformed PR was found to cross-react with a non-steroid-binding 59-kDa rabbit protein that was a component of several steroid receptor heterocomplexes.[18,19] Later, a human 56-kDa protein cross-reacting with the same antibody was detected that seemed to be overexpressed under heat shock conditions (Hsp56).[20] Independently, a 59-kDa protein associated with Hsp90 in the absence of steroid receptor was discovered and was termed HBI (heat shock binding immunophilin).[21] The deduced amino acid sequence of the rabbit[22] and human[23] cDNAs indicates a molecular mass of approximately 52,000 Da. Hence, this protein is now generally termed FKBP52 (see below).

FKBP52 belongs to the class of large immunophilins, which comprises two unrelated protein families: cyclophilins and FKBPs (FK506-binding proteins), both of which are peptidylprolyl *cis–trans*-isomerases (PPIases).[24–26] The PPIases catalyze slow steps in protein-folding reactions by accelerating conformational interconversions of prolyl peptide bonds in the polypeptide backbone. They bind with high affinity to immunosuppressants such as cyclosporin A, FK506, and rapamycin, which inhibit their prolyl isomerase activity.[27] The physiological role of PPIases remains enigmatic.

In conjunction with p23 and Hsp90, the large immunophilins (FKBP52,

[16] S. Bose, T. Weikl, H. Bügl, and J. Buchner, *Science* **274,** 1715 (1996).
[17] B. C. Freeman, D. O. Toft, and R. I. Morimoto, *Science* **274,** 1718 (1996).
[18] P. K. K. Tai and L. E. Faber, *Can. J. Biochem. Cell. Biol.* **63,** 41 (1985).
[19] P. K. K. Tai, Y. Maeda, K. Nakao, N. G. Wakim, J. L. Duhring, and L. E. Faber, *Biochemistry* **25,** 5269 (1986).
[20] E. R. Sanchez, *J. Biol. Chem.* **265,** 22067 (1990).
[21] N. Massol, M. C. Lebeau, J. M. Renoir, L. E. Faber, and E. E. Baulieu, *Biochem. Biophys. Res. Commun.* **187,** 1330 (1992).
[22] M. C. Lebeau, N. Massol, J. Herrick, L. E. Faber, J. M. Renoir, C. Radanyi, and E. E. Baulieu, *J. Biol. Chem.* **267,** 4281 (1992).
[23] D. A. Peattie, M. W. Harding, M. A. Fleming, M. T. DeCenzo, J. A. Lippke, D. J. Livingstone, and M. Benasutti, *Proc. Natl. Acad. Sci. U.S.A.* **89,** 10974 (1992).
[24] J. E. Kay, *Biochem. J.* **314,** 361 (1996).
[25] G. Fischer, *Angew. Chem., Int. Ed. Engl.* **33,** 1415 (1996).
[26] F. X. Schmid, *Annu. Rev. Biophys. Biomol. Struct.* **22,** 123 (1993).
[27] S. L. Schreiber, *Science* **251,** 283 (1991).

FKBP51, and Cyp40) play a crucial role in the activation of steroid hormone receptors and kinases. Data indicate that FKBP52 chaperones protein-folding and unfolding reactions by the accumulation of folding-competent intermediates.[16] Independently, Freeman and co-workers demonstrated a similar activity for Cyp40.[17] Using immunosuppressive agents to inhibit its catalytic activity, it was shown that the chaperone activity of FKBP52 is independent of its PPIase activity. Sequence and hydrophobic cluster analysis predicts that FKBP52 can be divided into three domains,[28] with the amino-terminal domain being responsible for its PPIase activity. Thus, the large immunophilins, in general, may have acquired additional domains responsible for the chaperone function.

*p23*

p23 was first observed as a component of the avian progesterone receptor complex.[2] Subsequently, monoclonal antibodies to p23 were shown to cross-react with a protein of the same size in organisms ranging from *Saccharomyces cerevisiae* to humans, suggesting that p23 is ubiquitous and its expression is not unique to cells that contain progesterone receptors.[29] Sequence comparisons demonstrated that the protein is highly conserved with the chicken p23 sharing 96.3% identity at the amino acid level with its human homolog.[29]

Human p23 consists of 160 amino acids and has a calculated molecular mass of 18,700 Da, but exhibits decreased mobility on sodium dodecyl sulfate (SDS)–polyacrylamide gels. Database searches did not reveal any significant homologies to proteins of known function. Thus, p23 is the only component of the mature progesterone receptor complex that does not belong to the family of heat shock proteins or immunophilins.

p23 was shown to be essential for correct assembly of mature progesterone receptor complexes.[8] Moreover, the presence of p23 in these complexes seems to play a major role in receptor activation, because disruption of p23–Hsp90 interactions in the complex results in the loss of the hormone-binding ability of the receptor.[12] Results demonstrating the chaperone properties of p23[16,17] suggest a participation of the protein in a cytosolic "superchaperone machinery" that guides maturation of different substrate proteins including steroid receptors.

[28] M. C. Lebeau, N. Binart, F. Cadepond, M. G. Catelli, B. Chambraud, N. Massol, C. Radanyi, G. Redeuilh, J. M. Renoir, M. Sabbah, G. Schweizer-Groyer, and E. E. Baulieu, "Steroid Hormone Receptors" (V. K. Milhoud, ed.). Birkhäuser, Boston, 1993.

[29] J. L. Johnson, T. G. Beito, C. J. Krco, and D. O. Toft, *Mol. Cell. Biol.* **14,** 1956 (1994).

*Hop/p60*

p60 was originally identified as a transformation-sensitive protein IEF SSP 3521 in MRC-5 fibroblasts.[30] It was found to have striking sequence similarity to yeast STI1, a protein reported to be a stress-inducible mediator of the heat shock response.[31] In 1993, Smith and co-workers[7] identified a 60-kDa protein from a variety of tissues that copurified together with Hsp70 and Hp90. Immunoblot analysis using a monoclonal antibody prepared against p60 derived from avian fibroblasts showed that p60 was present in cytosols from organisms ranging from *Xenopus* to humans.[7]

Human p60 consists of 543 amino acid residues and thus has a calculated molecular mass of approximately 63,000 Da. Two tetratricopeptide repeat (TPR) regions were identified[30] with clusters of three and six tandem, degenerate motifs, respectively.[13] These motifs are believed to be important for the interaction with Hsp70 and Hsp90 in the formation of a transient complex in the pathway of progesterone receptor formation. To account for these matchmaking properties, p60 was renamed Hop (Hsp70/Hsp90 organizing protein). Hop binds Hsp70 and Hsp90 independently and is thus thought to be a key mediator in the assembly of the two chaperone machineries.[13]

Studies on the chaperone functions of Hsp90-associated proteins indicated that, in contrast to FKBP52 and p23, Hop does not show chaperone activity.[16,17]

## Purification of Recombinant Rabbit FKBP52

### *Method*

The method described below allows the purification of rabbit FKBP52 expressed in *Escherichia coli.* Typically, a yield of 1 mg of pure protein per gram of cells is achieved.

*Protein Expression.* High-level expression of rabbit FKBP52 (GenBank Accession No. M84474) in *E. coli* is achieved by using the T7 polymerase/promoter system. The pET23 expression construct is a kind gift of D. Toft (Mayo Clinic, Rochester, MI). It contains the cDNA for rabbit FKBP52 cloned into the *Bam*HI restriction site of the multiple cloning site of the pET23 plasmid. This plasmid is transformed into *E. coli* BL21(DE3), which contains the T7 polymerase structural gene under control of the *lacUV5*

[30] B. Honoré, H. Leffers, P. Madsen, H. H. Rasmussen, J. Vanderkerckhove, and J. E. Celis, *J. Biol. Chem.* **267,** 8485 (1992).

[31] C. M. Nicolet and E. A. Craig, *Mol. Cell. Biol.* **9,** 3638 (1989).

promoter and can thus be induced with isopropyl-β-D-thiogalactopyranoside (IPTG).

For large-scale expression of FKBP52, Luria broth (LB) medium containing ampicillin (100 μg/ml) is inoculated with a stationary overnight culture equal to 1/50 the final culture volume. This culture is grown at 37° to an $OD_{600\,nm}$ of 0.4. Protein expression is induced with IPTG at a final concentration of 1 m*M*. Induction is performed at 25° for 3 hr with agitation (200 rpm).

The cells are harvested by centrifugation at 2000 *g* for 20 min at 4°. The bacterial cell pellet is washed three times by resuspending in a 1/10 volume of phosphate-buffered saline (PBS) and subsequent pelleting at 4000 *g* at 4°. Cell pellets may be stored at −80°.

*Purification.* Purification of recombinant FKBP52 is achieved using three chromatography steps including DE52, hydroxyapatite, and Resource-Q. All purification steps are performed at 4°. The buffers for column chromatographies are filtered and degassed before use.

1. The cell pellet from a 4-liter culture is resuspended in 40 ml of buffer A [10 m*M* Tris-HCl (pH 7.5), 10 m*M* glycerol, 5 m*M* dithioerythritol (DTE), 1 m*M* ethylenediaminetetraacetic acid (EDTA)]. In addition, the buffer contains the protease inhibitors Pefabloc (40 μg/ml), leupeptin (1 μg/ml), and pepstatin (1 μg/ml). The cells are lysed by using a French press operated at 18,000 psi (three times) and the crude extract is subsequently clarified by centrifugation (40,000 *g*, 60 min, 4°).
2. The supernatant is loaded onto a 50-ml DE52 column (column material: Whatman, Clifton, NJ), equilibrated in buffer A. After washing with two column volumes of buffer A to remove unspecifically bound protein, a KCl gradient (0–0.5 *M* KCl in 10 column volumes of buffer A) is applied. FKBP52 elutes at 0.1 *M* KCl.
3. The FKBP52-containing fractions are pooled and protease inhibitors are added (as described above). The pooled fractions are dialyzed overnight at 4° against buffer KP-1 (10 m*M* potassium phosphate buffer, pH 6.8). The pool is applied onto a 2-ml hydroxyapatite fast protein liquid chromatography (FPLC) column (Bio-Rad, Hercules, CA) previously equilibrated with five column volumes of KP-1 buffer. Unspecifically bound protein is removed by washing with two column volumes of KP-1 buffer. The remaining bound protein is eluted with increasing potassium phosphate concentrations up to 0.5 *M* over 10 column volumes. FKBP52 elutes between 0.1 and 0.2 *M* potassium phosphate.
4. The FKBP52-containing fractions are pooled and dialyzed against buffer A. The pooled fractions are applied onto a 1-ml Resource-Q column (Pharmacia, Piscataway, NJ) previously equilibrated with five column vol-

umes of buffer A. After washing the column in the same buffer, bound protein is eluted with a 0–0.3 *M* KCl gradient over 20 column volumes. FKBP52 is found in the fractions corresponding to KCl concentrations of 0.1–0.16 *M*. These fractions are pooled, dialyzed against 40 m*M* *N*-2-hydroxyethylpiperazine-*N'*-2-ethanesulfonic acid (HEPES)–KOH, pH 7.5, at 4° and concentrated to a concentration of 2–3 mg/ml by ultrafiltration using an Amicon (Danvers, MA) YM10 membrane. The pure protein is frozen in liquid nitrogen and stored at −80°.

*Comments*

1. The most efficient purification step is the hydroxyapatite chromatography.
2. A serious problem is that FKBP52 is sensitive to degradation by proteases. Thus, the presence of protease inhibitors in every step is crucial.
3. Degradation also occurs when the protein is frozen and thawed during purification.
4. Usually the protein concentration is determined using the Bradford protein assay with bovine serum albumin (BSA) as a standard.[32] Protein concentration can also be monitored using the theoretical extinction coefficient of 0.83 at 280 nm for a 1-mg/ml solution at a path length of 1 cm.
5. When run on an SDS–polyacrylamide gel, the apparent molecular mass of FKBP52 is consistent with the molecular mass calculated from the amino acid sequence.
6. The purification steps of the protein are performed by following the clearly visible bands of the overexpressed FKBP52 on an SDS–polyacrylamide gel[33] and collecting the appropriate fractions. The identity of the protein is further confirmed by Western blot analysis[34] using a specific antibody against FKBP52.

## Purification of Recombinant Human p23

### *Method*

The method described here allows the purification of recombinant human p23 expressed in *E. coli*. Typically, the purification yields 1 mg of pure protein per gram of cells.

*Protein Expression.* The expression plasmid for human p23 is a generous gift of D. Toft (Mayo Clinic). This construct contains the human p23 cDNA

[32] M. Bradford, *Anal. Biochem.* **72,** 248 (1976).
[33] U. K. Laemmli, *Nature* (*London*) **227,** 680 (1970).
[34] H. Towbin, T. Staehelin, and J. Gordon, *Proc. Natl. Acad. Sci. U.S.A.* **76,** 4350 (1979).

(GenBank Accession No. L24804) cloned into pET23 expression plasmid at the *Nde*I site. The p23 gene is therefore under the control of the T7 promoter. The plasmid is transformed into *E. coli* BL21(DE3)pLysS, which contains the T7 polymerase structural gene under control of the *lacUV5* promoter and can thus be induced with IPTG. In addition, the strain carries the gene for T7 lysozyme, so that cell lysis can be achieved by shock-freezing and thawing.

Growing of the cells and induction of protein expression are performed as described for FKBP52.

The cells are harvested by centrifugation at 4000 *g* for 5 min at 4°. The bacterial cell pellet is washed once by resuspending in a 1/10 volume of PBS and subsequent pelleting at 4000 *g* for 5 min at 4°.

*Purification.* Purification of p23 is accomplished in a four-step procedure, including precipitation, two different anion-exchange steps, and one gel-filtration step. All purification steps are performed at 4°. The buffers for column chromatographies are filtered and degassed before use.

1. The cell pellet is resuspended in 50 m*M* Tris-HCl (pH 7.5), 10 m*M* glycerol, 2 m*M* EDTA to 1/20 the original culture volume. The buffer also contains the protease inhibitors Pefabloc (40 $\mu$g/ml), leupeptin (1 $\mu$g/ml), and pepstatin (1 $\mu$g/ml).

The cells are lysed by shock-freezing of the cell suspension in liquid nitrogen and subsequent thawing at room temperature (alternatively, the frozen cell suspension can be stored at −80° for longer periods). The suspension is clarified by centrifugation at 40,000 *g* for 60 min at 4°.

2. The soluble fraction of the homogenate is loaded onto a 50-ml DE52-cellulose column (column material: Whatman) previously equilibrated in buffer B [50 m*M* Tris-HCl (pH 7.5), 10 m*M* glycerol, 5 m*M* DTE, 2 m*M* EDTA]. The column is washed with two column volumes of buffer B to remove unspecifically bound protein and then a 0–0.5 *M* KCl gradient in buffer B is applied to the column (500 ml). p23 elutes from the column between 0.3 and 0.4 *M* KCl. The p23-containing fractions are pooled and protease inhibitors are added.

3. The p23-containing pool is precipitated by the addition of ammonium sulfate (dissolved in buffer B) to a final concentration of 2.5 *M*. The precipitation reaction is incubated with gentle stirring overnight at 4°. Precipitated proteins are removed from the solution by centrifugation (40,000 *g*, 60 min, 4°). The p23-containing supernatant is concentrated 5- to 10-fold using an Amicon YM10 membrane. The concentrate is dialyzed against buffer B.

4. The dialyzed concentrate is applied to a 6-ml Resource-Q FPLC column (Pharmacia) equilibrated with buffer B. Elution of bound protein is achieved by a 0–0.5 *M* KCl gradient in buffer B (60 ml). p23 elutes from

the column between 0.35 and 0.42 *M* KCl. The p23-containing fractions are pooled and concentrated using an Amicon YM10 membrane.

5. The concentrate from step 3 is further purified on a 120-ml Superdex 75 preparation-grade column (Pharmacia) equilibrated in buffer C [40 m*M* HEPES–KOH (pH 7.5), 5 m*M* DTE] supplemented with 500 m*M* NaCl. Fractions containing pure p23 are pooled, dialyzed against buffer C supplemented with 10 m*M* glycerol, frozen in liquid nitrogen, and stored at −80°.

*Comments*

1. The most significant purification step is the first anion-exchange chromatography, which results in removal of the major contaminating components.

2. The p23 concentration is determined using either the Bradford protein assay[32] or a calculated extinction coefficient of 1.96 at 280 nm for a 1-mg/ml solution at a path length of 1 cm.

3. The protein can be concentrated to at least 15 mg/ml without significant aggregation.

4. As described before,[29] the migration pattern of human p23 on SDS–polyacrylamide gels suggests a protein of 23 kDa although the calculated molecular mass of human p23 is 18.7 kDa.

5. The purification steps of the protein are performed by following the clearly visible bands of the overexpressed p23 on an SDS–polyacrylamide gel[33] and collecting the appropriate fractions. The identity of the protein is further confirmed by Western blot analysis[34] using a specific antibody against p23.

## Purification of Recombinant Human Hop/p60

### *Method*

Following the expression and purification protocol described below, a yield of 6 mg of Hop per gram of *E. coli* cells can be achieved.

*Protein Expression.* Human Hop cDNA (GenBank Accession No. M86752) cloned into the expression plasmid pET28(a) under the control of a T7 promoter is a kind gift of D. Smith (University of Nebraska, Omaha, NE). The plasmid is transformed into *E. coli* BL21(DE3) using standard methods.

For large-scale expression of Hop/p60, LB medium containing kanamycin (40 $\mu$g/ml) is inoculated with a stationary overnight culture equal to 1/40 the final culture volume. The culture is grown at 37° to an $OD_{600\ nm}$ of 0.5.

After induction with IPTG to a final concentration of 0.4 m*M*, the cells are allowed to grow for another 3 hr at room temperature. The culture is harvested by centrifugation at 4000 *g* for 10 min at 4°. The cell pellets may be stored at −80°.

*Purification.* Purification of recombinant Hop is accomplished in three steps including Q-Sepharose, hydroxyapatite, and size-exclusion chromatography. All purification steps are performed at 4°. The buffers for column chromatographies are filtered and degassed before use.

1. The cell pellet from a 1-liter culture is resuspended in 3 ml of 10 m*M* Tris-HCl, pH 7.5, per gram of wet cell pellet containing 1 m*M* phenylmethylsulfonyl fluoride (PMSF), *p*-aminobenzoic acid (PABA, 520 $\mu$g/ml), and pepstatin A (1 $\mu$g/ml) as protease inhibitors. The cells are lysed in a French press at 18,000 psi for three cycles. The suspension is clarified by centrifugation at 40,000 *g* for 60 min at 4°.
2. The supernatant is loaded onto a 40-ml Q-Sepharose column (column material: Pharmacia) equilibrated with lysis buffer. Bound protein is eluted from the column with a linear gradient from 0 to 0.15 *M* NaCl over 10 column volumes. Hop elutes around 0.13 *M* NaCl. Hop-containing fractions are pooled and concentrated by ultrafiltration using an Amicon YM10 membrane. The concentrate is dialyzed overnight against 10 m*M* potassium phosphate buffer, pH 7.5.
3. The Hop pool is then loaded onto a preequilibrated 2-ml FPLC hydroxyapatite column (Bio-Rad) and eluted with a linear potassium phosphate gradient ranging from 10 to 300 m*M* over 15 column volumes (pH 7.5). Hop elutes between 80 and 220 m*M* potassium phosphate. The fractions containing Hop are collected, pooled, and concentrated as described above.
4. The concentrate from step 3 is loaded on a 120-ml Superdex-200 preparative-grade column (Pharmacia). The column is equilibrated with two column volumes of 10 m*M* Tris, 400 m*M* NaCl, pH 7.5. Hop-containing fractions are pooled, concentrated as described above to 5 mg/ml, and finally dialyzed against 40 m*M* HEPES–KOH, pH 7.5. The pure protein is frozen in liquid nitrogen and stored at −80°.

*Comments*

1. The most significant purification step is the hydroxyapatite chromatography, which results in the removal of most of the contaminating components.
2. The Hop concentration is determined using either the method described by Bradford[32] or the calculated extinction coefficient of 0.74 at 280 nm for a 1-mg/ml solution at a path length of 1 cm.
3. The protein could be concentrated to at least 20 mg/ml without significant aggregation.

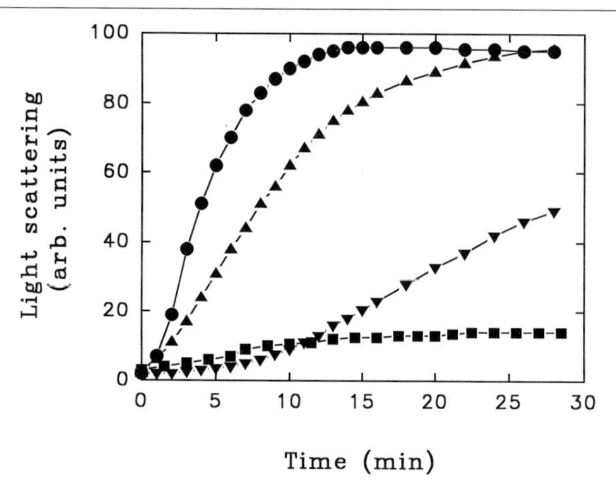


Fig. 1. Influence of Hsp90 partner proteins on the thermal aggregation of porcine heart citrate synthase (CS). The influence of FKBP52, p23, and Hop on the aggregation of the thermolabile enzyme CS at elevated temperatures was determined in order to examine their ability to bind nonnative proteins. The experimental procedure is described by Buchner *et al.* ([32] in this volume[34a]). The kinetics of aggregation were determined by light-scattering measurements. CS was diluted to a final concentration of 75 n$M$ into thermostatted (43°) buffer (40 m$M$ HEPES–KOH, pH 7.5) in the presence of 3 $\mu M$ control protein IgG (●), 0.45 $\mu M$ recombinant rabbit FKBP52 (■), 1.13 $\mu M$ recombinant human p23 (▼), and 0.75 $\mu M$ recombinant human Hop (▲). The concentrations for all proteins are based on dimers.

4. On SDS–PAGE the apparent molecular mass of Hop is consistent with that calculated from the amino acid sequence.

5. The purification steps of the protein are performed by following the clearly visible bands of overexpressed Hop on an SDS–polyacrylamide gel[33] and collecting the appropriate fractions. The identity of the protein is further confirmed by Western blot analysis[34] using a specific antibody against Hop.

## Chaperone and Prolyl Isomerase Assays

The chaperone activity of Hsp90 partner proteins can be tested using the citrate synthase (CS) unfolding assay (cf. [32] in this volume[34a]). Both p23 and FKBP52 suppress aggregation of CS at 43° in a concentration-dependent way (Fig. 1). Whereas FKBP52 seems to form apparently stable complexes with nonnative CS, p23 is only able to decelerate aggregation significantly. In addition to the aggregation assay, the influence of Hsp90

[34a] J. Buchner, S. Bose, C. Mayr, and U. Jakob, *Methods Enzymol.* **290,** Chap. 32, 1998 (this volume).

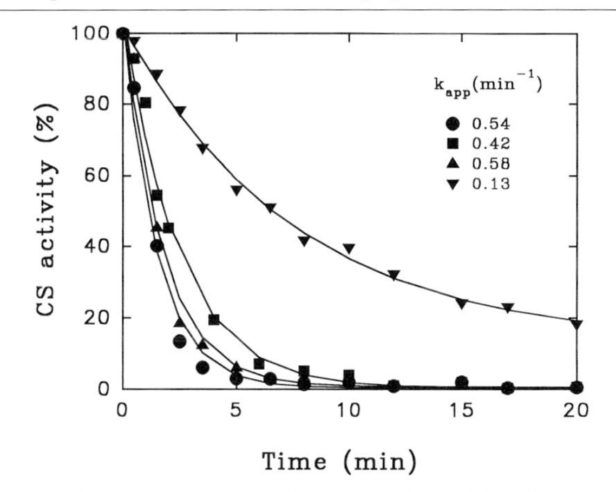


FIG. 2. Influence of Hsp90 partner proteins on the thermal inactivation of porcine heart citrate synthase (CS). In addition, the influence of FKBP52, p23, and Hop on the temperature-induced inactivation of CS was determined. CS was diluted to a final concentration of 75 n*M* into thermostatted (43°) buffer (40 m*M* HEPES–KOH, pH 7.5) and enzymatic activity was measured at different time points[34b] in the presence of 0.45 $\mu M$ FKBP52 (■), 0.15 $\mu M$ p23 (▼), 1.5 $\mu M$ Hop (▲), and 6.6 $\mu M$ IgG (●). The solid lines represent single exponential functions with apparent rate constants as shown in the inset. The concentrations of all proteins are based on dimers.

partner proteins on the inactivation process of CS at elevated temperatures can be tested. This allows the determination of whether structured folding intermediates are recognized and complexed (cf. [32] in this volume[34a]). In this assay, FKBP52 does not influence the inactivation process significantly; however, p23 is efficient in apparently stabilizing CS against inactivation (Fig. 2).

In both assays, p60/Hop is inactive. For Hop, the only activity to be determined is the binding to Hsp70 and Hsp90.[7,13]

The PPIase activity of FKBP52 can be measured in a coupled assay with chymotrypsin using the synthetic peptide *N*-Suc-Ala-Ala-Pro-Phe *p*-nitroanilide.[35] The assay is based on the fact that the *p*-nitroanilide can be cleaved only by chymotrypsin when the Ala-Pro bond is in *trans* configuration. The PPIase activity of FKBP52 can be inhibited by rapamycin.

[34b] P. A. Srere, H. Brazil, and L. Gonen, *Acta Chem. Scand.* **17,** 129 (1963).

[35] G. Fischer, H. Bang, and C. Mech. *Biomed. Biochim. Acta* **43,** 1101 (1984).

# [34] Purification and Properties of Hsp104 from Yeast

*By* ERIC C. SCHIRMER and SUSAN LINDQUIST

## Introduction

In this chapter we describe protocols for the purification of wild-type heat shock protein 104 (Hsp104) from *Saccharomyces cerevisiae* and a rapid procedure for the purification of a modified version of Hsp104, carrying an amino-terminal histidine extension, from *Escherichia coli.* We also describe systems for the expression of toxic HSP100 proteins and mutant Hsp104 proteins. In addition, methods for determining the ATPase activity of the purified protein and its oligomerization properties are provided. Requests for reagents employed may be made through the Web site http://http.bsd.uchicago.edu/~hsplab/index.html.

## Properties of Hsp104

### *Biological Properties*

Hsp104 is a heat-shock protein that promotes survival under extreme stresses such as heat and high concentrations of ethanol.[1,2] It appears to enhance survival by promoting the solubilization and reactivation of protein aggregates *in vivo.*[3] Hsp104 also functions in the maintenance and curing of a prion-like, protein conformation-based phenotype in yeast, referred to as [*PSI*$^+$],[4] in which the translation termination factor Sup35 is sequestered into aggregates causing ribosomes to read through stop codons.[5–8]

### *Biochemical Properties*

Hsp104 has a calculated relative molecular weight of 102,000 and an estimated *pI* of 5.14. On the basis of the similarities between these predicted

[1] Y. Sanchez and S. L. Lindquist, *Science* **248,** 1112 (1990).
[2] Y. Sanchez, J. Taulien, K. A. Borkovich, and S. Lindquist, *EMBO J.* **11,** 2357 (1992).
[3] D. A. Parsell, A. S. Kowal, M. A. Singer, and S. Lindquist, *Nature* (*London*) **372,** 475 (1994).
[4] Y. O. Chernoff, S. L. Lindquist, B.-i. Ono, S. G. Inge-Vechtomov, and S. W. Liebman, *Science* **268,** 880 (1995).
[5] B. Cox, *Curr. Biol.* **4,** 744 (1994).
[6] M. F. Tuite and I. Stansfield, *Nature* (*London*) **372,** 614 (1994).
[7] S. V. Paushkin, V. V. Kushnirov, V. N. Smirnov, and M. D. Ter-Avanesyan, *EMBO J.* **15,** 3127 (1996).
[8] M. M. Patino, J.-J. Liu, J. R. Glover, and S. Lindquist, *Science* **273,** 622 (1996).


METHODS IN ENZYMOLOGY, VOL. 290 0076-6879/98 $25.00

values and the observed migration of Hsp104 on two-dimensional sodium dodecyl sulfate-polyacrylamide gel electrophoresis (SDS–PAGE), we assume that Hsp104 does not undergo extensive posttranslational modifications. Hsp104 has no tryptophan residues, which makes tracking the protein during chromatographic procedures difficult. Hsp104 contains two nucleotide-binding domains, which demonstrate specificity for adenine nucleotides.[9,10] Interestingly, despite its two ATP-binding domains, Hsp104 does not bind ATP–agarose (either with a short or long arm spacer; Sigma, St. Louis, MO).[11] At low protein concentrations Hsp104 oligomerizes in the presence of ATP. These oligomers are most likely hexamers, on the basis of sizing chromatography, glutaraldehyde cross-linking, and scanning transmission electron microscopy (STEM).[9] Hsp104 is an ATPase with a $K_m$ of ~5 m$M$ and a $V_{max}$ of ~2 nmol min$^{-1}$ $\mu$g$^{-1}$.[10] ATPase activity is stimulated by certain proteins and peptides.[12]

## Purification of Hsp104 from Yeast

### *Plasmid and Strain Construction*

To increase yield in *S. cerevisiae, HSP104* can be expressed from high-copy vectors with strong promoters (Table I). Most commonly we employ a vector in which wild-type *HSP104* coding sequences are regulated by the highly inducible glucocorticoid response elements of the p2UG vector.[13] With this construct (p2UG104),[9] Hsp104 is induced by the addition of 10 $\mu M$ deoxycorticosterone to yeast cells, which also carry a plasmid encoding the mammalian glucocorticoid receptor (pG-N795).[14] To reduce degradation problems during purification, the protein is expressed in strain BJ5457 (A741; Table II), carrying a deletion of the pep4 and prb1 protease genes.[15] A variant carrying an *HSP104* deletion is employed to avoid contamination of wild-type Hsp104 when purifying mutant variants of Hsp104 or related HSP100 proteins from other organisms (A798, Table II).

Constructs for the expression of Hsp104 from the *GAL1-10* promoter are also available (104b-U; Table I). The growth of large-scale cultures in raffinose and galactose is costly, but a method to circumvent this can be

[9] D. A. Parsell, A. S. Kowal, and S. Lindquist, *J. Biol. Chem.* **269,** 4480 (1994).
[10] E. C. Schirmer, C. Queitsch, A. S. Kowal, D. A. Parsell, and S. Lindquist, submitted (1998).
[11] D. A. Parsell and S. Lindquist, unpublished observations (1990).
[12] E. C. Schirmer and S. Lindquist, unpublished observations (1995).
[13] D. Picard, M. Schena, and K. R. Yamamoto, *Gene* **86,** 257 (1990).
[14] M. Schena, D. Picard and K. R. Yamamoto, *Methods Enzymol.* **194,** 389 (1991).
[15] E. W. Jones, *Methods Enzymol.* **194,** 428 (1991).

TABLE I
PLASMIDS FOR EXPRESSION OF Hsp104 IN YEAST

| Accession number | Plasmid | Promoter | Selection | Copy number | Induction by: | Product | Refs. |
|---|---|---|---|---|---|---|---|
| 9029 | p2UG | *GRE-CYC1* | URA3, $Amp^R$ | High (2 $\mu$m) | GR/DOC[a] | — | [b] |
| 5316 | p2UG104 | *GRE-CYC1* | URA3, $Amp^R$ | High (2 $\mu$m) | GR/DOC[a] | Hsp104 | [c] |
| 5228 | pG-N795 | *GPD* | TRP1, $Amp^R$ | High (2 $\mu$m) | Constitutive | GR | [d] |
| 5800 | 104b-U | *GAL1–10* | URA3, $Amp^R$ | Low (CEN6, ARS4) | Galactose | Hsp104 | [e,f] |
| 5306 | pYS104 | *HSE* | URA3, $Amp^R$ | Low (CEN6, ARS4) | Heat | Hsp104 | [g] |
| 5632 | PLH102 | *GPD* | URA3, $Amp^R$ | High (2 $\mu$m) | Constitutive | Hsp104 | [h] |

[a] GR/DOC: Glucocorticoid receptor (GR), which is constitutively expressed in cells, is activated by addition of deoxycorticosterone (DOC).
[b] D. Picard, M. Schena, and K. R. Yamamoto, *Gene* **86,** 257 (1990).
[c] D. A. Parsell, A. S. Kowal, and S. Lindquist, *J. Biol. Chem.* **269,** 4480 (1994).
[d] M. Schena, D. Picard, and K. R. Yamamoto, *Methods Enzymol.* **194,** 389 (1991).
[e] E. C. Schirmer, S. Lindquist, and E. Vierling, *Plant Cell* **6,** 1899 (1994).
[f] E. C. Schirmer, C. Queitsch, A. S. Kowal, and S. Lindquist, submitted (1997).
[g] Y. Sanchez and S. L. Lindquist, *Science* **248,** 1112 (1990).
[h] L. Henninger and S. Lindquist, unpublished observations (1993).

found in Joshua-Tor *et al.*[16] Hsp104 can also be induced to high levels in wild-type cells with a heat shock of 37–39° for 90 min. (A plasmid containing Hsp104 behind its natural heat shock promoter is pYS104).[1] Hsp104 can be expressed constitutively at a high level from the GPD promoter (PHL102; Table I),[17] but this protein seems to have a lower specific activity in thermotolerance than Hsp104 induced by heat stress.[18] The reason for this is unclear; however, pending its resolution we recommend the GRE expression system. It yields the highest expression (with the exception of the GPD expression system): 5- to 10-fold greater than that observed in heat-treated cells.[9]

*Solutions and Equipment*

Deoxycorticosterone (Sigma), 10 m*M* (1000×) in ethanol
Yeast–peptone–dextrose (YPD, 10×), per liter: 100 of yeast extract, 200 g of Bacto-peptone, 200 g of dextrose, 0.4 g of adenine sulfate; components become soluble during autoclaving
KCl stock (2.5 *M*)
Buffer A (10×): 0.5 *M* Tris (pH 7.7), 20 m*M* ethylenediaminetetraacetic acid (EDTA), 100 m*M* $MgCl_2$, 50% (v/v) glycerol. Add 1.4 m*M*

[16] L. Joshua-Tor, H. E. Xu, S. A. Johnston, and D. C. Rees, *Science* **269,** 945 (1995).
[17] L. Henninger and S. Lindquist, unpublished observations (1993).
[18] S. Lindquist and G. Kim, *Proc. Natl. Acad. Sci. U.S.A.* **93,** 5301 (1996).

TABLE II
STRAINS FOR PURIFICATION OF Hsp104 FROM YEAST

| Accession number | Strain | Genotype | Refs. |
|---|---|---|---|
| A741 | BJ5457 | *α, ura3-52, trp1, lys2-801, leu2Δ1, his3Δ200, pep4::HIS3, prb1Δ1.6R, can1, GAL* | [a] |
| A798 | BJ5457HSP104::LEU2 | *α, ura3-52, trp1, lys2-801, leu2Δ1, his3Δ200, pep4::HIS3, prb1Δ1.6R, can1, GAL, hsp104::LEU2* | [b] |
| A750 | BJ5457/p2UG104, pG-N795 | *α, ura3-52, trp1, lys2-801, leu2Δ1, his3Δ200, pep4::HIS3, prb1Δ1.6R, can1, GAL,* carrying plasmids p2UG104 and pG-N795 | [b] |

[a] E. W. Jones, *Methods Enzymol.* **194,** 428 (1991).
[b] D. A. Parsell, A. S. Kowal, and S. Lindquist, *J. Biol. Chem.* **269,** 4480 (1994).

2-mercaptoethanol and 1 m*M* 4-(2-aminoethyl)benzenesulfonyl fluoride (AEBSF; protease inhibitor) fresh before use. *Note:* Owing to the high cost of AEBSF, it is replaced with 1 m*M* PMSF (phenylmethylsulfonyl fluoride) for dialysis.

Phosphate (pH 6.8) stock, 0.5 *M*: Roughly 49% (w/v) $Na_2HPO_4$ plus 51% (w/v) $NaH_2PO_4$

Buffer B: 50 m*M* potassium phosphate (pH 6.8), 5% (v/v) glycerol, 1.4 m*M* 2-mercaptoethanol, 1 m*M* AEBSF

Additional protease inhibitors: Added from concentrated stock solutions to the following final concentrations: 12 $\mu$g/ml (peptstatin A, in ethanol); 7 $\mu$g/ml (leupeptin, in $H_2O$); 2 $\mu$g/ml (aprotinin in 10 m*M* HEPES); 1 m*M* benzamidine (in $H_2O$); 1 $\mu M$ sodium metabisulfate (in $H_2O$)

Bead-Beater (BioSpec Products, Bartlesville, OK): Available with 350-ml cup or 15-ml cup. This apparatus is in essence a blender with a 4-mm-thick Teflon blade and an outer shell surrounding the lysis chamber to accommodate an ice–water bath. For much smaller preparations a Mini-Beadbeater-8 cell disrupter (BioSpec Products) accommodates 2-ml microcentrifuge tubes.

### *Growth of Cells for Protein Purification*

A high cell density and a high ratio of glass beads to cells in the lysis step greatly enhance the efficiency of breakage. The procedure detailed below is for 50 g (wet weight) of packed cells, but it can readily be scaled for smaller or larger preparations.

Strain A750 is grown at 25° in minimal medium containing dextrose (2%, w/v), ammonium sulfate (0.5%, w/v), yeast nitrogen base without amino acids (0.17%, w/v), adenine (10 mg/liter), arginine (50 mg/liter), lysine (50 mg/liter), methionine (20 mg/liter), phenylalanine (50 mg/liter), threonine (100 mg/liter), tyrosine (50 mg/liter), aspartic acid (70 mg/liter), leucine (100 mg/liter), and histidine (20 mg/liter). The absence of tryptophan and uracil forces maintenance of the GRE-HSP104 and GR (glucocorticoid receptor) expression plasmids. Reversion of the *trp* allele in this strain is sometimes problematic; strains should be monitored for Hsp104 expression competence before preparing large-scale cultures.

An initial overnight culture of cells in midlog phase ($\sim 2-5 \times 10^6$ ml$^{-1}$) is used to inoculate four 6-liter flasks, each containing 1.5 liters of medium. After dilution, the growth rate slows to a doubling time of ~3 hr; however, as cells again reach midlog phase, the doubling time reaches ~2.3 hr. When cultures reach midlog phase, Hsp104 is induced with deoxycorticosterone at a final concentration of 10 $\mu M$. After 8 additional hours of incubation, the medium is supplemented with 10× YPD and cells are incubated an additional 10–15 hr. The rich medium allows cells to reach a higher stationary-phase density, and plasmid loss in the absence of selection is not high during this short period. Cells (final density, $\sim 1-3 \times 10^8$ cells/ml) are collected by centrifugation (4500 rpm for 20 min at 4°). Cell pellets are resuspended in ice-cold water, combined, and subjected to an additional round of centrifugation. The gram weight of the cell pellet is measured, and cells are resuspended by adding <50% gram weight (i.e., for 160 g of cells add 80 ml) of buffer A. Protein can be purified with equal success from cells lysed immediately or cells frozen with liquid nitrogen. To freeze cells, the slurry (in buffer A) is dripped into a bath of liquid nitrogen. Frozen droplets are removed to a storage container (keep the lid loose initially to prevent explosions from the vaporizing nitrogen) and stored at −80° until use.

*Lysis of Cells*

Cells resuspended in buffer A are supplemented with additional protease inhibitors (see Solutions and Equipment). Densely resuspended cells (165 ml) are mixed with 150 ml of glass beads (425–600 $\mu$m in diameter; acid washed) and placed in a 350-ml Bead-Beater cup with an ice–water bath contained in the outer shell. Cells are blended 12 times for 30 sec each, with a 90-sec recovery period between pulses to recool the inner chamber. If frozen cells are used, they should be thawed by constant stirring of the frozen droplets in a warm water bath, making certain all cells are kept cold. Cells may be lysed when frozen clumps of cells are less than 0.5

cm in diameter (in this case the initial blending pulse can be extended to 2 min without significant heating of the lysate). Lysates are removed from the beads and the beads are washed with ice-cold $H_2O$ (one-third the lysate volume). Lysate and wash are accrued and subjected to centrifugation (18,000 rpm for 20 min at 4°). Supernatants are diluted 1:1 with buffer A before being applied to columns.

*Chromatography*

All columns should be maintained at 4° through the procedure and all buffers, equipment, and other reagents cooled before use. As Hsp104 has no tryptophan and produces a weak signal at 280 nm, all fractions should be kept until the quality of peak fractions has been confirmed by analysis by 10% (w/v) SDS–PAGE. The range indicated for elution of peak fractions covers the wider range observed from multiple preparations. In individual purifications, Hsp104 has sometimes eluted closer to one or the other end of this range. Hsp104 accumulates to 5–10% of the total cellular protein, and the first two columns each yield about 10-fold purification; so Hsp104 eluted from the first DEAE column is 90–95% pure. The last two columns increase this to >98%.

1. *Affi-Gel Blue:* Lysate from 50 g of cells is applied onto a 30-ml Affi-Gel Blue (Bio-Rad, Hercules, CA) column preequilibrated in buffer A with a flow rate of 70 ml/hr. At a similar flow rate, the column is washed with five column volumes of buffer A (150 ml) followed by buffer A containing 100 m*M* KCl (150 ml). Protein is eluted with 100 ml of buffer A containing 1 *M* KCl. The eluant is dialyzed for 4 hr to overnight against buffer A containing 1 m*M* PMSF and cleared by centrifugation.

2. *DEAE I:* The dialyzed eluant from the Affi-Gel Blue column is applied onto a 15-ml DEAE column (Pharmacia, Piscataway, NJ) preequilibrated in buffer A at a flow rate of 50 ml/hr. The column is washed with 75 ml of buffer A and protein is eluted with a 200-ml linear gradient of 0–500 m*M* KCl in buffer A. Fractions eluting between 70 and 140 m*M* KCl are enriched in Hsp104 and these are accrued and dialyzed against buffer B.

3. *Hydroxyapatite:* Accrued fractions from the DEAE column are applied onto a 15-ml hydroxyapatite column preequilibrated in buffer B at a flow rate of 20 ml/hr. After washing with five column volumes of buffer B, proteins are eluted with a linear gradient of buffer B from 50–400 m*M* potassium phosphate, pH 6.8. Fractions eluting between 135 and 180 m*M* phosphate are pooled and precipitated with addition of solid ammonium sulfate (crushed to powder with a mortar and pestle) to 70% of saturation.

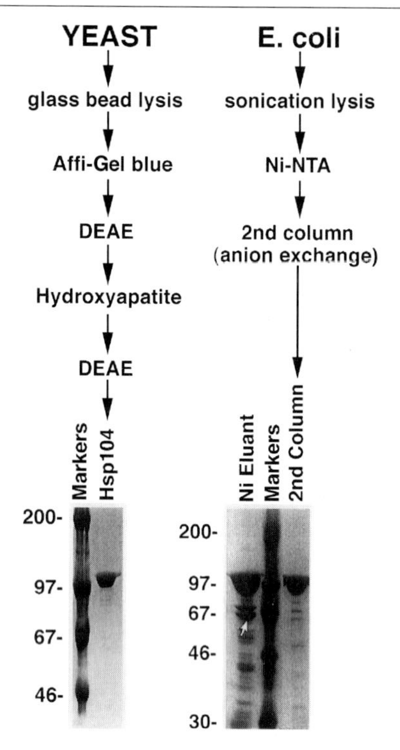


FIG. 1. Flow chart of Hsp104 purification procedures from yeast and *E. coli*. The purity of the final products is shown at the bottom. The arrow in the Ni eluant lane indicates common degradation products of Hsp104, which react with antibodies against Hsp104.

The precipitate is collected by centrifugation (15,000 rpm for 20 min at 4°), resuspended in buffer A, and dialyzed to completion against buffer A.

4. *DEAE II:* Insoluble material is removed by centrifugation and proteins are applied onto a second 15-ml DEAE column at a flow rate of 50 ml/hr. The column is washed with 200 ml of buffer A, followed by elution with a linear gradient of 50 to 300 m*M* KCl in buffer A. Hsp104 elutes from the column at a salt concentration of 105–150 m*M*. The purity of the preparation is shown in Fig. 1.

*Yields and Calculating Protein Concentration*

Every gram of packed cells yields roughly 0.5 mg of purified Hsp104 protein, as assessed by amino acid analysis and using a calculated extinction coefficient $\varepsilon = 31{,}900\ M^{-1}$ at 276 nm[19] (there is no difference between

[19] H. Edelhoch, *Biochemistry* **6,** 1948 (1967).

native and denatured Hsp104 at this wavelength). The Bradford assay gives values roughly twice this when using BSA (bovine serum albumin) as a standard.

### *Scaling for Larger or Smaller Preparations*

For large-scale cultures, a 20-liter polypropylene carboy is used to grow cultures. Sterile silastic tubing is connected through an adaptor in the lid. One end is attached to filtered house air, and the other end extends to the bottom of the carboy. Sampling of the culture density is effected through a second access point in the lid of the carboy, which also allows the exchange of $CO_2$. A 200-ml overnight culture is used to inoculate a 20-liter culture. The culture is grown in a 20-liter polypropylene carboy at room temperature; effort should be made to maintain the room at approximately 25° as the doubling time of the cells is much longer at lower temperatures. The rate of growth can be increased with a space heater next to the carboy. Aeration should be maintained at a level that prevents cells from settling, but does not cool them (house air is typically lower than room temperature).

For small-scale cultures a 15-ml Bead-Beater chamber is available to maintain high cell density in minimal volumes; other apparatuses exist for use with microcentrifuge tubes.

## Purification of HSP100 Proteins from Other Species in Yeast

### *Yeast Expression Systems for Toxic Proteins*

Expression of some HSP100 proteins (e.g., Hsp101 from *Arabidopsis thaliana,*[20] Hsp101 from *Glycine max,*[21] and Hsp100 from *Leishmania major*[22]) is apparently toxic to *E. coli.* To facilitate cloning in *E. coli,* a modified *HSP100* gene is employed to reduce basal expression. In this system, the sequence around the initiating AUG is changed to a context unfavorable for expression in *E. coli,* but still capable of strong induction in yeast. Specifically, the *HSP100* gene is modified to contain a polylinker site (*Bam*HI) followed by three guanine nucleotides directly in front of the initiating AUG, and this is inserted into a pRS313-based vector[23] carrying *URA3* as the selectable marker and the *GAL1-10* promotor in the polylinker (104b-U; Table I).[20] This plasmid also has reduced basal expression from the *GAL* promoter during growth in raffinose, allowing yeast cells carrying

[20] E. C. Shirmer, S. Lindquist, and E. Vierling, *Plant Cell* **6,** 1899 (1994).
[21] Y.-R. J. Lee, R. T. Nagao, and J. L. Key, *Plant Cell* **6,** 1889 (1994).
[22] A. Hubel, S. Brandau, A. Dresel, and J. Clos, *Mol. Biochem. Parasitol.* **70,** 107 (1995).
[23] R. S. Sikorski and P. Hieter, *Genetics* **122,** 19 (1989).

toxic HSP100 varaints to be grown in raffinose before galactose induction. For protein purification, a protease-disrupted strain carrying an *hsp104* deletion should be used (BJ5457HSP104::LEU2; Table II).

When expressing HSP100 proteins from other organisms using this system, attention should be paid to issues of codon usage and translation termination sequences. Because procedures for the purification of Hsp104 depend on anion-exchange chromatography and the middle region of HSP100 proteins is both highly charged and variable in size between subtypes, it is likely that the methods presented here will be applicable only for members of the same subtype. A method for purification of ClpA (an HSP100 protein that lacks the charged middle region) has been published by Maurizi *et al.*[24]

### *Vector to Facilitate Production of Mutant hsp104 Proteins*

A modified *HSP104* gene, $HSP104_R$, facilitates the cloning and analysis of Hsp104 mutants. This vector contains unique restriction sites approximately every 500 bp throughout the coding sequence of Hsp104 that do not change the encoded amino acids or significantly alter codon usage. For mutagenesis studies, for example, the segment of interest containing the mutation is sequenced, excised, and inserted into an unmutagenized version of $HSP104_R$ (104b-U; Table I) to ensure that no unintended mutations are present.

## Protein Purification from *Escherichia coli*

### *Plasmid and Strain Constructions*

Adding a short stretch of histidine residues to the end of a protein allows rapid purification in one step because histidine residues have a high affinity for nickel resins. Hsp104 protein carrying a six residue (6*x*) histidine extension at its amino terminus is as stable and functions in thermotolerance as well as wild-type Hsp104 in yeast (Hsp104 with a carboxy-terminal 6*x*-histidine expression functions in thermotolerance, but is less stable).[25] *In vitro,* Hsp104 protein with a 6*x*-histidine extension purified from *E. coli* exhibits ATP hydrolysis similar to that of wild-type Hsp104 purified from yeast, whether the histidine extension is cleaved from the protein or not.[12] The modified protein also assembles into oligomers indistinguishably from the wild-type yeast protein.[26]

[24] M. R. Maurizi, M. W. Thompson, S. K. Singh, and S.-H. Kim, *Methods Enzymol.* **244,** 314 (1994).

[25] D. A. Parsell and S. Lindquist, unpublished observations (1993).

[26] A. S. Kowal and S. Lindquist, unpublished observations (1996).

The plasmid employed for purification in *E. coli*, pETH6104b, is a modified pET28a expression vector (Novagen, Madison, WI), with the T7 epitope tag removed. pETH6104b contains $HSP104_R$ coding sequences under inducible control of the T7 promoter for expression using the system developed by Studier *et al.*[27] This plasmid produces an Hsp104 protein carrying a 6*x*-histidine extension at the amino terminus that can be cleaved from the protein using thrombin.[28] The selectable marker for the plasmid is kanamycin. The plasmid is transformed into pLysS cells, which contain a plasmid encoding T7 lysozyme that is selected for with chloramphenicol.[29] We have also had success purifying Hsp104 using the pJC45 vector system (see Clos and Brandau[30]).

Purification of Hsp104 from *E. coli* using pETH6104b is simple and fast. A disadvantage is that a higher level of degradation of Hsp104 occurs in *E. coli* than in yeast. The process of reducing these degradation products substantially reduces yields. Degradation products are minimized by inducing Hsp104 when cells are at a low density, for a short period (proteases are induced in late-log phase).

*Buffers and Solutions*

LB medium (per liter): 10 g of Bacto-tryptone, 5 g of NaCl, 5 g of yeast extract
Kanamycin, 50 mg/ml (1000×, in $H_2O$)
Chloramphenicol, 34 mg/ml (1000×, in ethanol)
Isopropyl-β-D-thiogalactopyranoside (IPTG, 1 *M*; Sigma) (1000× in $H_2O$)
Imidazole (pH 8.0) stock: 1 *M*
Nickel buffer (5×): 100 m*M* Tris (pH 8.0), 2 *M* NaCl. Imidazole and AEBSF are added before use
Nickel binding buffer: 20 m*M* Tris (pH 8.0), 400 m*M* NaCl, 0.01% (v/v) Triton X-100, 10 m*M* imidazole, 1 m*M* AEBSF
Buffer Q (10×): 200 m*M* Tris (pH 8.0), 5 m*M* EDTA, 50 m*M* $MgCl_2$
NaCl stock: 5 *M*

*Growth of Bacterial Cells Containing Hsp104 Proteins*

A 4-ml overnight culture of *E. coli* cells is used to inoculate 1 liter of LB medium containing kanamycin (50 μg/ml) and chloramphenicol (34 μg/

[27] F. W. Studier, A. H. Rosenberg, J. J. Dunn, and J. W. Dubendorf, *Methods Enzymol.* **185,** 60 (1990).
[28] E. C. Schirmer, J. R. Glover, and S. Lindquist, unpublished observations (1995).
[29] F. W. Studier, *J. Mol. Biol.* **219,** 37 (1991).
[30] J. Clos and S. Brandau, *Protein Expression Purif.* **5,** 133 (1994).

ml). Cells are grown to an $A_{595}$ of <0.4 (1-cm light path). Typically, this requires ~3 hr. Expression of Hsp104 protein is then induced by the addition of IPTG to a final concentration of 1 m*M*. Incubation is continued for 1 hr and cells are collected by centrifugation (in a cooled rotor as described above for yeast). Cells must be processed immediately because Hsp104 activity is lost when *E. coli* cells are frozen.

*Cell Lysis*

Cells are resuspended as a thick slurry in a minimal volume of cold nickel binding buffer and transferred to siliconized microcentrifuge tubes. (Hsp104 binds to the walls of polypropylene tubes.)[25] Cells are sonicated in a Branson sonicator (Branson Ultrasonic Corp., Danbury, CT) with a microtip adaptor using 3 cycles of 40 pulses each at a 90% duty cycle with cooling for several minutes on ice between cycles. Before applying to the nickel matrix, lysates are cleared of cellular debris by centrifugation (14,000 rpm for 20 min at 4°).

*Column Chromatography*

1. *Nickel–NTA:* Lysates containing 6*x*-histidine tagged proteins are applied onto a $Ni^{2+}$–NTA–agarose (nickel–nitrilotriacetic acid–agarose; Qiagen, Chatsworth, CA) column preequilibrated in nickel binding buffer. For every liter of cells ~0.75 ml of packed matrix is used. After washing with 20 column volumes of nickel binding buffer, proteins are eluted with the same buffer containing 220 m*M* imidazole. For every liter of cells, roughly 10 mg of Hsp104 is obtained from the nickel matrix purification step.

2. *Anion-exchange medium:* Several different anion-exchange media can be used for the second step in purification: POROS HQ columns can be resolved using FPLC, HPLC, or perfusion chromatography systems; Resource-Q columns can be resolved using a peristaltic pump or FPLC; and DEAE columns can be resolved with gravity flow or a peristaltic pump.

*POROS HQ:* Protein eluted from the nickel column is dialyzed against buffer Q. Before injection, protein is filtered through a low protein binding Millex-GV 0.22-μm pore size filter unit (Millipore, Bedford, MA). The POROS 20 HQ column (PerSeptive Biosystems, Framingham, MA) is first equilibrated with 20 column volumes of buffer *Q*. The sample is applied and the column is washed with 20-column volume of 60 m*M* NaCl in buffer Q. Hsp104 is eluted from the column using a 20-column volume linear gradient from 60 to 660 m*M* NaCl in buffer Q. Hsp104 elutes from the matrix at ~300 m*M* NaCl.[31]

[31] M. Ramakrishnan, D. Hattendorf, J. Glover, E. C. Schirmer, and S. Lindquist, unpublished observations (1997).

*Resource Q:* A Resource-Q (Pharmacia) FPLC column functions similarly to the POROS column, but using a gradient from 50 to 900 m*M*.

*DEAE:* Protein eluted from the nickel column is dialyzed against buffer A and applied onto a 5-ml DEAE column (Pharmacia) preequilibrated with buffer A. Hsp104 is eluted with a 50–300 m*M* KCl gradient in buffer A (as for the fourth column in the yeast purification protocol).

If the eluant from the nickel column is less clean than that shown in Figure 1, then a third column step may be necessary. Note: Some lower molecular weight contaminants may not be observed by SDS–PAGE unless acrylamide concentrations ≥12% are used. To remove these contaminants, the eluant from the anion exchange column is diluted in buffer Q and applied to a POROS-Heparin column. The column is washed with 10-column volumes of buffer Q and Hsp104 is eluted with a 0–450 m*M* NaCl gradient in buffer Q (Hsp104 elutes at ~300 m*M* salt).[31]

DEAE requires less specialized equipment, but Resource-Q and POROS yield similar results and provide better purification and recoveries than DEAE. Owing to the loss of material in separating contaminants and Hsp104 degradation products, the final yields of purified full-length Hsp104 from 1 liter of starting culture are typically only 1–4 mg.

The removal of the 6*x*-histidine extension using thrombin cleavage requires low temperatures and high salt concentrations to limit the activity of the enzyme because, under optimal conditions, it also cleaves Hsp104.

## Storage of Purified Hsp104 Protein

### *Concentration of Hsp104 Protein*

Hsp104 binds to many types of concentrating media such as Amicon Centricon and Centriprep concentrators, but Ultrafree-15 centrifugal filter devices with a molecular weight cutoff of 30,000 (Millipore) exhibit low binding and Hsp104 can be concentrated to >20 mg/ml in these devices without precipitating from solution. Hsp104 can also be concentrated by precipitation with ammonium sulfate without loss of activity as described earlier for the yeast protein purification following the hydroxyapatite column step.

### *Short-Term Storage of Purified Hsp104*

Hsp104 protein purified from yeast retains full ATPase activity when stored on ice in buffer A with 10% (v/v) glycerol for 1 month and loses only 50% activity after 9 months. Little if any degradation is observed over this time. However, if stored in ATPase assay buffer (see below), a physiological buffer lacking glycerol, wild-type Hsp104 loses ~50% of its ATPase activity within several weeks and mutant variants of Hsp104 typically lose an even greater percentage of activity.[12]

*Long-Term Storage of Purified Hsp104*

Hsp104 purified from yeast and flash-frozen in liquid nitrogen can be stored indefinitely at $-80°$ in buffer A containing 10% (v/v) glycerol with little or no loss of activity. Hsp104 retains ATPase activity when concentrated by lyophilization.[25] If shipped on dry ice, tubes should be sealed with Parafilm to prevent the introduction of $CO_2$, which precipitates Hsp104. Repeated freeze–thaw cycles reduce the activity of Hsp104 as previously noted for the related HSP100 protein, ClpA.[24]

## Assays to Test Activity of Purified Hsp104

*ATPase Assays*

Hsp104 hydrolyzes ATP under a variety of buffer and pH conditions. The ATPase activity reaches a maximum at pH 6.5, drops to half this level at pH 7.5, and increases again to an intermediate level at pH 9.[10] Activity is generally tested in 20 m*M* *N*-2-hydroxyethylpiperazine-*N*′-2-ethanesulfonic acid (HEPES, pH 7.5), 140 m*M* KCl, 15 m*M* NaCl, 10 m*M* $MgCl_2$, and ATP, pH 7.5. The assay should be performed in siliconized Eppendorf tubes, or in the presence of 0.02% (v/v) Triton X-100 to prevent Hsp104 binding to the walls of the tubes.[25] A reaction mix is made from 10× stocks of the individual buffer components and Hsp104. This is aliquoted to tubes on ice and peptides and other varied components are added. A typical assay volume is 25 $\mu$l and contains 1 $\mu$g of Hsp104. The tubes are preincubated at 37° for 1 min and ATP is added to start the reaction. The reaction is incubated at 37° for 7 min and then terminated by addition of 800 $\mu$l of Malachite Green reagent [0.034% (w/v) Malachite Green (Sigma), 1.05% (w/v) ammonium molybdate, 1 *M* HCl, filtered to remove insoluble material].[32] After 1 min at room temperature, color development is stopped by addition of 100 $\mu$l of 34% (w/v) citric acid. Two hundred microliters of the sample is removed to 96-well assay plates (Costar, Cambridge, MA) and the $A_{650}$ determined with a Molecular Devices (Palo Alto, CA) $V_{max}$ kinetic microplate reader with SoftMax software. Values are calibrated against $KH_2PO_4$ standards and corrected for phosphate released in the absence of Hsp104. Hsp104 cleaves ATP at a rate of ~2 nmol $min^{-1}$ $\mu g^{-1}$.[10]

[32] T. P. Geladopoulos, T. G. Sotiroudis, and A. E. Evangelopuolos, *Anal. Biochem.* **192,** 112 (1991).

Peptide-stimulated ATPase activity is assayed by replacing 1/10 of the reaction volume $H_2O$ with either peptide at 2 mg/ml or a control of the buffer in which the peptide is dissolved. At a final concentration of 200 $\mu$g/ml, poly (L-lysine) ($M_r$ 33,000 Sigma) should increase the ATPase activity of Hsp104 roughly sevenfold, and the oxidized chain B of insulin (Sigma) should increase the ATPase activity of Hsp104 roughly 50%.[12]

*Oligomerization*

Hsp104 is dialyzed against 20 m*M* HEPES (pH 7.5), 2 m*M* EDTA, 200 m*M* KCl, 10 m*M* $MgCl_2$, 2 m*M* dithiothreitol (DTT). Protein is diluted to a final concentration of 0.0425 mg/ml and incubated with 1–5 m*M* nucleotides or appropriate controls on ice for 10 min. Precipitated material is removed from reactions by centrifugation (14,000 rpm for 10 min at 4°) and the supernatant is aliquoted into tubes for each time point (100 $\mu$l/tube). Cross-linking is initiated by the addition of 4 $\mu$l of a 2.6% (v/v) glutaraldehyde stock (freshly prepared from an EM-grade stock 50% solution; Electron Microscopy Sciences, Ft. Washington, PA), and is allowed to proceed for 2 min to 1 hr. The reaction is quenched by the addition of 100 $\mu$l of 1 *M* glycine. Samples are then placed on ice, 75 $\mu$g of a carrier peptide is added (insulin chain B, oxidized; Sigma), and cross-linked proteins are precipitated with trichloroacetic acid (TCA; final concentration, 10%, v/v). No difference in precipitated protein is observed whether stopped reactions are incubated on ice for 1 or 12 hr. Samples are centrifuged

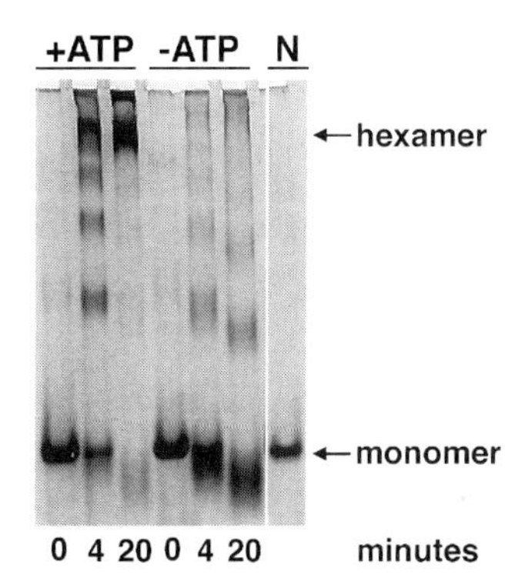


FIG. 2. Example of cross-linking gel to assess Hsp104 oligomerization. Hsp104 was mixed with nucleotide or buffer and cross-linked with glutaraldehyde as described in text. In the presence of ATP, most of the Hsp104 assembles into oligomers over time. In the absence of ATP, although some oligomers are observed, most of the Hsp104 remains monomeric. N indicates the migration of Hsp104 that was not incubated with glutaraldehyde. The expected migration Hsp104 monomers and hexamers is indicated on the basis of the migration of cross-linked phosphorylase *b* standards on the same gel.

(14,000 rpm for 30 min at 4°) and washed several times with 100% ethanol. The pellets are dried and suspended in sample buffer as described in Sigma Technical Bulletin MWS-877X (according to the method of Weber and Osborn[33]) and analyzed by 3.5% (w/v) SDS–PAGE, using Bio-Rad minigels, which require 3.5 hr to resolve samples (Fig. 2). Cross-linked phosphorylase *b* standards (Sigma) are resolved on the same gels to estimate molecular weights. Gels are stained with silver using the procedure of Morrissey.[34]

## Acknowledgments

The procedure for purification of Hsp104 from yeast was originally developed by Katherine Borkovitch, modified by Dawn Parsell,[9] and further modified by E. Schirmer and John Glover. Procedures for purification of Hsp104 from *E. coli* were developed by E. Schirmer, J. Glover, M. Ramakrishnan, and Doug Hattendorf. The cross-linking assay was developed by Anthony Kowal and we thank him for assistance in preparing this manuscript. This work was supported by the Howard Hughes Medical Institute and the Department of Energy Grant No. DE FG02 95E R20207.

[33] K. Weber and M. Osborn, *J. Biol. Chem.* **244,** 4406 (1969).
[34] J. H. Morrissey, *Anal. Biochem.* **117,** 307 (1981).

# [35] SecB: A Chaperone from *Escherichia coli*

*By* Linda L. Randall, Traci B. Topping, Virginia F. Smith, Deborah L. Diamond, and Simon J. S. Hardy

SecB is a molecular chaperone in *Escherichia coli* that is dedicated to the facilitation of the export of a number of proteins destined for the periplasmic space or the outer membrane. This role in export is demonstrated *in vivo* by the accumulation of pulse-labeled precursor species in a strain that is devoid of SecB[1] and *in vitro* by showing that SecB is required for translocation of precursors into inverted vesicles of cytoplasmic membrane.[2] Like all proteins classified as molecular chaperones, SecB has the ability to bind selectively and with high affinity to polypeptides that are in a nonnative state. SecB binds precursor polypeptides and maintains them in a state competent for translocation through the cytoplasmic membrane. Translocation cannot occur if the polypeptide is either folded or aggre-

[1] C. A. Kumamoto and J. Beckwith, *J. Bacteriol.* **163,** 267 (1985).
[2] J. B. Weiss, P. H. Ray, and P. J. Bassford, Jr., *Proc. Natl. Acad. Sci. U.S.A.* **85,** 8978 (1988).



0076-6879/98 $25.00

gated.[2–4] Because both folding and aggregation of proteins can be rapid, it is clear that the rate at which a chaperone binds must be high if it is to interfere with these processes and facilitate proper localization of its ligands. The molecular basis of such binding, which occurs rapidly with high selectivity and with high affinity in the absence of any consensus in sequence among the ligands, is particularly intriguing (see Randall and Hardy[5] for a review of the binding). SecB provides a good system for study because it is readily purified in large quantities.

## Purification of SecB Protein

SecB is purified from strains of *E. coli* harboring plasmids that contain the *secB* gene under strong, regulated promoters. Several plasmids have been constructed that result in high levels of expression of the *secB* gene.[2,6] Here we describe a purification scheme adapted from Weiss *et al.*[2] using plasmid pJW25, which has transcription of the *secB* gene under the control of the T7 promoter. The plasmid is harbored in *E. coli* strain BL21(DE3), which contains the cloned gene for T7 RNA polymerase under control of the inducible *lacUV5* promoter. An alternative method of purification is given in Kumamoto *et al.*[6] The purification described here is based on the quantity of protein obtained from a 1.5-liter culture.

### *Growth of Culture*

The strain harboring the plasmid is stored at −70° in 20% (v/v) glycerol. The stock is made by adding 1.8 ml of a dense culture to 0.6 ml of sterile 80% (v/v) glycerol. Each time the strain is to be used it is streaked from the glycerol stock onto a Luria broth (LB) plate containing 100 $\mu$g of ampicillin per milliliter to ensure that the plasmid is present. Between 10 and 20 colonies are picked from the plate and inoculated into separate liquid cultures of LB containing 50 $\mu$g of ampicillin per milliliter. These cultures are assessed to evaluate the level of expression of SecB. A portion of each culture is induced by addition of isopropyl-$\beta$-D-thiogalactoside (IPTG, 0.1 m$M$) and after growth for 2 hr analyzed by sodium dodecyl sulfate (SDS) gel electrophoresis as described below. Several of the uninduced cultures that are found to express SecB at a high level after induction are pooled and the pool is used to inoculate LB containing 50 $\mu$g of ampicillin per milliliter. After the culture has become turbid, 1.5 ml of the

[3] L. L. Randall and S. J. S. Hardy, *Cell* **46,** 921 (1986).
[4] H. de Cock, W. Overeem, and J. Tommassen, *J. Mol. Biol.* **224,** 369 (1992).
[5] L. L. Randall and S. J. S. Hardy, *Trends Biochem. Sci.* **20,** 65 (1995).
[6] C. A. Kumamoto, L. Chen, J. Fandl, and P. C. Tai, *J. Biol. Chem.* **264,** 2242 (1989).

culture is added to 150 ml of the same medium. This is grown to an optical density (OD) of 1.0 at 560 nm and transferred to a cold box for storage at 4°. The 150-ml culture is used the next day to inoculate 1.5 liters of LB (containing 50 μg of ampicillin per milliliter) distributed among five 2-liter flasks to ensure maximal aeration. The cultures are grown with shaking at 30° to OD 0.6. To induce the expression of SecB, IPTG is added to a final concentration of 0.1 m*M* and growth is continued for an additional 2 hr, at which time the OD will be approximately 1.4. The cells are harvested by centrifugation for 10 min at 10,000 rpm at 4° using a GSA (Sorvall, Norwalk, CT) rotor. The cell pellets are suspended in a total of 150 ml of 10 m*M* Tris-HCl, pH 7.6, containing 30 m*M* NaCl and centrifuged in one GSA centrifuge bottle for 10 min at 10,000 rpm at 4°. A typical yield from 2100 OD ml of cells is about 7 g (wet weight) of pellet. The cell pellet can be frozen at this stage and stored at −70° until needed.

Just before addition of the IPTG and again just before harvesting, samples are taken from the culture and used to check that good induction of SecB has occurred. The proteins in these samples are precipitated with 10% (w/w) trichloroacetic acid (TCA) and analyzed by SDS gel electrophoresis on a 14% (w/v) polyacrylamide gel. We usually apply samples to the gel that contain the protein from 0.05 OD ml of the culture (i.e., 85 μl of the culture at OD 0.6 and 35 μl of the culture at OD 1.4). There should be a prominent band of $M_r$ 17,000 in the sample taken after induction.

*Preparation of Cellular Extract*

Thirty-five milliliters of 20 m*M* Tris-HCl, pH 7.6, containing 2 m*M* ethylenediaminetetraacetic acid (EDTA), 0.1 m*M* phenylmethylsulfonyl fluoride (PMSF), and lysozyme (0.1 mg/ml) is added to the frozen cell pellet (total, 2100 OD ml) so that when it is suspended the OD is 60. The PMSF, which is stored as a stock at 0.1 *M* in ethanol, is diluted into the buffer just before use because it rapidly decomposes in $H_2O$ ($t_{1/2}$ is approximately 1 hr at room temperature)[7] and is no longer active as a protease inhibitor. The suspension is incubated for 20 min on ice (including the time for suspension). To disrupt the cells by sonication, the suspension is distributed among five plastic, 20-ml scintillation vials (approximately 8 ml in each). The sonicator probe must be of a large size so that when inserted into the vial a large proportion of the suspension is in contact with the surface of the probe to ensure efficient transfer of energy. The vials are held in a salt–ice bath and each subjected to three 5-sec bursts of sonication at 1-min intervals and then the viscosity is checked. If the lysate cannot be easily pipetted additional bursts of sonication are applied.

[7] G. T. James, *Anal. Biochem.* **86,** 574 (1986).

The sonicated sample is divided between two tubes and is centrifuged at 4° for 20 min at 15,000 rpm using the SS34 rotor (Sorvall). The supernatants are transferred to two clean tubes and centrifuged again (4°, 20 min, 15,000 rpm). A sample of the pellet and of the supernatant should be saved to follow the progress of purification. The supernatant is now ready for chromatography.

*Column Chromatography*

The purification involves passage of the sample over a Q-Sepharose ion-exchange column, followed by chromatography using a molecular sieve resin, Sephacryl S-300, and finally a Mono Q ion-exchange column. The Mono Q HR 16/10 ion-exchange column may be purchased from Pharmacia (Piscataway, NJ) as a prepacked column. The Sephacryl S-300 resin is purchased from Pharmacia and used to pack a column of bed volume 165 ml (1.6 × 82.5 cm). The Q-Sepharose resin is purchased from Pharmacia; the column volume is 25 ml (1.6 × 12.5 cm). Preparation of columns and all steps of purification are carried out at 4°. SecB in the eluted fractions is detected by SDS gel electrophoresis [14% (w/v) polyacrylamide]. Table I gives a summary of the purification procedure and Fig. 1 shows the extent of enrichment at each step.

The Q-Sepharose column is prepared for use by washing with four column volumes (100 ml) of 20 m*M* Tris-HCl (pH 7.6), 2 *M* NaCl and

TABLE I
CHROMATOGRAPHY FOR SecB PURIFICATION

| Column matrix | Column size (cm) | Equilibration buffer | Flow rate (ml/min) | Gradient | Protein applied[a] (mg) | SecB elution position[b] | Volume pooled (ml) |
|---|---|---|---|---|---|---|---|
| Q-Sepharose | 1.6 × 12.5 | 20 m*M* Tris-HCl (pH 7.6), 0.1 *M* NaCl | 2 | 0.1 to 0.6 *M* NaCl in 300 ml | 800 | 0.4 *M* NaCl | 50 |
| Sephacryl S-300 | 1.6 × 82.5 | 50 m*M* Tris-HCl (pH 7.6), 0.1 *M* NaCl | 0.4 | Isocratic | 100 | 110 ml | 25 |
| Mono Q | 1.6 × 10[c] | 50 m*M* Tris-HCl (pH 7.6), 0.1 *M* NaCl | 2 | 0.1 to 0.6 *M* NaCl in 500 ml | 85 | 0.4 *M* NaCl | 25 |

[a] Protein estimated by the Lowry assay [O. H. Lowry, N. J. Rosebrough, A. L. Farr, and R. J. Randall, *J. Biol. Chem.* **193,** 265 (1951)]. The values given apply to a purification from a 1.5-liter culture.

[b] The salt concentrations are approximate; they are calculated from the volume of the gradient that had been eluted where the peak of SecB appeared.

[c] This column has the capacity for a SecB preparation from 6 liters of culture. The Mono Q 10/10 column of dimensions 1.0 × 10 cm is adequate for a SecB preparation from 1.5 liters of culture.

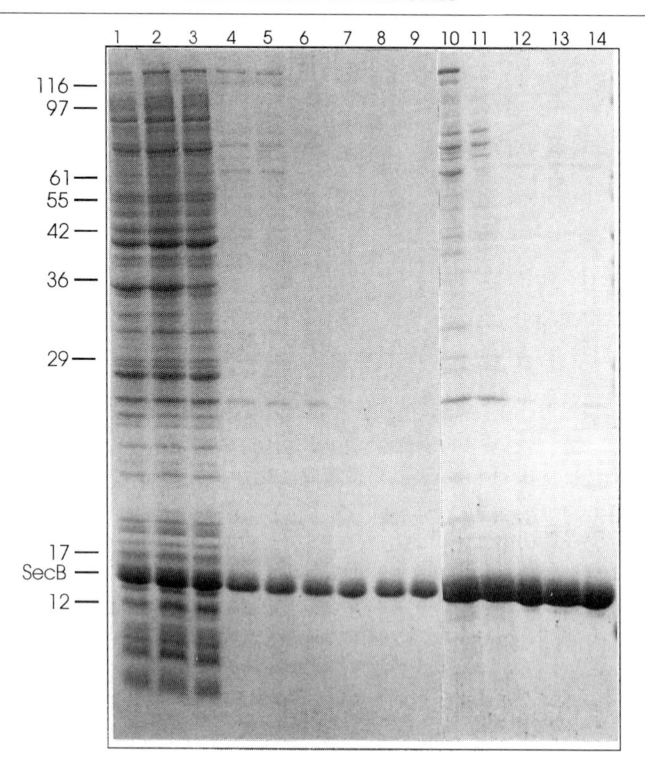


FIG. 1. Purification of SecB. The purity of SecB is assessed at each step by SDS–polyacrylamide (14%, w/v) gel electrophoresis. Lanes 1–9 contain 0.05 OD ml cell equivalents of the following: lane 1, whole cells; lane 2, sonicated cells; lane 3, supernatant; lanes 4 and 5, pooled fractions from Q-Sepharose column before (lane 4) and after (lane 5) concentration; lane 6, pooled fractions from Sephacryl S-300 column; lanes 7 and 8, pooled fractions from Mono Q column before (lane 7) and after (lane 8) concentration; lane 9, final dialyzed purified SecB. Lanes 10–14 contain 0.2 OD ml equivalents of the same sample as displayed in lanes 5–9, respectively. The position of migration for each molecular weight standard is indicated on the left-hand side of the gel by the corresponding molecular weight ($\times 10^{-3}$): 116, $\beta$-galactosidase; 97, phosphorylase *b*; 61, $\alpha$-amylase; 55, glutamate dehydrogenase; 42, actin; 36, glyceraldehyde-3-phosphodehydrogenase; 29, carbonate dehydratase; 17, myoglobin; 12, cytochrome *c*.

then equilibrating with four column volumes (100 ml) of 20 m*M* Tris-HCl (pH 7.6), 0.1 *M* NaCl at 2 ml/min. The sample is loaded onto the prepared column at 2 ml/min; the column is washed with 30 ml of 20 m*M* Tris-HCl (pH 7.6), 0.1 *M* NaCl and then a 300-ml gradient from 0.1 to 0.6 *M* NaCl in 20 m*M* Tris-HCl, pH 7.6, is developed. Fractions of 5 ml are collected and those containing SecB are pooled. The center of the peak containing SecB occurs at around 0.4 *M* NaCl. When purifying any protein it is good practice to pool all fractions across the peak containing the protein of interest, even though it might be relatively more pure in a subset of those fractions. In this way one avoids the danger of unknowingly enriching for a particular isoform.

The protein in the pooled fractions must be concentrated and equilibrated in the buffer used for the size-exclusion chromatography. The protein is concentrated using an Amicon (Danvers, MA) concentrator (50 ml) with a PM10 membrane to reduce the sample volume to 5 ml. The sample can be dialyzed against 50 m*M* Tris-HCl (pH 7.6), 0.1 *M* NaCl or after concentration to 5 ml it can be subjected to two cycles of dilution with the buffer to 50 ml, followed by concentration to 5 ml. Precipitates that form during the concentration and dialysis procedures are removed by centrifugation for 5 min at 12,000 *g* using an Eppendorf centrifuge.

The sample is loaded onto a S-300 column equilibrated with 50 m*M* Tris-HCl (pH 7.6), 0.1 *M* NaCl at a flow rate of 0.4 ml/min. The fractions containing SecB, eluting at about 110 ml, are pooled and applied directly to the Mono Q 16/10 column, which has been equilibrated in 50 m*M* Tris-HCl (pH 7.6), 0.1 *M* NaCl. After application of the sample, the column is washed with 20 ml of the same buffer. A salt gradient from 0.1 to 0.6 *M* NaCl in 50 m*M* Tris-HCl, pH 7.6, is developed over 500 ml. The fractions containing SecB are pooled, concentrated, and dialyzed twice against 1 liter of 10 m*M* Tris-HCl (pH 7.6), 0.3 *M* NaCl for longer than 8 hr each time. The solution is clarified by centrifugation using an Eppendorf centrifuge and the purified protein is stored frozen at −70°. Stored in this way, SecB retains activity for at least 1 year. The high ionic strength of the buffer is necessary because circular dichroism studies show that prolonged exposure to low ionic strength causes SecB to lose structure.[8] A typical yield from 1.5 liters of culture is 60 mg of SecB.

### *Parameters*

The extinction coefficient for SecB at 280 nm is 0.72 ml $mg^{-1}$ $cm^{-1}$. SecB is soluble to at least 30 mg/ml. We find that determination of the concentration of SecB using the Lowry assay[9] with bovine serum albumin (BSA) as a standard routinely overestimates the concentration by a factor of approximately 1.2. SecB purified as described here has been shown to exist as a tetramer using electrospray mass spectrometry.[10] The monomers exist in two species, one with a mass of 17,146 Da (indicating that the N-terminal methionine has been removed) and a larger species with a mass

[8] K. Park, L. L. Randall and G. D. Fasman, unpublished results (1993).

[9] O. H. Lowry, N. J. Rosebrough, A. L. Farr, and R. J. Randall, *J. Biol. Chem.* **193,** 265 (1951).

[10] V. F. Smith, B. L. Schwarz, L. L. Randall, and R. D. Smith, *Protein Science* **5,** 448 (1996).

of 17,189 Da (consistent with N-acetylation of either the amino terminus or the ε-$NH_2$ of a lysine). The sequence of the *secB* gene and the deduced sequence of amino acids in the protein can be found in GenBank (accession code M24490) or in Kumamoto and Nault.[11]

## Assays of Function

### *Formation of Complexes between SecB and Polypeptide Ligands*

SecB binds selectively and with high affinity to nonnative proteins to form complexes that have dissociation constants that vary from 1 to 100 n*M*. The complexes are formed by rapidly diluting a denaturant, either guanidinium chloride or urea, away from an unfolded protein in the presence of SecB or by adding SecB within a few seconds of dilution of the denaturant. In many cases, complexes have sufficiently high stability to be isolated using size-exclusion chromatography. We have successfully used both fast protein liquid chromatography (FPLC) with the Superose 12 column from Pharmacia (Piscataway, NJ) and high-performance liquid chromatography (HPLC) with the TSK3000.SW column from TosoHaas (Montgomeryville, PA). Both columns readily resolve SecB, which elutes aberrantly at the position characteristic of a macromolecule having a molecular mass of 90 kDa or higher, from ligands that have molecular masses of 50 kDa or less. The complex, however, is not well resolved from free SecB. Thus, a change in the absorbance profile at $A_{280}$ is often not sufficient to demonstrate that a specific complex has been formed and it is necessary to analyze fractions eluted from the column by SDS gel electrophoresis to show that the putative ligand coelutes with SecB. Absorbance profiles are shown for the complex between SecB and maltose-binding protein ($M_r$ 41,000) using the TSK3000.SW column (Fig. 2). Necessary controls include a demonstration that the unfolded protein when applied to the column does not elute at the position of SecB and that when SecB is added after the protein has been allowed to refold or when the protein has never been unfolded no complex is formed. Recovery is often poor when a protein is applied to a column in a denatured state in the absence of SecB.

For conventional protein–ligand complexes one can predict the success of isolation of the complex on the basis of the concentration of components during the procedure relative to the dissociation constant of the complex. However, in the case of a complex between a chaperone and a nonnative ligand other factors come into play. The polypeptides that are ligands are competent to interact with chaperones only as long as they are nonnative.

[11] C. A. Kumamoto and A. K. Nault, *Gene* **75**, 167 (1989).

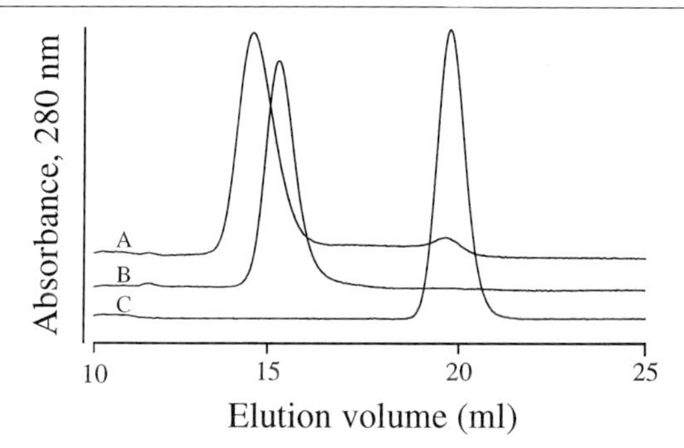


FIG. 2. Complex between SecB and nonnative ligand demonstrated by size-exclusion chromatography. High-performance liquid chromatography using a TSK3000.SW column was carried out as described in text. The absorbance profiles are as follows: (A) mixture of SecB (10 $\mu$g) and unfolded maltose-binding protein (4 $\mu$g); (B) SecB only (20 $\mu$g); (C) folded maltose-binding protein (5 $\mu$g).

In addition, in the specific case of the chaperone SecB, although the binding is of a high affinity, it is characterized by high rate constants for both association and dissociation and thus the ligand cycles rapidly between the bound and the free state. When the ligand is free, there is a certain probability, related to the rate constant of folding for the particular polypeptide, that it will fold and thereby lose the ability to bind to SecB. During any isolation procedure the population of the ligand will be continuously partitioning between binding and folding. Because the stability of most folded proteins would exceed the stability of the complex, at equilibrium no complex would remain. Thus it is crucial that precautions be taken to ensure that rebinding is kinetically favored over folding during the isolation procedure. The temperature should be as low as possible to retard folding and if possible agents should be included to block formation of stable structure of the ligand. For example, because reducing agents and chelating agents do not affect the activity of SecB, they may be included if the ligand of interest is a protein that is stabilized by disulfide bonds or metal ions. The solution used during formation of the complex as well as during chromatography should contain at least 50 m*M* salt in addition to a buffer. The reason for this is discussed below in the following section describing the fluorescence assay.

*Fluorescence Assay of Interaction with SecB*

SecB binds to its natural ligands to maintain them in a nonnative and unaggregated state so that they are competent for translocation through the cytoplasmic membrane. This interaction can be detected by using changes in intrinsic fluorescence to monitor folding if the ligand polypeptides contain either tryptophanyl or tyrosinyl residues. For ligands containing tryptophan we use excitation and emission wavelengths of 295 nm (bandwidth, 2 nm) and 344 nm (bandwidth, 5 nm), respectively. Tyrosine fluorescence is observed using an excitation wavelength at 280 nm (bandwidth, 2 nm) and an emission wavelength of 303 nm (bandwidth, 5 nm). The concentration of denaturant to be used for unfolding and the final concentration after dilution that will allow refolding should be determined for each protein by doing an equilibrium study of the reversible folding transition. For this study native protein is added to varying concentrations of denaturant in 3 ml of buffer [usually 10 m*M* *N*-2-hydroxyethylpiperazine-*N'*-2-ethanesulfonic acid (HEPES, pH 7.6)] and the fluorescence intensity of each sample is measured after equilibrium for several hours at 25°. The observed fluorescence amplitudes are converted to apparent fraction of unfolded protein, $F_{app}$, such that

$$F_{app} = (Y_{obs} - Y_N)/(Y_U - Y_N)$$

where $Y_{obs}$ refers to the observed fluorescence intensity at a particular denaturant concentration, and $Y_N$ and $Y_U$ refer to the calculated values for the native and unfolded forms, respectively, at the same denaturant concentration. Generally, a linear dependence of $Y_{obs}$ on the denaturant concentration will be observed in the pre- and posttransition baseline regions, so the values for $Y_N$ and $Y_U$ are obtained by extrapolation of the baselines into the region of the unfolding transition.

Maltose-binding protein contains eight tryptophanyl residues and is thus amenable to the fluorescence assay. The purified polypeptide is denatured by incubation for 2 hr at room temperature in buffered solutions containing either 2 *M* guanidinium chloride or 6 *M* urea. Refolding of the protein is initiated by rapid dilution of the denaturant directly into 2.7 ml of buffer contained in a cuvette that is held in the spectrophotometer. This is achieved by placing approximately 75 $\mu$l of the solution of denaturant containing the protein onto a plastic plunger, designed for simultaneous addition and mixing, and rapidly inserting and withdrawing it from the cuvette (Fig. 3). The rate of refolding is determined from the rate of change in the amplitude of fluorescence. The change in fluorescence intensity ($\Delta F$) is the absolute value of the difference between the intensity of fluorescence at a given time and the intensity when the system has reached equilibrium (Fig. 4, left-hand side). A plot of the log of this value vs time (Fig. 4, right-hand side) allows one to extract the relaxation time, $\tau$. Interaction with SecB is detected as a change in the rate of appearance of folded protein when SecB is present in the cuvette into which the protein is diluted. If conditions are such that the kinetic partitioning favors binding over folding, for example

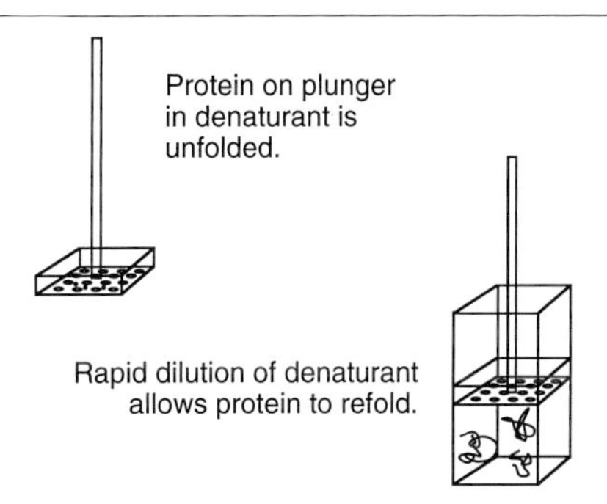


FIG. 3. Refolding of protein in fluorescence assay.

low temperature or the presence of a slow-folding ligand, observed effects can be extreme, resulting in an apparent blockage of folding over the time of a usual assay (30 to 45 min). If the rate of folding of the polypeptide and the rate of interaction with SecB, which is a function of the pseudo first-order rate constant of association, i.e., $[\text{SecB}]k_{on}$, are similar then an effect on $\tau$ will be observed, but the entire population will fold. Effects on the folding can be reported quantitatively as increases of $\tau$ or $t_{1/2}(\tau = t_{1/2}/0.693)$.

It is crucial to note that if urea is used as the denaturant in this assay, the solution in the cuvette must contain at least 50 m$M$ salt. We have observed that at low ionic strength SecB cannot block the folding of maltose-binding protein and has only a small effect on $\tau$ even though using another assay (see Peptide-Binding Assays, below) it can be demostrated that SecB

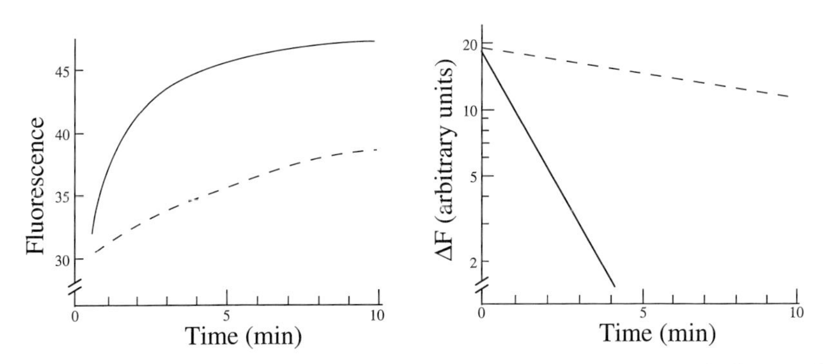


FIG. 4. Effect of SecB on rate of folding. Refolding of unfolded maltose-binding protein is initiated by dilution of GuHCl from 2 $M$ to 55 m$M$. The left-hand side shows the increase in fluorescence of maltose-binding protein alone (solid line) and maltose-binding protein diluted in the presence of SecB (dashed line). The right-hand side displays the data in the graphic representation that is used to extract the relaxation time (or $t_{1/2}$).

interacts with peptides under conditions of low ionic strength. If guanidinium chloride is used as the denaturant other salts can be omitted provided that the residual guanidinium chloride in the assay is at least 50 m*M*. Nevertheless, we prefer to use 10 m*M* HEPES, pH 7.6, containing 150 m*M* potassium acetate to approximate physiological ionic strength. The cuvette holder must be thermostatted because the temperature affects the rate of folding of the ligands. In addition, the cuvette must be mounted over a stirring motor so that the solution can be stirred continuously. A convenient and relatively inexpensive instrument that is suitable for these studies is the Shimadzu RF-540 fluorescence spectrophotometer (Shimadzu, Columbia, MD). An excellent introduction to fluorescence spectroscopy is found in Ref. 11a.

This assay as described can be used directly to demonstrate interaction of SecB with a ligand only if the ligand undergoes a change in fluorescence on folding. However, the binding of other ligands can be assessed indirectly if they compete with maltose-binding protein for the binding to SecB. For an example, see the study of Hardy and Randall.[12] The putative competitor is denatured and placed on the plunger with the denatured maltose-binding protein. If the competitor interacts with SecB then the effects on the changes in fluorescence of maltose-binding protein are reduced because less SecB is available. Such competition assays can be performed only with competitors that contain no tryptophan, because competitors with tryptophan would interfere with the fluorescence signal from maltose-binding protein.

### *Peptide-binding Assays*

*Protection of SecB from Proteolysis.* An assay for binding of short peptide ligands, which have affinities that are too low to allow isolation of complexes, is based on the resistance to proteolysis of a complex comprising SecB and ligand relative to the proteolysis of the free components. At low ionic strength, when uncomplexed with ligand, SecB is quantitatively cleaved by proteinase K. This cleavage leads to a shift in the position of migration of SecB on SDS–polyacrylamide gels from 17 kDa to approximately 12 kDa[12] even though analysis of the cleavage product by mass spectrometry reveals that only 14 aminoacyl residues are removed from the C terminus.[10] When bound to peptide ligands SecB is resistant to the proteolysis. For some ligands such as carboxamidomethylated bovine pancreatic trypsin inhibitor (R-BPTI), the free ligand is resistant to proteolysis but is rendered sensitive on binding to SecB. If the ligand to be tested

[11a] D. A. Harris and C. L. Bashford (eds.), "Spectrophotometry and Spectrofluorimetry: A Practical Approach. IRL Press, Washington, DC, 1987.

[12] S. J. S. Hardy and L. L. Randall, *Science* **251,** 439 (1991).

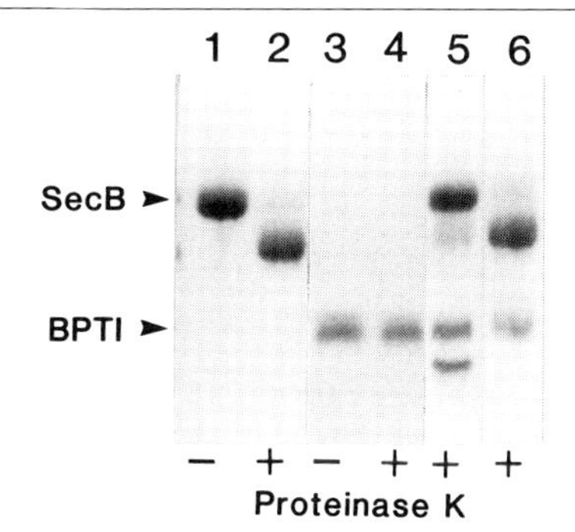


FIG. 5. Protection of SecB from proteolysis by ligand. Incubation mixtures (0.4 ml, 10 m*M* HEPES, pH 7.6) contain purified protein as follows: lanes 1 and 2, SecB (4 μg); lanes 3 and 4, R-BPTI (1 μg); lane 5, SecB (4 μg) and R-BPTI (1 μg); lane 6, SecB (4 μg) and native BPTI (2 μg). The mixtures were incubated on ice either with (+) or without (−) proteinase K (0.3 μg/ml) and processed as described in text. [Reprinted with permission from L. L. Randall, *Science* **257,** 241 (1992). Copyright © 1992, American Association for the Advancement of Science.]

exists in a stably folded form a control can be done to show that the folded ligand does not render SecB resistant to proteolysis. Figure 5 shows an example of the assay using R-BPTI as ligand. The assay is carried out in Eppendorf centrifuge tubes (1.5-ml size) containing 4 μg of purified SecB in 0.4 ml of 10 m*M* HEPES, pH 7.6 (0.15 μ*M* tetrameric SecB). The peptide to be tested is added at varying concentrations and the mixtures incubated on ice for 20 min with and without proteinase K at a final concentration of 0.3 μg/ml (add 20 μl of a 0.006-mg/ml solution). Proteolysis is terminated by addition of trichloroacetic acid to a final concentration of 10% (w/w). The precipitates are collected by centrifugation in an Eppendorf centrifuge for 15 min. The pellets are washed once with acetone and suspended in SDS gel electrophoresis sample buffer. After incubation at 100° for 5 min, the samples are analyzed by electrophoresis on an SDS-15% (w/v) polyacrylamide gel or if one wants to resolve small peptide ligands as well as the SecB a peptide gel systcm can be used.[13] The amount of SecB that is rendered resistant can be quantified by densitometric scanning of the gels and the data can be represented graphically. Examples of this quantitative assay can be found in the publication by Randall.[14]

The proteinase K (EC 3.4.21.64, endopeptidase K) used in this assay is obtained from Sigma (St. Louis, MO). We purchase 5 mg and suspend the entire amount in 10 m*M* HEPES, pH 7.6, at 2.5 mg/ml, dispense it in 50-

[13] H. Schägger and G. von Jagow, *Anal. Biochem.* **166,** 368 (1987).
[14] L. L. Randall, *Science* **257,** 241 (1992).

μl aliquots, and store the aliquots frozen at −20°. An aliquot is thawed just before use and diluted to 0.006 mg/ml in 10 m*M* HEPES, pH 7.6. Once thawed, the aliquots are not reused.

The volume of the assay can be reduced if the peptides tested have low affinities. However, it must be remembered that on saturation of the peptide-binding sites SecB undergoes a conformational change to expose a hydrophobic patch[14] and at high concentration in small volumes precipitation will become a problem. Caution must also be taken when testing ligands of low affinity because addition of large quantities of peptide might render SecB apparently resistant to proteolysis, when what has happened is that the proteolytic capacity of proteinase K has been exceeded.

*Enhancement of 1-Anilinonaphthalene 8-Sulfonate Fluorescence.* The fluorescent compound 1-anilinonaphthalene 8-sulfonate (ANS), which binds to hydrophobic clusters of aminoacyl residues,[15] can be used to detect hydrophobic sites on SecB that are exposed as a result of binding to peptide ligands. When ANS is bound the intensity of fluorescence increases and the emission maximum is shifted from ~520 nm to shorter wavelengths. ANS at 15 μ*M* is held in a thermostatted cuvette containing 2.7 ml of 10 m*M* HEPES, pH 7.6, at 5°. The excitation wavelength used is 370 nm (slit width, 2 nm) and emission is scanned from 400 to 600 nm (slit width, 5 nm) to establish the spectrum of ANS alone. SecB is then added at 0.15 μ*M* tetramer and the spectrum scanned again. Free SecB does not bind ANS and thus no difference between the spectra should be observed. As illustrated in Fig. 6, addition of a peptide ligand results in an increase in fluorescence intensity and a shift in maximum of emission to 472 nm. The intensity of fluoresence that is observed when SecB is saturated with various peptide ligands varies from two- to threefold higher than the fluorescence of free ANS. The binding can be quantitatively assessed by adding the saturating amount of the ligand in 6–10 separate steps while following emission at 472 nm. In these experiments it is crucial to determine whether the ligand alone binds ANS. It is important to remember that at saturation SecB will have a tendency to aggregate as a result of exposure of hydrophobic surfaces; therefore, one must be aware of possible artifacts resulting from light scatter.

The solution of ANS should be made fresh daily. We usually make a solution of 10 m*M* ANS in 10 m*M* HEPES, pH 7.6, to be able to weigh the ANS accurately. From that solution we prepare a 1 m*M* ANS solution, which is used for addition to the cuvette, by dilution with 10 m*M* HEPES, pH 7.6. We obtain ANS from Sigma. We have not observed a detectable

[15] G. V. Semisotnov, N. A. Rodionova, O. I. Razgulyaev, V. N. Uversky, A. F. Gripas', and R. I. Gilmanshin, *Biopolymers* **31,** 119 (1991).

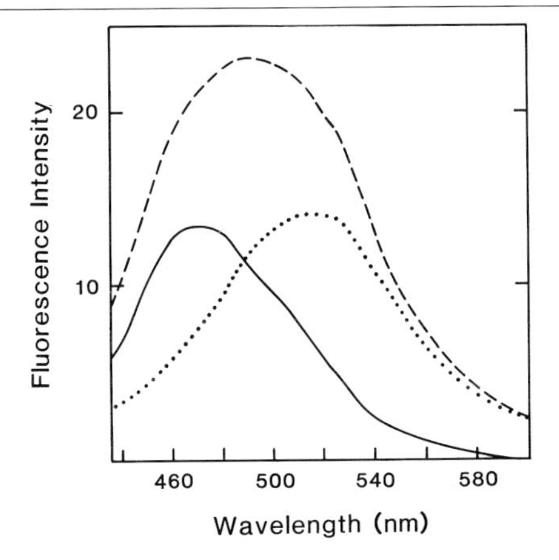


FIG. 6. Enhancement of ANS fluorescence as an assay of ligand binding. Fluorescence measurements are made as described in text. The emission spectra shown are for (···) ANS (15 $\mu M$) and SecB (0.15 $\mu M$ tetramer); (---) ANS, SecB, and peptide S4 (5.4 $\mu M$); and (—) the difference spectrum. S4 is a peptide of molecular weight 2856 from human immunodeficiency virus (HIV) gp41 protein. [Reprinted with permission from L. L. Randall, *Science* **257,** 241 (1992). Copyright © 1992, American Association for the Advancement of Science.]

difference between ANS from Sigma and that from Molecular Probes (Eugene, OR).

## Determination of Binding Frame within Ligands

The binding frame for SecB within two physiologic ligands, maltose-binding protein and galactose-binding protein, has been determined.[16,17] In each case, the area in direct contact with SecB covers approximately 160 aminoacyl residues and is positioned in the center of the primary sequence of each polypeptide. There is no apparent consensus in sequence or similarity of structural elements that can explain selection of the central region. It should be possible to identify the binding frame within other ligands of known sequence if they bind tightly enough to be isolated in complex with SecB. The approach is to subject complexes to proteolytic digestion and determine which fragments of the ligand are protected from degradation and remain bound to SecB. The protease used is proteinase K, which cleaves after aromatic and aliphatic residues.[18] The conditions of digestion must be optimized for each ligand such that the ligand is degraded, but the SecB

[16] T. B. Topping and L. L. Randall, *Protein Sci.* **3,** 730 (1994).
[17] V. J. Khisty, G. R. Munske, and L. L. Randall, *J. Biol. Chem.* **270,** 25920 (1995).
[18] E. Kraus, H.-H. Kiltz, and U. F. Femfert, *Hoppe-Seyler's Z. Physiol. Chem.* **357,** 233 (1976).

remains intact. A good starting point is 0.01 mg of proteinase K per milliliter for 10 min on ice. To have sufficient material for sequence determination of the peptides isolated as bound to SecB one must start with approximately 1.0 mg of ligand in complex with SecB in 1 ml. We routinely have the SecB tetramer in at least a twofold molar excess over the ligand to ensure that all ligand is bound. Complexes between SecB and a ligand can be formed in one of two ways: (1) denatured ligand is added to SecB directly from the GuHCl solution; or (2) the denatured ligand is diluted from the GuHCl in the absence of SecB, allowing the polypeptide to collapse from the fully denatured state, and SecB is added within 10 sec. In the studies we have carried out both techniques gave the same results. In both cases the solutions are held on ice. An example of the final concentrations is as follows: 40 $\mu M$ SecB tetramer, 20 $\mu M$ ligand, 100 m$M$ NaCl, 50 m$M$ Tris-HCl (pH 7.0), and guanidinium hydrochloride at a concentration dependent on the volume of denatured protein added (we keep it below 0.1 $M$, but this may not be necessary). After incubation for 10 min on ice, proteinase K is added to 0.01 mg/ml. After an additional incubation for 10 min on ice, phenylmethylsulfonyl fluoride is added to 1 m$M$ to stop proteolysis. It is important that the concentration of guanidium hydrochloride necessary to denature the protein be determined for each ligand.

One-third of the proteolyzed sample is used for analysis of the complete mixture of peptides generated from the ligand (termed *total*). This pattern should be compared to a pattern of peptides generated from digestion of the denatured ligand in the absence of SecB. The remaining two-thirds of the sample containing the proteolyzed complex are subjected to HPLC on a TSK3000.SW size-exclusion column to separate peptides bound to SecB from those that are free. For good resolution it is necessary to make several runs of the column, injecting 200 $\mu$l each time. Chromatography is carried out at 10° in 100 m$M$ NaCl, 50 m$M$ Tris-HCl, pH 7.0, at a flow rate of 1 ml/min. Fractions containing SecB eluting between 13.5 and 16.0 min, are pooled, and the peptides they contain are referred to as *bound peptides.* Analysis of a portion of each fraction by gel electrophoresis shows the position of elution of the free peptides, usually between 19 and 26 min. To ensure complete recovery of all peptides, the fractions from 16.5 to 30 min are taken and the peptides therein are referred to as *free peptides.* The total, bound, and free peptides are processed for analysis by reversed-phase HPLC using a Vydac (Hesperia, CA) $C_4$ column at room temperature. To expose all peptides to the same conditions, the unfractionated sample (total) is diluted 7.5-fold with the TSK3000.SW buffer to mimic passage through the column. The following additions are made to the total and to the bound peptide samples at room temperature to disrupt the complex between SecB and peptides and to precipitate the SecB: solid guanidinium hydrochloride to 3 *M*, acetonitrile to 20% (v/v), and trifluoroacetic acid (TFA) to give a pH of 2.0. The fractions containing the free peptides are lyophilized and suspended in 18% (v/v) acetonitrile–0.1% (v/v) TFA. We often lyophilize the fractions separately to avoid any risk of introducing proteinase K by pooling the fractions. Each of the three samples, total, bound, and free, is clarified by centrifugation (10,000 *g* for 15 min) before application to the $C_4$ column. This step removes the SecB, which will have been precipitated. After sample application and washing the salts through with 18% (v/v) acetonitrile in water containing 0.1% (v/v) TFA, a gradient of acetonitrile in water containing 0.1% (v/v) TFA is developed from 18 to 54% (v/v) over 70 ml at a flow rate of 1 ml/min. Fractions of 1.5 ml are collected and taken to dryness in a SpeedVac (Savant, Hicksville, NY). The fragments of ligand present in each fraction are resolved using a peptide gel system.[13] When the gel is run with the intention of determining the sequence of amino acid residues, 0.1 m*M* thioglycolate is added to the running buffer and the peptides are transferred to an Immobilon $P^{SQ}$ polyvinylidene fluoride membrane (Millipore, Bedford, MA). The Coomassie blue-stained bands are excised and the sequence of the first six to eight aminoacyl residues is determined using an automated sequencer such as the Applied Biosystems (Foster City, CA) 475A sequencing system with pulsed liquid update. The sequence information along with the molecular weights estimated from the position of migration on peptide gels allows one to locate the peptides within the sequence of the ligand.

## Acknowledgments

This work was supported by a grant from the National Institutes of Health (GM29798) to L. L. R.

# Author Index

Numbers in parentheses are footnote reference numbers and indicate that an author's work is referred to although the name is not cited in the text.

## A



## B

C

D

## E

## F



## G

H

## L

## M

## Q



## R

## S

## T

U



V

W

X



Y

## Z

# Subject Index

## A



## B

## C

## D



## E

## F

## H

## I

## R

ISBN 0-12-182191-9